SOLUTION MANUAL

# Student Notes and Problems

S N A P

# MATH 30 (PURE)

Castle Rock Research Corp

Canadian Cataloguing in Publication Data

Rao, Gautam, 1961 –
***STUDENT NOTES AND PROBLEMS*** – Math 30 (Pure)
Solution Manual

1. Mathematics – Juvenile Literature. I. Title

**Published by**
Castle Rock Research Corporation
2340 Manulife Place
10180 – 101 Street
Edmonton, AB T5J 3S4

7 8 9 FP 09 08 07

Printed in Canada

**Publisher**
Gautam Rao

**Contributors**
Allan Ashmore
Bob Frizzell
Brett Hanneson
Shirley Knull
Ken Kulka
Moses Renert

Castle Rock Research Corp

*Dedicated to the memory of Dr. V. S. Rao*

# STUDENT NOTES AND PROBLEMS WORKBOOKS

Student Notes and Problems (SNAP) workbooks are a series of support resources in mathematics for students in grades 3 to 12 and in science for students in grades 9 to 12. SNAP workbooks are 100% aligned with curriculum. The resources are designed to support classroom instructions and provide students with additional examples, practice exercises, and tests. SNAP workbooks are ideal for use all year long at school and at home.

The following is a summary of the key features of all SNAP workbooks.

## UNIT OPENER PAGE

- summarizes the curriculum outcomes addressed in the unit in age-appropriate language
- identifies the lessons by title
- lists the prerequisite knowledge and skills the student should know prior to beginning the unit

## LESSONS

- provide essential teaching pieces and explanations of the concepts
- include example problems and questions with complete, detailed solutions that demonstrate the problem-solving process

## NOTES BARS

- contain key definitions, formulas, reminders, and important steps or procedures
- provide space for students to add their own notes and helpful reminders

## PRACTICE EXERCISES

- include questions that relate to each of the curriculum outcomes for the unit
- provide practice in applying the lesson concepts

## REVIEW SUMMARY

- provides a succinct review of the key concepts in the unit

## PRACTICE TEST

- assesses student learning of the unit concepts

## ANSWERS AND SOLUTIONS

- demonstrate the step-by-step process or problem-solving method used to arrive at the correct answer

Answers and solutions are provided in each workbook for the odd-numbered. A SNAP *Solutions Manual* that contains answers and complete solutions for all questions is also available.

# CONTENTS

## Transformations and Functions



## Exponents and Logarithms



## Trigonometry

## Permutations and Combinations



## Statistics and Probability

## Conics

# TRANSFORMATIONS

## Lesson 1 Review of Functions

### CLASS EXERCISES ANSWERS AND SOLUTIONS

**1.** $f(2) = 2^2 - 2 - 5$

$= 4 - 2 - 5$

$= -3$

This corresponds to the point (2, –3).

**2.** Let $x = 0$:

$f(0) = 0^2 - 2(0) + 6$

$= 6$

The point is (0, 6).

**3.** Let $y = 0$:

$0 = \sqrt{x+3} - 4$

$4 = \sqrt{x+3}$

$16 = x + 3$ square both sides

$13 = x$

The $x$-intercept is 13, which corresponds to the point (13, 0).

### PRACTICE EXERCISES ANSWERS AND SOLUTIONS

**1.** $f(x) = x^2 - x - 6$

Factor:

$= (x-3)(x+2)$

$x$-intercepts occur when $y = 0$

$0 = (x-3)(x+2)$

$x = 3, -2$

$y$-intercepts occur when $x = 0$

$y = 0^2 - 0 - 6$

$y = -6$

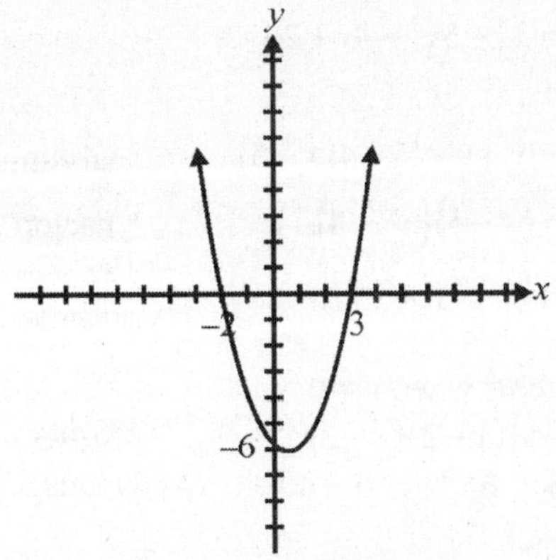


**2.** $f(x) = \sqrt{x-4}$

$x$-intercept when $y = 0$

$0 = \sqrt{x-4}$

$0 = x - 4$ • Square both sides

$4 = x$

$y$-intercept when $x = 0$

$y = \sqrt{0-4}$

$y = \sqrt{-4}$ • not a real number

$\therefore$ there is no $y$-intercept.

Notice that $y = \sqrt{x-4}$ is $y = \sqrt{x}$ translated 4 units to the right.

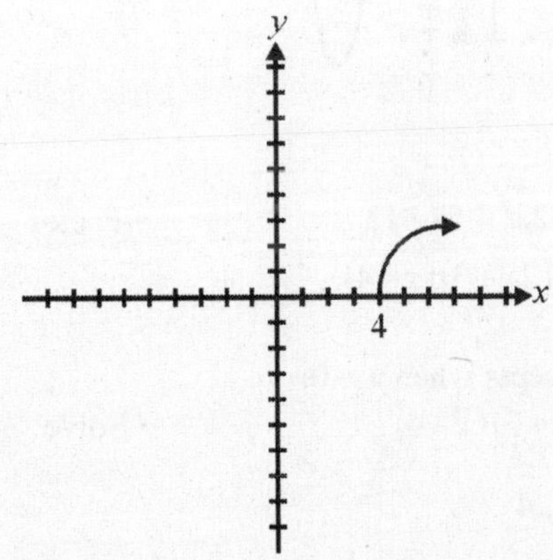

**3.** $f(x) = x^3 - 5x^2 - 4x + 20$

Factor:

$= x^2(x-5) - 4(x-5)$ • Group

$= (x-5)(x^2-4)$ • Factor

$= (x-5)(x+2)(x-2)$

$x$-intercepts when $y = 0$

$0 = (x-5)(x+2)(x-2)$ • Solve

$x = 5,\ 2,\ -2$

$y$-intercept when $x = 0$

$y = 0^3 - 5(0) - 4(0) + 20$

$y = 20$

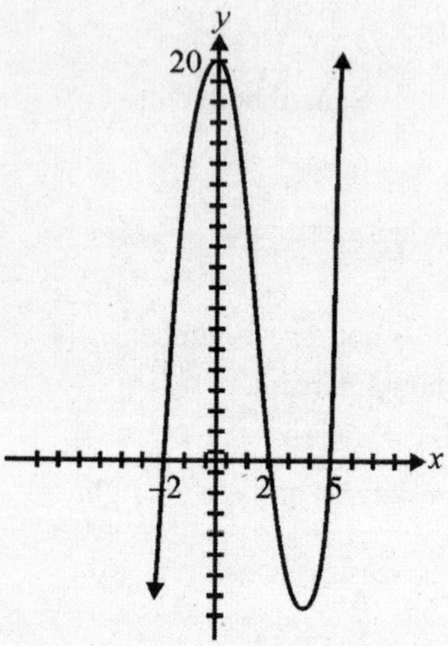


**4.** $f(x) = 2x^2 - 5x - 12$ • Factor

$= (2x+3)(x-4)$

$x$-intercepts when $y = 0$:

$0 = (2x+3)(x-4)$ • Solve

$x = \frac{-3}{2},\ 4$

$y$-intercept when $x = 0$

$y = 2(0)^2 - 5(0) - 12$

$y = -12$

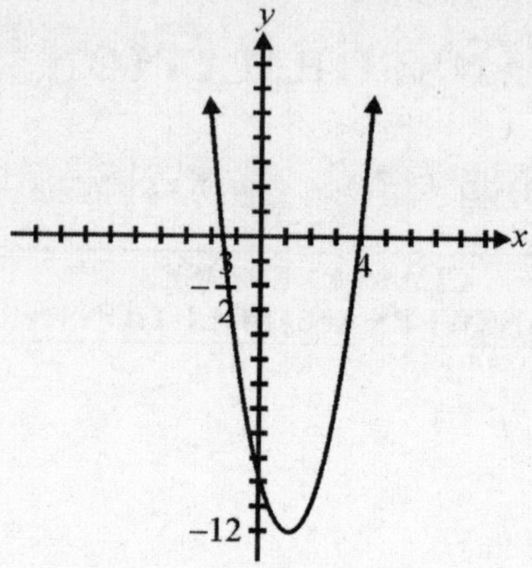


**5.** $f(x) = \frac{1}{x-3}$

$x$-intercept when $y = 0$:

$0 = \frac{1}{x-3}$

There is no solution, therefore, there is no $x$-intercept.

$y$-intercept when $x = 0$:

$y = \frac{1}{0-3} = -\frac{1}{3}$

Notice that the graph of $y = \frac{1}{x-3}$ is the graph of $y = \frac{1}{x}$ translated 3 units right.

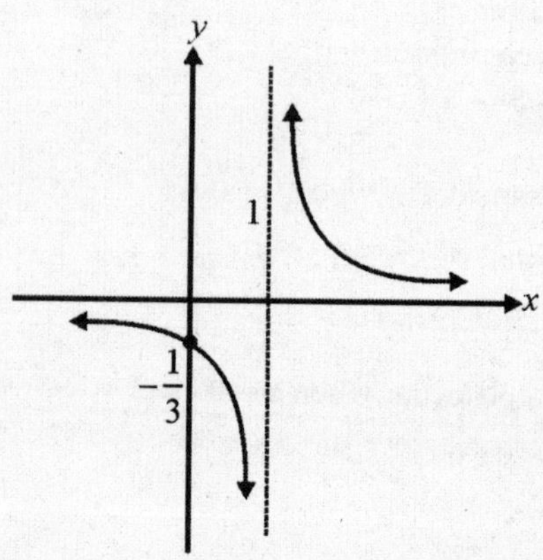


 

**6.** $f(x)=4x+1$

$x$-intercept when $y=0$:
$0=4x+1$
$x=-\frac{1}{4}$

$y$-intercept when $x=0$
$y=4(0)+1$
$=1$

Notice that the graph of $y=4x+1$ is $y=x$, stretched vertically about the $x$-axis by a factor of 4 and translated 1 unit up.

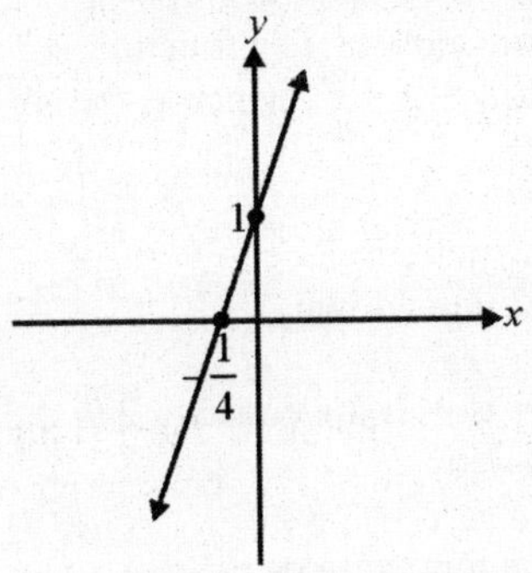


**7.** $f(x)=|x|+3$

$x$-intercept when $y=0$:
$0=|x|+3$
$-3=|x|$

There is no solution, therefore, there are no $x$-intercepts.
$y$-intercept when $x=0$:
$y=|0|+3$
$y=3$

Notice that the graph of $y=|x|+3$ is the graph of $y=|x|$ translated 3 units up.

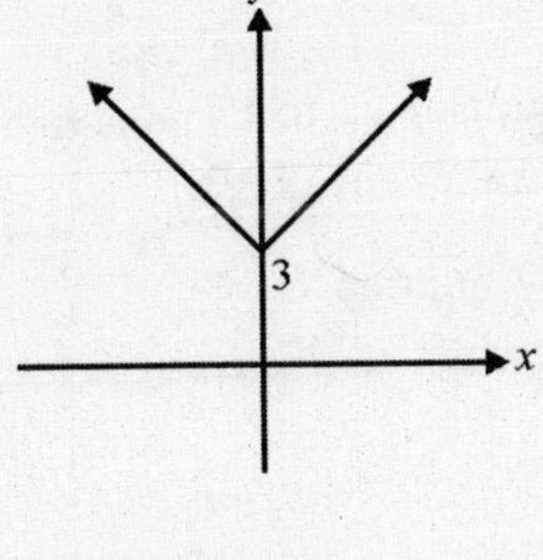


**8.** $f(x)=x^2+3$
$f(4)=4^2+3$
$f(4)=19$

**9.** $f(x)=-x^3+3x^2-x$
$f(-2)=-(-2)^3+3(-2)^2-(-2)$
$=-(-8)+3(4)+2$
$=22$

**10.** $f(x)=\sqrt{x+5}-5$
$f(4)=\sqrt{4+5}-5$
$=\sqrt{9}-5$
$=3-5$
$=-2$

**11.** $f(x)=\frac{1}{x-4}+5$
$f(5)=\frac{1}{5-4}+5$
$=6$

**12.** $f(3)=7$ (constant function)

**13. A**
$3x^2-7x+2=0$
$(3x-1)(x-2)=0$
$x=\frac{1}{3}, 2$

**14. B**
$f(-2)=-(-2)^3+2(-2)-1$
$=8-4-1$
$=3$

**15.**
$3x^2+5x-28=0$
$(3x-7)(x+4)=0$
$x=\frac{7}{3}, -4$
$x \doteq 2.33$

| 2 | . | 3 | 3 |
|---|---|---|---|

**16.**

$$0=\sqrt{3x-1}-2$$
$$\sqrt{3x-1}=2$$
$$3x-1=4$$
$$3x=5$$
$$x=\frac{5}{3}\doteq 1.67$$

| 1 | . | 6 | 7 |
|---|---|---|---|

## *Lesson 2 Horizontal and Vertical Translations*

### CLASS EXERCISES ANSWERS AND SOLUTIONS

**1.**

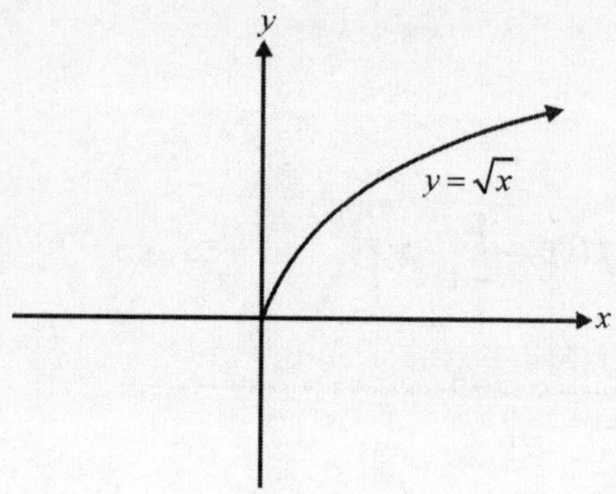


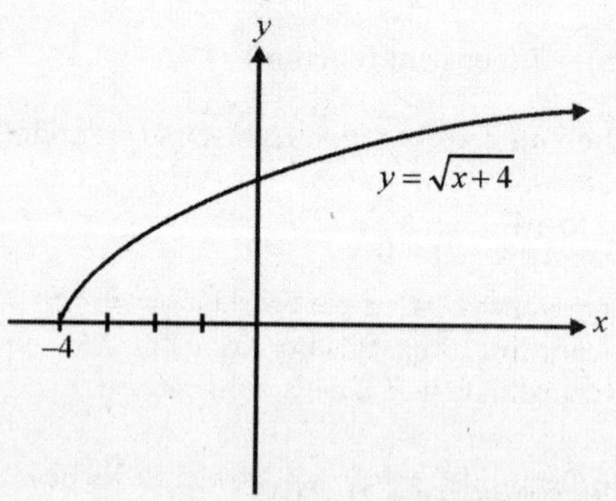


The graph of $y=\sqrt{x+4}$ is the graph of $y=\sqrt{x}$ translated 4 units to the left.

To follow the pattern of changing $y = f(x)$ to $y = f(x - h)$, consider $y = \sqrt{x+4}$ as $y=\sqrt{x-(-4)}$ and then you can see that $h = -4$. Thus, the horizontal translation has been 4 units to the left.

In general, always consider $y = f(x)$ being changed to $y = f(x - h)$ where $h$ is the horizontal translation.

**2.** The graph of $y = x^2$ contains point (3, 9). The graph of $y - 2 = x^2$ or $y = x^2 + 2$ contains point (3, 11). Point (3, 9) was translated vertically 2 units up.

**3.** The graph of $y = \sqrt{x} + 5$ is the graph of $y = \sqrt{x}$ translated 5 units up. The function $y = \sqrt{x} + 5$ is equivalent to $y - 5 = \sqrt{x}$ . Here, $k = 5$ and the vertical translation was 5 units up. In the form $y = \sqrt{x} + 5$, the constant 5 is the vertical translation.

**4.** The graph of $y = x^2 - 2$ is the graph of $y = x^2$ translated 2 units down. The function $y = x^2 - 2$ is equivalent to $y + 2 = x^2$, which is the same as $y - (-2) = x^2$.

Here, $k = -2$, which produces a vertical translation of 2 units down.

In the form $y = x^2 + 2$, the constant 2 is the vertical translation.

Thus, for the general case of $y = f(x) + k$, $k$ is the vertical translation.

### PRACTICE EXERCISES ANSWERS AND SOLUTIONS

**1.** 3 units left

**2.** $y=|x+2|-4$

**3.** 3 units right and 6 units up

**4.** $(x+4)^2+(y-2)^2-25=0$

**5.** 5 units to the right

**6.** $y=\frac{1}{x-2}+5$

**7.** 4 units down

**8.** The original graph is translated 5 units right so, $x$-intercept are at (8, 0) and (2, 0).

**9.** 4 units right and 3 units up

**10.** $y=\frac{1}{x-4}$

 

**11.** Translate the graph 7 units left, so replace $x$ with $x + 7$.

**12.** Translate the graph 2 units up, so the equation is

$$\frac{(y-2)^2}{16}-\frac{x^2}{4}=1$$

**13.** **B**
The points are translated 2 units left.

**14.** **B**

**15.** **C**

**16.**

$$y=(x-5-2)^2$$

$$y=(x-7)^2 \qquad \boxed{h=7}$$

| 7 | | | |
|---|---|---|---|

**17.**

$$y=(x-3)^2+3$$

$$y\geq 3$$

| 3 | | | |
|---|---|---|---|

## *Lesson 3 Reflecting Graphs*

### CLASS EXERCISES ANSWERS AND SOLUTIONS

**1.** You should see that the second graph is the first graph reflected in the $y$-axis.

Notice that the $x$-intercept changed from 2 to –2 but the $y$-intercept remained at –8.

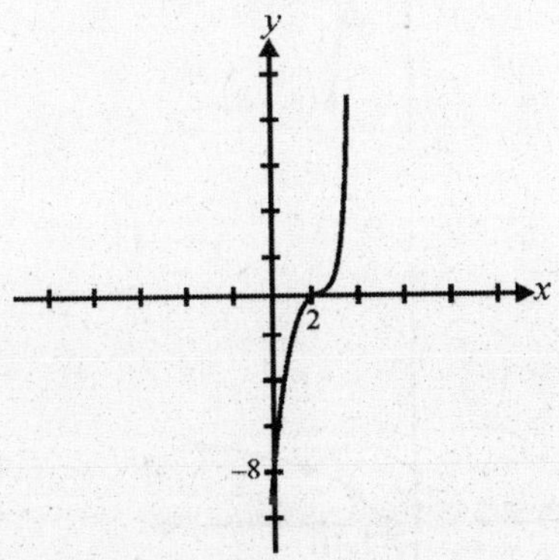


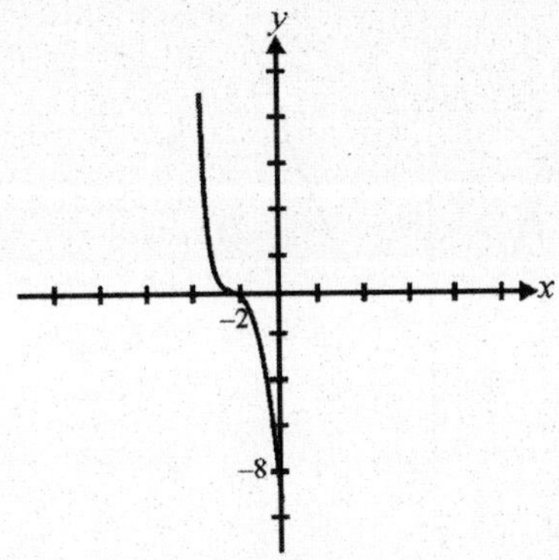


**2.**

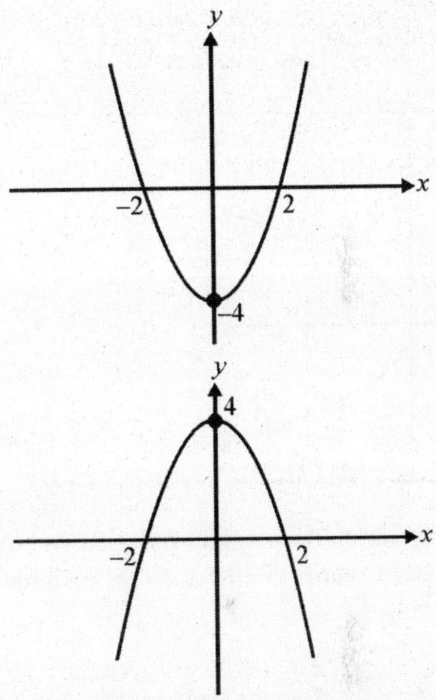


The $x$-intercepts remain constant (invariant points), but the $y$-intercept changes from –4 to +4.

For every point on the original graph, the $x$-coordinate remains the same but the $y$-coordinate will have a sign change.

For example, $y = x^2 - 4$ contains the point (3, 5) and $y = -x^2 + 4$ contains the point (3, –5)

> **In general, the graph of $-y = f(x)$ or $y = -f(x)$ is the graph of $y = f(x)$ reflected in the $x$-axis.**

Next, consider changing the function $y = f(x)$ to $x = f(y)$. That is, from the original function, interchange the $x$- and $y$-variables.

**3.**

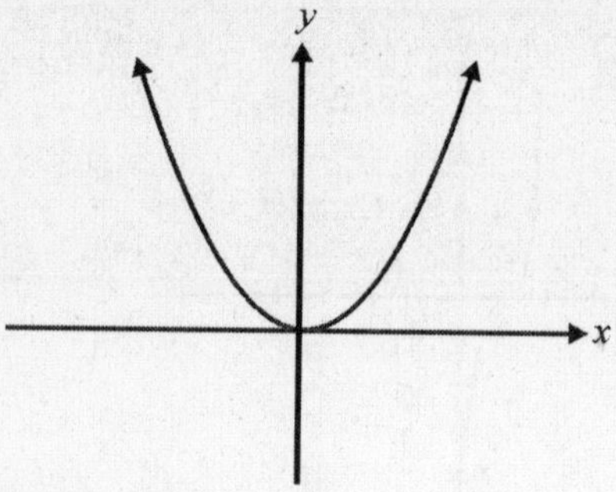


Now, graph $x = y^2$.
Solve for $y$:
$\pm\sqrt{x} = y$, which is $\sqrt{x} = y$ or $-\sqrt{x} = y$

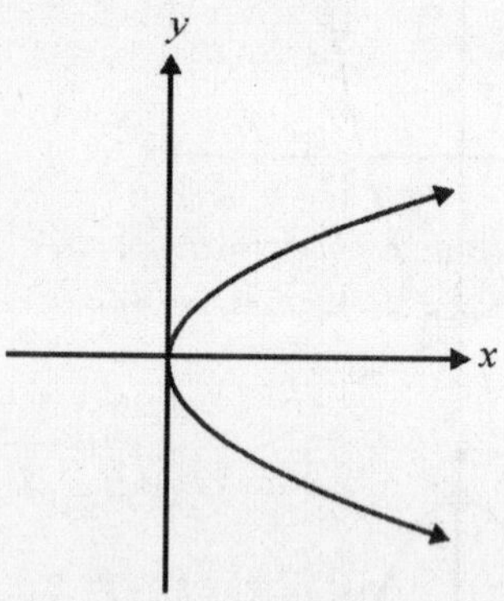


Now graph both the original function and its inverse in the same $xy$-plane, along with the line $y = x$.

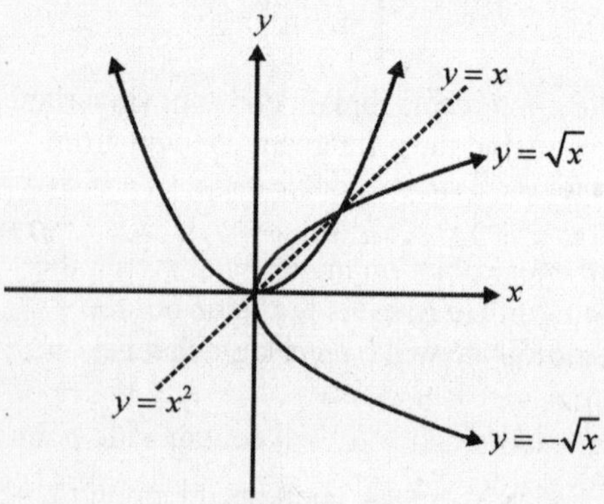


From these graphs you can see that the second graph, $y = \pm\sqrt{x}$ is the graph of $y = x^2$ reflected in the line $y = x$.

This inverse relation is not a function because there are $x$-coordinates paired with more than one $y$-coordinate. For example, (4, 2) and (4, –2). To enter this inverse relation in your calculator you need to enter it as two separate functions: $y_1 = \sqrt{x}$ and $y_2 = -\sqrt{x}$.

## PRACTICE EXERCISES ANSWERS AND SOLUTIONS

**1.** $y$-axis

**2.** $y = (-x)^2 - (-x) - 3$
$y = x^2 + x - 3$

**3.** $x$-axis

**4.** $y = -\left(x^2 - 5x - 3\right)$
$y = -x^2 + 5x + 3$

**5. a)**

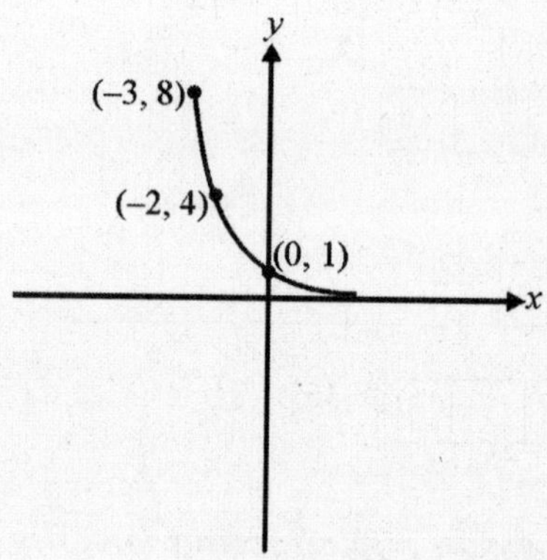


**b)**

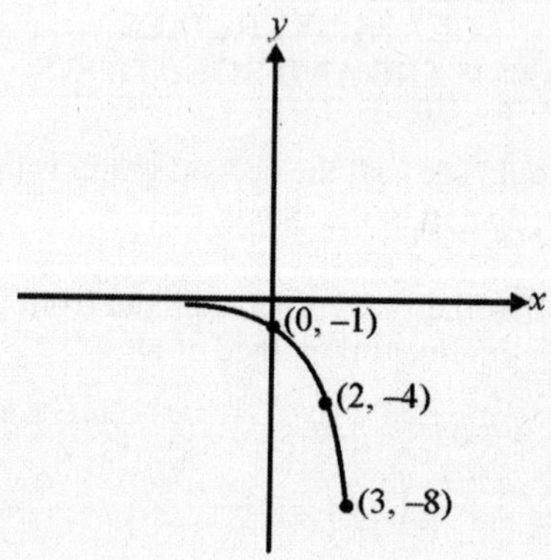


**c)**

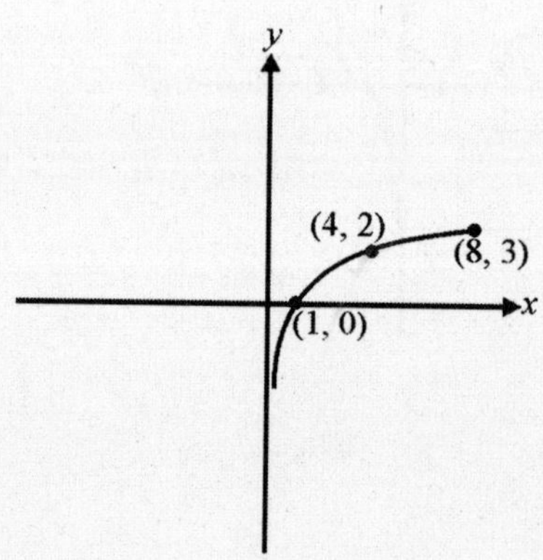


 

**6.** $y=(-x)^3-8$
$y=-x^3-8$

**7.** $x$-axis

**8.** $y=-\sqrt{2x+5}$

**9.** $y=(-x)^2$ Their graphs are the same.

**10.** A reflection in the $x$-axis gives $y=-(x^3)$.

A reflection in the $y$-axis gives $y=(-x)^3$.

Both are equivalent to $y=-x^3$.

**11.** $x=3y-7$
$3y=x+7$
$y=\frac{x+7}{3}$ $\quad f^{-1}(x)=\frac{x+7}{3}$

**12.** $(y-2)^2+(x+3)^2=36$
centre: (–3, 2)

**13. C**
(–3, –5) reflection in the $y$-axis

**14. D**
(–2, 3) reflection in the $x$-axis

**15.** $y=-(-x^3+9)$
$0=+x^3-9$
$x^3=9$
$x=\sqrt[3]{9}\doteq 2.08$

| | | | |
|---|---|---|---|
| 2 | . | 0 | 8 |

**16.** $x=3^y-9$
$0=3^y-9$
$y=2$

| | | | |
|---|---|---|---|
| 2 | | | |

## Lesson 4 Horizontal and Vertical Stretches

### CLASS EXERCISES ANSWERS AND SOLUTIONS

**1.**

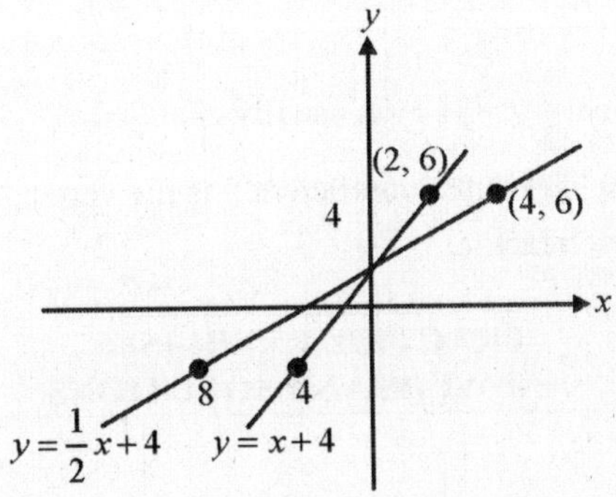


From the first graph, if you keep the $y$-coordinate the same, you will notice that the $x$-coordinate becomes twice the original.

For example, (2, 6) is a point on the graph of $y=x+4$. If the $y$-coordinate is 6 for the new function, then

$$6=\frac{1}{2}x+4$$
$$12=x+8$$
$$x=4$$

Thus, (4, 6) is a point on the graph of $y=\frac{1}{2}x+4$.

Also notice that the $x$-intercept has moved from –4 to –8. Thus, the point (–4, 0) has changed to (–8, 0).

In this case, the original graph was stretched horizontally about the $y$-axis by a factor of 2.

**2.** The graph of $y=f(-3x)$ is the graph of $y=f(x)$ stretched horizontally about the $y$-axis by a factor of $\frac{1}{3}$ and reflected in the $y$-axis.

**3.** Since $\frac{1}{a}=\frac{1}{3}$, the vertical stretch factor is 3. Thus, the graph of $y=f(x)$ is stretched vertically about the $x$-axis by a factor of 3 to give the graph of $\frac{1}{3}y=f(x)$.

Since $\frac{1}{3}y=f(x)$ is usually changed to $y=3f(x)$, the coefficient 3 is the vertical stretch factor.

## PRACTICE EXERCISES ANSWERS AND SOLUTIONS

**1.** horizontally, $\frac{1}{3}$

**2.** vertically, 2

**3. a)** $y=(2x)^2+3(2x)\rightarrow y=4x^2+6x$

**b)** $y=\frac{1}{2}(x^2+3x)\rightarrow y=\frac{1}{2}x^2+\frac{3}{2}x$

**4. a)** (2, 3), horizontal stretch by a factor of $\frac{1}{3}$

**b)** (6, –6), vertical stretch by a factor of 2 and reflection in the $x$-axis

**5.** $y=\left|-\frac{1}{2}x+3\right|$

**6.** vertically, 5

**7.** horizontally, 2

**8.** The circle is stretched horizontally and vertically by a factor of 2.

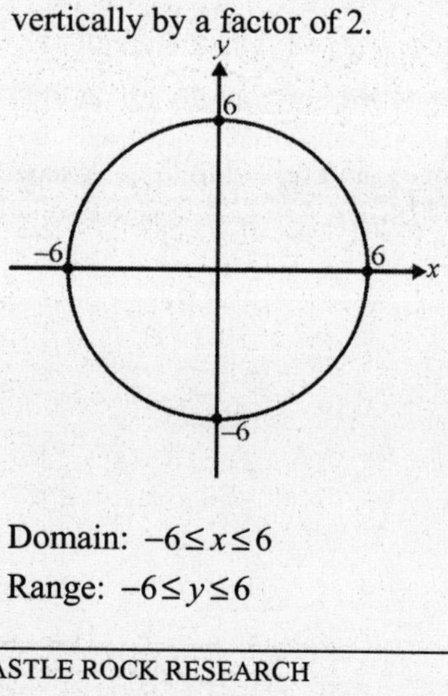


Domain: $-6\le x\le 6$

Range: $-6\le y\le 6$

**9.** $x$-axis, $\frac{1}{|k|}$

**10.** $y=3\sqrt{\frac{1}{2}x}$

**11. a)** 8

**b)** $\frac{1}{2}$  $y=8x^3\rightarrow y=(2x)^3$

**12.** $\left(\frac{1}{3}a,2b\right)$

**13.** The circle is stretched vertically by a factor of $\frac{1}{2}$.

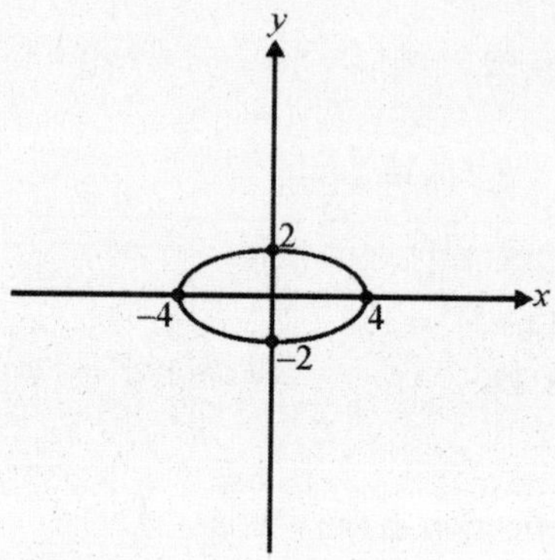


Domain: $-4\le x\le 4$

Range: $-2\le y\le 2$

**14.** $(a,-2b)$

**15.** $y=\sqrt{-\frac{1}{4}x}\rightarrow y=\frac{\sqrt{-x}}{2}$

**16.** $y=-3(|x|-2)\rightarrow y=-3|x|+6$

**17.** Doing the transformations in the order given results in:

**i)** $y=(2x)^3+3$

**ii)** $y=2\left[(2x)^3+3\right]$

**iii)** $y=-2\left[(2x)^3+3\right]$

Simplifying:

$y=-2\left[8x^3+3\right]$

$y=-16x^3-6$

 

**18.** In order, the transformations give:

**i)** $y = \frac{3}{x-2}$

**ii)** $y = \frac{3}{\frac{1}{2}x - 2}$

**iii)** $y = \frac{3}{\frac{1}{2}x - 2} \rightarrow y = \frac{-6}{x+4}$

**19.** Vertical stretch by a factor of 2; reflection in the $x$-axis

**20.** Horizontal stretch by a factor of 3; reflection in the $x$-axis

**21.** Vertical stretch by a factor of $\frac{1}{4}$; reflection in the $x$-axis

**22.** Horizontal stretch by a factor of $\frac{1}{3}$; vertical stretch by a factor of $\frac{1}{2}$; reflection in the $x$-axis

**23.** Notice that $y = \frac{2}{3^x}$ can be written $y = (2)(3^{-x})$.

Vertical stretch by a factor of 2; reflection in the $y$-axis

**24.** Horizontal stretch by a factor of $\frac{1}{3}$; vertical stretch by a factor 2; reflection in the $x$-axis

**25.** $y = 3x^2 - 3$

**26.** $y = \left|\frac{1}{2}x\right| + 1$

**27.** $y = -4x^2 + 8$

**28.** $y = \sin\left(\frac{1}{2}x\right)$

**29.** C

**30.** A

**31.** For $f(x) = (3x-2)(2x+3)(2x-1)$, the largest $x$-intercept is $\frac{2}{3}$.

For $y = f(4x)$, this intercept is $\frac{1}{4} \times \frac{2}{3} = \frac{1}{6}$.

$\frac{1}{6} \doteq 0.17$

| 0 | . | 1 | 7 |
|---|---|---|---|

**32.** Graph is stretched horizontally by a factor of 2.

The value of $k$ is $\frac{1}{2}$.

| 0 | . | 5 | 0 |
|---|---|---|---|

## *Practice Quiz*

### ANSWERS AND SOLUTIONS

**1.**
$$\begin{aligned} f(-3) &= -2(-3)^3 + 4(-3) - 7 \\ &= -2(-27) - 12 - 7 \\ &= 54 - 12 - 7 \\ &= 35 \end{aligned}$$

**2.** $f(x) = 4x^2 - 1$

$x$-intercepts;
$$0 = 4x^2 - 1$$
$$1 = 4x^2$$
$$x^2 = \frac{1}{4}$$
$$x = \pm\frac{1}{2}$$

$y$-intercept
$$y = (0)^2 - 1$$
$$y = -1$$

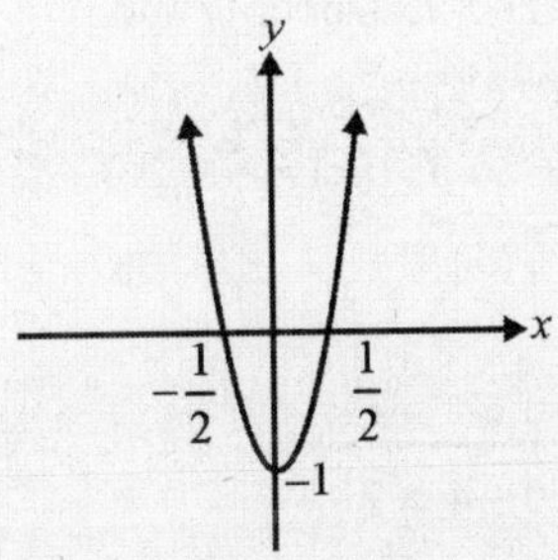

**3.** $y = f(x) = \dfrac{x-2}{6x^2-13x-5}$

Non-permissible values when

$6x^2-13x-5=0$

$(3x+2)(2x-5)=0$

$x = \dfrac{-2}{3}, \dfrac{5}{2}$

**4.** $y=\dfrac{1}{x} \rightarrow y=\dfrac{1}{x-3}+4$

Translations of 3 units right, 4 units up.

**5.** Vertices are translated 3 units left, 1 unit up.

Thus, the equation is $\dfrac{(x+3)^2}{25}-\dfrac{(y-1)^2}{4}=1.$

**6.** Translate 3 left, 2 up.

Thus, the equation is

$(x+3)^2+(y-2)^2-25=0.$

**7.** $y=\sqrt{4x+2}$

Reflected in the $y$-axis $\rightarrow$ $y=\sqrt{-4x+2}$

**8.** $f(x)=\dfrac{1}{x-2}$

Inverse:

$x=\dfrac{1}{y-2}$

$(y-2)(x)=1$

$y-2=\dfrac{1}{x}$

$y=\dfrac{1}{x}+2 \quad \therefore f^{-1}(x)=\dfrac{1}{x}+2$

**9.** $f(x)=\sqrt{x+4}$

$y$-intercept: $y=\sqrt{0+4}=2$

Thus, the $x$-intercept for $y=f^{-1}(x)$ is 2.

**10.** $(x-4)^2+(y+1)^2=25$

Reflected in

$y=0: (-x-4)^2+(y+1)^2=25$

$(x+4)^2+(y+1)^2=25$

The centre of the new circle is (–4, –1).

**11.** The graph of $y=5x^2-5$ is the graph of $y=x^2-1$ stretched <u>vertically</u> about the <u>$x$</u>-axis by a factor of <u>5</u>.

**12.** The graph is stretched vertically about the $x$-axis by a factor of 3 and reflected in the $x$-axis. Thus, (2, –3) becomes (2, 9).

**13.** The graph is stretched vertically about the $x$-axis by a factor of 3 and horizontally about the $y$-axis by a factor of $\dfrac{1}{4}$. Thus, $(a, b)$ becomes $\left(\dfrac{a}{4}, 3b\right)$.

**14.** $y=\sqrt{x-3}$

$\rightarrow y=\dfrac{1}{2}\sqrt{x-3}$ vertical stretch by $\dfrac{1}{2}$

$\rightarrow y=\dfrac{1}{2}\sqrt{\dfrac{x}{4}-3}$ horizontal stretch by 4

$\rightarrow y=-\dfrac{1}{2}\sqrt{\dfrac{x}{4}-3}$ reflection in the $x$-axis

**15.** $y=|2x| \rightarrow y=\left|\dfrac{1}{2}(x-1)\right|+3$

**i)** Horizontal stretch about the $y$-axis by a factor of 4: $y=\left|\dfrac{2}{4}x\right|$

**ii)** Translations of 1 unit right and 3 units up:

$y=\left|\dfrac{1}{2}(x-1)\right|+3$

## Lesson 5 Combining Transformations

### CLASS EXERCISES ANSWERS AND SOLUTIONS

**1.** The results of the transformations are:

**i)** Stretched vertically be a factor of 2.

$\frac{1}{2}y=\frac{1}{x}\rightarrow y=2\left(\frac{1}{x}\right)$

**ii)** Reflected in the $x$-axis.

$y=-2\left(\frac{1}{x}\right)$

**iii)** Translated 3 units left and 1 unit down.

$y=-2\left(\frac{1}{x+3}\right)-1$

Simplifying:

$y=-2\left(\frac{1}{x+3}\right)-1$

$y=\frac{-2}{x+3}-1$

$y=\frac{-2-1(x+3)}{x+3}$

$y=\frac{-2-1x-3}{x+3}$

$y=\frac{-x-5}{x+3}$

A sketch from the calculator of the two functions gives:

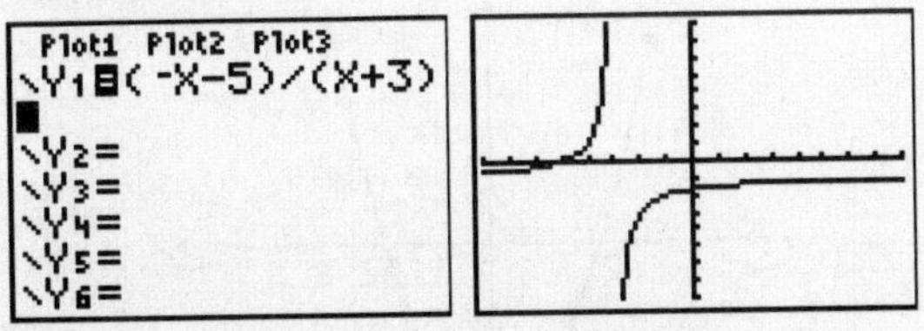


### PRACTICE EXERCISES ANSWERS AND SOLUTIONS

**1.** $y=2\sqrt{x-3}$

**2.** Horizontal stretch by a factor of $\frac{1}{2}$; translation 3 right $(4,7)\rightarrow(2,7)\rightarrow(5,7)$

**3.** $(3,6)\rightarrow(6,6)\rightarrow(6,10)$

**4.** $y=f(3x-9)\rightarrow y=f\left[3(x-3)\right]$
$(3,7)\rightarrow(1,7)\rightarrow(4,7)$

**5.** **i)** $y=2x^3$

**ii)** $y=2\left(\frac{1}{3}x\right)^3\rightarrow y=\frac{2}{27}x^3$

**6.** $(-2,8)\rightarrow(-2,4)\rightarrow(-6,4)$

**7.** $y=(2x-4)^2\rightarrow y=\left[2(x-2)\right]^2$
Horizontal stretch about the $y$-axis by a factor of $\frac{1}{2}$; translate 2 units right

**8.**

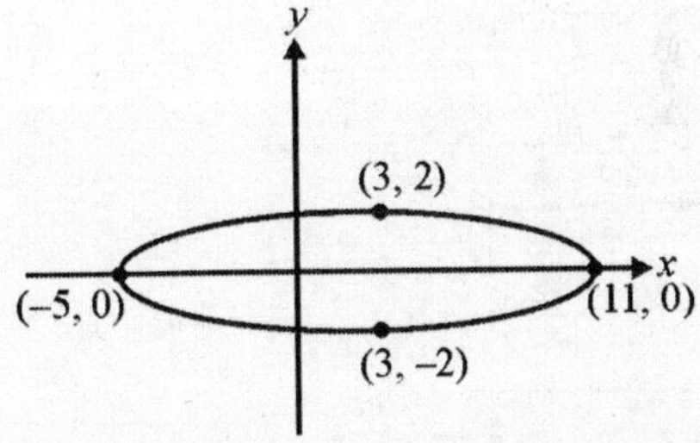


**9.** $(3,7)\rightarrow(3,14)\rightarrow(3,-14)\rightarrow(3,-11)$

**10.** $(a,b)\rightarrow(2a,b)\rightarrow(2a+4,b)$

**11.** **i)** $y=3|x|$

**ii)** $y=3|2x|$

**iii)** $y=3|2x|-5$

**12.** $y=\frac{1}{2x+6}\rightarrow y=\frac{1}{2(x+3)}$

The graph of $y=\frac{1}{x}$ has been stretched vertically (or horizontally) by a factor of $\frac{1}{2}$ and translated 3 units left.

**13.** **i)** $y=3(x^2-1)$

**ii)** $y=-3(x^2-1)$

**iii)**

$y=-3\left(\left(x-2\right)^2-1\right)+4$

$y=3\left(x^2-4x+4-1\right)+4$

$y=3x^2-12x+9+4$

$y=3x^2-12x+13$

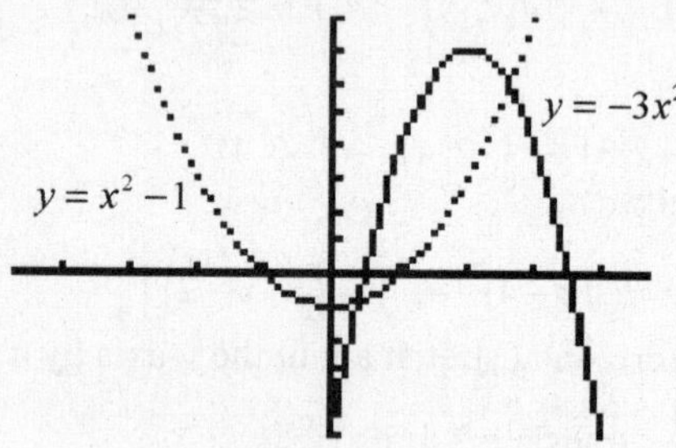


**14. i)** $y=\sqrt{\frac{1}{2}x+2}$

**ii)** $y=\sqrt{-\frac{1}{2}x+2}$

**iii)**

$y=\sqrt{-\frac{1}{2}(x-3)+2}+1$

$y=\sqrt{-\frac{1}{2}x+\frac{3}{2}+2}+1$

$y=\sqrt{-\frac{1}{2}x+\frac{7}{2}}+1$

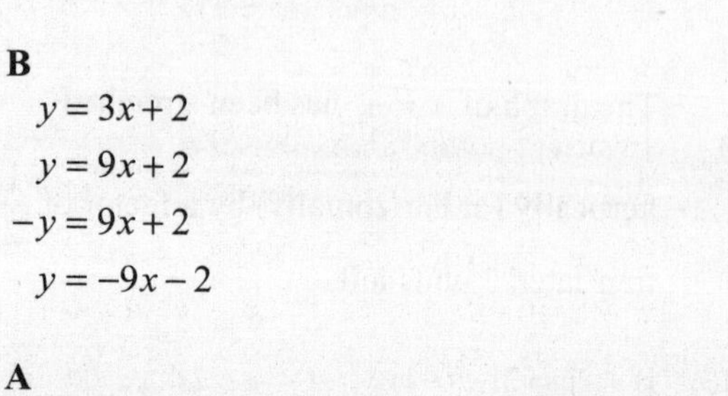


**15. B**

$y=3x+2$

$y=9x+2$

$-y=9x+2$

$y=-9x-2$

**16. A**

$(a,b)\rightarrow(b,a)\rightarrow(b,a+2)$

**17. C**

$y=6x^3$

$y=6(-x)^3$

$y=6\left(-(x+1)\right)^3-2$

$y=6(-x-1)^3-2$

**18.** $(6,8)\rightarrow\left(6,\frac{8}{3}\right)\rightarrow\left(7,\frac{2}{3}\right)$

| | | | |
|---|---|---|---|
| 0 | . | 6 | 7 |

**19.** $(1,2)\rightarrow\left(\frac{1}{3},2\right)\rightarrow\left(\frac{10}{3},2\right)$

| | | | |
|---|---|---|---|
| 3 | . | 3 | 3 |

**20.** $(4,6)\rightarrow\left(\frac{4}{3},3\right)\rightarrow\left(\frac{13}{3},3\right)$

$\frac{13}{3}\doteq 4.33$

| | | | |
|---|---|---|---|
| 4 | . | 3 | 3 |

## *Lesson 6 Graphing Reciprocal Functions*

### CLASS EXERCISES ANSWERS AND SOLUTIONS

1. First, sketch $y=x^2-4$.

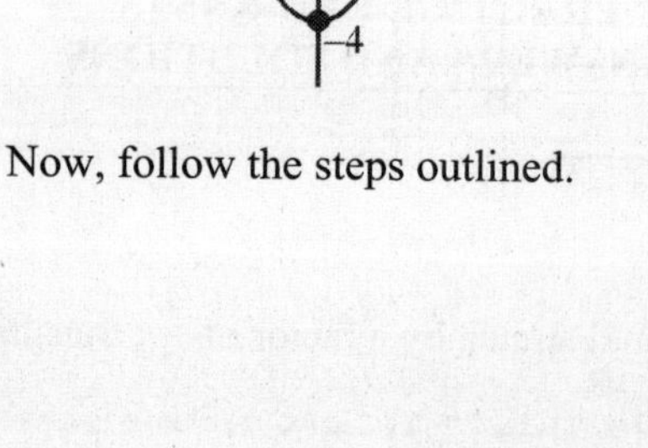


Now, follow the steps outlined.

**Step 1**

Show vertical asymptotes

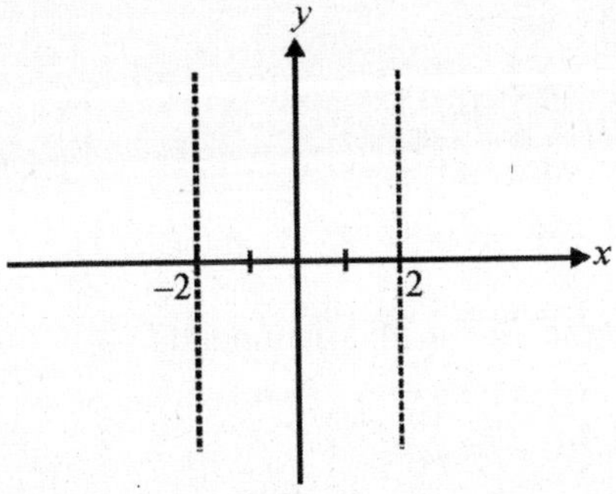


**Step 2**

Find invariant points ($y = 1$ or $y = -1$ on the graph of $y = f(x) = x^2 - 4$)

If $y = 1$:

$1 = x^2 - 4$

$5 = x^2$

$\pm\sqrt{5} = x$

If $y = -1$:

$-1 = x^2 - 4$

$3 = x^2$

$\pm\sqrt{3} = x$

These points are $(\sqrt{5}, 1), (-\sqrt{5}, 1), (\sqrt{3}, -1)$, and $(-\sqrt{3}, -1)$.

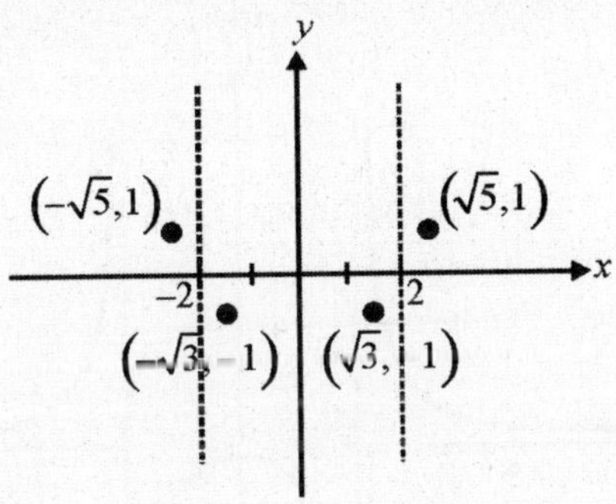


**Step 3**

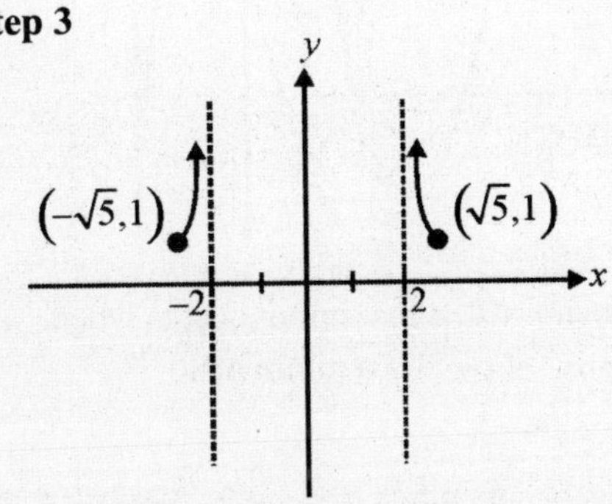


**Step 4**

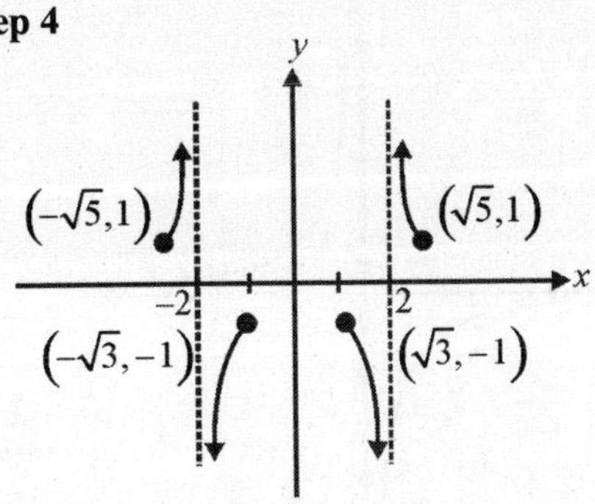


**Step 5**

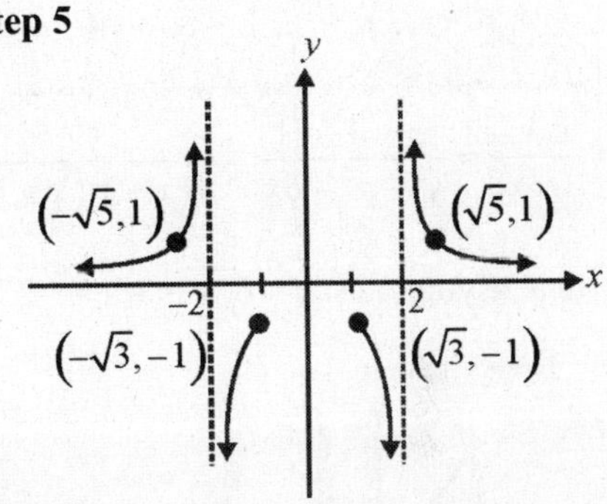


**Step 6**

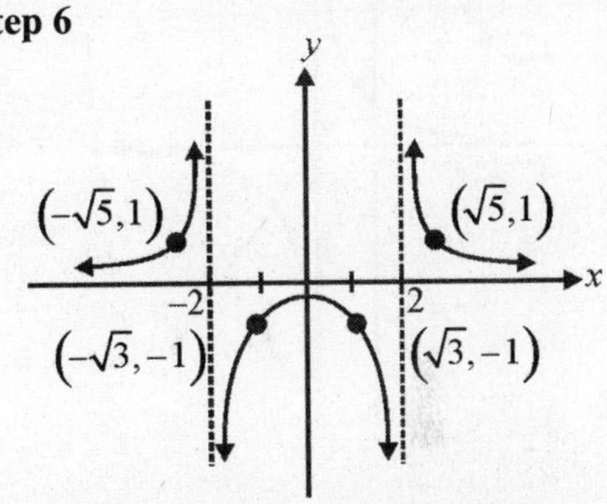


## PRACTICE EXERCISES ANSWERS AND SOLUTIONS

1. $y = \dfrac{1}{f(x)} = \dfrac{1}{x^2 - x - 12} = \dfrac{1}{(x-4)(x+3)}$

   vertical asymptotes: $x = 4, x = -3$
   horizontal asymptote: $y = 0$

2. $x = -3, x = 1, x = 2$

3. Invariant points when $y = \pm 1$

   When $y = 1$:

   $x^2 - x - 5 = 1$

   $x^2 - x - 6 = 0$

   $(x-3)(x+2) = 0$

   $x = 3, -2$

   When $y = -1$:

   $x^2 - x - 5 = -1$

   $x^2 - x - 4 = 0$

   $x = \dfrac{1 \pm \sqrt{1 - 4(1)(-4)}}{2(1)}$

   $x = \dfrac{1 \pm \sqrt{17}}{2}$

**4.**

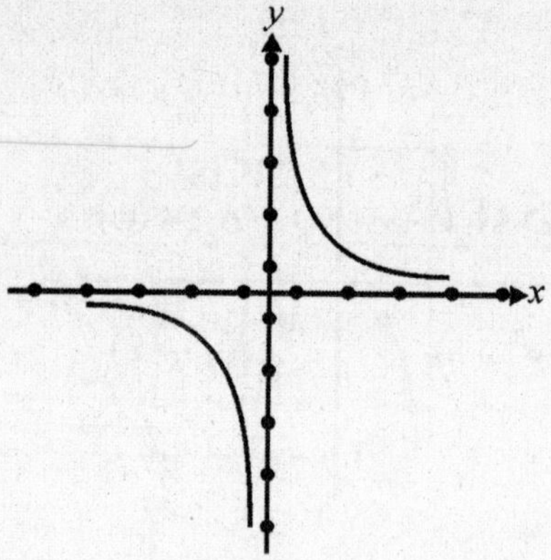


**5.** $y=\dfrac{1}{f(x)}=\dfrac{1}{x^3-x^2-6x}=\dfrac{1}{x(x-3)(x+2)}$

Vertical asymptotes: $x=0$, $x=3$, $x=-2$

**6.**

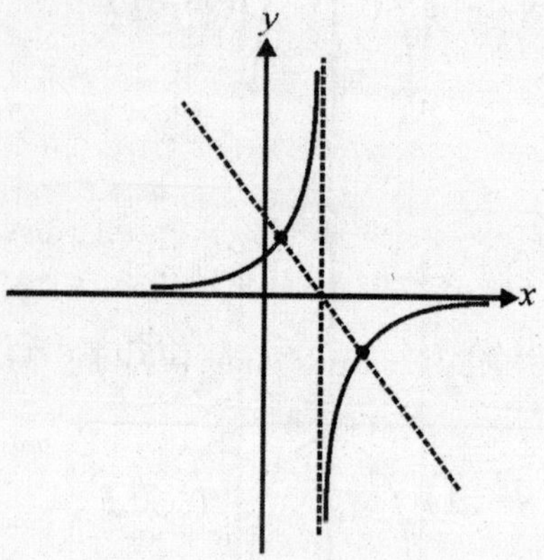


**7.** $4, \dfrac{1}{4}$

**8.**

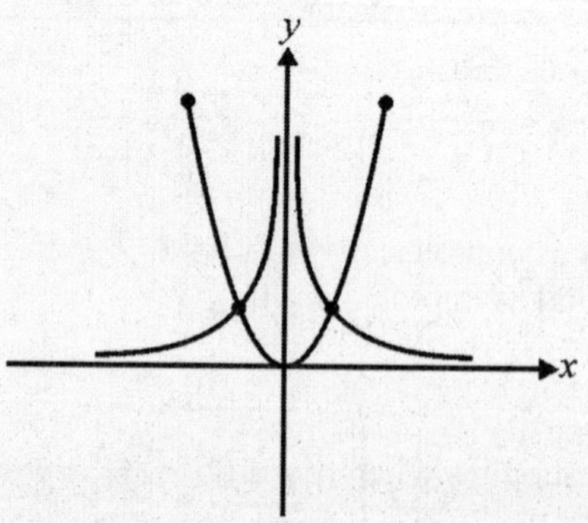


**9.** The vertical asymptotes are defined by the equations $x=4$, and $x=-3$.

**10.**

$$x+3=\frac{1}{x+3}$$
$$(x+3)^2=1$$
$$x+3=\pm 1$$
$$x=-4,\ -2$$

The points are (–4, –1) and (–2, 1).

These are the invariant points.

**11.**

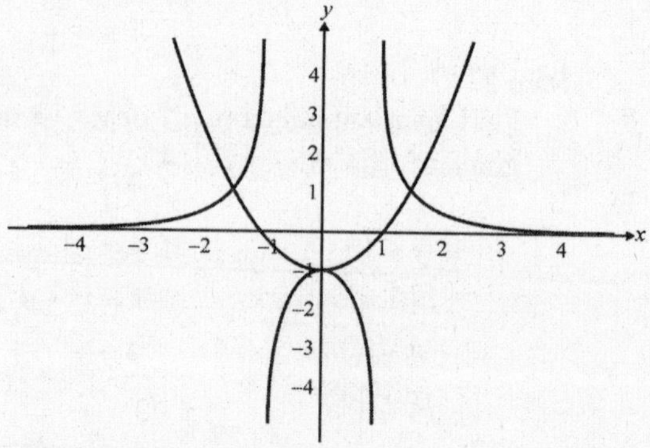


**12.** From the calculator, the graph of

$y=\dfrac{1}{f(x)}=\dfrac{1}{x^2-x-5}$ is

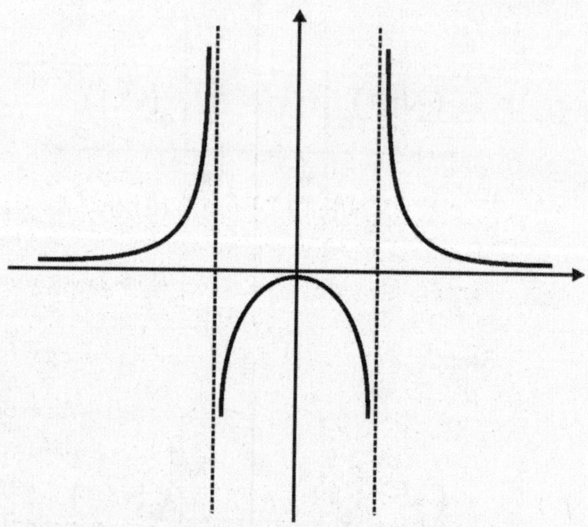

When $y<0$, a maximum occurs when $y=x^2-x-5$ is a minimum.

This occurs when $x=\dfrac{-b}{2a}=\dfrac{-5}{2(1)}=\dfrac{1}{2}$.

When $x=\frac{1}{2}$, $y=\left(\frac{1}{2}\right)^2-\frac{1}{2}-5=-5.25$.

Therefore, for $y<0$, the maximum is $\frac{1}{-5.25}=-\frac{4}{21}$.

The range is $y>0$ or $y\le -\frac{4}{21}$.

**13. D**

$$x^2-4x-12=0$$
$$(x-6)(x+2)=0$$
$$x=6,-2$$

**14. B**

$$x^2-3x-17=1$$
$$x^2-3x-18=0$$
$$(x-6)(x+3)=0$$
$$x=6,-3$$

The points are (6, 1) and (–3, 1).

**15.** $y=\frac{1}{x^2-4}$ has a horizontal asymptote of $y=0$, so $y=\frac{1}{x^2-4}+3$ has a horizontal asymptote of $y=3$.

| 3 | | | |
|---|---|---|---|

**16.** Invariant points when $-\frac{1}{2}x^2+2x-1=\pm1$

| | |
|---|---|
| $-\frac{1}{2}x^2+2x-1=1$ | $-\frac{1}{2}x^2+2x-1=-1$ |
| $-\frac{1}{2}x^2+2x-2=0$ | $-\frac{1}{2}x^2+2x=0$ |
| $x^2-4x+4=0$ | $x^2-4x=0$ |
| $(x-2)^2=0$ | $x(x-4)=0$ |
| $x=2$ | $x=0,4$ |

The sum of the $x$-coordinates is 6.

| 6 | | | |
|---|---|---|---|

## *Lesson 7 Stretching Graphs About Lines Parallel to the x- and y-axes*

### CLASS EXERCISES ANSWERS AND SOLUTIONS

**1. a)**

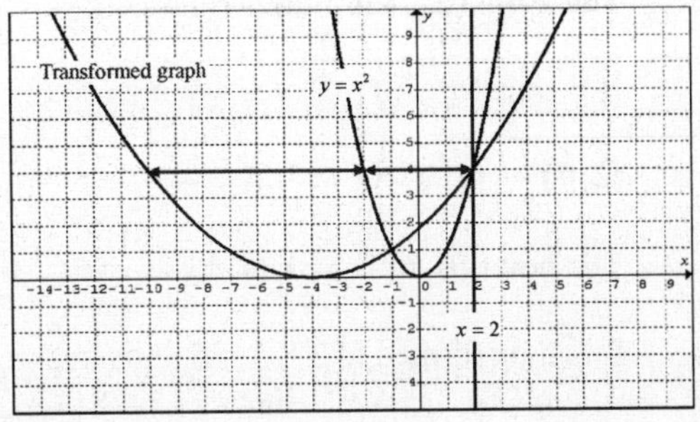


Notice that any point on the transformed function is three times as far as the original from $x=2$. If the point is on the negative (left) side of $x=2$, the transformed point is three times as far away in the negative direction.

**b)** Consider the point (3, 9) on the original graph. The point (3, 9) is one unit from $x=2$. Since the stretch factor is 3, this distance must become 3 units from $x=2$.

Thus, (3, 9) is transformed to (5, 9). The point (5, 9) is 3 units from $x=2$.

### PRACTICE EXERCISES ANSWERS AND SOLUTIONS

**1.** The point (3, 5) is 4 units above $y=1$.

It is transformed to $4\times4=16$ units above $y=1$.

$(3, 5)\rightarrow(3, 17)$

**2.**

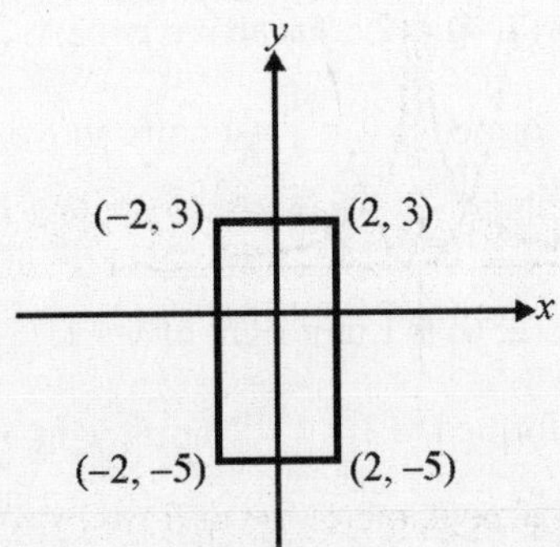


**3.** The point (1, 1) is 1 unit left of $x=2$.

It is transformed to $3\times1=3$ units left of $x=2$.

$(1, 1)\rightarrow(-1, 1)$

**4.** The point (–2, –8) is 9 units below $y = 1$.

It is transformed to $2 \times 9 = 18$ units below $y = 1$.

(–2, –8) → (–2, –17)

**5.** The point (2, 7) is 3 units right of $x = -1$.

It is transformed to $2 \times 3 = 6$ units right of $x = -1$.

(2, 7) → (5, 7)

**6.** The point (6, 6) is 3 units above $y = 3$.

It is transformed to $2 \times 3 = 6$ units above $y = 3$.

(6, 6) → (6, 9)

**7.** The point (4, 16) is 2 units right of $x = 2$.

It is transformed to $3 \times 2 = 6$ units right of $x = 2$.

(4, 16) → (8, 16)

**8.** The $x$-intercept is –2 at (–2, 0), and the $y$-intercept is 2 at (0, 2).

The point (–2, 0) is 5 units left of $x = 3$.

It is transformed to $2 \times 5 = 10$ units left of $x = 3$.

(–2, 0) → (–7, 0)

The point (0, 2) is 3 units left of $x = 3$.

It is transformed to $2 \times 3 = 6$ units left of $x = 3$.

(0, 2) → (–3, 2)

**9.** The point (2, 4) is 2 units above $y = 2$.

It is transformed to $2 \times 2 = 4$ units above $y = 2$.

(2, 4) → (2, 6)

The point (2, 6) is 1 unit right of $x = 1$.

It is transformed to $3 \times 1 = 3$ units right of $x = 1$.

(2, 6) → (4, 6)

**10.** The point (–2, 3) is 3 units right of $x = -5$.

It is transformed to $4 \times 3 = 12$ units right of $x = -5$.

(–2, 3) → (7, 3)

**11.** The original $y$-intercepts are 5 and –5.

The $y$-intercept of 5 is 3 units above $y = 2$.

It is transformed to $2 \times 3 = 6$ units above $y = 2$.

The new $y$-intercept is 8.

The $y$-intercept of –5 is 7 units below $y = 2$.

It is transformed to $2 \times 7 = 14$ units below $y = 2$.

The new $y$-intercept is –12.

Thus, the new $y$-intercepts are 8 and –12.

**12.**

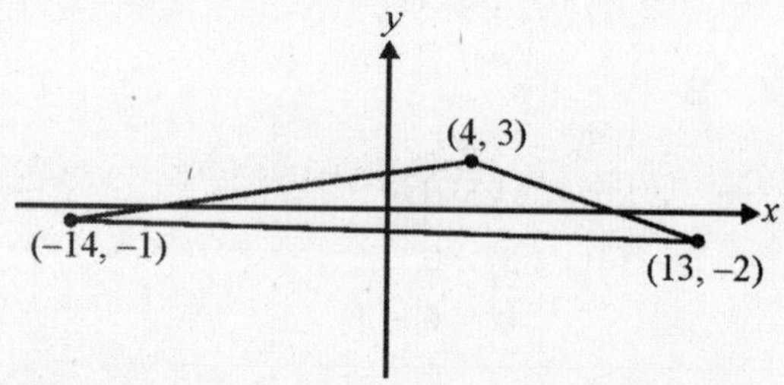


**13. a)**

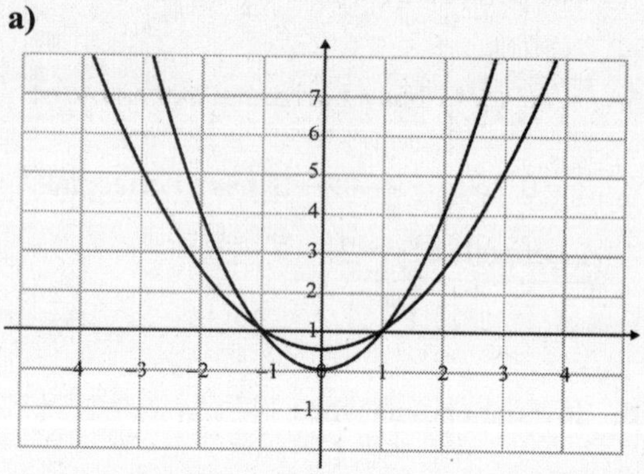


**b)** The point (2, 4) is transformed to $\left(2, 2\frac{1}{2}\right)$.

**c)** (–4, 16) is 15 units above $y = 1$.

It is transformed to $7\frac{1}{2}$ units above $y = 1$.

(–4, 16) is transformed to $\left(-4, 8\frac{1}{2}\right)$.

**14. A**

The minimum occurs at (2, 5).

The point (2, 5) is 1 unit right of $x = 1$.

It is transformed to $3 \times 1 = 3$ units right of $x = 1$.

The point (2, 5) is transformed to (4, 5).

**15. D**
The range of $y = x^2 - 6x + 10$ is $y \geq 1$.

The line $y = 1$ is 3 units above $y = -2$.

The distance is stretched by a factor of 4 so that the smallest range value will be $4 \times 3 = 12$ units above $y = -2$.

Thus, the new range is $y \geq 10$.

**16.** The point (4, 5) is 2 units right of $x = 2$ and the point (10, 5) is 8 units right of $x = 2$.

The value of $k$ is 4.

| 4 | | | |
|---|---|---|---|

## *Practice Test*

### ANSWERS AND SOLUTIONS

**1.** $f(-2) = \dfrac{-3}{6(-2)-2} = \dfrac{-3}{-14} = \dfrac{3}{14}$

**2.** $f(x) = \sqrt{x+4} - 3$

*x*-intercept:
$0 = \sqrt{x+4} - 3$
$3 = \sqrt{x+4}$
$9 = x + 4$
$x = 5$

*y*-intercept:
$y = \sqrt{0+4} - 3$
$y = 2 - 3$
$y = -1$

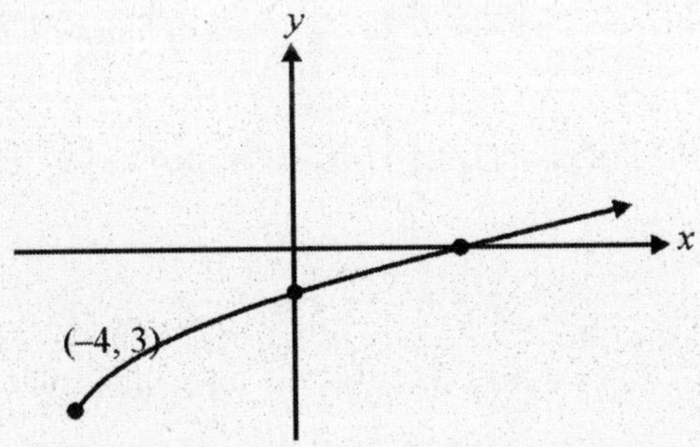


**3.** $y = \dfrac{1}{2x^2 + 7x - 4}$

Vertical asymptotes occur when $2x^2 + 7x - 4 = 0$.

$(2x-1)(x+4) = 0$

Thus, $x = \dfrac{1}{2}$ and $x = -4$ are the vertical asymptotes.

**4.** $y = \dfrac{1}{x}$

$y = \dfrac{3}{x}$, vertical stretch by factor of 3

$y = \dfrac{3}{x+6}$, horizontal translation 6 units left

**5.** $f(x) = \dfrac{1}{2x-1}$

Find the inverse:

$$x = \frac{1}{2y-1}$$
$$(2y-1)x = 1$$
$$2y - 1 = \frac{1}{x}$$
$$2y = \frac{1}{x} + 1$$
$$y = \frac{1}{2x} + \frac{1}{2}$$
$$y = \frac{1+x}{2x}$$
$$\therefore f^{-1}(x) = \frac{1+x}{2x}$$

**6.** $y = -2f(3x)$

$(a,b) \rightarrow \left(\frac{1}{3}a, 2b\right)$

- Stretched vertically by a factor of 2 and horizontally by a factor of $\frac{1}{3}$

$\left(\frac{1}{3}a, 2b\right) \rightarrow \left(\frac{1}{3}a, -2b\right)$

- Reflected in the *x*-axis

**7.** $y = x^3$

$y = \frac{1}{3}x^3$ • Stretched vertically by a factor of $\frac{1}{3}$

$y = \frac{1}{3}\left(\frac{1}{5}x\right)^3$ • Stretched horizontally by a factor of 5

$y = \frac{1}{3}\left(\frac{1}{5}x\right)^3 + 2$ • Translated 2 units up

$y = \frac{1}{375}x^3 + 2$ • Simplified

**8.** $y = x^2 \rightarrow y = (2x-6)^2 + 4$

$y = [2(x-3)]^2 + 4$

The graph would be:

**i)** stretched horizontally about the $y$-axis by a factor of $\frac{1}{2}$

**ii)** translated 3 units right and 4 units up

**9.** $y = \sqrt{x}$

**i)** Stretched vertically about the $x$-axis by a factor of 2: $y = 2\sqrt{x}$

**ii)** Reflected in the $y$-axis: $y = 2\sqrt{-x}$

**iii)** Translated 1 unit left, 4 units up: $y = 2\sqrt{-(x+1)} + 4$ or $y = \sqrt{-x-1} + 4$

**10.**

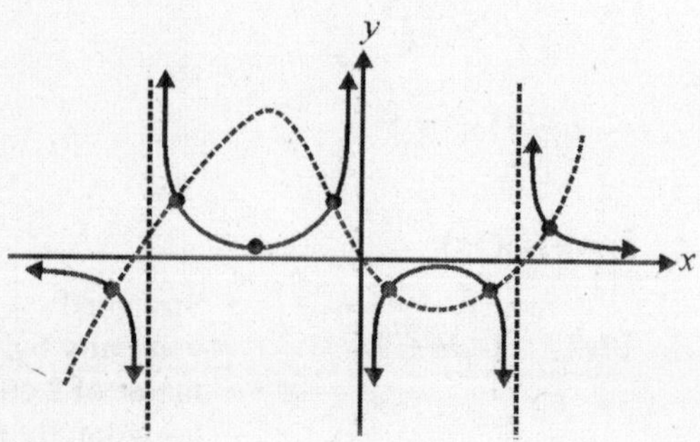


-- $y = f(x)$

— $y = \frac{1}{f(x)}$

**11.** The point (2, 6) moves to a location half as far from $x = -4$.

It is 6 units from $x = -4$.
It moves to 3 units from $x = -4$.

Thus, (2, 6) is transformed to (–1, 6).

**12.** $x^2 + y^2 = 16$

The $x$-intercepts are +4 and –4.

The $x$-intercept of +4 is 3 units from $x = 1$.

It moves to 9 units from $x = 1$, thus it is at 10.

The $x$-intercept of –4 is 5 units in the negative direction from $x = 1$.

It is now 15 units from $x = 1$, thus it is at –14.

Thus, the new $x$-intercepts are 10 and –14.

**13.** $y = |3x| \rightarrow y = \left|\frac{3}{2}x - 1\right| - 2$

$y = \left|3\cdot\frac{1}{2}\left(x - \frac{2}{3}\right)\right| - 2$ • Write in $y = af[b(x-h)] + k$ from

**i)** Stretched horizontally about the $y$-axis by a factor of 2. (Notice that the original was $y = |3x|$.)

**ii)** Translated $\frac{2}{3}$ units right and 2 units down.

**14.** For $f(x) = \dfrac{1}{f(x)}$:

$$x - 4 = \frac{1}{x-4}$$
$$(x-4)^2 = 1$$
$$x - 4 = \pm 1$$
$$x = 5, 3$$

For $x = 5$:
$y = f(5) = 5 - 4 = 1$

For $x = 3$:
$y = f(3) = 3 - 4 = -1$

Thus, the points are (5, 1) and (3, – 1).

These are the invariant points for the graphs of the reciprocal functions,
$y = f(x)$ and $y = \dfrac{1}{f(x)}$.

**15. C**

$$y = x^2 - 2$$
$$\rightarrow y = 3(x^2 - 2)$$
$$\rightarrow y = 3((x-4)^2 - 2)$$
$$-3 = 3((x-4)^2 - 2)$$
$$-1 = (x-4)^2 - 2$$
$$1 = (x-4)^2$$
$$\pm 1 - x - 4$$
$$x = 5, 3$$

**16. B**
The domain of $f(x) = \sqrt{x-5}$ is $x \geq 5$.

$g(x) = -f(x) = -\sqrt{x-5}$ has the same domain as $f(x) = \sqrt{x-5}$.

**17.** (1, 3) is

**i)** Stretched horizontally about the *y*-axis by a factor of $\dfrac{1}{2}$: $\left(\dfrac{1}{2}, 3\right)$

**ii)** Translated 5 units right and 2 units up:
$\left(\dfrac{1}{2} + 5, 3 + 2\right)$

Thus, the *x*-coordinate is 5.5.

| | | | |
|---|---|---|---|
| 5 | . | 5 | |

**18.**

$$y = \frac{1}{x} \rightarrow y = \frac{2}{5x - 12}$$
$$y = \frac{2}{5\left(x - \frac{12}{5}\right)}$$

The horizontal translation is $\dfrac{12}{5}$ units or 2.4 units.

| | | | |
|---|---|---|---|
| 2 | . | 4 | |

# EXPONENTS AND LOGARITHMS

## *Lesson 1 Review of Exponents and Solving Exponential Equations*

### CLASS EXERCISES ANSWERS AND SOLUTIONS

1.

$$\left(\frac{27}{125}\right)^{2n-5} = \left(\left(\frac{5}{3}\right)^{-3}\right)^{2n-5}$$
- Replace $\frac{27}{125}$ with $\left(\frac{5}{3}\right)^{-3}$

$$= \left(\frac{5}{3}\right)^{-3(2n-5)}$$
- Apply the Power of a Power Law

$$= \left(\frac{5}{3}\right)^{-6n+15}$$
- Simplify

2.

$$\frac{8^{3x+7}}{16^{x-1}} = \frac{\left(2^3\right)^{3x+7}}{\left(2^4\right)^{x-1}}$$
- Replace with base *2*

$$= \frac{2^{9x+21}}{2^{4x-4}}$$
- Apply the Power of a Power Law

$$= 2^{(9x+21)-(4x-4)}$$
- Apply the Quotient Law

$$= 2^{5x+25}$$
- Simplify

3. To solve for $x$, raise both sides of the equation to the power of $\frac{5}{2}$, the reciprocal of the exponent $\frac{2}{5}$.

$$\left(x^{\frac{2}{5}}\right)^{\frac{5}{2}} = \left(\frac{1}{9}\right)^{\frac{5}{2}}$$
- Raise both sides to the power of $\frac{5}{2}$

$$x = \left(\frac{1}{9}\right)^{\frac{5}{2}}$$
- Apply the Power of a Power Law

$$x = \left(\sqrt[2]{\frac{1}{9}}\right)^5$$
- Apply the Rational Exponent Property

$$x = \left(\frac{1}{3}\right)^5$$
- Evaluate

$$x = \frac{1}{243}$$
- Evaluate

The solution is $x = \frac{1}{243}$.

4. In this problem, the powers do not have the same base. However, although the bases are different, we can rewrite both sides with a common base of 2.

$$\left(2^3\right)^{3x+1} = \left(2^5\right)^{2-x}$$
- Replace 8 with $2^3$ and replace 32 with $2^5$

$$2^{9x+3} = 2^{10-5x}$$
- Apply the Power of a Power Law

$$9x + 3 = 10 - 5x$$
- Equal bases means the exponents can be equated

$$14x = 7$$
- Solve for $x$

$$x = \frac{1}{2}$$

The solution is $x = \frac{1}{2}$.

### PRACTICE EXERCISES ANSWERS AND SOLUTIONS

1. $\left(\frac{k^3 \times k^2 \times k}{k^4}\right)^5 = \left(\frac{k^6}{k^4}\right)^5 = \left(k^2\right)^5 = k^{10}$

2. $\left(2b^6c^{m+2}\right)^5 = 2^5b^{30}c^{5m+10} = 32b^{30}c^{5m+10}$

3. $\left(\frac{4m^5n^8}{6m^2n^9}\right)^3 = \left(\frac{2m^3}{3n}\right)^3 = \frac{8m^9}{27n^3}$

4. $\left(\sqrt[4]{x}\right)^6 \times \frac{1}{\sqrt{x}} = x^{\frac{6}{4}} \times x^{-\frac{1}{2}} = x^{\frac{3}{2}-\frac{1}{2}} = x$

5. $\left(\frac{2h^6}{5n^{-3}}\right)^{-2} = \frac{2^{-2}h^{-12}}{5^{-2}n^6} = \frac{5^2}{2^2h^{12}n^6} = \frac{25}{4h^{12}n^6}$

6. $\left(\frac{8x^3y}{6x^2}\right) \times \left(\frac{3x^2y^3}{2y^2}\right)^3 = \left(\frac{4xy}{3}\right) \times \left(\frac{3x^2y}{2}\right)^3$

$= \left(\frac{4xy}{3}\right)\left(\frac{27x^6y^3}{8}\right) = \frac{9x^7y^4}{2}$

7. $\left(\left(2^2\right)^5\right)^7 = \left(x^{10}\right)^7 = x^{70}$

8. $\left(\sqrt{\left(x^4\right)^6}\right)^{-\frac{1}{3}} = \left(\sqrt{x^{24}}\right)^{-\frac{1}{3}} = \left(x^{12}\right)^{-\frac{1}{3}} = x^{-4} = \frac{1}{x^4}$

**9.** $\left(3x^{-3}y^{4}\right)^{4} = 3^{4}x^{-12}y^{16} = \dfrac{81y^{16}}{x^{12}}$

**10.** $\left(\left(\dfrac{1}{x^{4}}\right)^{2}\right)^{-3} = \left(\left(x^{-4}\right)^{2}\right)^{-3} = x^{24}$

**11.** $\dfrac{\left(2x^{3}y^{-3}\right)\left(6x^{3}\right)}{\left(6x^{-1}y^{2}\right)\left(8xy\right)} = \dfrac{12x^{6}y^{-3}}{48y^{3}} = \dfrac{x^{6}}{4y^{6}}$

**12.** $\left(\sqrt{x}\right)^{7}\cdot\sqrt[5]{x} = x^{\frac{7}{2}}\cdot x^{\frac{1}{5}} = x^{\frac{35}{10}+\frac{2}{10}} = x^{\frac{37}{10}}$

**13.** $9^{-2} = \dfrac{1}{9^{2}} = \dfrac{1}{81}$

**14.** $16^{\frac{5}{4}} = \left(\sqrt[4]{16}\right)^{5} = 2^{5} = 32$

**15.** $25^{\frac{-3}{2}} = \dfrac{1}{25^{\frac{3}{2}}} = \dfrac{1}{\sqrt{25^{3}}} = \dfrac{1}{125}$

**16.** $-17^{0} = -1$

**17.** $\left(\dfrac{25}{81}\right)^{-\frac{1}{2}} = \left(\dfrac{81}{25}\right)^{\frac{1}{2}} = \sqrt{\dfrac{81}{25}} = \dfrac{9}{5}$

**18.** $2^{-7} = \dfrac{1}{2^{7}} = \dfrac{1}{128}$

**19.** $81^{-0.25} = 81^{-\frac{1}{4}} = \dfrac{1}{81^{\frac{1}{4}}} = \dfrac{1}{\sqrt[4]{81}} = \dfrac{1}{3}$

**20.** $\left(-\dfrac{1}{7}\right)^{0} = 1$

**21.** $\dfrac{\left(5^{6}\right)^{2}}{\left(5^{-10}\right)\left(\dfrac{1}{5^{4}}\right)} = \dfrac{5^{12}}{\dfrac{1}{5^{14}}} = 5^{12}\times 5^{14} = 5^{26}$

**22.** $\left(-\dfrac{2}{3}\right)^{-3} = \left(-\dfrac{3}{2}\right)^{3} = -\dfrac{27}{8}$

**23.** $25^{-\frac{3}{2}} = \dfrac{1}{25^{\frac{3}{2}}} = \dfrac{1}{\sqrt{25^{3}}} = \dfrac{1}{5^{3}} = \dfrac{1}{125}$

**24.** $(-8)^{\frac{4}{3}} = \left(\sqrt[3]{-8}\right)^{4} = (-2)^{4} = 16$

**25.** $\left(\dfrac{49}{25}\right)^{\frac{3}{2}} = \left(\sqrt{\dfrac{49}{25}}\right)^{3} = \left(\dfrac{7}{5}\right)^{3} = \dfrac{343}{125}$

**26.** $\dfrac{\left(27^{\frac{2}{3}}\right)^{2}}{81^{\frac{1}{2}}} = \dfrac{27^{\frac{4}{3}}}{81^{\frac{1}{2}}} = \dfrac{\sqrt[3]{27^{4}}}{\sqrt{81}} = \dfrac{81}{9} = 9$

**27.** $(-27)^{-\frac{4}{3}} = \left(\dfrac{1}{-27}\right)^{\frac{4}{3}} = \left(\sqrt[3]{\dfrac{1}{-27}}\right)^{4} = \left(-\dfrac{1}{3}\right)^{4} = \dfrac{1}{81}$

**28.** $(49)^{3x} = \left(7^{2}\right)^{3x} = 7^{6x}$

**29.** $64^{2-x} = \left(4^{3}\right)^{2-x} = 4^{6-3x}$

**30.** $8^{3x+2} = \left[\left(\sqrt{2}\right)^{6}\right]^{3x+2} = \left(\sqrt{2}\right)^{18x+12}$

**31.**

$$\left(\frac{1\,000}{27}\right)^{x+7} = \left(\left(\frac{10}{3}\right)^{3}\right)^{x+7}$$

$$= \left(\left(\frac{3}{10}\right)^{-3}\right)^{x+7} = \left(\frac{3}{10}\right)^{-3x-21}$$

**32.** $\left(\dfrac{1}{36}\right)^{2x} = \left(6^{-2}\right)^{2x} = 6^{-4x}$

**33.** $\left(\dfrac{16}{81}\right)^{3x} = \left[\left(\dfrac{2}{3}\right)^{4}\right]^{3x} = \left(\dfrac{2}{3}\right)^{12x}$

**34.** $\left(\dfrac{1}{9}\right)^{6x+1} = \left(3^{-2}\right)^{6x+1} = 3^{-12x-2}$

**35.** $10\,000^{x-5} = \left(\left(\dfrac{1}{10}\right)^{-4}\right)^{x-5} = \left(\dfrac{1}{10}\right)^{-4x+20}$

**36.** $\left(\sqrt[4]{8}\right)^{3x} = \left(8^{\frac{1}{4}}\right)^{3x} = 8^{\frac{3x}{4}} = \left(2^{3}\right)^{\frac{3x}{4}} = 2^{\frac{9x}{4}}$

**37.** $\dfrac{9^{3x}}{81^{2x+3}} = \dfrac{\left(3^{2}\right)^{3x}}{\left(3^{4}\right)^{2x+3}} = \dfrac{3^{6x}}{3^{8x+12}} = 3^{6x-(8x+12)} = 3^{-2x-12}$

**38.** $\frac{10\,000^{x+7}}{1\,000^{3-5x}}=\frac{\left(10^4\right)^{x+7}}{\left(10^3\right)^{3-5x}}=\frac{10^{4x+28}}{10^{9-15x}}=10^{19x+19}$

**39.** $\frac{16^{2x}}{32^{4-x}}=\frac{\left(2^4\right)^{2x}}{\left(2^5\right)^{4-x}}=\frac{2^{8x}}{2^{20-5x}}=2^{8x-(20-5x)}=2^{13x-20}$

**40.** $\frac{6^{3x-1}}{36^{2+x}}=\frac{6^{3x-1}}{\left(6^2\right)^{2+x}}=\frac{6^{3x-1}}{6^{4+2x}}=6^{x-5}$

**41.**

$$\left(\frac{8}{27}\right)^{2x}=\left(\frac{9}{4}\right)^{x+1}$$
$$\left[\left(\frac{2}{3}\right)^3\right]^{2x}=\left[\left(\frac{2}{3}\right)^{-2}\right]^{x+1}$$
$$\left(\frac{2}{3}\right)^{6x}=\left(\frac{2}{3}\right)^{-2x-2}$$
$$6x=-2x-2$$
$$8x=-2$$
$$x=-\frac{1}{4}$$

**42.**

$$2x^{\frac{1}{4}}=10\rightarrow x^{\frac{1}{4}}=5$$
$$\left(x^{\frac{1}{4}}\right)^4=(5)^4$$
$$x=625$$

**43.**

$$x^{\frac{3}{2}}=\frac{1}{27}$$
$$\left(x^{\frac{3}{2}}\right)^{\frac{2}{3}}=\left(\frac{1}{27}\right)^{\frac{2}{3}}$$
$$x=\sqrt[3]{\left(\frac{1}{27}\right)^2}$$
$$x=\left(\frac{1}{3}\right)^2$$
$$x=\frac{1}{9}$$

**44.**

$$\frac{9^{x+3}}{81^{x-4}}=27^{2x+7}$$
$$\frac{\left(3^2\right)^{x+3}}{\left(3^4\right)^{x-4}}=\left(3^3\right)^{2x+7}$$
$$\frac{3^{2x+6}}{3^{4x-16}}=3^{6x+21}$$
$$3^{-2x+22}=3^{6x+21}$$
$$-2x+22=6x+21$$
$$-8x=-1$$
$$x=\frac{1}{8}$$

**45.**

$$x^{-\frac{2}{5}}=\frac{4}{9}$$
$$\left(x^{-\frac{2}{5}}\right)^{-\frac{5}{2}}=\left(\frac{4}{9}\right)^{-\frac{5}{2}}$$
$$x=\frac{1}{\left(\frac{4}{9}\right)^{\frac{5}{2}}}$$
$$x=\frac{1}{\left(\sqrt{\frac{4}{9}}\right)^5}$$
$$x=\frac{1}{\left(\frac{2}{3}\right)^5}$$
$$x=\frac{1}{\frac{32}{243}}$$
$$x=\frac{343}{32}$$

**46.**

$$\left(27^{2x+1}\right)^3=\left(\sqrt[4]{81^{4x+8}}\right)^3$$
$$\left(\left(3^3\right)^{2x+1}\right)^3=\left[\left(\left(3^4\right)^{4x+8}\right)^{\frac{1}{4}}\right]^3$$
$$\left(3^9\right)^{2x+1}=\left(3^{4x+8}\right)^3$$
$$3^{18x+9}=3^{12x+24}$$
$$18x+9=12x+24$$
$$6x=15$$
$$x=\frac{15}{6}=\frac{5}{2}$$

 

**47.**

$$5x^{\frac{3}{2}} = 135$$
$$x^{\frac{3}{2}} = 27$$
$$\left(x^{\frac{3}{2}}\right)^{\frac{2}{3}} = (27)^{\frac{2}{3}}$$
$$x = \left(\sqrt[3]{27}\right)^2 = 3^2 = 9$$

**48.**

$$27^x = 9^{2x-1}$$
$$\left(3^3\right)^x = \left(3^2\right)^{2x-1}$$
$$3^{3x} = 3^{4x-2}$$
$$3x = 4x - 2$$
$$-x = -2$$
$$x = 2$$

**49.**

$$6^{3x-6} = 1$$
$$6^{3x-6} = 6^0$$
$$3x - 6 = 0$$
$$3x = 6$$
$$x = 2$$

**50.**

$$4^{8x} = \frac{1}{16}$$
$$\left(2^2\right)^{8x} = 2^{-4}$$
$$2^{16x} = 2^{-4}$$
$$16x = -4$$
$$x = -\frac{1}{4}$$

**51.**

$$8(2x-1)^3 = 125$$
$$(2x-1)^3 = \frac{125}{8}$$
$$\left((2x-1)^3\right)^{\frac{1}{3}} = \left(\frac{125}{8}\right)^{\frac{1}{3}}$$
$$2x - 1 = \sqrt[3]{\frac{125}{8}}$$
$$2x - 1 = \frac{5}{2}$$
$$2x = \frac{7}{2}$$
$$x = \frac{7}{4}$$

**52.**

$$8^{\frac{1}{4}} \times \left(\frac{1}{4}\right)^{\frac{x}{2}} = 16^{\frac{3}{4}}$$
$$\left(2^3\right)^{\frac{1}{4}} \times \left(2^{-2}\right)^{\frac{x}{2}} = \left(2^4\right)^{\frac{3}{4}}$$
$$2^{\frac{3}{4}} \times 2^{-x} = 2^3$$
$$2^{\frac{3}{4}-x} = 2^3$$
$$\frac{3}{4} - x = 3$$
$$-x = 3 - \frac{3}{4}$$
$$-x = \frac{9}{4}$$
$$x = -\frac{9}{4}$$

**53.**

$$\frac{2^{3x}}{4^x} = \frac{3^{2x}}{9^x}$$
$$\frac{2^{3x}}{2^{2x}} = \frac{3^{2x}}{3^{2x}}$$
$$2^x = 1$$
$$x = 0$$

**54.**

$$2^x + 2^x = 2^4$$
$$(2)\left(2^x\right) = 2^4$$
$$2^{x+1} = 2^4$$
$$x + 1 = 4$$
$$x = 3$$

**55. C**

$$4^{2x-3} = \left(\sqrt{2}\right)^{3x+6}$$
$$\left(2^2\right)^{2x-3} = \left(2^{\frac{1}{2}}\right)^{3x+6}$$
$$2^{4x-6} = 2^{\frac{3}{2}x+3}$$
$$4x - 6 = \frac{3}{2}x + 3$$
$$\frac{5}{2}x = 9$$
$$x = 9 \times \frac{2}{5} = \frac{18}{5} = 3.6$$

**56. D**

$$27\left(\frac{1}{3}\right)^{x-2} = \sqrt[3]{9^{2x}}$$

$$3^3\left(3^{-1}\right)^{x-2} = \left(9^{2x}\right)^{\frac{1}{3}}$$

$$3^3\left(3^{-x+2}\right) = \left(\left(3^2\right)^{2x}\right)^{\frac{1}{3}}$$

$$3^{-x+5} = 3^{\frac{4x}{3}}$$

$$-x+5 = \frac{4x}{3}$$

$$-3x+15 = 4x$$

$$7x = 15$$

$$x = \frac{15}{7}$$

**57.** $\frac{1}{2}\times 2^{1\,000} = 2^{1\,000-1}$

$= 2^{999}$

| | | | |
|---|---|---|---|
| 9 | 9 | 9 | |

**58.**

$$\left(4^{x+1}\right)\left(8^x\right) = 32$$

$$\left(2^2\right)^{x+1}\left(2^3\right)^x = 2^5$$

$$\left(2^{2x+2}\right)\left(2^{3x}\right) = 2^5$$

$$2^{5x+2} = 2^5$$

$$5x+2 = 5$$

$$5x = 3$$

$$x = \frac{3}{5} = 0.6$$

| | | | |
|---|---|---|---|
| 0 | . | 6 | |

## *Lesson 2 Logarithms*

### CLASS EXERCISES ANSWERS AND SOLUTIONS

**1.**

| Exponential Form | Logarithmic Form |
|---|---|
| $2^3 = 8$ | $3 = \log_2 8$ |
| $10^{-2} = 0.01$ | $-2 = \log 0.01$ |
| $3^4 = 81$ | $\log_3 81 = 4$ |
| $10^4 = 10\,000$ | $\log 10000 = 4$ |
| $(e)^0 = 1$ | $\ln 1 = 0$ |
| $h^i = k$ | $\log_h k = j$ |
| $2^{3y} = x+1$ | $3y = \log_2(x+1)$ |
| $2^{-6} = \frac{1}{64}$ | $\log_2 \frac{1}{64} = -6$ |
| $b^3 = 8$ | $6 = 2\log_b 8$ |
| $a = 2(10)^y$ | $y = \log\left(\frac{a}{2}\right)$ |
| $a = b(c)^d$ | $d = \log_c \frac{b}{d}$ |
| $2\,000 = 1\,000(1.05)^n$ | $n = \log_{1.05} 2$ |
| $600\,000 = 500(2)^{\frac{t}{80}}$ | $\frac{t}{80} = \log_2 1200$ |

**2.** You can think of this as solving the equation, $\log_2\left(\frac{1}{8}\right) = x$, which is $2^x = \frac{1}{8}$ in exponential form. Writing in base 2 gives $2^x = 2^{-3}$ and thus, $x = -3$.

**3.** If $\log_3 81 = x$ and $\log_3\left(\frac{1}{27}\right) = y$, then we require $x + y$.

Since $3^4 = 81$ and $3^{-3} = 27$, $x = 4$ and $y = -3$.

Thus, $x + y = 4 + (-3) = 1$.

**4.** The exponential form is

$x^{-2} = \left(\frac{1}{49}\right) \to x^{-2} = (7)^{-2}$.

Thus, $x = 7$.

**5.** The exponential form is $\left(\frac{1}{3}\right)^{-4} = m \to 3^4 = m$.

Thus, $m = 3^4 = 81$.

 

6. The exponential form is $7^x = 49^{50}$.

$7^x = (7^2)^{50}$

$7^x = 7^{100}$

Therefore, $x = 100$.

7. The exponential form is $3^4 = 2x - 3$.

$2x - 3 = 81$

$2x = 84$

Thus, $x = 42$.

## PRACTICE EXERCISES ANSWERS AND SOLUTIONS

1. $3^2 = 9 \rightarrow \log_3 9 = 2$

2. $5^3 = 125 \rightarrow \log_5 125 = 3$

3. $m^x = a + 4 \rightarrow \log_m (a+4) = x$

4. $b^{x-2} = 8 \rightarrow \log_b 8 = x - 2$

5. $150 = 2^{x+1} \rightarrow \log_2 150 = x + 1$

6. $\left(\frac{1}{2}\right)^{-6} = 64 \rightarrow \log_{\frac{1}{2}} 64 = -6$

7. $\left(\frac{3}{2}\right)^{x} = y \rightarrow \log_{\frac{3}{2}} y = x$

8. $2^{\log_2 2} = x \rightarrow \log_2 x = \log_2 2$

9. $10^{\log(0.001)} = y \rightarrow \log y = \log(0.001)$

10. $\log_4\left(\frac{1}{64}\right) = -3 \rightarrow 4^{-3} = 164$

11. $-3 = \log_{\left(\frac{1}{2}\right)} 8 \rightarrow \left(\frac{1}{2}\right)^{-3} = 8$

12. $x = \log_b 25 \rightarrow b^x = 25$

13. $\log_p x = 7 \rightarrow p^7 = x$

14. $\frac{a}{2} = \log_2 32 \rightarrow 2^{\frac{a}{2}} = 32$

15. $\log(x-6) = \frac{1}{4} \rightarrow 10^{\frac{1}{4}} = x - 6$

16. $x - 2 = \log_3 (3^5) \rightarrow 3^{x-2} = 3^5$

17. $\log_2 (y-2) = x - 3 \rightarrow 2^{x-3} = y - 2$

18. $-1 = \log_{\frac{b}{a}}\left(\frac{a}{b}\right) \rightarrow \left(\frac{b}{a}\right)^{-1} = \frac{a}{b}$

19. $\log_{25} 5 = x$

$25^x = 5$

$(5^2)^x = 5^1$

$2x = 1$

$x = \frac{1}{2}$

20. $\log_{27} 3 = x$

$27^x = 3$

$(3^3)^x = 3$

$3x = 1$

$x = \frac{1}{3}$

21. $2\log_2 128 = x$

$\log_2 128 = \frac{x}{2}$

$2^{\frac{x}{2}} = 128$

$2^{\frac{x}{2}} = 2^7$

$\frac{x}{2} = 7$

$x = 14$

22. $4\log_6\left(\frac{1}{36}\right) = x$

$\log_6 \frac{1}{36} = \frac{x}{4}$

$6^{\frac{x}{4}} = \frac{1}{36}$

$6^{\frac{x}{4}} = 6^{-2}$

$\frac{x}{4} = -2$

$x = -8$

**23.** $\log_2 64 - \log_2 16$

If $\log_2 64 = x$, then $2^x = 64$, and $x = 6$.

If $\log_2 16 = y$, then $2^y = 16$, and $y = 4$.

$\log_2 64 - \log_2 16 = x - y = 6 - 4 = 2$

**24.** $\log_5 1 - \log_5\left(\frac{1}{125}\right)$

If $\log_5 1 = x$, then $5^x = 1$, and $x = 0$.

If $\log_5\left(\frac{1}{125}\right) = y$, then $5^y = \frac{1}{125}$, and $y = -3$.

$\log_5 1 - \log_5\left(\frac{1}{125}\right) = x - y = 0 - (-3) = 3$

**25.** $\log_2 x = -5$

$x = 2^{-5} = \frac{1}{2^5} = \frac{1}{32}$

**26.** $\log_x\left(\frac{1}{4}\right) = -2$

$x^{-2} = \frac{1}{4}$

$x^{-2} = 2^{-2}$

$x = 2$

**27.** $\log_x\left(\frac{4}{9}\right) = -2$

$x^{-2} = \frac{4}{9}$

$x^{-2} = \left(\frac{3}{2}\right)^{-2}$

$x = \frac{3}{2}$

**28.** $\log_5(2x+1) = 2$

$5^2 = 2x + 1$

$25 = 2x + 1$

$24 = 2x$

$x = 12$

**29.** $\log_3(x+5) = \log_5 1$

$\log_3(x+5) = 0$

$x + 5 = 3^0$

$x + 5 = 1$

$x = -4$

**30.** $\log_{\frac{1}{2}} 16 = 2x - 3$

$\left(\frac{1}{2}\right)^{2x-3} = 16$

$\left(2^{-1}\right)^{2x-3} = 2^4$

$2^{-2x+3} = 2^4$

$-2x + 3 = 4$

$-2x = 1$

$x = -\frac{1}{2}$

**31.** $\log_2\left(x^2 + x\right) = 1$

$x^2 + x = 2^1$

$x^2 + x - 2 = 0$

$(x+2)(x-1) = 0$

$x = -2, 1$

Both are valid solutions.

**32.** $\log_6\left(x^2 - x\right) = 1$

$x^2 - x = 6^1$

$x^2 - x - 6 = 0$

$(x-3)(x+2) = 0$

$x = 3, -2$

Both are valid solutions.

**33. A**

$\log_b b^{4b} = x$

$b^x = b^{4b}$

$x = 4b$

**34. B**

$$\log_5(25) = x+2y \rightarrow 5^{x+2y} = 25 \rightarrow 5^{x+2y} = 5^2$$

$$\log_{25}\left(\frac{1}{5}\right) = x-y \quad 25^{x-y} = \frac{1}{5} \rightarrow 5^{2x-2y} = 5^{-1}$$

$$\begin{array}{lcl} x+2y=2 & \rightarrow & 2x+4y=4 \\ \underline{2x-2y=-1} & & \underline{2x-2y=-1} \\ & & 6y=5 \\ & & y=\frac{5}{6} \end{array}$$

**35.**

$$\begin{array}{ll} \log_{\sqrt{3}}(9)=x & \log_{\sqrt{3}}(3\sqrt{3})=y \\ \left(\sqrt{3}\right)^x=9 & \left(\sqrt{3}\right)^y=3\sqrt{3} \\ \left(\sqrt{3}\right)^x=\left(\sqrt{3}\right)^4 & \left(\sqrt{3}\right)^y=\left(\sqrt{3}\right)^3 \\ x=4 & y=3 \\ x\times y=12 & \end{array}$$

| 1 | 2 | | |
|---|---|---|---|

**36.**

$$\log_{\frac{1}{4}}\left(\frac{1}{8}\right)=b$$

$$\left(\frac{1}{4}\right)^b=\frac{1}{8}$$

$$\left(\frac{1}{2}\right)^{2b}=\left(\frac{1}{2}\right)^3$$

$$b=\frac{3}{2}=1.5$$

| 1 | . | 5 | |
|---|---|---|---|

## *Lesson 3 Laws of Logarithms*

### CLASS EXERCISES ANSWERS AND SOLUTIONS

1. The Product Law is still applicable when there are more than two multiplicands.

$$\begin{aligned}\log_3(9\times81\times27)&=\log_3 9+\log_3 81+\log_3 27\\&=2+4+3\\&=9\end{aligned}$$

2. We need to simplify the log of the product 25 multiplied by $a$. We use the Product Law.

$$\begin{aligned}\log_5 25a&=\log_5 25+\log_5 a\\&=2+\log_5 a\end{aligned}$$

No further simplification is possible.

3. Once again, we have the sum of logs. All the logs have the same base, $x$. We can use the Product Law to rewrite this expression as

$$\begin{aligned}\log_x 3+\log_x 5+\log_x 2m&=\log_x(3\times5\times2m)\\&=\log_x 30m\end{aligned}$$

4. The values of the two individual logarithms are not rational numbers.

To what power do we raise 3 to get 72? To what power should 3 be raised to get 8?

Since we are subtracting logs with the same base—base 3—we can use the Quotient Law in reverse to rewrite the expression as a single log.

$\log_3 72-\log_3 8=\log_3\frac{72}{8}$ • Rewrite as a single log

$=\log_3 9$ • Simplify the fraction

$=2$ • Evaluate

5. There is no way that you would want to first expand $125^8$ and then find the $\log_5$ of the result. With the Power Law, we can easily evaluate the expression.

$\log_5(125^8)=8\log_5 125$ • Slide the exponent in front (Power Law)

$=8\times3$ • Evaluate

$=24$

6.

$\log_2(16^{3k+1})=(3k+1)\log_2 16$ • Slide the exponent in front (Power Law)

$=(3k+1)4$ • Evaluate $\log_2 16$

$=12k+4$ • Simplify

**7.** Here, we will apply the Power Law in reverse. Since we have a coefficient of 3, we can slide that back so that it becomes the exponent in the log expression.

$$\begin{aligned}3\log_7 2 &= \log_7 2^3\\ &= \log_7 8\end{aligned}$$

**8.** Since we have the same base throughout the expression, we can combine the terms. We can begin with $\log_3 A + \log_3 B$. Using the Product Law, this can be rewritten as $\log_3 AB$.

$$\log_3 A + \log_3 B - \log_3 C = \log_3 AB - \log_3 C$$

Since we now have the difference of two logs, we can apply the Quotient Law to get:

$$\begin{aligned}\log_3 A + \log_3 B - \log_3 C &= \log_3 AB - \log_3 C\\ &= \log_3 \frac{AB}{C}\end{aligned}$$

**9.**

$$\begin{aligned}\log_3 m^2 - 3\log_3 k &= \log_3 m^2 - \log_3 k^3 && \text{• Apply Power Law}\\ &= \log_3\left(\frac{m^2}{k^3}\right) && \text{• Apply Quotient Law}\end{aligned}$$

**10.** Since we are given that $\log_b 9 = M$, we want to rewrite $\log_b \frac{1}{81}$ so that it includes $\log_b 9$. If we do that, we can later substitute $M$ for $\log_b 9$.

$$\begin{aligned}\log_b \frac{1}{81} &= \log_b 9^{-2} && \text{• Rewrite } \tfrac{1}{81} \text{ as an exponent with base 9}\\ &= -2\log_b 9 && \text{• Apply Power Law}\\ &= -2(M) && \text{• Substitute } M \text{ for } \log_b 9\\ &= -2M && \text{• Simplify}\end{aligned}$$

**11. a)** Since $\log_2 3 = x$, we want to rewrite $\log_2 48$ so that it includes $\log_2 3$.

$$\begin{aligned}\log_2 48 &= \log_2(3\times16) && \text{• Rewrite 48 as } 3\cdot16\\ &= \log_2 3 + \log_2 16 && \text{• Apply Product Law}\\ &= x + 4 && \text{• Replace } \log_2 3 \text{ with } x \text{ and evaluate } \log_2 16\end{aligned}$$

**b)** We want to rewrite $\log_2 \sqrt[4]{9^3}$ so that it includes $\log_2 3$.

$$\begin{aligned}\log_2 \sqrt[4]{9^3} &= \log_2 9^{\frac{3}{4}} && \text{• Rewrite } \sqrt[4]{9^3} \text{ as a power}\\ &= \log_2 \left(3^2\right)^{\frac{3}{4}} && \text{• Rewrite 9 as a power of 3}\\ &= \log_2 3^{\frac{3}{2}} && \text{• Apply Power Law of Exponents}\\ &= \frac{3}{2}\log_2 3 && \text{• Apply Power Law of Logarithms}\\ &= \frac{3}{2}x && \text{• Replace } \log_2 3 \text{ with } x\end{aligned}$$

## PRACTICE EXERCISES ANSWERS AND SOLUTIONS

**1.** $\log_2 120 - \log_2 15 = \log_2 \frac{120}{15} = \log_2 8 = 3$

**2.** $\log_5 4 + \log_5 \frac{1}{20} = \log_5\left(4\cdot\frac{1}{20}\right) = \log_5 \frac{1}{5} = -1$

**3.** $\log_3 6 - \log_3 54 = \log_3 \frac{6}{54} = \log_3 \frac{1}{9} = -2$

**4.**

$$\begin{aligned}\log_7 14 + \log_7 \frac{1}{2} + \log_7 \frac{1}{7} &= \log_7\left(14\cdot\frac{1}{2}\cdot\frac{1}{7}\right)\\ &= \log_7 1 = 0\end{aligned}$$

**5.** $\log_3 9^{20} = 20\log_3 9 = 20(2) = 40$

**6.** $\log_8 2^{\frac{3}{2}} = \frac{3}{2}\cdot\log_8 2 = \frac{3}{2}\cdot\left(\frac{1}{3}\right) = \frac{1}{2}$

**7.** $\log_5\left(\frac{1}{25}\right)^{60} = 60\log_5\left(\frac{1}{25}\right) = 60(-2) = -120$

**8.** $\log_{\frac{1}{2}} 8^{11} = 11\cdot\log_{\frac{1}{2}} 8 = 11\cdot(-3) = -33$

**9.** $3\log_m k = \log_m k^3$

**10.** $\frac{1}{2}\log_7 9 = \log_7 9^{\frac{1}{2}}$

**11.** $\log_3 45 - \log_3 15 = \log_3 \frac{45}{15} = \log_3 3 = 1$

 

**12.**

$$\log_5\left(\sqrt{5}\right)^{25} = \log_5\left(5^{\frac{1}{2}}\right)^{25} = \frac{25}{2}\cdot\log_5 5$$
$$= \frac{25}{2}(1) = 12.5$$

**13.**

$$\log_4\frac{1}{12}+\log_4\frac{3}{16} = \log_4\left(\frac{1}{12}\times\frac{3}{16}\right)$$
$$= \log_4\frac{1}{64} = -3$$

**14.** $\log_9 3^{10} = 10\log_9 3 = 10\left(\frac{1}{2}\right) = 5$

**15.** $\log_2 48 - \log_2 3 = \log_2\frac{48}{3} = \log_2 16 = 4$

**16.**

$$\log_3 9^5 - \log_6 6^{30} + \log 6^{10}$$
$$= 5(\log_3 9) - 30(\log_6 6) + 10(\log_6 6)$$
$$= (5\cdot 2) - (30\cdot 1) + (10\cdot 1) = 10 - 30 + 10$$
$$= -10$$

**17.** $\log_6 2 + \log_6 18 = \log_6(36) = 2$

**18.** $\log_4 64^{50} = 50\cdot\log_4 64 = 50\cdot 3 = 150$

**19.**

$$\log_2\frac{1}{12}+\log_2\frac{3}{8} = \log_2\left(\frac{1}{12}\times\frac{3}{8}\right)$$
$$= \log_2\frac{1}{32} = -5$$

**20.**

$$\log_{10} 100000^{100} = 100\cdot\log_{10} 100000$$
$$= 100\cdot 5 = 500$$

**21.** $\log_3 2 - \log_3 54 = \log_3\frac{2}{54} = \log_3\frac{1}{27} = -3$

**22.**

$$\log_3 18 + \log_3 6 + \log_3 3^{\frac{3}{4}} = \log_3\left(18\cdot 6\cdot\frac{3}{4}\right)$$
$$= \log_3 81 = 4$$

**23.**

$$\log_{10} 5 + \log_{10} 4 - \log_{10} 200 = \log_{10}\left(\frac{5\times 4}{200}\right)$$
$$= \log_{10}\left(\frac{1}{10}\right) = -1$$

**24.**

$$2\log_{\sqrt{3}} 6 - 2\log_{\sqrt{3}} 2 - 5\log_{\sqrt{3}} 1$$
$$= \log_{\sqrt{3}} 6^2 - \log_{\sqrt{3}} 2^2 - \log_{\sqrt{3}} 1^5$$
$$= \log_{\sqrt{3}}\frac{36}{4} - \log_{\sqrt{3}} 1 = \log_{\sqrt{3}}\frac{9}{1} = 4$$

**25.**

$$\log_x E + \log_x F - \log_x E - \log_x H$$
$$= \log_x\left(\frac{EF}{EH}\right)$$

**26.**

$$5\log_2 x^2 + 2\log_2 x - \frac{1}{2}\log_2 x^6$$
$$= \log_2\left(x^2\right)^5 + \log_2 x^2 - \log_2\left(x^6\right)^{\frac{1}{2}}$$
$$= \log_2\left(\frac{x^{10}\cdot x^2}{x^3}\right) = \log_2\left(\frac{x^{12}}{x^3}\right) = \log_2 x^9$$

**27.**

$$3\log_2 m - \log_2 n + \frac{1}{2}\log_2 p$$
$$= \log_2 m^3 - \log_2 n + \log_2 p^{\frac{1}{2}}$$
$$= \log_2\frac{m^3 p^{\frac{1}{2}}}{n} \text{ or } \log_2\frac{m^3\sqrt{p}}{n}$$

**28.**

$$\frac{1}{n}\log_c G - 2\log_c H = \log_c G^{\frac{1}{n}} - \log_c H^2$$
$$= \log_c\frac{G^{\frac{1}{n}}}{H^2} = \log_c\left(\frac{\sqrt[n]{G}}{H^2}\right)$$

**29.** $\log_6 E + \log_6 H - \log_6 J = \log_6\left(\frac{EH}{J}\right)$

**30.** $4\log_3 K - \log_3 L = \log_3 K^4 - \log_3 L = \log_3\left(\frac{K^4}{L}\right)$.

**31.**

$$\log_6 2A + 3\log_6 B - \frac{1}{2}\log_6 C$$
$$= \log_6 2A + \log_6 B^3 - \log_6 C^{\frac{1}{2}}$$
$$= \log_6 \left(\frac{2AB^3}{C^{\frac{1}{2}}}\right) \text{ or } \log_6 \left(\frac{2AB^3}{\sqrt{C}}\right)$$

**32.**

$$4\log_5 K - (\log_5 L + 5\log_5 M)$$
$$= \log_5 K^4 - (\log_5 L + \log_5 M^5)$$
$$= \log_5 \left(\frac{K^4}{LM^5}\right)$$

**33.**

$$\frac{1}{2}\log_3 J + 7\log_3 K - \log_3 (LM)$$
$$= \log_3 J^{\frac{1}{2}} + \log_3 K^7 - \log_3 (LM)$$
$$= \log_3 \left(\frac{J^{\frac{1}{2}}K^7}{LM}\right) \text{ or } \log_3 \left(\frac{\sqrt{J}\,K^7}{LM}\right)$$

**34.**

$$\frac{2}{3}\log_7 5A - \frac{7}{9}\log_7 B = \log_7 (5A)^{\frac{2}{3}} - \log_7 B^{\frac{7}{9}}$$
$$= \log_7 \left(\frac{(5A)^{\frac{2}{3}}}{B^{\frac{7}{9}}}\right)$$

**35.**

$$\log_x \sqrt{5}x = \log_x \sqrt{5} + \log_x x$$
$$= \log_x 5^{\frac{1}{2}} + 1$$
$$= \frac{1}{2}\log_x 5 + 1$$
$$= \frac{1}{2}k + 1$$

**36.**

$$\log_x 15x^2 - \log_x 3$$
$$= \log_x 15 + \log_x x^2 - \log_x 3$$
$$= \log_x 15 + 2(1) - \log_x 3 = \log_x \frac{15}{3} + 2$$
$$= \log_x 5 + 2 = k + 2$$

**37.** $\log_5 (XYZ) = \log_5 X + \log_5 Y + \log_5 Z$

**38.**

$$\log_2 \left(\frac{4A}{B^2}\right) = \log_2 4 + \log_2 A - \log_2 B^2$$
$$= \log_2 4 + \log_2 A - 2\log_2 B$$

**39.**

$$\log_7 \left(\frac{\sqrt[3]{X^7}}{YZ}\right) = \log_7 \sqrt[3]{X^7} - \log_7 Y - \log_7 Z$$

or

$$\log_7 X^{\frac{7}{3}} - \log_7 Y - \log_7 Z$$

**40.**

$$\log\left(\frac{WX}{\sqrt{Y}}\right) = \log W + \log X - \log\sqrt{Y}$$

or

$$\log W + \log X - \frac{1}{2}\log Y$$

**41.**

$$\log_x \left(\frac{3A+2}{4B^2}\right)$$
$$= \log_x (3A+2) - \log 4B^2$$
$$= \log_x (3A+2) - (2\log_x B + \log_x 4)$$
$$= \log_x (3A+2) - 2\log_x B - \log_x 4$$

**42.**

$$\log_a \left(\frac{\sqrt[3]{M+2N}}{4\sqrt{P^7}}\right)$$
$$= \log_a (M+2N)^{\frac{1}{3}} - \log_a (P)^{\frac{7}{4}}$$
$$= \frac{1}{3}\log_a (M+2N) - \frac{7}{4}\log_a P$$

**43.** $\log_x 64 = \log_x (8^2) = 2\log_x 8 = 2A$

**44.**

$$\log_x 4 + \log_x 12x - \log_x 6x = \log_x \left(\frac{4\cdot 12x}{6x}\right)$$
$$= \log_x 8 = A$$

**45.**
$$2\log_x \sqrt{8} + \log_x 8\sqrt{8}$$
$$= \log_x \left(\sqrt{8}\right)^2 + \log_x 8 + \log_x \sqrt{8}$$
$$= \log_x 8 + \log_x 8 + \log_x 8^{\frac{1}{2}}$$
$$= \log_x 8 + \log_x 8 + \frac{1}{2}\log_x 8$$
$$= A + A + \frac{1}{2}A \quad = \frac{5}{2}A$$

**46.**
$$\log_x 6x^3 - \log_x 3 = \log_x \frac{6x^3}{3} = \log_x 2x^3$$
$$= \log_x 8^{\frac{1}{3}} x^3$$
$$= \frac{1}{3}\log_x 8 + 3\log_x x = \frac{1}{3}A + 3$$

**47.**
$$\log_7 2 + \log_7 \frac{1}{98} = \log_7 \left(2 \times \frac{1}{98}\right)$$
$$= \log_7 \left(\frac{1}{49}\right) = -2$$

**48.** $\log_5 50 - \log_5 10 = \log_5 \frac{50}{10} = \log_5 5 = 1$

**49.**
$$\log_3 4a + \log_3 \frac{a}{36} - 2\log_3 a$$
$$= \log_3 4a + \log_3 \frac{a}{36} - \log_3 a^2$$
$$= \log_3 \frac{4a \times \frac{a}{36}}{a^2} = \log_3 \left(\frac{\frac{a^2}{9}}{a^2}\right)$$
$$= \log_3 \left(\frac{a^2}{9} \times \frac{1}{a^2}\right) = \log_3 \frac{1}{9} = -2$$

**50.**
$$\frac{\log_{\frac{1}{2}} \frac{1}{4} + \log_{\frac{1}{2}} 64 - \log_{\frac{1}{2}} 16}{\log_{\frac{1}{2}} 2 - \log_{\frac{1}{2}} 4 - \log_{\frac{1}{2}} 8} = \frac{\log_{\frac{1}{2}} \left(\frac{\frac{1}{4} \times 64}{16}\right)}{\log_{\frac{1}{2}} \left(\frac{2}{4}\right) - \log_{\frac{1}{2}} 8}$$
$$= \frac{\log_{\frac{1}{2}} 1}{\log_{\frac{1}{2}} \frac{1}{16}} = \frac{0}{4} = 0$$

**51.**
$$\log_b 135 = \log_b \left(3^3 \times 5\right) = \log_b 3^3 + \log_6 5$$
$$= 3\log_b 3 + \log_b 5 = 3A + B$$

**52.**
$$\log_b \left(\frac{\sqrt{15}}{27}\right)$$
$$= \log_b \frac{3^{\frac{1}{2}} \times 5^{\frac{1}{2}}}{3^3} = \log_b \frac{5^{\frac{1}{2}}}{3^{\frac{5}{2}}}$$
$$= \log_b 5^{\frac{1}{2}} - \log_b 3^{\frac{5}{2}} = \frac{1}{2}\log_b 5 - \frac{5}{2}\log_b 3$$
$$= \frac{1}{2}B - \frac{5}{2}A$$

**53.**
$$\log_5 x = k - \log_5 y$$
$$\log_5 x + \log_5 y = k$$
$$\log_5 (xy) = k$$
$$xy = 5^k$$
$$y = \frac{5^k}{x}$$

**54. D**
$$\log_3 \left[\left(27\sqrt{3}\right)\left(3^x\right)\right] = \log_3 27\sqrt{3} + \log_3 3^x$$
$$= \log_3 27 + \frac{1}{2}\log_3 3 + \log_3 3^x$$
$$-3 + \left(\frac{1}{2} \times 1\right) + \log_3 3^x = 3.5 + x$$

**55. C**
$$A = B^3C \rightarrow C = \frac{A}{B^3}$$
$$\log_5 C = \log_5 A - 3\log_5 B$$
$$= \log_5 A - 3\log_5 B$$

**56.**
$$\frac{\log x^8 - \log x^2}{\log x^7 + \log x^4} = \frac{\log \frac{x^8}{x^2}}{\log\left(x^7\right)\left(x^4\right)} = \frac{\log x^6}{\log x^{11}}$$
$$= \frac{6 \log x}{11 \log x} = \frac{6}{11} \doteq 0.55$$

| 0 | . | 5 | 5 |
|---|---|---|---|

**57.**

$$\log_{27} 3^8 = \log_{27}\left(27^{\frac{1}{3}}\right)^8 = \log_{27} 27^{\frac{8}{3}}$$

$$= \frac{8}{3}\log_{27} 27 = \frac{8}{3}(1) \doteq 2.67$$

| 2 | . | 6 | 7 |
|---|---|---|---|

## *Lesson 4 Logarithms—Special Cases, Base Conversion*

### CLASS EXERCISES ANSWERS AND SOLUTIONS

**1.** We need to solve for $x$ using exponents, such that $\left(\frac{1}{3}\right)^x = 1$.

The value of $x$ is 0, therefore, $\log_{\frac{1}{3}} 1 = 0$.

**2.** Since $3.7^1 = 3.7$, $\log_{3.7} 3.7 = 1$.

**3.** Since $2.1^5 = 2.1^5$, we know that $\log_{2.1}(2.1^5) = 5$.

**4.**

$\left(\frac{1}{10}\right)^{\log_{\frac{1}{10}} 10000} = \left(\frac{1}{10}\right)^{-4}$ • Evaluate the exponent

$= 10000$ • Evaluate the power

**5.** Notice that in this question, the exponent cannot be evaluated in the same way as the previous two questions because 29 is not a power of 2. So, we will take a slightly different approach.

Let $2^{\log_2 29} = y$. • We wish to solve for $y$

$\log_2 y = \log_2 29$ • Change the exponential form to logarithmic form

$y = 29$ • Equivalent logarithmic expressions

$2^{\log_2 29} = 29$ • By substitution, $y = 29$

**6.** When changing the base, we should choose a new base such that 32 and $\frac{1}{8}$ are both powers of that new base. What would be a good base to use?

A new base of 2 would work because 32 and $\frac{1}{8}$ are both powers of 2.

$\log_{32} \frac{1}{8} = \frac{\log_2 \frac{1}{8}}{\log_2 32}$ • Apply Change of Base Law

$= \frac{-3}{5}$ • Evaluate top and bottom

$= -\frac{3}{5}$ • Simplify

**7.** Notice that we have a rational expression where both logs have the same base: base 6. This allows us to use base conversion to rewrite the expression as a single log.

$\frac{\log_6 1000}{\log_6 10} = \log_{10} 1000$ • Evaluate

$= 3$

### PRACTICE EXERCISES ANSWERS AND SOLUTIONS

**1.** $\log_4 1 + \log_8 1$
$= 0 + 0 = 0$

**2.** $3^{\log_3 81} = 81$

**3.** $7^{\log_5 1} + 8^{\log_3 3} = 7^0 + 8^1 = 1 + 8 = 9$

**4.** $8^{\log_8\left(\frac{1}{16}\right)} = \frac{1}{16}$

**5.** $\log_3 1^8 - \log_3 3^8 = 8\log_3 1 - 8\log_3 3$
$= 8(0) - 8(1) = 0 - 8 = -8$

**6.** $\log_3(\log_6(\log_2 64)) = \log_3(\log_6(6))$
$= \log_3(1) = 0$

**7.** $\frac{(\log_{10} 12) + (\log_{10} 3)}{\log_{10} 6} = \frac{\log_{10} 36}{\log_{10} 6} = \log_6 36 = 2$

**8.** $\log_3 7^{\log_7(3^{12})} = \log_3\left(7^{\log_7(3^{12})}\right) = \log_3 3^{12} = 12$

**9.** $\log_{\frac{1}{5}}(5^6) = 6\log_{\frac{1}{5}} 5 = 6(-1) = -6$

**10.** $4^{\log_2 2^{-1}} = 4^{-1\log_2 2} = 4^{-1(1)} = \frac{1}{4}$

**11.** $\log_3 3^5 + \log_5 5^3 = 5\log_3 3 + 3\log_5 5$
$= 5(1) + 3(1) = 8$

 

**12.**
$$\log_8 \frac{1}{16} = \frac{\log_2 \frac{1}{16}}{\log_2 8} = \frac{-4}{3}$$

Use base conversion with base 2, since $\frac{1}{16}$ and 8 are powers of 2.

**13.**
$$\frac{\log_7 \frac{1}{81}}{\log_7 9} = \log_9 \frac{1}{81} = -2$$

**14.**
$$\frac{\log_{14} \sqrt{7}}{\log_{14} 7} = \log_7 \sqrt{7} = \log_7 7^{\frac{1}{2}} = \frac{1}{2} \times \log_7 7$$
$$= \frac{1}{2} \times 1 = \frac{1}{2}$$

**15.**
$$\log_{\frac{1}{81}} \sqrt{27} = \frac{\log_3 \sqrt{27}}{\log_3 \frac{1}{81}} = \frac{\frac{1}{2} \times \log_3 27}{\log_3 \frac{1}{81}}$$
$$= \frac{\frac{1}{2} \times 3}{-4} = -\frac{3}{8}$$

**16.** $\log_{81}\left(\frac{1}{27}\right) = \frac{\log_3 \frac{1}{27}}{\log_3 81} = -\frac{3}{4}$

**17.** $\frac{\log_7 2^4}{\log_7 2} = \frac{4 \log_7 2}{\log_7 2} = 4$

**18.** $\log_{\sqrt{5}} \sqrt[5]{5^2} = \frac{\log_5 5^{\frac{2}{5}}}{\log_5 5^{\frac{1}{2}}} = \frac{\frac{2}{5}}{\frac{1}{2}} = \frac{2}{5} \times \frac{2}{1} = \frac{4}{5}$

**19.**
$$\log_8 \sqrt[5]{2^7} = \log_8 2^{\frac{7}{5}} = \frac{7}{5} \log_8 2$$
$$= \frac{7}{5}\left(\frac{1}{3}\right) = \frac{7}{15} \qquad \left(\text{Note: } \log_8 2 = \frac{1}{3}\right)$$

**20.**
$$\log_3 4 - \log_3 36 = \log_3\left(\frac{4}{36}\right) = \log_3\left(\frac{1}{9}\right)$$
$$= \log_3\left(\frac{1}{3^2}\right) = -2$$

**21.** $\log_2\left(\frac{4}{3}\right) + \log_2 768 = \log_2 1024 = 10$

**22.** $\log_a 1 + \log_a 1 + \log_a 1 = \log_a 1 = 0$

**23.**
$$\log_2\left(2^8\right) - \log_2\left(\frac{1}{8}\right)^2$$
$$= 8 - \log_2\left(2^{-3}\right)^2 = 8 - (-6) = 14$$

**24.**
$$\log_3 3 + \log_3 2 - \log_3 27 - \log_3 6$$
$$= \log_3 6 - \log_3 27 - \log_3 6$$
$$= -\log_3 27 = -3$$

**25.** $\log_2 \sqrt[3]{16} = \frac{1}{3} \times \log_2 16 = \frac{1}{3} \times 4 = \frac{4}{3}$

**26.** $\frac{\log_{11} \sqrt{2}}{\log_{11} \sqrt{8}} = \log_{\sqrt{8}} \sqrt{2} = \frac{\log_2 \sqrt{2}}{\log_2 \sqrt{8}} = \frac{\frac{1}{2}}{\frac{3}{2}} = \frac{1}{3}$

**27.** $\log_{81} 27 = \frac{\log_3 27}{\log_3 81} = \frac{3}{4}$

**28.**

| LHS | RHS |
|---|---|
| $\frac{\log_2 10}{\log_2 2}$ | $\log_2 5$ |
| $= \log_2 10$ | Thus, the original equation is not true. |

**29.**

| LHS | RHS |
|---|---|
| $\log_6 \sqrt{2}$ | $\frac{\log_6 2}{2}$ |
| $= \log_6 2^{\frac{1}{2}}$ | $= \frac{1}{2} \log_6 2$ |
| $= \frac{1}{2} \log_6 2$ | |
| = R.S. | Thus, the original equation is true. |

**30.**

| LHS | RHS |
|---|---|
| $4^{\log_3 14}$ | $14$<br>$= 4^{\log_4 14}$<br>Since<br>$\log_3 14 \neq \log_4 14$,<br>the original equation is not correct. |

**31.**

| LHS | RHS |
|---|---|
| $\log_2 \frac{1}{8}$<br>$= -3$ | $\frac{1}{\log_2 8}$<br>$= \frac{1}{3}$<br>Thus, the original equation is not true. |

**32.**

| LHS | RHS |
|---|---|
| $\log_5 8$ | $3\log_5 2$<br>$= \log_5 2^3$<br>$= \log_5 8$<br>Thus, the original equation is true. |

**33.**

| LHS | RHS |
|---|---|
| $2^{\frac{\log_7 8}{\log_7 2}}$<br>$= 2^{\log_2 8}$<br>$= 2^3$ | $2^4$<br><br>$2^4$<br>Thus, the original equation is not true. |

**34.**

| LHS | RHS |
|---|---|
| $\log_b\left(5^{-3}\right)$ | $-3\log_b 5$<br>$= \log_b 5^{-3}$<br>Thus, the original equation is true. |

**35.**

| LHS | RHS |
|---|---|
| $\log_b 0$<br>Does not exist | $1$<br>Thus, the original equation is not true. |

**36.**

| LHS | RHS |
|---|---|
| $\frac{\log_b y}{\log_b x}$ | $\log_b y - \log_b x$<br>$= \log_b\left(\frac{y}{x}\right)$<br>Thus, the original equation is not true. |

**37.**

| LHS | RHS |
|---|---|
| $\log_b 27$<br>$= \log_b 3^3$ | $\frac{3}{4}\log_b 81$<br>$= \log_b 81^{\frac{3}{4}}$<br>$= \log_b 3^3$<br>Thus, the original equation is true. |

**38.**

| LHS | RHS |
|---|---|
| $\log_b 2 - \log_b \sqrt{2}$<br>$= \log_b \frac{2}{\sqrt{2}}$ | $\log_b \sqrt{2}$<br>Thus, the original equation is not true. |

**39.**

| LHS | RHS |
|---|---|
| $(\log_b y)(\log_b x)$ | $\log_b y + \log_b x$<br>$= \log_b(yx)$<br>Thus, the original equation is not true. |

**40.**

| LHS | RHS |
|---|---|
| $\frac{1}{3}\log_b 11$<br>$= \log_b 11^{\frac{1}{3}}$ | $\log_b\left(\frac{11}{3}\right)$<br>Thus, the original equation is not true. |

**41.**

| LHS | RHS |
|---|---|
| $\log_b \frac{1}{5} - \log_b 5$<br>$= \log_b \frac{\frac{1}{5}}{5}$<br>$= \log_b \frac{1}{25}$ | $-\log_b 25$<br>$= \log_b 25^{-1}$<br>$= \log_b \frac{1}{25}$<br>Thus, the original equation is true. |

 

**42.**

| LHS | RHS |
|---|---|
| $\log_b 5$ | $\frac{1}{2}\log_b 10$<br>$= \log_b 10^{\frac{1}{2}}$<br>Thus, the original equation is not true. |

**43. D**

$$\log_{\frac{1}{d}} x = \frac{\log_d(x)}{\log_d\left(\frac{1}{d}\right)} = \frac{\log_d x}{-1} = -\log_d x$$

**44. D**

$$3\log_A 2A - \log_A(1.6A)$$
$$= \log_A(2A)^3 - \log_A(1.6A)$$
$$= \log_A\left(\frac{8A^3}{1.6A}\right) = \log_A(5A^2) = \log_A 5 + \log_A A^2$$
$$= \log_A 5 + 2$$

**45.** $-\log_{27}\left(\frac{1}{9}\right) = \log_{27} 9 = \frac{\log_3 9}{\log_3 27} = \frac{2}{3} \doteq 0.67$

| 0 | . | 6 | 7 |
|---|---|---|---|

**46.** $\log_{27}\left(\sqrt[5]{3^2}\right) = \log_{27} 3^{\frac{2}{5}} = \frac{\log_3 3^{\frac{2}{5}}}{\log_3 27} = \frac{\frac{2}{5}}{3} = \frac{2}{15} \doteq 0.13$

| 0 | . | 1 | 3 |
|---|---|---|---|

## *Lesson 5 Using the Calculator to Evaluate Logarithms*

### CLASS EXERCISES ANSWERS AND SOLUTIONS

**1.** Method 1: On most scientific calculators, enter [“78”] and then press [LOG].

Method 2: On most graphing calculators, press [LOG][7][8][ENTER].

Using either method, you should get $\log 78 \doteq 1.892$.

You can also use your calculator to verify that this evaluation is correct.

If $\log 78 \doteq 1.892$, then it must follow that $10^{1.892} \doteq 78$. Check to see if this is correct.

**2.** Method 1: On most scientific calculators, enter [“500”] and then press [LN].

Method 2: On most graphing calculators, press [LN][5][0][0][ENTER].

Using either method, you should get $\ln 500 \doteq 6.215$.

You can use your calculator to verify that this is correct.

If $\ln 500 \doteq 6.215$, then we can conclude that $e^{6.215} \doteq 500$. Check to see if this is correct.

**3.** First, we must rewrite the expression using the base conversion formula. When we rewrite the expression, we can choose any number as our new base. We will choose 10 so that we can evaluate the new expression on a calculator.

$\log_3 \frac{1}{4} = \frac{\log \frac{1}{4}}{\log 3}$ • Rewrite using the base conversion formula

$\doteq -1.262$ • From the calculator

With a graphing calculator, the entire expression can be evaluated by entering $\log(1 \div 4) \div \log 3$, and then pressing [ENTER].

**4.** Before evaluating $\log_5 12$, $\log_5 4$, and $\log_5 8$ individually, notice that all three terms have the same base: base 5. This indicates that they can be combined into a single logarithm by using the laws of logarithms.

$$\log_5 12 + \log_5 4 - \log_5 8 = \log_5\left(\frac{12 \times 4}{8}\right)$$
$$= \log_5 6$$

This simplifies the calculation.

To evaluate $\log_5 6$, we rewrite the expression by completing a base conversion. When we rewrite the expression, we will choose a new base of 10 or $e$, so that the expression can be evaluated using a calculator. Using base 10 or base $e$ works equally well for this evaluation. Using base $e$ gives:

$\log_5 6 = \frac{\ln 6}{\ln 5}$

$\doteq 1.113$ • From the calculator. Using a graphing calculator, enter $\ln(6) \div \ln(5)$, then press [ENTER]

So, $\log_5 12 + \log_5 4 - \log_5 8 \doteq 1.113$.

5. Does this expression look familiar? The log in the numerator has the same base as the log in the denominator. This means that if we choose, we can use the reverse of the base conversion formula to rewrite this expression in a simpler form.

The result is $\frac{\log_3 55}{\log_3 10} = \log_{10} 55$.

Since $\log_{10} 55$ is a common logarithm, we can also write it as log 55.

Using a calculator, we find that $\log 55 \doteq 1.740$.

Thus, $\frac{\log_3 55}{\log_3 10} \doteq 1.740$.

## PRACTICE EXERCISES ANSWERS AND SOLUTIONS

1. $\log 100 = 2.000$ (exact)

2. $\log 78422 \doteq 4.894$

3. $\log \frac{1}{100000} = -5.000$ (exact)

4. $\log 0.0027 \doteq -2.569$

5. $\log 15 + \log 46 = \log(15 \times 46) \doteq 2.839$

6. $\ln 12 \doteq 2.485$

7. $\ln 100 \doteq 4.605$

8. $\ln e = 1$ (exact)

9. $\log 5000 \doteq 3.699$

10. $\log \sqrt{10} = \log 10^{\frac{1}{2}} = \frac{1}{2} \log 10 = \frac{1}{2}(1)$
$= 0.500$ (exact)

11. $\ln\left(\frac{1}{e}\right) = \ln 1 - \ln e = 0 - 1 = -1.000$ (exact)

12. $\ln 543 - \ln 17 = \ln\left(\frac{543}{17}\right) \doteq 3.464$

13. $\log_{11} 80 = \frac{\log 80}{\log 11} \doteq 1.83$

14. $\log_e 678 = \ln 678 \doteq 6.52$

15. $\log_4\left(26^{100}\right) = 100 \times \log_4 26$
$= 100 \times \frac{\log 26}{\log 4} \doteq 235.02$

16. $\log_3 25 + \log_3 4 - \log_3 100$
$= \log_3 \frac{25 \times 4}{100}$
$= \log_3 1 = 0$ (exact)

17. $\frac{\log_7 63}{\log_7 5} = \log_5 63 = \frac{\log 63}{\log 5} \doteq 2.57$

18. $e^{\ln 100} = e^{\log_e 100} = 100.00$ (exact)

19. $\frac{\log_e 81}{\ln 3} = \frac{\ln 81}{\ln 3} = \frac{\ln 3^4}{\ln 3} = \frac{4 \ln 3}{\ln 3} = 4.00$ (exact)

20. $\log_{\sqrt{3}} 98 = \frac{\log 98}{\log \sqrt{3}} \doteq 8.35$

21. $\frac{\log_5 600}{\log_5 e} = \log_e 600 = \ln 600 \doteq 6.40$

22. $\ln e^{\log_2 10} = \log_2 10 \ln e = \log_2 10 = \frac{\log 10}{\log 2} \doteq 3.32$

23. $\frac{\log_{13} 200}{\log_{13} 6} = \log_6 200 = \frac{\log 200}{\log 6} \doteq 2.96$

24. $\log_2 18 - \log_2 5 + \log_2\left(\frac{20}{9}\right)$
$= \log_2\left(\frac{18 \times \frac{20}{9}}{5}\right)$
$= \log_2 8 = 3.00$ (exact)

25. $\log_5 123 = \frac{\log 123}{\log 5} \doteq 2.990$

 

**26.** $\log_2 12 + \log_2 3 - \log_2 9$
$= \log_2\left(\frac{12\times 3}{9}\right)$
$= \log_2 4 = 2.00$ (exact)

**27.** $\log_e 10000 = \ln 10000 \doteq 9.210$

**28.** $\log_{\sqrt{5}} 9 = \frac{\log 9}{\log\sqrt{5}} \doteq 2.730$

**29.**
$\log_5 \frac{\sqrt{7}}{\sqrt{35}} = \log_5\sqrt{\frac{7}{35}} = \log_5\sqrt{\frac{1}{5}} = \log_5\left(\frac{1}{5}\right)^{\frac{1}{2}}$
$= \frac{1}{2}\log_5\left(\frac{1}{5}\right) = \frac{1}{2}(-1) = -\frac{1}{2} = -0.500$ (exact)

**30.** $\log_3 9^{80} = 80\log_3 9 = 80\times 2 = 160.000$ (exact)

**31.** $\log_8 5 - \log_8 15 + \log_8 12$
$= \log_8\left(\frac{5\times 12}{15}\right)$
$= \log_8(4) = \log_8(8)^{\frac{2}{3}} = \frac{2}{3}\log_8 8 = \frac{2}{3}(1)$
$\doteq 0.667$

**32.** $\log_5 2 + \log_2 5 = \frac{\log 2}{\log 5} + \frac{\log 5}{\log 2} \doteq 2.753$

**33.** $6^{\log_6 22} = 22.000$ (exact)

**34.**
$\log_{\sqrt{2}} 6 - \log_{\sqrt{2}} 3 = \log_{\sqrt{2}}\left(\frac{6}{3}\right) = \log_{\sqrt{2}} 2$
$= \log_{\sqrt{2}}\left(\sqrt{2}\right)^2 = 2.00$ (exact)

**35.** $\ln 64 + \ln 34 + \ln 7 = \ln(64\times 34\times 7)$
$= \ln 15232 \doteq 9.631$

**36. C**
$\frac{\log_4 17}{\log_4 21} = \log_{21} 17$

**37. D**
$\log_5 19 = \frac{\log_x 19}{\log_x 5}$ where $x > 0$, $x \neq 1$

**38.** $\log_4 18 = \frac{\log 18}{\log 4} \doteq 2.08$

| 2 | . | 0 | 8 |
|---|---|---|---|

**39.**
$\log_5 7^2 + \log_7 5^2 = 2\times\frac{\log 7}{\log 5} + 2\times\frac{\log 5}{\log 7}$
$\doteq 2.418 + 1.654 \doteq 4.07$

| 4 | . | 0 | 7 |
|---|---|---|---|

## *Lesson 6 Solving Exponential and Logarithmic Equations*

### CLASS EXERCISES
### ANSWERS AND SOLUTIONS

**1.** We can write 4 and 8 as powers of 2.

$(2^2)^{x-1} = (2^3)^{5x}$ • Rewrite 4 and 8 with base 2
$2^{2x-2} = 2^{15x}$ • Apply Product Law of Exponents
$2x - 2 = 15x$ • By comparison, the exponents are equal
$-2 = 13x$ • Subtract $2x$ from both sides
$\frac{-2}{13} = x$ • Solve for $x$

**2.** This time, we will take the natural log of both sides. (Common log will also work).

$3^{x+3} = 8^{2x-1}$
$\ln\left(3^{x+3}\right) = \ln\left(8^{2x-1}\right)$ • Take the natural log of both sides
$(x+3)\ln 3 = (2x-1)\ln 8$ • Apply the Power Law
$(x+3)(1.099) \doteq (2x-1)(2.079)$ • Evaluate ln 3 and ln 8
$1.099x + 3.297 \doteq 4.158x - 2.079$ • Expand
$5.376 \doteq 3.059x$ • Isolate $x$
$1.757 \doteq x$ • Solve for $x$

**3.**

$\log_2\left(\frac{6x+2}{x-1}\right) = 3$ • Quotient Law
$2^3 = \frac{6x+2}{x-1}$ • Rewrite in exponential form
$8x - 8 = 6x + 2$ • Cross products are equal
$2x = 10$ • Solve for $x$
$x = 5$

Verify that $x = 5$.

$\log_2(6(5)+2)-\log_2(5-1)$

$\log_2 32-\log_2 4=5-2=3$ = right side of the equation

There is no log of zero or a negative number and $x = 5$ is verified.

The final solution is $x = 5$.

**4.**

$\log_3 x+\frac{\log_3 x}{\log_3 9}=5$ • Base conversion, using base 3

$\log_3 x+\frac{\log_3 x}{2}=5$ • Evaluate $\log_3 9$

$2\log_3 x+\log_3 x=10$ • Multiply both sides by 2

$3\log_3 x=10$ • Simplify

$\log_3 x=\frac{10}{3}$ • Simplify

$3^{\frac{10}{3}}=x$ • Rewrite in exponential form

$38.9 \doteq x$ • Evaluate

Verification: $x \doteq 38.9$

LHS: $\log_3 38.9+\log_9 38.9 \doteq 5$ (using the calculator) = RHS

$x \doteq 38.9$ is verified as the solution.

## PRACTICE EXERCISES ANSWERS AND SOLUTIONS

**1.**

$$5^{x-2}=125^{x+2}$$
$$5^{x-2}=\left(5^3\right)^{x+2}$$
$$5^{x-2}=5^{3x+6}$$
$$x-2=3x+6$$
$$-2x=8$$
$$x=-4$$

**2.**

$$81^{4x}=27^{x+2}$$
$$\left(3^4\right)^{(4x)}=\left(3^3\right)^{(x+2)}$$
$$3^{16x}=3^{3x+6}$$
$$16x=3x+6$$
$$13x=6$$
$$x=\frac{6}{13}$$

**3.**

$8^{x+4}\times16^{x-5}=64^{3-x}$

$\left(2^3\right)^{x+4}\left(2^4\right)^{x-5}=\left(2^6\right)^{3-x}$ • Write 8, 16, and 64 with a base of 2

$2^{3x+12}\times2^{4x-20}=2^{18-6x}$ • Laws of exponents

$2^{7x-8}=2^{18-6x}$

$7x-8=18-6x$ • Compare

$13x=26$

$x=2$

**4.**

$$27^{2x+1}=9^x$$
$$\left(3^3\right)^{2x+1}=\left(3^2\right)^x$$
$$3^{6x+3}=3^{2x}$$
$$6x+3=2x$$
$$4x=-3$$
$$x=-\frac{3}{4}$$

**5.** $\log_2 x-\log_2(x+2)=3$

Write left side as a single log:

$$\log_2\left(\frac{x}{x+2}\right)=3$$

Write in exponential form:

$$2^3=\frac{x}{x+2}$$
$$8=\frac{x}{x+2}$$
$$8x+16=x$$
$$7x=-16$$
$$x=\frac{-16}{7}$$

To satisfy the equation $x > 0$ and $x + 2 > 0$.

Thus, there is no solution since $-\frac{16}{7}$ is not greater than 0.

 

**6.** $\log x - \log 12 = \log \frac{1}{3}$

$\log\left(\frac{x}{12}\right) = \log\left(\frac{1}{3}\right)$

$\frac{x}{12} = \frac{1}{3}$

$x = 4$

| LHS | RHS |
|---|---|
| $\log 4 - \log 12$ | $\log \frac{1}{3}$ |
| $\log\left(\frac{4}{12}\right)$ | |
| $\log \frac{1}{3}$ | |

**7.** $\ln(2x-1) = \ln(4x-3) - \ln x$

$\ln(2x-1) = \ln\left(\frac{4x-3}{x}\right)$ • Write right side as a single log

$2x - 1 = \frac{4x-3}{x}$ • Compare

$2x^2 - 1 = 4x - 3$

$2x^2 - 5x + 3 = 0$ • To solve a quadratic equation make one side equal to zero

$(2x-3)(x-1) = 0$ • Factor

$2x - 3 = 0$ or $x - 1 = 0$

$2x = 3 \quad x = 1$

$x = \frac{3}{2}$

For $x$ to be a solution, $2x - 1 > 0$, $4x - 3 > 0$, and $x > 0$. Thus, $x > \frac{1}{2}$, $x > \frac{3}{4}$, and $x > 0$.

$\therefore x > \frac{3}{4}$

Thus, the solutions are $x = \frac{3}{2}$ and $x = 1$ since both satisfy the original condition of $x > \frac{3}{4}$.

Verify $x = \frac{3}{2}$:

| LHS | RHS |
|---|---|
| $\ln\left(2\left(\frac{3}{2}\right) - 1\right)$ | $\ln\left(4\left(\frac{7}{2}\right) - 3\right) - \ln\left(\frac{3}{2}\right)$ |
| $\ln 2$ | $\ln(3) - \ln\left(\frac{3}{2}\right)$ |
| | $\ln\left(\frac{3}{\frac{1}{2}}\right)$ |
| | $\ln 2$ |

Verify $x = 1$:

| LHS | RHS |
|---|---|
| $\ln(2(1))$ | $\ln(4(1) - 3)\ln(1)$ |
| $\ln 1$ | $\ln 1 - \ln 1$ |
| 0 | 0 |

**8.** $2\log_2(x+1) - \log_2 x - 2 = 0$

$\log_2\left[\frac{(x+1)^2}{x}\right] = 2$

$2^2 = \frac{(x+1)^2}{x}$

$4x = x^2 + 2x + 1$

$x^2 - 2x + 1 = 0$

$(x-1)(x-1) = 0$

$x = 1$

$x = 1$ satitfies $(x + 1) > 0$ and $x > 0$

| LHS | RHS |
|---|---|
| $2\log_2(1+1) - \log_2 1 - 2$ | 0 |
| $2(1) - 0 - 2 = 0$ | |

**9.** $\log_5\left(\frac{x+6}{x-2}\right) = 1$

$5^1 = \frac{x+6}{x-2}$

$5x - 10 = x + 6$

$4x = 16$

$x = 4$

$x = 4$ is a solution since $x = 4$ satitfies $\frac{x+6}{x-2} > 0$

| LHS | RHS |
|---|---|
| $\log_5\left(\frac{4+6}{4-2}\right)$ | 1 |
| $\log_5(5)$ | |
| 1 | |

**10.** $\log_3 x + \log_3(2x-9) = 4$

$$\log_3(x(2x-9)) = 4$$
$$\log_3(2x^2-9x) = 4$$
$$3^4 = 2x^2 - 9x$$
$$2x^2 - 9x - 81 = 0$$
$$(x-9)(2x+9) = 0$$
$$x = 9 \text{ or } x = \frac{-9}{2} = -4.5$$

For $x$ to be a solution, $x > 0$ and $2x - 9 > 0$. Thus, $x > 4.5$.

Thus, $x = 9$, since $-4.5$ is not $> 4.5$.

| LHS | RHS |
|---|---|
| $\log_3(9)+\log_3(2(9)-9)$ | 4 |
| $2+\log_3 9$ | |
| $2+2$ | |
| $=4$ | |

**11.** $2\log_2(x+1) - \log_2 x - 2 = 0$

$$\log_2 \frac{(x+1)^2}{x} = 2$$
$$2^2 = \frac{(x+1)^2}{x}$$
$$4x = x^2 + 2x + 1$$
$$0 = x^2 - 2x + 1$$
$$0 = (x-1)(x-1)$$

Thus, $x = 1$.
$x = 1$ satisfies $(x + 1) > 0$ and $x > 0$.

| LHS | RHS |
|---|---|
| $2\log_2(1+1) - \log_2 1 - 2$ | 0 |
| $2\log_2(2) - 0 - 2$ | |
| $2(1) - 0 - 2$ | |
| 0 | |

**12.** $\log_8 37 = x$

$$\frac{\log 37}{\log 8} = x$$
$$x \doteq 1.74$$

**13.**

$$\log_4\left[\log_{\frac{1}{16}} x\right] = -1,$$

$$4^{-1} = \log_{\frac{1}{16}} x$$

- Rewrite in exponential form

$$\frac{1}{4} = \log_{\frac{1}{16}} x$$

$$\left(\frac{1}{16}\right)^{\frac{1}{4}} = x$$

- Again, rewrite in exponential form
- Evaluate

$$\sqrt[4]{\frac{1}{16}} = x$$

$$x = \frac{1}{2}$$

This satisfies the original condition of $x > 0$.

**14.**

$$\log 5 = 2\log x - \log(x+2)$$
$$\log 5 = \log\left(\frac{x^2}{x+2}\right)$$
$$5 = \frac{x^2}{x+2}$$
$$5x + 10 = x^2$$
$$x^2 - 5x - 10 = 0$$
$$x = \frac{-b \pm \sqrt{b^2 - 4ac}}{2a}$$
$$= \frac{5 \pm \sqrt{25 - 4(1)(-10)}}{2(1)}$$
$$x = \frac{5 \pm \sqrt{65}}{2} \doteq -1.53 \text{ or } 6.53$$

- $-1.53$ is extraneous, it is not $> 0$

$\therefore x \doteq 6.53$ (satisfies $x > 0$)

**15.**

$$7^{2x+3} = 99$$
$$\log 7^{2x+3} = \log 99$$

- Take common log of both sides

$$(2x+3)\log 7 = \log 99$$

- Use the power law on left side

$$(2x+3)(0.845) \doteq 1.996$$
$$1.69x + 2.535 \doteq 1.996$$
$$1.69x \doteq -0.539$$
$$x \doteq -0.32$$

 

**16.**
$$\log_2\left[3\left(2^x\right)\right] = \log_2\left(5^{x-3}\right)$$
$$\log_2 3 + \log_2 2^x = \log_2 5^{x-3}$$
$$\log_2 3 + x = (x-3)\log_2 5$$
$$\log_2 3 + x = x\log_2 5 - 3\log_2 5$$
$$\log_2 3 + 3\log_2 5 = x\log_2 5 - x$$
$$\log_2 3 + \log_2 5^3 = x\left(\log_2 5 - 1\right)$$
$$x = \frac{\log_2\left(3\times 5^3\right)}{\log_2 5 - 1} = \frac{\log_2(375)}{\log_2 5 - 1}$$
$$x \doteq 6.47$$

**17.** $\log_2 8 - \log_2 4 = \log_2 3x + \log_2 2$
$$\log_2 \frac{8}{4} = \log_2(3x\cdot 2)$$
$$\log_2 2 = \log_2 6x$$
$$2 = 6x$$
$$x = \frac{2}{6} = \frac{1}{3} \doteq 0.33 \text{ (satisfies } x > 0)$$

**18.** $\log_2(x+1) - \log_2(x-1) = 4$
$$\log_2\left(\frac{x+1}{x-1}\right) = 4$$
$$2^4 = \frac{x+1}{x-1}$$
$$16x - 16 = x + 1$$
$$15x = 17$$
$$x = \frac{17}{15} \doteq 1.13$$

This solution satisfies $(x + 1) > 0$ and $(x - 1) > 0$.

**19.** $\log\left(x^3+11\right) = 3\ln\left(x^2+2\right) - 8$

To solve, graph:
$Y_1 = \log\left(x^3+11\right)$, $Y_2 = 3\ln\left(x^2+2\right) - 8$

The solution is $x \doteq 5.29$.

**20.** $e^{\log\left(x^3\right)} = 3^{x-1} - 4$

To solve, graph:
$Y_1 = e^{\log\left(x^3\right)}$, $Y_2 = 3^{x-1} - 4$

The solution is $x \doteq 2.89$.

**21.** $\log(2x) = x^2 - 5$

To solve, graph:
$Y_1 = \log(2x)$, $Y_2 = x^2 - 5$

The solution is $x \doteq 2.38$.

**22.**
$$\log_3 x = \left(\frac{1}{2}\right)^{x-5}$$

To solve, graph:
$Y_1 = \log_3 x = \dfrac{\log x}{\log 3}$, $Y_2 = \left(\dfrac{1}{2}\right)^{x-5}$

The solution is $x \doteq 4.54$.

**23.** $(\log x)(\ln x) = \dfrac{1}{5}x^2 - 7$

To solve, graph:
$Y_1 = (\log x)(\ln x)$, $Y_2 = \dfrac{1}{5}x^2 - 7$

The solution is $x = 6.53$.

**24.** $\log x^3 = \sqrt{4-x}$

To solve, graph:
$Y_1 = \log x^3$, $Y_2 = \sqrt{4-x}$

The solution is $x \doteq 2.53$.

**25.** $(\ln x)^2 + \ln x = 12$

Let $u = \ln x$. The equation now becomes:
$$u^2 + u = 12$$
$$u^2 + u - 12 = 0$$
$$(u+4)(u-3) = 0$$
$$u + 4 = 0 \text{ or } u - 3 = 0$$
$$u = -4 \text{ or } u = 3$$

If $u = -4$, $-4 = \ln x$ –4
$$-4 = \log_e x$$
$$e^{-4} = x$$
$$2.718^{-4} \doteq x$$
$$0.018 \doteq x$$

If $u = 3$, $3 = \ln x$

$3 = \log_e x$

$e^3 = x$

$2.718^3 \doteq x$

$20.08 \doteq x$

Thus $x \doteq 0.02$ or $x \doteq 20.08$.

(both solutions satisfy $x > 0$).

**26.** $2(\log x)^2 - 3 - 5\log x = 0$

Let $u = \log x$.

The equation now becomes:

$2u^2 - 5u - 3 = 0$

$(u-3)(2u+1)$

$u = 3$ or $u = -0.5$

If $u = 3$, $\log x = 3$.

$10^3 = x$

$x = 1\ 000$

If $u = -0.5$, $\log x = -0.5$.

$10^{-0.5} = x$

$x \doteq 0.316$

Thus, $x = 1\ 000$ or $x \doteq 0.32$.

(both solutions satisfy $x > 0$).

**27.** $(\log_3 x)^2 + \log_3 x - 2 = 0$

Let $u = \log_3 x$.

The equation becomes $u^2 + u - 2 = 0$.

$(u+2)(u-1) = 0$

$u + 2 = 0$ or $u - 1 = 0$

$u = -2$ or $u = 1$

If $u = -2$, $\log_3 x = -2$.

$x = 3^{-2}$

$x = \frac{1}{9}$

If $u = 1$, $\log_3 x = 1$.

$x = 3^1$

$x = 3$

Thus, $x = \frac{1}{9}$ or $x = 3$ (both satisfy $x > 0$).

**28.** $2(\log_9 x)^2 - 7\log_9 x = 4$

Let $u = \log_9 x$.

$2u^2 - 7u - 4 = 0$

$(u-4)(2u+1)$

$u = 4$ or $u = -\frac{1}{2}$

If $u = 4$, $\log_9 x = 4$.

$x = 9^4$

$x = 6\ 561$

If $u = -\frac{1}{2}$, $\log_9 x = -\frac{1}{2}$.

$x = 9^{-\frac{1}{2}} = \frac{1}{\sqrt{9}}$

$x = \frac{1}{3}$

Thus, $x = 6.561$ or $\frac{1}{3}$ (both satisfy $x > 0$).

**29.**

$\log_4 x + \log_8 x = 5,$

$\log_4 x + \frac{\log_4 x}{\log_4 8} = 5$ • Base conversion using base 4

$\log_4 x + \frac{\log_4 x}{1.5} = 5$ • Evaluate $\log_4 8$

$1.5\log_4 x + \log_4 x = 7.5$ • Multiply both sides by 1.5

$2.5\log_4 x = 7.5$ • Simplify

$\log_4 x = 3$ • Divide both sides by 2.5

$4^3 = x$ • Rewrite in exponential form

$64 = x$ • Evaluate

The solution is $x = 64$ (satisfies $x > 0$).

**30.** $\log_5 x - \log_{25} x = 2$

$\log_5 x - \frac{\log_5 x}{\log_5 25} = 2$

$\log_5 x - \frac{\log_5 x}{2} = 2$

$2\log_5 x - \log_5 x = 4$

$\log_5 x = 4$

$x = 5^4 = 625$ (satisfies $x > 0$).

 

**31.**
$$\log_4 x - \frac{\log_4 x}{\log_4 8} = 1$$
$$\log_4 x - \frac{\log_4 x}{1.5} = 1$$
$$1.5\log_4 x - \log_4 x = 1.5$$
$$0.5\log_4 x = 1.5$$
$$\log_4 x = 3$$
$$x = 4^3$$
$$x = 64 \text{ (satisfies } x > 0\text{)}.$$

**32.**
$$\log_9 x + 3\log_{27} x = 6$$
$$\log_9 x + \frac{\log_9 x^3}{\log_9 27} = 6$$
$$\log_9 x + \frac{\log_9 x^3}{1.5} = 6$$
$$1.5\log_9 x + \log_9 x^3 = 9$$
$$1.5\log_9 x + 3\log_9 x = 9$$
$$4.5\log_9 x = 9$$
$$\log_9 x = 2$$
$$x = 9^2$$
$$x = 81 \text{ (satisfies } x > 0\text{)}.$$

**33. D**
$$\log_2(x-3) + \log_2(x-1) = 3$$
$$\log_2((x-3)(x-1)) = 3$$
$$\log_2\left(x^2 - 4x + 3\right) = 3$$
$$2^3 = x^2 - 4x + 3$$
$$x^2 - 4x - 5 = 0$$
$$(x-5)(x+1) = 0$$
$$x = 5 \text{ or } x = -1$$

Thus, $x = 5$, since $x = -1$ does not satisfy $x > 3$.

**34. A**
$$\log_3(x+2) + \log_3(x-4) = 3$$
$$\log_3((x+2)(x-4)) = 3$$
$$3^3 = x^2 - 2x - 8$$
$$x^2 - 2x - 35 = 0$$
$$(x-7)(x+5) = 0$$
$$x = 7 \text{ or } x = -5$$

Thus, $x = 7$, since $-5$ does not satisfy $(x - 4) > 0$.

**35.** $\log_7(x-6) + \log_7(x+6) = 2$
$$\log_7(x-6)(x+6) = 2$$
$$\log_7\left(x^2 - 36\right) = 2$$
$$7^2 = x^2 - 36$$
$$x^2 - 85 = 0$$
$$x = \pm\sqrt{85} \doteq 9.22 \text{ or } -9.22$$

Thus, $x = 9.22$, since $-9.22$ does not satisfy $(x - 6) > 0$.

| 9 | . | 2 | 2 |
|---|---|---|---|

**36.**
$$\log_{(x-4)}\left(x^2 - 2x - 61\right) = 2$$
$$x^2 - 2x - 61 = (x-4)^2$$
$$x^2 - 2x - 61 = x^2 - 8x + 16$$
$$6x = 77$$

$x = \frac{77}{6} \doteq 12.8$ (satisfies the conditions of $x - 4 > 0$, and $x^2 - 2x - 61 > 0$)

| 1 | 2 | . | 8 |
|---|---|---|---|

## *Lesson 7 Exponential and Logarithmic Graphs*

### PRACTICE EXERCISES ANSWERS AND SOLUTIONS

**1.** $f(x) = \left(\frac{1}{5}\right)^{x-3} + 7$

This is the same as $y = \left(\frac{1}{5}\right)^{x-3} + 7$.

The inverse is:

$x = \left(\frac{1}{5}\right)^{y-3} + 7$ • Interchange $x$ and $y$

$x - 7 = \left(\frac{1}{5}\right)^{y-3}$ • Subtract 7 from each side

$\log_{\frac{1}{5}}(x-7) = y - 3$ • Rewrite in log form

$\log_{\frac{1}{5}}(x-7) + 3 = y$ • Solve for $y$

Thus, $f^{-1}(x) = \log_{\frac{1}{5}}(x-7) + 3$

**2.** $f(x) = \log_5(x+2) - 4$

This is the same as $y = \log_5(x+2) - 4$.

The inverse is:

$$x = \log_5(y+2) - 4$$
$$x + 4 = \log_5(y+2)$$
$$5^{(x+4)} = (y+2)$$
$$5^{x+4} - 2 = y$$

Thus, $f^{-1}(x) = 5^{x+4} - 2$.

**3.** $f(x) = \log_{\frac{3}{8}}(x+6)$

This is the same as $y = \log_{\frac{3}{8}}(x+6)$.

The inverse is:

$$x = \log_{\frac{3}{8}}(y+6)$$

- Interchange $x$ and $y$

$$\left(\frac{3}{8}\right)^x = y + 6$$

- Rewrite in exponential form

$$\left(\frac{3}{8}\right)^x - 6 = y$$

- Solve for $y$

Thus, $f^{-1}(x) = \left(\frac{3}{8}\right)^x - 6$.

**4.** $f(x) = 4^{3-x} + 9$

This is the same as $y = 4^{3-x} + 9$.

The inverse is:

$$x = 4^{3-y} + 9$$
$$x - 9 = 4^{3-y}$$
$$\log_4(x-9) = 3 - y$$
$$y = -\log_4(x-9) + 3$$

Thus, $f^{-1}(x) = -\log_4(x-9) + 3$.

**5.** $y = \log_3(x-4)$

This will be the graph of $y = \log_3 x$, translated 4 units right.

$y = \log_3(x-4)$

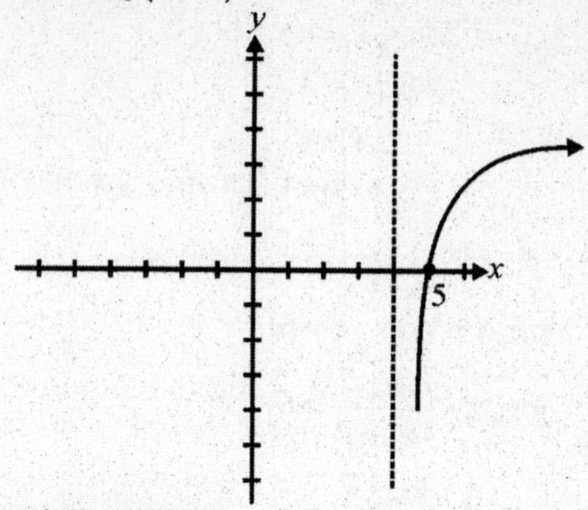


$x$-intercept is 5
Vertical asymptote: $x = 4$
Domain: $x > 4$
Range: $y \in R$

**6.** $y = 5^{x-4} + 1$

This will be the graph of $y = 5^x$, translated 4 units right and 1 unit up.

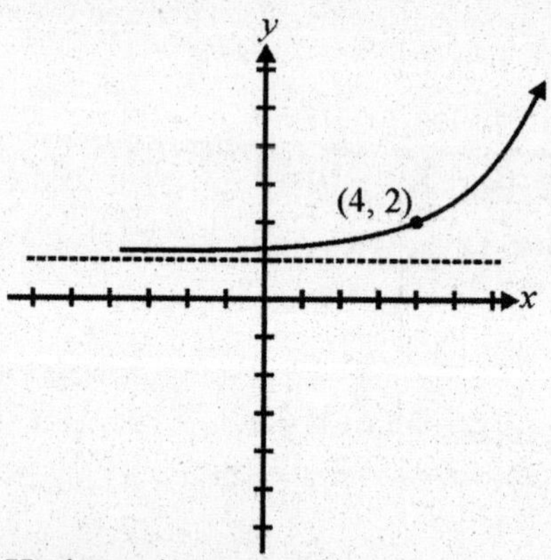


Horizontal asymptote: $y = 1$
Domain: $x \in R$
Range: $y > 1$

**7.** $y = \log_{\frac{1}{4}}(x-1) - 4$

This will be the graph of $y = \log_{\frac{1}{4}}(x)$ translated

1 unit right and 4 units down

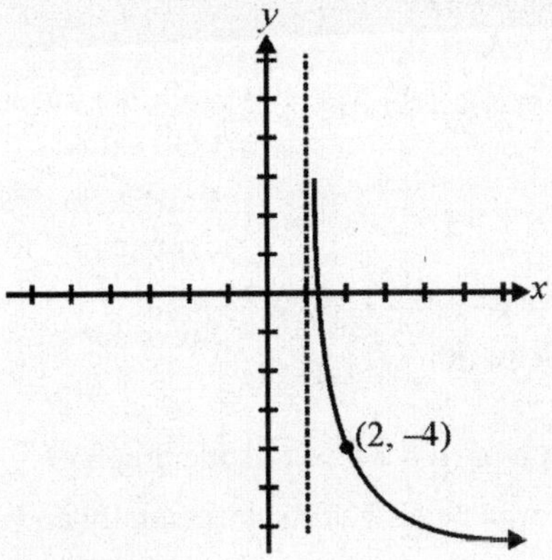


Asymptote: $x = 1$
Domain: $x > 1$
Range: $y \in R$

**8.** $y = \left(\frac{1}{7}\right)^x + 2$

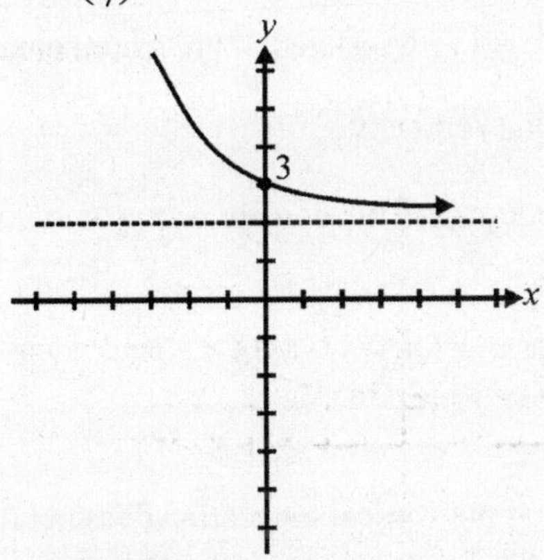


$y$-intercept is 3
Horizontal asymptote: $y = 2$
Domain: $x \in R$
Range: $y > 2$

**9.** $f(x) = 5^{x+2} - 1$

This is the same as $y = 5^{x+2} - 1$.

The inverse is:

$x = 5^{y+2} - 1$ • Interchange $x$ and $y$

$x + 1 = 5^{y+2}$ • Add 1 to each side

$\log_5(x+1) = y + 2$ • Rewrite in log form

$\log_5(x+1) - 2 = y$ • Solve for $y$

Thus, $f^{-1}(x) = \log_5(x+1) - 2$.

**10.** $f(x) = \ln(x-2)$

This is the same as $y = \ln(x-2)$.

The inverse is:

$x = \ln(y-2)$
$x = \log_e(y-2)$
$e^x = y - 2$
$y = e^x + 2$

Thus, $f^{-1}(x) = e^x + 2$.

**11.** $f(x) = \log_2 x + 3$

This is the same as $y = \log_2 x + 3$.

The inverse is:

$x = \log_2 y + 3$ • Interchange $x$ and $y$

$x - 3 = \log_2 y$ • Subtract 3 from both sides

$2^{x-3} = y$ • Rewrite in exponential form

Thus, $f^{-1}(x) = 2^{x-3}$.

**12.** $f(x) = \left(\frac{1}{5}\right)^x - 6$

This is the same as $y = \left(\frac{1}{5}\right)^x - 6$.

The inverse is:

$x = \left(\frac{1}{5}\right)^y - 6$

$x - 6 = \left(\frac{1}{5}\right)^y$

$\log_{\frac{1}{5}}(x-6) - y$

Thus, $f^{-1}(x) = \log_{\frac{1}{5}}(x-6)$.

**13.**

$f(x)=\left(\frac{1}{4}\right)^{x+2}-5$

This is the same as $y=\left(\frac{1}{4}\right)^{x+2}-5$.

The inverse is:

$x=\left(\frac{1}{4}\right)^{y+2}-5$ • Interchange $x$ and $y$

$x+5=\left(\frac{1}{4}\right)^{y+2}$ • Add 5 to both sides

$\log_{\frac{1}{4}}(x+5)=y+2$ • Rewrite in log form

$\log_{\frac{1}{4}}(x+5)-2=4$ • Solve for $y$

Thus, $f^{-1}(x)=\log_{\frac{1}{4}}(x+5)-2$.

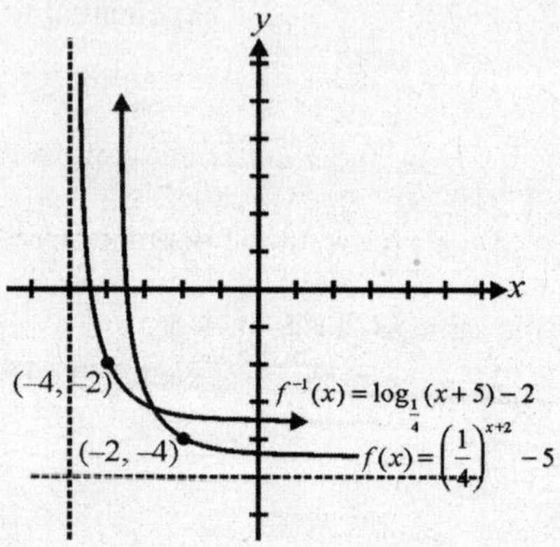


**14.** $f(x)=\log_2(x-1)+5$

This is the same as $y=\log_2(x-1)+5$.

The inverse is:

$x=\log_2(y-1)+5$

$x-5=\log_2(y-1)$

$2^{(x-5)}=y-1$

$2^{x-5}+1=y$

Thus, $f^{-1}(x)=2^{x-5}+1$.

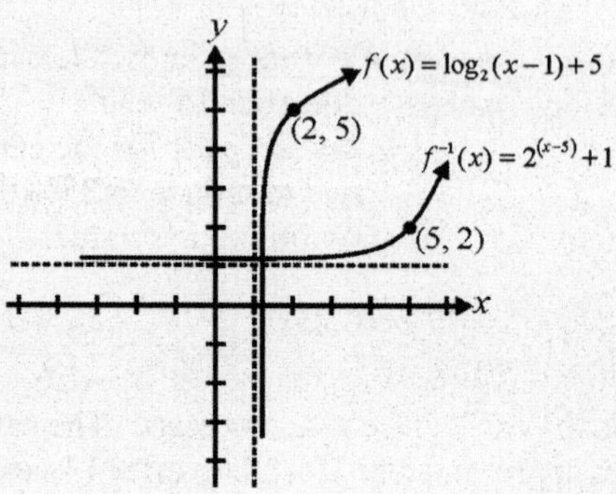


**15.** For $y=\log_{\frac{2}{3}}(x+3)+7$ the $x$-intercept occurs when $y=0$

Solve:

$0=\log_{\frac{2}{3}}(x+3)+7$

$-7=\log_{\frac{2}{3}}(x+3)$ • Subtract 7 from both sides

$\left(\frac{2}{3}\right)^{-7}=x+3$ • Rewrite in exponential form

$17.09\doteq x+3$

$x\doteq 14.09$ • Solve for $x$

**16.** The graph of $y=2^{x-b}+c$ is the graph of $y=2^x$, translated $b$ units horizontally and $c$ units vertically.

For $y=2^x$, the range is $y>0$, thus, for $y=2^{x-b}+c$, the range is $y>c$.

**17.** The graph of $y=\log 45(x-M)+j$ is the graph of $y=\log_{\frac{1}{5}}(x)$ translated $M$ units horizontally and $j$ units vertically.

For $y=\log_{\frac{1}{5}}(x)$, the domain is $x>0$.

Thus, for $y=\log_{\frac{1}{5}}(x-M)+j$, the domain is $x>M$.

(Note that the vertical translation does not affect the domain.)

**18. C**

$y=\log(5x)$

The inverse is

$x=\log(5y)$

$10^x=5y$

$y=\frac{10^x}{5}$

Thus, $f^{-1}(x)=\frac{10^x}{5}$

**19. D**

The graph of $y=3^{x+2}-3$ is the graph $y=3^x$, translated 2 units left and 3 units down. The range of $y=3^x$ is $y>0$, thus, the range for $y=3^{x+2}-3$, is $y>-3$.

**20.** $y=\log_2 x+2$.

The $x$- intercept occurs when $y=0$.

$0=\log_2 x+2$
$-2=\log_2 x$
$2^{-2}=x$

The $x$-intercept is $\frac{1}{4}=0.25$.

| 0 | . | 2 | 5 |
|---|---|---|---|

**21.** $f(x)=3^{x-1}$ is the same as $y=3^{x-1}$

The inverse is

$x=3^{y-1}$
$\log_3 x=y-1$
$f^{-1}(x)=\log_3 x+1$
$f^{-1}(9)=\log_3 9+1=2+1=3$

| 3 | | | |
|---|---|---|---|

## Lesson 8 Applications

### CLASS EXERCISES ANSWERS AND SOLUTIONS

**1.** $t=40 \rightarrow P=50(2^{\frac{t}{3}})$

$=50(2^{\frac{40}{3}})$
$\doteq 50(10\ 321.27)$
$\doteq 516\ 063.7$

The population after 40 hours will be approximately 516 064 ants.

**2.** In this question, we are given the population, $P$, and need to find the time, $t$, to reach this population. So, we substitute 10 000 000 for $P$ and solve for $t$.

$P=10\ 000\ 000 \rightarrow 10\ 000\ 000=50(2^{\frac{t}{3}})$

$200\ 000=2^{\frac{t}{3}}$ • Divide both sides by 50

$\log 200000=\log(2^{\frac{t}{3}})$ • Take the common log of both sides

$\log 200000=\frac{t}{3}\log 2$ • Apply the Power Law

$\frac{\log 200000}{\log 2}=\frac{t}{3}$ • Divide both sides by log2

$\frac{3\log 200000}{\log 2}=t$ • Multiply both sides by 3

$52.83 \doteq t$ • Evaluate

The population will reach 10 000 000 in approximately 53 hours.

**3.** Since this is a problem involving exponential growth, we can use the formula $N_t=N_0\times R^{\frac{t}{p}}$. First, we need to identify the variables whose values are given.

$N_0=30$ • At time zero, the population is 30.

$R=4$ • If the population quadruples, the growth rate is 4.

$t=100$ • We want the population after 100 days.

$p=9$ • It takes 9 days for the population to quadruple, so the period is 9.

$\rightarrow N_{100}=30\times 4^{\frac{100}{9}}$
$\doteq 146\ 783\ 323$

After 100 days, the spider population is approximately 146 783 323.

**4.** Once again, we need to identify the variables whose values are given.

$N_0=50\ 000$ • At time 0, the car's value is \$50 000.

$R=0.85$ • If the car depreciates at 15%, the growth rate is 0.85. In other words, each year the car is worth 85% of what its value was in the previous year.

$t=14$ • We want to know the car's value after 14 years.

$p=1$ • It takes 1 year for the car's value to reduce to 85% the previous year's value.

$\rightarrow N_{14}=50\ 000\times(0.85)^{\frac{14}{1}}$
$\doteq 5\ 138.48$

$N_{14}$ is the car's value after 14 years. The car is worth approximately \$5 138.48 after 14 years.

**5.**

$N_0 = 300$ • At time 0, the mass is 300 g.

$R = \frac{1}{2}$ • The substance is reduced to $\frac{1}{2}$ its mass during each period.

$t = 100$ • We want the mass in 100 years.

$p = 7$ • It takes 7 years for the substance to reduce to $\frac{1}{2}$ its mass.

$$\rightarrow N_{100} = 300 \times (0.5)^{\frac{100}{7}}$$
$$\doteq 0.015$$

The mass of the substance will be approximately 0.015 grams in 100 years.

## PRACTICE EXERCISES ANSWERS AND SOLUTIONS

**1.** Use $N_t = N_o \times R^{\frac{t}{p}}$:

$$1\,000\,000 = 400(1.25)^{\frac{t}{5}}$$
• Substitute values into formula

$$2\,500 = 1.25^{\frac{t}{5}}$$
• Divide both sides by 400

$$\log 2500 = \log 1.25^{\frac{t}{5}}$$
• Take the log of each side to solve the exponential equation

$$\log 2500 = \frac{t}{5}\log 1.25$$
• Power Law

$$\frac{\log 2500}{\log 1.25} = \frac{t}{5}$$
• Divide both sides by log 1.25

$$5\left(\frac{\log 2500}{\log 1.25}\right) = t$$
• Multiply both sides by 5

$$175.3 \doteq t$$

Thus, the time is approximately 175 minutes.

(Note $t$ is in minutes since $P$ was given in minutes.)

**2.** Use $N_t = N_o \times R^{\frac{t}{p}}$:

$$100 = 4\,000 \times (0.94)^{\frac{t}{2}}$$
$$0.025 = 0.94^{\frac{t}{2}}$$
$$\log 0.025 = \log 0.94^{\frac{t}{2}}$$
$$\log 0.025 = \frac{t}{2}\log 0.94$$
$$t = \frac{2 \cdot \log 0.025}{\log 0.94}$$
$$t \doteq 119.24$$

It will take approximately 119 years for the population to reduce to 100.

**3.** Use $N_t = N_o \times R^{\frac{t}{p}}$:

$$8\,000 = 20\,000(R)^{\frac{6}{1}}$$
• Substitute values into the formula

$$0.4 = R^6$$
• Divide both sides by 20 000

$$\sqrt[6]{0.4} = R$$
• Take the 6th root of both sides

$$0.86 \doteq R$$

If the value is 86% of the previous year, the depreciation is 14% (100% – 86%).

**4.** Use $N_t = N_o \times R^{\frac{t}{p}}$:

$$8 = 500 \times (0.5)^{\frac{t}{125}}$$
$$0.016 = 0.5^{\frac{t}{125}}$$
$$\log 0.016 = \log 0.5^{\frac{t}{125}}$$
$$\frac{125 \log 0.016}{\log 0.5} = t$$
$$t = 745.72$$

It will take approximately 745.7 years for the substance to decay to 8 grams.

 

**5.** Let the intensity of the 7.4 earthquake be $I_1$.
$I_1 = I_r(10)^{7.4}$

Let the intensity of the 3.3 earthquake be $I_2$.
$I_2 = I_r(10)^{3.3}$

Compare the intensities by dividing $\frac{I_1}{I_2}$:

$$\frac{I_1}{I_2} = \frac{I_r(10)^{7.4}}{I_r(10)^{3.3}}$$

$= 10^{7.4-3.3}$ • Reduce
$= 10^{4.1}$ • Simplify
$\doteq 12\ 589.3$

Thus, the magnitude 7.4 earthquake is approximately 125 896 times as intense as the magnitude 3.3 earthquake.

**6.** Let $I_A$ be the intensity of earthquake $A$.
Let $I_B$ be the intensity of earthquake $B$.

$I_A = I_r(10)^m \quad I_B = I_r(10)^{6.4}$

You know that $\frac{I_B}{I_A} = 722$.

Thus,

$$\frac{I_r(10)^{6.4}}{I_r(10)^m} = 722$$

$$\frac{10^{6.4}}{10^m} = 722$$

$$10^{(6.4-m)} = 722$$

$$\log 722 = (64 - m)$$

$$m = 64 - \log 722 \doteq 3.54$$

Earthquake $A$ has a magnitude of approximately 3.5.

**7.** Use $N_t = N_0 \times R^{\frac{t}{p}}$:

$$50\ 000 = N_0(3)^{\frac{27}{5}}$$

$$\frac{50\ 000}{3^{\frac{27}{5}}} = N_0$$ • Divide both sides by $3^{\frac{27}{5}}$

$$133 \doteq N_0$$

Thus, the original population was 133.

**8.** Use $N_t = N_0 \times R^{\frac{t}{p}}$

$$N_{40} = 1\ 000\ 000 \times (0.93)^{\frac{40}{3}}$$

$$N_{40} = 379\ 990.42$$

In 40 years, the population will be 379 990.

**9.** For an orchestra:
$98 = 10\log L_o \Rightarrow 9.8 = \log L_o \Rightarrow L_o = 10^{9.3}$

For a vacuum cleaner:
$80 = 10\log L_v \Rightarrow 8 = \log L_v \Rightarrow L_v = 10^8$

The relative loudness of an orchestra compared to a vacuum cleancr is:

$$\frac{L_o}{L_v} = \frac{10^{9.3}}{10^8} = 10^{1.8} \doteq 63.10$$

Thus, an orchestra is approximately 63 times louder then the vacuum cleaner.

**10.** Use $N_t = N_o \times R^{\frac{t}{p}}$:

$5\ 000 = N_o(0.88)^{\frac{8}{1}}$ • Substitute values into the formula

$\frac{5\ 000}{(0.88)^8} = N_o$ • Divide both sides by $(0.88)^8$

$1\ 390.3 \doteq N_o$ • Evaluate

Thus, the original cost was 13 903.

**11.** Use $N_t = N_o \times R^{\frac{11}{p}}$:

$$47 = 50 \times (0.5)^{\frac{11}{p}}$$

$$0.94 = 0.5^{\frac{11}{p}}$$

$$\log 0.94 = \log 0.5^{\frac{11}{p}}$$

$$\log 0.94 = \frac{11}{P}\log 0.5$$

$$p = \frac{11\log 0.5}{\log 0.94}$$

$$p \doteq 123.23$$

The half-life of substance $Z$ is approximately 123 days.

**12.** Let the intensity of the 7.1 earthquake be $I_1$.

$I_1 = I_r(10)^{7.1}$

Let the intensity of the 5.1 earthquake be $I_2$.

$I_2 = I_r(10)^{5.1}$

Compare the intensities by dividing:

$$\frac{I_1}{I_2} = \frac{I_r(10)^{7.1}}{I_r(10)^{5.1}}$$

$$= 10^2 = 100$$

Thus, the magnitude 7.1 earthquake is 100 times as intense as the magnitude 5.1 earthquake.

**13.** For eggs: $\left[H^+\right] = \frac{1}{10^{7.8}}$

For milk: $\left[H^+\right] = \frac{1}{10^{5.9}}$

Thus, the ratio of alkalinity for eggs to milk is:

$$\frac{\frac{1}{10^{5.9}}}{\frac{1}{10^{7.8}}} = \frac{10^{7.8}}{10^{5.9}} = 10^{1.9} \doteq 79.4$$

Thus, eggs are approximately 79 times as alkaline as milk.

**14.** For the liquid: $\left[H^+\right] = \frac{1}{10^{\text{pH}}}$

For seawater:

$$\left[H^+\right] = \frac{1}{10^{8.5}}$$

$$\frac{\frac{1}{10^{\text{pH}}}}{\frac{1}{10^{8.5}}} = 7\ 500$$

$$\frac{10^{8.5}}{10^{\text{pH}}} = 7\ 500$$

$$10^{8.5-\text{pH}} = 7500 \log 7500 = 8.5 - \text{pH}$$

$$\text{pH} = 8.5 - \log 7500$$

$$\text{pH} \doteq 4.6$$

Thus, the liquid has a pH of approximately 4.6.

**15.** Use $N_t = N_o \times R^{\frac{t}{p}}$ :

$$300\ 000 = 5\ 000 \times R^{\frac{70}{1}}$$

$$60 = R^{70}$$

$$R = 60^{\frac{1}{70}} = 1.06$$

Thus, the annual growth rate is approximately 6%.

**16.** Use $N_t = N_o \times R^{\frac{t}{p}}$ :

$$N_t = 1(0.5)^{\frac{300}{50}}$$

- Substitute values into the formula (Note that $t$ and $p$ must be in the same units.)

$$N_t \doteq 0.016$$

Thus, the mass after 5 min will be approximately 0.2 kg.

**17.** Use $N_t = N_o \times R^{\frac{t}{p}}$ :

The population grows from $N_o$ to $3N_o$.

$$3N_o = N_o \times R^{\frac{t}{50}}$$

$$3 = 1 \times (2)^{\frac{t}{50}}$$

$$\log 3 = \log 2^{\frac{t}{50}}$$

$$\log 3 = \frac{t}{50} \log 2$$

$$\frac{50 \log 3}{\log 2} = t$$

$$t = 79.25$$

It will take approximately 79 days for the population to triple.

 

**18.** Use $N_t = N_o \times R^{\frac{t}{p}}$ :

$$30\ 000\ 000 = 7\times(3)^{\frac{t}{55}}$$
$$\frac{30\ 000\ 000}{7} = 3^{\frac{t}{55}}$$
$$\log\left(\frac{30\ 000\ 000}{7}\right) = \log 3^{\frac{t}{55}}$$
$$\log\left(\frac{30\ 000\ 000}{7}\right) = \frac{t}{55}\log 3$$
$$\frac{55\log\left(\frac{30\ 000\ 000}{7}\right)}{\log 3} = t$$

$t \doteq 764.5044$ seconds

$t \doteq 12$ minutes 45 seconds

The population will reach 30 000 000 at approximately 10:12:45 A.M.

**19.** For each trumpet: $10\log L = 60 \rightarrow L = 10^6$

For the trombone: $10\log L = 55 \rightarrow L = 10^{5.5}$

For 2 trumpets and 1 trombone, the total loudness is $L = 2\times10^6 + 10^{5.5}$.

The decibel level is $10\log\left(2\times10^6 + 10^{5.5}\right)$.

$\doteq 10(6.365) \doteq 64$

**20.** The acidity should be so that $\left[H^+\right] = 10^{-0.8}$.

The acidity is so that $\left[H^+\right] = \frac{1}{5}\times10^{-0.8}$.

$$\text{pH} = -\log\left(\frac{1}{5}\times10^{-0.8}\right)$$
$$= -\left[\log\frac{1}{5} + \log10^{-0.8}\right]$$
$$= -\log\frac{1}{5} + 0.8$$
$$\doteq 0.7 + 0.8 = 1.5$$

**21. D**

$$30 = 100\times\left(\frac{1}{2}\right)^{\frac{t}{235}}$$
$$0.3 = \left(\frac{1}{2}\right)^{\frac{t}{235}}$$
$$\log 0.3 = \log\left(\frac{1}{2}\right)^{\frac{t}{235}}$$
$$\log 0.3 = \frac{t}{235}\log\left(\frac{1}{2}\right)$$
$$\frac{235\log 0.3}{\log\left(\frac{1}{2}\right)} = t$$

$t \doteq 408$ days.

**22. C**

$$6\ 000 = 10\ 000\left(10^{-0.035y}\right)$$
$$0.6 = 10^{-0.035y}$$
$$\log 0.6 = -0.035y$$
$$y \doteq 6.34 \text{ years.}$$

$1998 + 6$ years $= 2004$

**23.** $92 = 10\log L$, where L is the relative loudness of the orchestra

$9.2 = \log L \rightarrow L = 10^{9.2}$

For two orchestras: $2L = 2\times10^{9.2}$

decibel level $= 10\log 2L$

$$= 10\log\left(2\times10^{9.2}\right)$$
$$= 10\left(\log 2 + \log 10^{9.2}\right)$$
$$= 10(0.30 + 9.2)$$
$$= 95$$

| 9 | 5 | | |
|---|---|---|---|

**24.**

$$N_{10} = 1\times(.95)^{\frac{10}{1}}$$
$$N_{10} \doteq 0.598\ 7$$

Approximately 60.0% of the light passes through 10 layers of glass.

| 6 | 0 | . | 0 |
|---|---|---|---|

## Lesson 9 Exponential Regression

### PRACTICE EXERCISES ANSWERS AND SOLUTIONS

**1.** Enter the data as shown.

Following the procedure, you will get
$y=(0.125)(5.03)^x$.

**2.** Enter the data as shown.

Following the procedure, you will get
$y=(817.29)(0.91)^x$.

**3.** Enter the data as shown.

Following the procedure, you will get:

**a)** $y=(90\ 296.6)(0.945)^x$

**b)** For 1950, $x$ will be 50.
$y=(90\ 296.6)(0.945)^{50}$
$y=5\ 336.6$
$y\doteq 5\ 337$

The population in 1950 was approximately 5 337.

**c)** For 2040, $x$ will be 140.
$y=(90\ 296.6)(0.945)^{140}$
$y\doteq 32.8$
or $y\doteq 33$

The population in 2040 will be approximately 33.

**d)** Here, $y=100$ and you will solve for $x$.
$100=(90\ 296.6)(0.945)^x$

$0.001\,1=0.945^x$ • Divide both sides by 90 296.6

$\log(0.0011)=\log(0.945)^x$ • Take the log of each side

$\log(0.0011)=x\log(0.945)$ • Power Law

$\dfrac{\log(0.0011)}{\log(0.945)}=x$ • Divide both sides by log (0.945)

$120.4\doteq x$ • Simplify

Since $x=120$, this will be the year 2020.

**4.** Enter the data as shown.

Following the procedure, you will get
$y=(8.38)(1.30)^x$.

**5.** Enter the data as shown.

Following the procedure, you will get
$y=(415.36)(0.79)^x$.

**6.** Enter the data as shown.

Following the procedure, you will get

**a)** $y=(119.5)(0.99)^x$

**b)** Let $x=40$.
$y=(119.5)(0.99)^{40}$
$\doteq 79.9$

After 40 minutes the mass is approximately 79.9 kg.

**c)** How long does it take the mass to become $\frac{1}{2}$ of the original mass. That is, when does the mass become 60 kg?

Here, $y=60$. Solve for $x$:
$60=(119.5)(0.99)^x$

$0.502=0.99^x$ • Divide both sides by 119.5

$\log(0.502)=\log(0.99)^x$ • Take the log of each side

$\log(0.502)=x\log(0.99)$ • Power Law

$\dfrac{\log(0.502)}{\log(0.99)}=x$ • Divide both sides by log (0.99)

$68.6\doteq x$ • Simplify

Thus, the half-life of substance $K$ is approximately 68.6 minutes.

**7. a)** Enter the data as shown.
Following the procedure, you will get
$T=(80.54)(0.97)^t$

**b)** When $t=30$ minutes,
$T=(80.54)(0.97)^{30}$
$T=32.30$

The temperature will be approximately 32°C.

**c)** When $T = 22°C$,

$$22 = (80.54)(0.97)^t$$
$$0.2732 \doteq 0.97^t$$
$$\log 0.2732 \doteq \log 0.97^t$$
$$\log 0.2732 \doteq t \log 0.97$$

$$t \doteq \frac{\log 0.2732}{\log 0.97}$$
$$t \doteq 42.99$$

It will take approximately 43 min for the tea to cool down to 22°C.

**8.** Enter the data as shown.

Following the procedure, you will get $y = (11.26)(2.19)^x$.

**9.** Enter the data as shown.

Following the procedure, you will get $y = (340.91)(0.73)^x$.

## *Practice Quiz*

### ANSWERS AND SOLUTIONS

**1.**
$$\left(\frac{16x^5y}{9x^2}\right)\times\left(\frac{3x^2y^4}{2y}\right)^3 = \left(\frac{16x^3y}{9}\right)\times\left(\frac{3x^2y^3}{2}\right)^3$$
$$= \frac{16x^3y}{9}\times\frac{27x^6y^9}{8} = 6x^9y^{10}$$

**2.**
$$\frac{\left(\sqrt{3}\right)^{2x-4}}{27^{2+x}} = \frac{\left(3^{\frac{1}{2}}\right)^{(2x-4)}}{\left(3^3\right)^{2+x}} = \frac{3^{x-2}}{3^{6+3x}} = 3^{-2x-8}$$

**3.**
$$\frac{4^{x-3}}{16^x} = 8^{x-1}$$
$$\frac{\left(2^2\right)^{x-3}}{\left(2^4\right)^x} = \left(2^3\right)^{x-1}$$
$$\frac{2^{2x-6}}{2^{4x}} = 2^{3x-3}$$
$$2^{-2x-6} = 2^{3x-3}$$
$$-2x-6 = 3x-3$$
$$-5x = 3$$
$$x = -\frac{3}{5}$$

**4.**
$$16(x+1)^4 = 81$$
$$(x+1)^4 = \frac{81}{16}$$
$$(x+1) = \pm\frac{3}{2}$$
$$x = -1\pm\frac{3}{2}$$
$$x = -\frac{5}{2}, \frac{1}{2}$$

**5.**
$$x-2 = \log_2\left(4^3\right)$$
$$2^{(x-2)} = 64$$
$$2^{x-2} = 2^6$$
$$x-2 = 6$$
$$x = 8$$

**6.**
$$\log_3 x + \log_3(x-2) = 1$$
$$\log_3\left(x^2-2x\right) = 1$$
$$3^1 = x^2-2x$$
$$x^2-2x-3 = 0$$
$$(x-3)(x+1) = 0$$

$x = 3$ or $x = -1$ (extraneous, as it does not satisfy $x > 2$).

Thus, $x = 3$.

**7.**

$$4\log_5 m - \log_5 n + \frac{1}{2}\log_5 p$$

$$= \log_5\left(\frac{m^4 \times p^{\frac{1}{2}}}{n}\right)$$

**8.**

$$\log\left(\frac{W^2\sqrt{X}}{3YZ}\right)$$

$$= \log W^2 + \log\sqrt{X} - (\log 3YZ)$$

$$= 2\log W + \frac{1}{2}\log X - (\log 3 + \log Y + \log Z)$$

$$= 2\log W + \frac{1}{2}\log X - \log 3 - \log Y - \log Z$$

**9.**

$$\log_b 3 = A$$

$$\log_b 5 = B$$

$$\log_b 135 = \log_b(27 \times 5) = \log_b(3^3 \times 5)$$

$$= \log_b(3)^3 + \log_b 5$$

$$= 3\log_b 3 + \log_b 5$$

$$= 3A + B$$

**10.** $\log_4 120 = \dfrac{\log 120}{\log 4} \doteq 3.5$

**11.**

$$f(x) = 2^{5-x} + 4$$

$$y = 2^{5-x} + 4$$

inverse: $x = 2^{5-y} + 4$

$$x - 4 = 2^{5-y}$$

$$\log_2(x-4) = 5 - y$$

$$y = -\log_2(x-4) + 5$$

Thus, $f^{-1}(x) = -\log_2(x-4) + 5$.

**12.** $y = \log_4 x - 2$

This is the graph of $y = \log_4 x$ translated 2 units down.

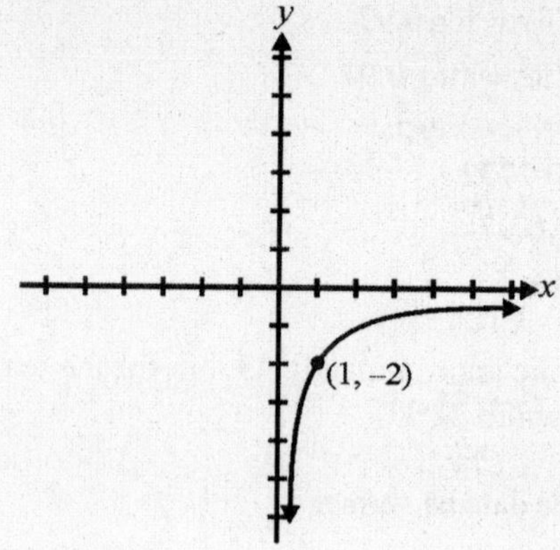


Domain: $x > 0$
Range: $y \in R$
Asymptote: $x = 0$

**13.** $y = \left(\frac{1}{2}\right)^{x+2}$

This is the graph of $y = \left(\frac{1}{2}\right)^x$ translated 2 units left.

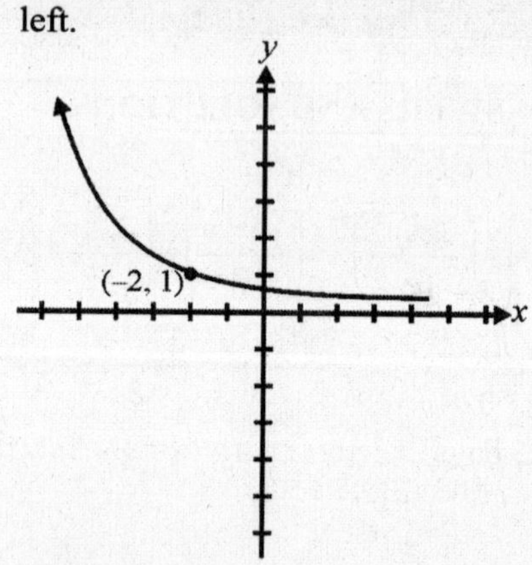


Domain: $x \in R$
Range: $y > 0$
Asymptote: $y = 0$

 

**14.** Use $N_t = N_0(R)^{\frac{t}{p}}$ :

$$10 = 100(0.5)^{\frac{t}{110}}$$

$$0.1 = 0.5^{\frac{t}{110}}$$

$$\log 0.1 = \frac{t}{110}\log 0.5$$

$$t = \frac{110(-1)}{\log 0.5}$$

$$t \doteq 365.4$$

It takes approximately 365.4 years for the 100 g sample to decay to 10 g.

**15.** Earthquake 1: $I_1 = I_r(10)^{7.4}$

Earthquake 2: $I_2 = I_r(10)^m$

$$\frac{I_r(10)^{7.4}}{I_r(10)^m} = 150 \qquad 10^{7.4-m} = 150$$

$$\log 150 = 7.4 - m$$

$$m = 7.4 - \log 150$$

$$m \doteq 5.22$$

The magnitude of the second earthquake is approximately is 5.2.

**16. a)** Enter the data as shown.
You will get: $P \doteq 20.30(1.17)^t$.

**b)** When $t = 25$,
$P \doteq 20.30(1.17)^{25}$
$P \doteq 1\,028.35$
In 25 days, the population of fruit flies will be approximately 1 028.

## *Lesson 10 Introduction to Geometric Sequences and Series*

### CLASS EXERCISES ANSWERS AND SOLUTIONS

**1.** To find $t_{11}$, we substitute 11 for $n$ in the formula.
$t_{11} = 2(3)^{11-1} = 2(3)^{10} = 2(59\,049) = 118\,098$

**2.** To find the common ratio, divide any term by the previous term.

If we divide term 2 by term 1, we get $\frac{t_2}{t_1} = \frac{-3}{1} = -3$.

If we divide term 3 by term 2, we get $\frac{t_3}{t_2} = \frac{9}{-3} = -3$.

In either case, the result is –3. Thus, the common ratio is –3.

**3.** We are dealing with a geometric sequence, so $\frac{t_2}{t_1} = \frac{t_3}{t_2}$.

This gives the equation $\frac{x+2}{x} = \frac{x+5}{x+2}$.

Solve for $x$.

$$\frac{x+2}{x} = \frac{x+5}{x+2}$$

$$(x+2)(x+2) = (x)(x+5)$$ • If the two ratios are equal, then the cross-products are equal

$$x^2 + 4x + 4 = x^2 + 5x$$ • Expand

$$4x + 4 = 5x$$ • Subtract $x^2$ from both sides

$$4 = x$$ • Solve

We were asked to find the first 3 terms of the sequence. They are $x$, $x + 2$, and $x + 5$.

Since $x = 4$, the terms are 4, 6, 9.

The common ratio of the sequence is $\frac{t_2}{t_1} = \frac{6}{4} = \frac{3}{2}$.

**4.** This is a finite sequence. To find the number of terms, we need to determine the position of the last term, 78 732.

In this sequence, $a = 4$ and $r = \frac{t_2}{t_1} = \frac{12}{4} = 3$. We use the formula $t_n = ar^{n-1}$ to find the general term. It is $t_n = 4(3)^{n-1}$.

The last term of the sequence is 78 732. So, we know that for some value of $n$, $t_n = 78\ 732$.

If we can find this value of $n$, we will know how many terms are in the sequence.

When we substitute $t_n = 78\ 732$ into the formula $t_n = 4(3)^{n-1}$, we get $78\ 732 = 4(3)^{n-1}$.

$78\ 732 = 4(3)^{n-1}$

$(3)^{n-1} = 19\ 683$ • Divide both sides by 4

$\log(3)^{n-1} = \log 19683$ • Take logs of both sides

$(n-1)\log 3 = \log 19683$ • Apply the Power Law of Logarithms

$n-1 = \dfrac{\log 19683}{\log 3}$ • Divide both sides by $\log 3$

$n = \dfrac{\log 19683}{\log 3} + 1$ • Add 1 to both sides

$n = 10$ • Evaluate

The sequence has 10 terms, and 78 732 is the 10th term.

## PRACTICE EXERCISES ANSWERS AND SOLUTIONS

1. $r = \dfrac{t_2}{t_1} = \dfrac{1}{-1} = -1$

2. $r = \dfrac{t_2}{t_1} = \dfrac{3\sqrt{2}}{\sqrt{6}} = \dfrac{3}{\sqrt{3}}$ or $\dfrac{3}{\sqrt{3}} \cdot \dfrac{\sqrt{3}}{\sqrt{3}} = \dfrac{3\sqrt{3}}{3} = \sqrt{3}$

3. $r = \dfrac{t_2}{t_1} = \dfrac{-21}{-3} = 7$

4. $r = \dfrac{t_2}{t_1} = \dfrac{2\pi^2}{2\pi} = \pi$

5. $r = \dfrac{t_2}{t_1} = \dfrac{-35}{7} = -5$

6. $r = \dfrac{t_2}{t_1} = \dfrac{3k}{3k^2} = \dfrac{1}{k}$

7. Common ratio, $r = \dfrac{t_2}{t_1} = \dfrac{t_3}{t_2}$

$r = \dfrac{-12}{4} = -3$

$t_5 = ar^{5-1} = ar^4$

$t_5 = (4)(-3)^4 = 324$

8. Since the sequence is geometric, $\dfrac{t_2}{t_1} = \dfrac{t_3}{t_2}$.

This gives $\dfrac{2x+2}{x-1} = \dfrac{4x+16}{2x+2}$.

$(2x+2)(2x+2) = (x-1)(4x+16)$

• cross-products are equal.

$4x^2 + 8x + 4 = 4x^2 + 12x - 16$ • Multiply

$8x + 4 = 12x - 16$ • Subtract $4x^2$ from both sides

$4 = 4x - 16$

$20 = 4x$

$5 = x$ • Solve for $x$

Since $x = 5$, the terms $x-1$, $2x+2$, and $4x+16$ are 4, 12, and 36.

9. As the sequence is geometric, then $\dfrac{t_2}{t_1} = \dfrac{t_3}{t_2}$.

$\dfrac{3^{x+2}}{3} = \dfrac{3^{6x+5}}{3^{x+2}}$

$3^{x+1} = 3^{5x+3}$ • Simplify

$x + 1 = 5x + 3$ • Both sides have the same base, thus the exponents can be equated

$-4x = 2$

$x = -\dfrac{1}{2}$ • Solve for $x$

10. $r = \dfrac{t_2}{t_1} = \dfrac{112}{64} = \dfrac{7}{4}$

$t_5 = ar^4$

$t_5 = (64)\left(\dfrac{7}{4}\right)^4$

$t_5 = 600.25$

 

**11.** Since the sequence is geometric, $\frac{t_2}{t_1}=\frac{t_3}{t_2}$.

This gives $\frac{2x-8}{x-7}=\frac{4x+2}{2x-8}$.

$(2x-8)(2x-8)=(x-7)(4x+2)$ • Cross-products are equal

$4x^2-32x+64=4x^2-26x-14$ • Multiply

$-32x+64=-26x-14$ • Subtract $4x^2$ from both sides

$64=6x-14$

$78=6x$

$13=x$ • Solve for $x$

If $x=13$, then the terms $x-7$, $2x-8$, and $4x+2$, are 6, 18, and 54.

**12.** $t_n=640\left(\frac{1}{2}\right)^{n-1}$

$t_8=640\left(\frac{1}{2}\right)^{8-1}$ • Substitute 8 for $n$

$t_8=640\left(\frac{1}{2}\right)^{7}$

$t_8=5$ • Simplify

**13.** $t_n=-6\,161(-0.235)^{n-1}$

$t_1=-6\,161(-0.235)^{1-1}$ • Substitute 1 for $n$

$t_1=-6\,161$ • Simplify

**14.** For the geometric sequence $\frac{3}{16},\frac{1}{4},3,\ldots,$

$r=\frac{t_2}{t_1}$.

$r=\frac{1}{4}\div\frac{3}{16}$

$r=\frac{1}{4}\times\frac{16}{3}$

$r=\frac{4}{3}$

$t_n=ar^{n-1}$

$t_n=\frac{3}{16}\left(\frac{4}{3}\right)^{n-1}$ • Substitute the values for $a$ and $r$ into $t_n$

**15.** $a=5$

$r=\frac{5\sqrt{2}}{5}=\sqrt{2}$

$t_n=2\,560$

Thus, using

$t_n=a(r)^{n-1}$,

$2\,560=(5)\left(\sqrt{2}\right)^{n-1}$

$512=\left(\sqrt{2}\right)^{n-1}$

$2^9=\left(2^{\frac{1}{2}}\right)^{n-1}$

$2^9=2^{\frac{1}{2}n-\frac{1}{2}}$

$9=\frac{1}{2}n-\frac{1}{2}$

$9\frac{1}{2}=\frac{1}{2}n$

$19=n$

Thus, there are 19 terms in this sequence

Or solve $512=\left(\sqrt{2}\right)^{n-1}$ by using logarithms:

$\log(512)=\log\left(\sqrt{2}\right)^{n-1}$

**16.** $t_6=486\rightarrow ar^5=486$

$t_4=54\rightarrow ar^3=54$

Dividing these two equations gives $\frac{ar^5}{ar^3}=\frac{486}{54}$.

$r^2=9$ • The *as* divide out

$r=\pm3$ • Take the square root of both sides

If

$r=3$

$ar^3=54$

$a(3)^3=54$

$ar^3=54$

$27a=54$

$a=2$

This sequence is 2, 6, 18,…

$t_n=2(3)^{n-1}$ • Substitute the values for $a$ and $r$ into $t_n$

If

$$r = -3$$
$$ar^3 = 54$$
$$a(-3)^3 = 54$$
$$-27a = 54$$
$$a = -2$$

This sequence is –2, 6, –18,...

$t_n = ar^{n-1}$

$t_n = (-2)(-3)^{n-1}$

**17.** In the first week, you earn \$8.00/hour and in the second week, you will earn \$8.80/hour.

This is an increase of $8.00 \times 10\% = 0.80$.

In the third week, you will earn \$9.68/hour, as $8.80 \times 10\% = 0.88$.

These hourly wages form the geometric sequence 8.00, 8.80, 9.68,...

Here, $r = \frac{t_2}{t_1} = \frac{8.80}{8.00} = 1.1$.

Using $t_n = ar^{n-1}$, we need to determine when $t_n > 50$.

Thus, $ar^{n-1} > 50$.

$(8.00)(1.1)^{n-1} > 50$ • Substitute for $a$ and $r$

$(1.1)^{n-1} > 6.25$ • Divide both sides by 8.00

$(n-1)\log 1.1 > \log 6.25$ • Take the log of both sides and apply the Power Law

$n - 1 > 19.23$ • Divide both sides by log (1.1)

$n > 20.23$ • Solve for $n$

Thus, it will take 21 weeks before you start to earn more than \$50.00/hour.

**18.** $t_n = 54\left(\frac{1}{3}\right)^{n-1}$,

To find the first three terms, substitute 1, 2, and 3 for $n$.

$$t_1 = 54\left(\frac{1}{3}\right)^{1-1}$$
$$= 54\left(\frac{1}{3}\right)^{0}$$
$$= 54$$

$$t_2 = 54\left(\frac{1}{3}\right)^{2-1}$$
$$= 54\left(\frac{1}{3}\right)$$
$$= 18$$

$$t_3 = 54\left(\frac{1}{3}\right)^{3-1}$$
$$= 54\left(\frac{1}{3}\right)^{2}$$
$$= 54\left(\frac{1}{9}\right)$$
$$= 6$$

Thus, the first three terms are 54, 18, and 6.

**19.** For the geometric sequence $2\sqrt{3}, 6, 6\sqrt{3}, \ldots$ first find $r$.

$$r = \frac{t_3}{t_2} = \left(\frac{6\sqrt{3}}{6}\right) = \sqrt{3}$$

$t_n = ar^{n-1}$

$t_n = 2\sqrt{3}\left(\sqrt{3}\right)^{n-1}$ • Substitute for $a$ and $r$

$t_{11} = 2\sqrt{3}\left(\sqrt{3}\right)^{11-1}$ • Substitute 11 for $n$

$= 2\sqrt{3}\left(\sqrt{3}\right)^{10}$

$= 2\sqrt{3}(243) = 486\sqrt{3}$ • Simplify

 

**20.** $a = -512$

$r = \frac{128}{-512} = -\frac{1}{4}$

$t_n = (-512)\left(-\frac{1}{4}\right)^{n-1}$

$t_7 = (-512)\times\left(-\frac{1}{4}\right)^{7-1}$

$t_7 = -\frac{1}{8}$

**21.** For the geometric sequence 8, –8, 8, –8, 8,…, first find $r$.

$r = \frac{t_2}{t_1} = \frac{-8}{8} = -1$

$t_n = ar^{n-1}$

$t_n = 8(-1)^{n-1}$ • Substitute for $a$ and $r$

**22.** $a = 64$

$r = \frac{96}{64} = \frac{3}{2}$

$t_n = 1\,093.5$

Using $t_n = a(r)^{n-1}$:

$1\,093.5 = (64)\left(\frac{3}{2}\right)^{n-1}$

$\frac{1\,093.5}{64} = \left(\frac{3}{2}\right)^{n-1}$

$\log\left(\frac{1\,093.5}{64}\right) = \log\left(\frac{3}{2}\right)^{n-1}$

$\frac{\log\left(\frac{1\,093.5}{64}\right)}{\log\left(\frac{3}{2}\right)} = n-1$

$n = 8$

Thus, there are 8 terms in the sequence.

**23.** For the geometric sequence $\frac{128\sqrt{2}}{2}, 64, \frac{64\sqrt{2}}{2}, \ldots, \frac{1}{2}$, first simplify $t_1$ and $t_3$ to give the sequence $64\sqrt{2}, 64, 32\sqrt{2}, \ldots, \frac{1}{2}$.

$r = \frac{64}{64\sqrt{2}} = \frac{1}{\sqrt{2}}$

Using $t_n = ar^{n-1}$:

$\frac{1}{2} = 64\sqrt{2}\left(\frac{1}{\sqrt{2}}\right)^{n-1}$ • Substitute for $t_n$, $a$ and $r$

$\frac{1}{128\sqrt{2}} = \left(\frac{1}{\sqrt{2}}\right)^{n-1}$ • Divide both sides by $64\sqrt{2}$

$\frac{1}{\left(\sqrt{2}\right)^{14}\cdot\sqrt{2}} = \left(\frac{1}{\sqrt{2}}\right)^{n-1}$ • Write both sides with the same base

$\frac{1}{\left(\sqrt{2}\right)^{15}} = \frac{1}{\left(\sqrt{2}\right)^{n-1}}$

$15 = n-1$ • Solve for $n$

$n = 16$

**24.** If $t_5 = 5$, then $ar^4 = 5$.

If $t_8 = 40$, then $ar^7 = 40$.

$\frac{\not{a}r^7}{\not{a}r^4} = \frac{40}{5}$

$r^3 = 8$

$r = 2$

Solve for $a$

$a(2)^4 = 5$

$a = \frac{5}{16}$

The general term, $t_n = a(r)^{n-1}$, is:

$t_n = \left(\frac{5}{16}\right)(2)^{n-1}$

**25.** $a = 12$

$r = \frac{t_2}{t_1} = \frac{16}{12} = \frac{4}{3}$

The general term, $t_n = a(r)^{n-1}$, is:

$t_n = (12)\left(\frac{4}{3}\right)^{n-1}$

**26.** For the geometric sequence $\sqrt{3}, 3\sqrt{2}, 6\sqrt{3}, \ldots 7\ 776\sqrt{3}$, first find $r$.

$r = \frac{t_2}{t_1} = \frac{3\sqrt{2}}{\sqrt{3}}$

$r = \frac{3\sqrt{2}}{\sqrt{3}} \cdot \frac{\sqrt{3}}{\sqrt{3}}$ • Rationalize the denominator

$r = \frac{3\sqrt{6}}{3}$

$r = \sqrt{6}$ • Simplify

Now find *n*, using $t_n = ar^{n-1}$:

$7\ 776\sqrt{3} = \sqrt{3}\left(\sqrt{6}\right)^{n-1}$ • Substitute for $t_n$, $a$, and $r$

$7\ 776 = \left(\sqrt{6}\right)^{n-1}$ • Divide both sides by $\sqrt{3}$

$\log 7776 = (n-1)\log\sqrt{6}$ • Take the log of both sides and apply the Power Law

$10 = n - 1$ • Divide both sides by $\log\sqrt{6}$

$11 = n$ • Solve for *n*

There are 11 terms.

**27.** $a = 512$

$r = \frac{t_2}{t_1} = \frac{-256}{512} = \frac{-1}{2}$

$t_n = -1$

Using $t_n = a(r)^{n-1}$:

$-1 = (512)\left(\frac{-1}{2}\right)^{n-1}$

$\frac{-1}{512} = \left(\frac{-1}{2}\right)^{n-1}$

$\left(-\frac{1}{2}\right)^9 = \left(-\frac{1}{2}\right)^{n-1}$

$9 = n - 1$

$n = 10$

There are 10 terms.

**28.** If $t_8 = ar^{8-1}$, then $ar^7 = 9\ 375\sqrt{5}$.
If $t_4 = ar^{4-1}$, then $ar^3 = 15\sqrt{5}$.

$\frac{ar^7}{ar^3} = \frac{9\ 375\sqrt{5}}{15\sqrt{5}}$

$r^4 = 625$ • The *a*s divide out

$r = \pm 5$ • Take the 4th root of both sides

If $r = 5$:

$ar^3 = 15\sqrt{5}$

$a(5)^3 = 15\sqrt{5}$

$125a = 15\sqrt{5}$

$a = \frac{15\sqrt{5}}{125} = \frac{3\sqrt{5}}{25}$

The general term, $t_n = ar^{n-1}$, is:

$t_n = \frac{3\sqrt{5}}{25}(5)^{n-1}$

$= \frac{3\sqrt{5}}{5^2}(5)^{n-1}$

$= 3\sqrt{5}(5)^{n-3}$

If $r = -5$:

$ar^3 = 15\sqrt{5}$

$a(-5)^3 = 15\sqrt{5}$

$-125a = 15\sqrt{5}$

$a = \frac{15\sqrt{5}}{-125} = \frac{-3\sqrt{5}}{25}$

The general term, $t_n = ar^{n-1}$, is

$t_n = \frac{-3\sqrt{5}}{(-5)^2}(-5)^{n-1} = \frac{-3\sqrt{5}(-5)^{n-1}}{25}$

$= -3\sqrt{5}(-5)^{n-3}$

**29.** If $t_3 = 10\ 000$, then $ar^2 = 10\ 000$.
If $t_6 = 5\ 120$, then $ar^5 = 5\ 120$.

$\frac{\not{a}r^5}{\not{a}r^2} = \frac{5\ 120}{10\ 000}$

$r^3 = 0.512$

$r = 0.8$

 

Solve for $a$:

$a(0.8)^2 = 10\ 000$
$a = 15\ 625$

The general term is:

$t_n = (15\ 625)(0.8)^{n-1}$
$t_8 = (15\ 625)(0.8)^{8-1}$
$t_8 = 3\ 276.8$

Thus, $t_8$ is 3 276.8.

**30.** The sequence representing the height after each bounce is $12, 9, \frac{27}{4}, \ldots$

$r = \frac{t_2}{t_1} = \frac{9}{12} = \frac{3}{4}$

The general term is $t_n = ar^{n-1}$.

$t_n = 12\left(\frac{3}{4}\right)^{n-1}$ • Substituting for $a$ and $r$

We want to determine when $t_n < 50$ cm.

$t_n < 0.5$ m • Convert centimetres to metres

$12\left(\frac{3}{4}\right)^{n-1} < 0.5$

$\left(\frac{3}{4}\right)^{n-1} < \frac{1}{24}$ • Divide both sides by 12

$(n-1)\log\frac{3}{4} < \log\frac{1}{24}$ • Take the log of both sides and apply the Power Law

$n-1 > \dfrac{\log\frac{1}{24}}{\log\frac{3}{4}}$ • Divide both sides by $\log\frac{3}{4}$

(Note that $\log\frac{3}{4}$ is a negative number and thus, the direction of the inequality changes)

$n-1 > 11.05$
$n > 12.05$ • Solve for $n$

Thus, since $n > 12.05$, there will be 13 bounces before the ball rises to less than 50 cm.

**31. C**

A geometric sequence is one that has a common ratio $\left(\frac{t_2}{t_1} = \frac{t_3}{t_2} = \frac{t_4}{t_3}\right)$.

Only alternative C has a common ratio, with $r = \sqrt{5}$.

**32. A**

$a = t_1 = 32y^6$

$t_{n+1} = \frac{t_n}{4y^2}$

Since

$t_{n+1} = t_n \cdot r$, and $t_{n+1} = t_n\left(\frac{1}{4y^2}\right)$ then $r = \frac{1}{4y^2}$

$t_4 = ar^3 = (32y^6)\left(\frac{1}{4y^2}\right)^3$

$= \frac{32y^6}{64y^6} = \frac{1}{2}$

## *Lesson 11 Geometric Series*

### CLASS EXERCISES
### ANSWERS AND SOLUTIONS

**1.** In this series, the first term is 64, so we know that $a = 64$.

To find the common ratio, $r$, we divide $\frac{t_2}{t_1}$ to get $\frac{96}{64} = \frac{3}{2}$. So, $r = \frac{3}{2}$.

Now, which of the two formulas do we need to use to find the sum of this geometric series?

In this question, we are not told the value of $n$, the number of terms. However, we do know the value of the last term in the series; it is 729. So, we know that $t_n = 729$.

Since we know the value of the last term, we use the formula $S_n = \frac{rt_n - a}{r-1}, r \neq 1$.

Now, substitute the values $a = 64$, $r = \frac{3}{2}$, and $t_n = 729$ into the formula:

$$S_n = \frac{\left(\frac{3}{2}\right)(729) - 64}{\frac{3}{2} - 1}$$
$$= \frac{1\,029.5}{0.5}$$
$$= 2\,059$$

The sum of the finite geometric series is 2 059.

**2.** In this series, the first term is 3, so we know that $a = 3$.

To find the common ratio, $r$, we divide $\frac{t_2}{t_1}$ to get $\frac{-6}{3} = -2$. So, $r = -2$.

In this question, we are told that the sum of the first $n$ terms is $-1\,023$. So, $S_n = -1\,023$.

However, we do not know the value of the last term, $t_n$ and we do not know the number of terms, $n$. So, which of the two formulas should we use?

Since we are trying to find the value of $n$, we use the formula that includes $n$:

$$S_n = \frac{a(r^n - 1)}{r-1}$$

Now, we substitute the values $a = 3$, $r = -2$, and $S_n = -1\,023$ into the formula.

$$-1\,023 = \frac{3\left((-2)^n - 1\right)}{(-2) - 1}.$$

We need to solve this equation for $n$.

$$-1\,023 = \frac{3\left((-2)^n - 1\right)}{-3}$$ • Simplify the denominator

$$-1\,023 = -\left((-2)^n - 1\right)$$ • Simplify

$$1\,023 = (-2)^n - 1$$ • Multiply both sides by $-1$

$$1\,024 = (-2)^n$$ • Add 1 to both sides

$$n = 10$$ • $1\,024 = (-2)^{10}$

Thus, $n = 10$.

**3.** Once again, we are given the formula for the sum of terms. To find the value of $t_8$, we will use the formula $t_n = S_n - S_{n-1}$.

This means that $t_8 = S_8 - S_7$.

Let's find the values of $S_8$ and $S_7$ by substituting $n = 8$ and $n = 7$ into the formula for the sum of terms:

$$S_n = 3n^2 + n - 2$$
$$S_8 = 3(8)^2 + 8 - 2$$
$$= 198$$
$$S_7 = 3(7)^2 + 7 - 2$$
$$= 152$$

Since $t_8 = S_8 - S_7$, then $t_8 = 198 - 152 = 46$.

So, $t_8 = 46$.

 

## PRACTICE EXERCISES ANSWERS AND SOLUTIONS

1. For the series $7-14+28-56+...-3584$, first find $r$.

$$r=\frac{t_2}{t_1}=\frac{-14}{7}=-2$$

Since we know the value of $t_n$, use

$$S_n=\frac{rt_n-a}{r-1}.$$

$$S_n=\frac{(-2)(-3\ 584)-7}{-2-1}$$

- Substitute the values into the formula

$$S_n=-2\ 387$$

2. $a=3$

$$r=\frac{t_2}{t_1}=\frac{3\sqrt{2}}{3}=\sqrt{2}$$

$$n=8$$

Thus, use $S_n=\frac{a(r^n-1)}{r-1}$.

$$S_8=\frac{3\left(\left(\sqrt{2}\right)^8-1\right)}{\sqrt{2}-1}=\frac{3(2^4-1)}{\sqrt{2}-1}=\frac{45}{\sqrt{2}-1}$$

$$S_8=\frac{45}{\sqrt{2}-1}\cdot\frac{\left(\sqrt{2}+1\right)}{\left(\sqrt{2}+1\right)}=\frac{45\sqrt{2}+45}{2-1}=45\sqrt{2}+45$$

The sum of the first 8 terms is $45\sqrt{2}+45$.

3. For the geometric series, $-1+\frac{3}{4}-\frac{9}{16}+\frac{27}{64}-...$, first find $r$.

$$r=\frac{t_2}{t_1}=\frac{\frac{3}{4}}{-1}=-\frac{3}{4}$$

Since we want $S_{11}$, we know $n=11$, thus use

$$S_n=\frac{a(r^n-1)}{r-1}.$$

$$S_{11}=\frac{-1\left(\left[-\frac{3}{4}\right]^{11}-1\right)}{\frac{-3}{4}-1}$$

- Substitute the values into the formula

$$S_{11}=-0.60$$

- Simplify

Thus, the sum of the first 11 terms is approximately $-0.60$.

4. $t_n=a(r)^{n-1}$

$$t_n=-4(-3)^{n-1}$$

$$t_8=-4(-3)^{8-1}$$

Thus:

$a=-4$

$r=-3$

$t_8=8\ 748$

Use $S_n=\frac{rt_n-a}{r-1}$:

$$S_8=\frac{(-3)(8\ 748)-(-4)}{(-3)-1}$$

$$S_8=6\ 560$$

The sum of the first 8 terms is 6 560.

5. The series is $-2-10-50$, …, and $S_n=-7\ 812$.

Use $S_n=\frac{a(r^n-1)}{r-1}$ to solve for $n$.

$$-7\ 812=\frac{-2(5^n-1)}{5-1}$$

- Substitute the given values into the formula

$$-7\ 812=\frac{-2(5^n-1)}{4}$$

- Simplify the denominator

$$-7\ 812=\frac{5^n-1}{-2}$$

- Reduce

$$15\ 624=5^n-1$$

- Multiply both sides by $-2$

$$15\ 625=5^n$$

- Add 1 to both sides

$$\log 15625=n\log 5$$

- Take the log of both sides and apply the Power Law

$$6=n$$

- Divide both sides by log 5

Thus, there are 6 terms.

**6.** $S_n = 2n^3 - n^2 + 4$

$t_n = S_n - S_{n-1}$

$S_1 = 2(1)^3 - (1)^2 + 4$ • We want $t_1$, $t_2$, $t_3$

$S_1 = 5$

$t_1 = 5$

$S_2 = 2(2)^3 - (2)^2 + 4$

$S_2 = 16$

$t_2 = S_2 - S_1$

$t_2 = 16 - 5$

$t_2 = 11$

$S_3 = 2(3)^3 - (3)^2 + 4$

$S_3 = 49$

$t_3 = S_3 - S_2$

$t_3 = 49 - 16$

$t_3 = 33$

Thus, the first 3 terms are 5, 11, and 33.

The series is geometric if $\frac{t_2}{t_1} = \frac{t_3}{t_2}$

$\frac{t_2}{t_1} = \frac{11}{5} = 2.2$ and $\frac{t_3}{t_2} = \frac{33}{11} = 3.$

Thus, the series is not geometric because there is no common ratio.

**7.** $S_n = 3(2^n - 1)$

$S_1 = 3(2^1 - 1) = 3(1) = 3$

$S_2 = 3(2^2 - 1) = 3(3) = 9$

$S_3 = 3(2^3 - 1) = 3(7) = 21$

$t_1 = S_1 = 3$

$t_2 = S_2 - S_1$

$t_2 = 9 - 3 = 6$

$t_3 = S_3 - S_2$

$t_3 = 21 - 9$

$t_3 = 12$

Thus, the series is $3 + 6 + 12 + \ldots$

This is a geometric series, where

$r = \frac{t_2}{t_1} = \frac{6}{3} = 2.$

Using $t_n = ar^{n-1}$, the general term is:

$t_n = 3(2)^{n-1}$

**8.** Using $S_n = \frac{a(r^n - 1)}{r - 1}$, solve for $a$.

$2\,862 = \frac{a(3.5)^4 - 1}{3.5 - 1}$ • Substitute the given values into the formula

$2\,862 = \frac{a(149.062\,5)}{2.5}$ • Simplify

$7\,155 = a(149.062\,5)$ • Multiply both sides by 2.5

$48 = a$ • Divide both sides by 149.062 5

Thus, the first term is 48.

**9.** $a = 4$

$r = 3$

$t_n = 78\,732$

Use

$S_n = \frac{rt_n - a}{r - 1}$:

$S_n = \frac{(3)(78\,732) - (4)}{(3) - 1}$

$S_n = 118\,096$

**10.** Since

$t_6 = ar^{6-1}$, then $ar^5 = 24$.

$t_y = ar^{4-1}$, then $ar^3 = 54$.

$\frac{ar^5}{ar^3} = \frac{24}{54}$

$r^2 = \frac{4}{9}$ • The *a*s divide out

$r = \pm\frac{2}{3}$ • Take the square root of both sides

If $r = \frac{2}{3}$, then using $ar^3 = 54$:

$a\left(\frac{2}{3}\right)^3 = 54$

$a\left(\frac{8}{27}\right) = 54$

$a = 182.25$

 

Now, using $S_n = \dfrac{a(r^n - 1)}{r-1}$:

$$S_8 = \frac{182.25\left(\left[\frac{2}{3}\right]^8 - 1\right)}{\frac{2}{3} - 1}$$

- Substitute the values into the formula

$S_8 = 525.42$

- Simplify

If $r = -\frac{2}{3}$, then using $ar^3 = 54$:

$$a\left(-\frac{2}{3}\right)^3 = 54$$

$$a\left(-\frac{8}{27}\right) = 54$$

$$a = -182.25$$

Now, using $S_n = \dfrac{a(r^n - 1)}{r-1}$:

$$S_8 = \frac{-182.25\left(\left[-\frac{2}{3}\right]^8 - 1\right)}{-\frac{2}{3} - 1}$$

- Substitute the values in the formula

$S_8 = -105.08$

- Simplify

**11.** $S_1 = 18,\ S_2 = 6,\ S_3 = 14$

To find $t_n = a(r)^{n-1}$ we must find $a$, and $r$.

$a = t_1 = S_1 = 18$

$r = \dfrac{t_2}{t_1}$

$t_2 = S_2 - S_1$

$t_2 = 6 - 18$

$t_2 = -12$

$r = \dfrac{-12}{18} = \dfrac{-2}{3}$

The general term is $t_n = (18)\left(\dfrac{-2}{3}\right)^{n-1}$.

**12.** For the geometric series $-32 + 48 - 72 + \ldots$,

$r = \dfrac{t_2}{t_1} = \dfrac{48}{-32} = -1.5.$

Using $S_n = \dfrac{a(r^n - 1)}{r-1}$:

$$S_7 = \frac{-32\left((-1.5)^7 - 1\right)}{-1.5 - 1}$$

$S_7 = -231.5$

- Substitute the values into the formula
- Simplify

**13.** $r = -2,\ S_8 = -510$

$t_n = a(r)^{n-1}$

Use $S_n = \dfrac{a(r^n - 1)}{r-1}$:

$$-510 = \frac{a\left((-2)^8 - 1\right)}{(-2) - 1}$$

$$-510 = \frac{a(255)}{-3}$$

$$\frac{1\,530}{255} = a$$

$$a = 6$$

The general term is $t_n = 6(-2)^{n-1}$.

**14.** The given series is 3 125 + 2 500 + 2 000 + … + 819.2.

$r = \dfrac{t_2}{t_1} = \dfrac{2\,500}{3\,125} = 0.8.$

Using $S_n = \dfrac{rt_n - a}{r-1}$:

- Use this formula since we know $t_n$, but not $n$

$$S_n = \frac{(0.8)(819.2) - 3\,125}{0.8 - 1}$$

- Substitute the values into the formula

$S_n = 12\,348.2$

**15.** $r=-2,\ S_6=-105$

$t_1=?\ \ t_1=a$

Solve for $a$

Use $S_n=\dfrac{a(r^n-1)}{r-1}$:

$$-105=\frac{a((-2)^6-1)}{(-2)-1}$$
$$315=63a$$
$$a=5$$
$$t_1=5$$

The first term is 5.

**16.** The sum is given by $S_n=n-5n^2$.

$S_1$, where $n=1=1-5(1)^2=1-5=-4$
$S_2$, where $n=2=2-5(2)^2=2-20=-18$
$S_3$, where $n=3=3-5(3)^2=3-45=-42$

$t_1=S_1=-4$
$t_2=S_2-S_1,$
$t_2=-18-(-4)=-14$

$t_3=S_3-S_2,$
$t_3=-42-(-18)$
$t_3=-24$

Thus, the series is –4 –14 –24 …

This is not a Geometric series because $\frac{t_2}{t_1}\neq\frac{t_3}{t_2}$.

**17. C**

$a=16,\ r=-\frac{1}{2},\ n=7$

Use $S_n=\dfrac{a(r^n-1)}{r-1}$:

$$S_7=\frac{16\left(\left(-\frac{1}{2}\right)^7-1\right)}{\left(-\frac{1}{2}\right)-1}$$

$S_7=10.75.$

**18. C**

$t_2=8,\ t_7=256,\ S_8=\dfrac{a(r^8-1)}{r-1}$

We need to find $a$ and $r$:

$ar=8$
$ar^6=256$

Thus, $\dfrac{\not{a}r^6}{\not{a}r}=\dfrac{256}{8}$

$r^5=32$
$r=2$

Solve for $a$:
$t_2=a(2)=8$
$a=4$

$$S_8=\frac{4((2)^8-1)}{(2)-1}$$

$S_8=1\,020.$

**19.** $S_n=9(2^n)+9$ $\qquad t_4=S_4-S_3$

$S_4=9(2^4)+9=153$
$S_3=9(2^3)+9=81$
$t_4=153-81=72$

| 7 | 2 | | |
|---|---|---|---|

**20.** $a=1,\ r=2,\ t_n=2^k$

You need to determine when $S_n\geq 1\,000\,000$.

$$\frac{rt_n-1}{r-1}\geq 1\,000\,000$$
$$\frac{2(.2^k)-1}{2-1}\geq 1\,000\,000$$
$$2^k\geq 500\,000.5$$
$$k\log z\geq \log 500000.5$$
$$k\geq\frac{\log 500000.5}{\log^2}$$
$$k\geq 18.9$$
$$k=19$$

| 1 | 9 | | |
|---|---|---|---|

 

## Lesson 12 Sigma Notation

### CLASS EXERCISES ANSWERS AND SOLUTIONS

1. In this question, the formula for generating the terms of the series is $5j+2$. To generate these terms, we will substitute values of *j* from 3 to 6.

   First, substitute $j = 3$ to get $5(3)+2=17$. The first term of the series is 17.

   Next, substitute $j = 4$ to get $5(4)+2=22$. The second term of the series is 22.

   Substitute $j = 5$ to get $5(5)+2=27$. The third term of the series is 27.

   Next, substitute $j = 6$ to get $5(6)+2=32$. The fourth term of the series is 32.

   So, the series is $17+22+27+32$.

2. We will use the formula $4(-3)^{i-7}$ to generate the terms of the series.

   We will use the values of *i* from 9 to 11.

   First, substitute $i = 9$ to get $4(-3)^{9-7} = 4(-3)^2 = 4(9) = 36$. The first term of the series is 36.

   Next, substitute $i = 10$ to get $4(-3)^{10-7} = 4(-3)^3 = 4(-27) = -108$. The second term of the series is –108.

   Substitute $i = 11$ to get $4(-3)^{11-7} = 4(-3)^4 = 4(81) = 324$. The third term of the series is 324.

   So, the series is $36+(-108)+324$. We can also write this as $36-108+324$.

3. We will begin by using the formula $3(2)^{k-5}$ to generate the first four terms of the series.

   The sigma notation tells us to use the values of *k* from 4 to 12.

   First, substitute $k = 4$ to get $3(2)^{4-5} = 3(2)^{-1} = 3\left(\frac{1}{2}\right) = \frac{3}{2}$.

   The first term of the series is $\frac{3}{2}$.

   Next, substitute $k = 5$ to get $3(2)^{5-5} = 3(2)^0 = 3(1) = 3$.

   The second term of the series is 3.

   Substitute $k = 6$ to get $3(2)^{6-5} = 3(2)^1 = 3(2) = 6$. The third term is 6.

   Substitute $k = 7$ to get $3(2)^{7-5} = 3(2)^2 = 3(4) = 12$. The fourth term is 12.

   Although we could keep going and determine all of the terms using this method, this is not necessary.

   We have the first 4 terms of the series: $\frac{3}{2}+3+6+12+...$

   We can see that the series is geometric by noticing that there is a common ratio between consecutive terms—each term can be obtained by multiplying the previous term by 2. The common ratio is 2.

   We can also tell that it is geometric from its formula $3(2)^{k-5}$. It is of the same basic form as the general formula for a geometric series, $t_n = a(r)^{n-1}$. From $3(2)^{k-5}$, it is clear that each term is obtained by multiplying the previous term by 2.

   To find the sum of this series, we will use the formula $S_n = \frac{a(r^n - 1)}{r-1}$.

In this series, the first term is $\frac{3}{2}$, so we know that $a = \frac{3}{2}$. We also know that $r = 2$.

We now need to determine the value of $n$—the number of terms in the series. The number of terms = upper limit – lower limit + 1, so we have

$12 - 4 + 1 = 9$. The series has 9 terms ($n = 9$).

Now, substituting $a = \frac{3}{2}$, $r = 2$, and $n = 9$, into the formula gives:

$$S_9 = \frac{\frac{3}{2}(2^9 - 1)}{2-1}$$
$$= 766.5$$

The sum of the series is 766.5.

## PRACTICE EXERCISES ANSWERS AND SOLUTIONS

**1.**

$$t_1 = k_5 = 3(5)^2 - 5 = 70$$
$$t_2 = k_6 = 3(6)^2 - 6 = 102$$
$$t_3 = k_7 = 3(7)^2 - 7 = 140$$
$$t_4 = k_8 = 3(8)^2 - 8 = 184$$
$$t_5 = k_9 = 3(4)^2 - 9 = 234$$

$$\sum_{k=5}^{9} 3k^2 - k = 70 + 120 + 140 + 184 + 234$$

**2.** $n =$ upper limit – lower limit +1

$n = 97 - 22 + 1 = 76$

Thus, there are 76 terms in the series.

**3.** $\sum_{i=3}^{9} 5(-3)^{i-3} = 5 - 15 + 45 \cdots\cdots 5(-3)^{9-3}$

This series is geometric, where
$a = 5(-3)^{3-3} = 5$.

$$r = -3$$
$$n = 9 - 3 + 1 = 7$$
$$S_7 = \frac{5\left((-3)^7 - 1\right)}{-3-1}$$
$$S_7 = 2\ 735$$

**4.**

$$t_1 = -8\left(\frac{3}{4}\right)^{1-1} = -8$$
$$t_2 = -8\left(\frac{3}{4}\right)^{2-1} = -6$$

Thus, $r = \frac{-6}{-8} = \frac{3}{4}$.

You can also notice that in the general term form, $ar^{n-1}$, $r = \frac{3}{4}$.

**5.** $\sum_{k=4}^{6} (k-7)x = 72$

Thus
$$(4-7)x + (5-7)x + (6-7)x = 72$$
$$-3x + -2x + -x = 72$$
$$-6x = 72$$
$$x = -12$$

**6.** Notice that the series is geometric, and that $r = 3$. Thus, the general term is of the form $t_n = ar^{n-1}$.

$$a = 4$$
$$r = \frac{12}{4} = 3$$

Thus, $t_n = 4(3)^{n-1}$.

Solve for $n$:
$$78\ 732 = 4(3)^{n-1}$$
$$19\ 683 = 3^{n-1}$$
$$\log 19683 = (n-1)\log 3$$
$$\frac{\log 19683}{\log 3} = n - 1$$
$$n = 10$$

Thus $\sum_{n=1}^{10} 4(3)^{n-1}$.

**7.** $n =$ upper limit – lower limit +1

$n = 75 - 5 + 1 = 71$

There are 71 term in the series.

 

**8.** $\sum_{j=4}^{22} 5(3)^{j-6}$

The 4th term will occur when $j = 7$.

$t_7 = 5(3)^{7-6} = 15$

The 4th term is 15.

**9.** $\sum_{j=2}^{10} \left(-\frac{3}{4}\right)^{j-1}$

The series is geometric with

$a = \frac{-3}{4},\ r = \frac{-3}{4},\ n = 10 - 2 + 1 = 9$

Use $S_n = \frac{a(r^n - 1)}{r - 1}$:

$$S_9 = \frac{\frac{-3}{4}\left(\left(\frac{-3}{4}\right)^9 - 1\right)}{\frac{-3}{4} - 1} \doteq 0.46$$

**10.**

$$\sum_{k=1}^{9} \log\left(\frac{k+1}{k}\right) = \log 2 + \log\frac{3}{2} + \log\frac{4}{3} + \ldots + \log\frac{10}{9}$$

$$= \log\left(2 \times \frac{3}{2} \times \frac{4}{3} \times \frac{5}{4} \times \frac{6}{5} \times \frac{7}{6} \times \frac{8}{7} \times \frac{9}{8} \times \frac{10}{9}\right)$$

$= \log 10$

$= 1$

**11.** Notice that the series is geometric with $a = 5$, $r = \sqrt{2}$, and $t_n = 640$.

$t_n = ar^{n-1} = 5\left(\sqrt{2}\right)^{n-1}$

$640 = 5\left(\sqrt{2}\right)^{n-1}$

$128 = \left(\sqrt{2}\right)^{n-1}$

$2^7 = \left(2^{\frac{1}{2}}\right)^{n-1}$

$2^7 = 2^{\frac{1}{2}n - \frac{1}{2}}$

$7 = \frac{1}{2}n - \frac{1}{2}$

$14 = n - 1$

$n = 15$

Thus, the sigma notation is $\sum_{n=1}^{15} 5\left(\sqrt{2}\right)^{n-1}$.

**12.** $\sum_{k=1}^{x} 3(-5)^{k-1} = 39\ 063$

As the series is geometric, use $S_n = \frac{a(r^n - 1)}{r - 1}$:

$39\ 063 = \frac{3\left((-5)^n - 1\right)}{-5 - 1}$

$39\ 063 = \frac{3\left((-5)^n - 1\right)}{-6}$

$-78\ 125 = (-5)^n$

$(-5)^7 = (-5)^n$

There are 7 terms and $1 + x - 1 = 7$.

Thus, $x = 7$.

**13.**

$t_1 = j_6 = 3(6-4)^2 = 12$

$t_2 = j_7 = 3(7-4)^2 = 27$

$t_3 = j_8 = 3(8-4)^2 = 48$

$t_4 = j_9 = 3(9-4)^2 = 75$

$\sum_{j=6}^{9} 3(j-4)^2 = 12 + 27 + 48 + 75$

**14.**

$\sum_{j=1}^{1} 4\left(\frac{2}{3}\right)^{1-j}$

$r = \frac{t_2}{t_1}$

$t_1 = 4\left(\frac{2}{3}\right)^{1-1} = 4$

$t_2 = 4\left(\frac{2}{3}\right)^{1-2} = 6$

$r = \frac{6}{4} = \frac{3}{2}$

You can also notice that in the general term form, $t_n = a(r)^{n-1}$, $r = \frac{3}{2}$.

**15.**

$$t_1 = k_3 = 2(3+1)^{3-2} = 8$$
$$t_2 = k_4 = 2(4+1)^{4-2} = 50$$
$$t_3 = k_5 = 2(5+1)^{5-2} = 432$$
$$\sum_{k=3}^{5} 2(k+1)^{k-2} = 8+50+432$$

**16.** $\sum_{k=2}^{8} 7(-4)^{k-3}$

Notice that the series is geometric, and that $r = -4$.

Use $S_n = \dfrac{a(r^n - 1)}{r-1}$ $\quad n = 8-2+1$, $n = 7$

$$a = t_1 = k_2 = 7(-4)^{2-3} = \frac{-7}{4}$$
$$S_7 = \frac{\frac{-7}{4}\left((-4)^7 - 1\right)}{-4-1}$$
$$S_7 = -5\ 734.75$$

**17. A**

$$\log_3\left(\frac{1}{3}\right) + \log_3 1 + \log_3 3 + \log_3 9 + \log_3 27$$
$$= -1+0+1+2+3$$
$$\sum_{n=1}^{5}(n-2)$$

**18. D**

$$\sum_{k=1}^{15} 3^{k-2}$$
$$t_1 = k_1 = 3^{1-2} = \frac{1}{3}$$
$$t_2 = k_2 = 3^{2-2} = 1$$
$$r = \frac{1}{\left(\frac{1}{3}\right)} = 3, \quad a = \frac{1}{3}$$

Thus $\sum_{k=1}^{15} 3^{k-2}$ is equal to

$$S_n = \frac{a(r^n - 1)}{r-1} = \frac{\frac{1}{3}(3^{15} - 1)}{3-1}$$
$$= \frac{3^{15} - 1}{6}$$

**19.** $n$ = upper limit – lower limit + 1

$$n = (3k+6) - (2k) + 1$$
$$42 = 3k + 6 - 2k + 1$$
$$42 = k + 7$$
$$k = 35$$

| 3 | 5 | | |
|---|---|---|---|

**20.**

$$\sum_{k=a-2}^{a+15} 3^{k-1}$$
$$n = (a+15) - (a-2) + 1$$
$$n = a + 15 - a + 2 + 1$$
$$n = 18$$

| 1 | 8 | | |
|---|---|---|---|

## *Lesson 13 Applications*

### CLASS EXERCISES
### ANSWERS AND SOLUTIONS

**1.** Here, we have a geometric series where we do not know the value of the first term, *a*. However, we are given the number of terms, the sum, and the common ratio.

We are given the sum of Alex's salary over 5 years, so the number of terms is 5.

$$n = 5$$

The sum of the 5 terms is 155 680.05. So, $S_5 = 155\ 680.05$.

Finally, we know the common ratio. Each year, Alex receives a 10% raise. To calculate the salary for each year, we multiply the previous year's salary by 1.10. This means that the common ratio is 1.10, thus $r = 1.10$.

Since we know the values of $n$, $r$, and $S_n$, we will use the formula $S_n = \dfrac{a(r^n - 1)}{r-1}$ to find the value of $a$.

When we substitute the values of $n$, $r$, and $S_n$, we get $155\ 680.05 = \dfrac{a(1.10^n - 1)}{1.10 - 1}$.

Solving for $a$:

$$155\ 680.05 = \frac{a(1.10^5 - 1)}{1.10 - 1}$$

$155\ 680.05 = 6.105\ 1a$ • Evaluate

$25\ 500 = a$ • Divide both sides by 6.105 1

Since $a = 25\ 500$, we know that Alex's starting annual salary was \$25 500.

**2.** Total Distance Falling:
When the ball falls for the first time, it drops 9 m. So, $t_1 = 9$. The ball then rises to a point that is $\frac{2}{3}$ of the distance that it just fell. So, the ball rises to a height of 6 m $\left(\frac{2}{3} \times 9 = 6\right)$. So, on its second fall, the ball falls 6 m. This means that $t_2 = 6$. We calculate the value of $t_3$ by multiplying, the height of the last fall by $\frac{2}{3}$.

So, $t_3 = \frac{2}{3} \times 6 = 4\text{ m}$.

The first three terms of the series are 9 + 6 + 4, where each term represents the distance, in metres, that the ball falls each time.

In this series, $a = 9$, and $r = \frac{2}{3}$.

We are interested in finding the sum of the first seven terms in the series. So, $n = 7$.

When we substitute these values into the formula $S_n = \frac{a(r^n - 1)}{r - 1}$, we get the following result.

$$S_7 = \frac{9\left(\left(\frac{2}{3}\right)^7 - 1\right)}{\frac{2}{3} - 1} \doteq 25.42$$

The total distance that the ball falls is approximately 25.42 m.

Total Distance Rising:
We will now create a new series for the distance the ball rises with each bounce.

After the ball initially drops 9 m, it rises to a height of 6 m because $\frac{2}{3} \times 9 = 6$. So, in our new series, $t_1 = 6$.

After the ball falls again, it rises to a height that is $\frac{2}{3}$ the distance that it just rose. So, the ball rises to a height of 4 m, since $\frac{2}{3} \times 6 = 4$. So, $t_2 = 4$ is the second term of the series.

We calculate the value of $t_3$ by multiplying $t_2$ by $\frac{2}{3}$. We get $\frac{2}{3} \times 4 = \frac{8}{3}$. So, $t_3 = \frac{8}{3}$.

The first three terms of this series are $6 + 4 + \frac{8}{3}$, where each term represents the distance, in metres, that the ball rises each time.

In this series, $a = 6$, and $r = \frac{2}{3}$.

If we want to find the sum of all of the distances that the ball rises before it hits the ground for the 7th time, how many terms must we add?

When the ball hits the ground for the first time, it has not had the chance to rise yet. When the ball hits the ground for the second time, it has only had the chance to rise once.

So, when the ball hits the ground for the seventh time, it will have risen only six times.

Therefore, we want to find the sum of the first 6 terms of the series.

$$6 + 4 + \frac{8}{3} + \ldots$$

This means that $n = 6$.

When we substitute these values into the formula $S_n = \dfrac{a(r^n-1)}{r-1}$, we get the following result.

$$S_6 = \frac{6\left(\left(\frac{2}{3}\right)^6 - 1\right)}{\frac{2}{3}-1} \doteq 16.42$$

The total distance that the ball rises is approximately 16.42 m.

To calculate the total distance travelled, we add the total distance the ball fell and the total distance the ball rose.

We get 25.42 + 16.42 = 41.84 m. The ball travelled a total of 41.8 m (to the nearest tenth of a metre).

## PRACTICE EXERCISES ANSWERS AND SOLUTIONS

**1.** The series is
$600+600(1.05)+600(1.05)^2+...+t_{52}$

$t_1 = 600$ or $a = 600$
$r = 1.05$, $n = 52$

You are looking for $S_{52}$, so use $S_n = \dfrac{a(r^n-1)}{r-1}$.

$$S_{52} = \frac{600(1.05^{52}-1)}{1.05-1} \doteq 134\ 713.699\ 2$$

Over 52 weeks Lana received approximately \$139 713.70.

**2.** The series is
$1\ 100+1\ 100(0.5)+\ldots+t_{10}$
$t_1 = 1\ 100$ or $a = 1\ 100$
$n = 10$

$t_1 = 1\ 100$
$t_2 = 1\ 100 \times 0.5 = 550$
$t_3 = 550 \times 0.5 = 275$

$$r = \frac{t_3}{t_2} = \frac{275}{550} = 0.5$$

You are looking for $S_{10}$, so use $S_n = \dfrac{a(r^n-1)}{r-1}$.

$$S_{10} = \frac{1\ 100\left((0.5)^{10}-1\right)}{0.5-1} \doteq 2\ 197.851\ 6$$

Jeff took a total of approximately 2 197.9 mg of medication over the 10-day period.

**3.** $t_1 = 0.7$ or $a = 0.7$
The series is $0.7+0.7(1.04)+...t_{12}$.

$r = 1.04$
$n = 12$

You arc looking for $S_{12}$, so use $S_n = \dfrac{a(r^n-1)}{r-1}$.

$$S_{12} = \frac{0.7\left((1.04)^{12}-1\right)}{1.04-1} \doteq 10.518\ 1$$

Over the 12-week period, the puppy will eat approximately 10.5 kg of food.

**4.** The series is 14 + 7 + 7 + 3.5 + ….

Total Distance Falling:
$t_1 = 14$
$t_2 = 14(0.5) = 7$
$t_3 = 7(0.5) = 3.5$

Thus, $a = 14$ and $r = 0.5$.

 

You are looking for $S_9$, so use $S_n = \dfrac{a(r^n - 1)}{r - 1}$.

$$S_9 = \frac{14((0.5)^9 - 1)}{0.5 - 1} \doteq 27.945\ 3$$

Total Distance Rising:
$t_1 = 7,\ t_2 = 3.5,\ t_3 = 1.75$

Thus, $a = 7$ and $r = 0.5$.

You are looking for $S_8$, so use $S_n = \dfrac{a(r^n - 1)}{r - 1}$.

$$S_8 = \frac{7((0.5)^8 - 1)}{0.5 - 1} \doteq 13.945\ 3$$

Total Distance Travelled:
= total rising + total falling
≐ = 13.945 3 + 27.945 3
≐ = 41.890 6

The ball travelled a total of approximately 41.9 m.

**5.** The series is $25 + 25(1.15) + 25(1.15)^2 + \ldots + t_7$
$r = 1.15$
$n = 7$

You are looking for $S_7$, so use $S_n = \dfrac{a(r^n - 1)}{r - 1}$.

$$S_7 = \frac{25((1.15)^7 - 1)}{1.15 - 1} \doteq 276.670\ 0$$

The total amount of wheat harvested over the 7 days was approximately 277 acres.

**6.** The series is $150 + 150(0.94) + 150(0.94)^2 + \ldots$

**a)** $t_1 = 150$ or $a = 150$
$r = 0.94$
$n = 15$

You are looking for $S_{15}$, so use $S_n = \dfrac{a(r^n - 1)}{r - 1}$.

$$S_{15} = \frac{150((0.94)^{15} - 1)}{0.94 - 1} \doteq 1\ 511.770\ 5$$

In 15 minutes the balloon travels approximately 1 511.8 m.

**b)** 2 km = 2 000 m.

$$S_n = 2\ 000$$
$$2\ 000 = \frac{150((0.94)^n - 1)}{0.94 - 1}$$
$$-0.8 = ((10.94)^n - 1)$$
$$0.2 = (0.94)^n$$
$$\log 0.2 = n \log 0.94$$
$$n = \frac{\log 0.2}{\log 0.94} \doteq 26.01$$

It will take approximately 26 min for the balloon to reach a height of 2 km.

**7.** The series is $a + a(1.25) + a(1.25)^2 + \ldots t_6$.
$r = 1.25$
$S_6 = 262\ 625$
$n = 6$

You are looking for $a$, so use $S_n = \dfrac{a(r^n - 1)}{r - 1}$.

$$262\ 625 = \frac{a(1.25)^6 - 1}{1.25 - 1}$$
$$a = \frac{(262\ 625)(0.25)}{(1.25)^6 - 1}$$
$$a \doteq 23\ 326.22$$

Her starting salary was $23 326.22.

**8.** The number of people who know the secret is $3 + 9 + 27 + \ldots$

Each term is the number of people who are told the secret in each successive hour. You require the sum of 7 terms, or $S_7$
$a = 3,\ r = 3$

Use $S_n = \dfrac{a(r^n - 1)}{r - 1}$:

$$S_7 = \frac{3(3^7 - 1)}{3 - 1}$$
$$S_7 = 1\ 093$$

In 7 hours, 1 093 people other than you know the secret.

9. The series is $0.01+0.02+0.04+...t_n$.

$t_1 = 0.01$ or $a = 0.01$
$r = 2$
$S_n = 1\,000\,000$

You are looking for $n$, so use $S_n = \dfrac{a(r^n-1)}{r-1}$.

$$\frac{0.01(2^n-1)}{2-1} > 1\,000\,000$$
$$2^n - 1 > 100\,000\,001$$
$$2^n > 100\,000\,001$$
$$n\log 2 > \log 100000001$$
$$n > 26.57$$

By 27th week of the year, the total allowance will be over \$1 000 000.

## Practice Test

### ANSWERS AND SOLUTIONS

1.
$$\frac{(3)(9)^{2x-4}}{27^x} = \frac{(3)(3^2)^{2x-4}}{(3^3)^x}$$
$$= \frac{(3)^1(3)^{4x-8}}{(3)^{3x}} = \frac{3^{4x-7}}{3^{3x}} = 3^{x-7}$$

2. $\log_9 36 + \log_9 18 - \log_9 24$
$$= \log_9\left(\frac{(36\times18)}{24}\right) = \log_9 27 \quad = \frac{\log_3 27}{\log_3 9} = \frac{3}{2}$$

3.
$$4^{x-2} = \left(\frac{1}{8}\right)^{x+1}$$
$$(2^2)^{x-2} = \left(\frac{1}{2^3}\right)^{x+1}$$
$$2^{2x-4} = 2^{-3x-3}$$
$$2x-4 = -3x-3$$
$$5x = 1$$
$$x = \frac{1}{5}$$

4.
$$\log_2 x + \log_2(x+7) = 3$$
$$\log_2(x^2+7x) = 3$$
$$x^2+7x-8 = 0$$
$$(x+8)(x-1) = 0$$
$x = -8$ or $x = 1$
But $x > 0$, so $x = 1$.

5. $y = 3^{2x-2}$
The inverse is $x = 3^{2y-2}$.

$$\log_3 x = 2y-2$$
$$y = \frac{\log_3 x + 2}{2}$$

6.
$$y = 5\log(2x-4)+1$$
$$y = 5\log(2(x-2))+1$$

The transformations in order are:

1. Vertical stretch about the $x$-axis by a factor of 5.
2. Horizontal stretch about the $y$-axis by a factor of $\frac{1}{2}$.
3. Translations 2 right, 1up

7.
$$\log y - 2\log x - \frac{1}{2}\log(x-1) = 0$$
$$\log y = 2\log x + \frac{1}{2}\log(x-1)$$
$$\log y = \log x^2 + \log(x-1)^{\frac{1}{2}}$$
$$\log y = \log\left(x^2(x-1)^{\frac{1}{2}}\right)$$
$$y = x^2(x-1)^{\frac{1}{2}}$$

From the original form, you know that $x > 0$ and $(x-1) > 0$.

Thus, the domain is $x > 1$.

8.
$$\log_b 2 = A,\ \log_b 49 = B$$
$$\log_b 392 = \log_b(8\times49) = \log_b 8 + \log_b 49$$
$$= \log_b 2^3 + \log_b 49 = 3\log_b 2 + \log_b 49$$
$$= 3A + B$$

 

**9.** As it is a geometric sequence, there is a common ratio.

$$\frac{x+2}{x}=\frac{x+6}{x+2}$$
$$x^2+4x+4=x^2+6x$$

$$4x+4=6x$$
$$4=2x$$
$$2=x$$

Therefore, the terms are 2, 4, and 8.

**10.** $t_n=a(r)^{n-1}$

$a=-3$

$$r=\frac{t_2}{t_1}=\frac{2}{3}=\frac{-2}{3}$$

$$t_n=-3\left(\frac{-2}{3}\right)^{n-1}$$

**11.** $t_n=-3(-2)^{n-1}$

$$S_n=\frac{a\left(r^n-1\right)}{r-1}$$

$a=-3, r=-2, n=10$

$$S_{10}=\frac{-3\left((-2)^{10}-1\right)}{-2-1}$$
$$S_{10}-1\ 023$$

**12.** If $S_1=36$, then $a=36$.

$$S_2=24=a+t_2=a+ar$$
$$24=36+36r$$
$$r=\frac{-12}{3}=-\frac{1}{3}$$

Thus, $t_n=ar^{n-1}\ =36\left(-\frac{1}{3}\right)^{n-1}$.

**13.** If $t_3+t_4=50$, then $ar^2+ar^3=50$.
If $t_4+t_5=100$, then $ar^3+ar^4=100$.

$$\frac{ar^3+ar^4}{ar^2+ar^3}=\frac{100}{50}$$
$$r=2$$

Solve for $a$:

$$a(2)^2+a(2)^3=50$$
$$12a=50$$
$$a=\frac{50}{12}=\frac{25}{6}$$
$$t_1=\frac{25}{6}$$
$$t_2=\frac{25}{6}\times 2=\frac{25}{3}$$

**14.** $a=24,\ r=\sqrt{2},\ S_n=360+168\sqrt{2},\ n=?$

You are looking for $n$, so use $S_n=\dfrac{a\left(r^n-1\right)}{r-1}$.

$$360+168\sqrt{2}=\frac{24\left(\left(\sqrt{2}\right)^n-1\right)}{\sqrt{2}-1}$$
$$\left(15+7\sqrt{2}\right)\left(\sqrt{2}-1\right)=\left(\left(\sqrt{2}\right)^n-1\right)$$
$$15\sqrt{2}-15+14-7\sqrt{2}=\left(\sqrt{2}\right)^n-1$$

$$8\sqrt{2}-1=\left(\sqrt{2}\right)^n-1$$
$$8\sqrt{2}=\left(\sqrt{2}\right)^n$$
$$\left(\sqrt{2}\right)^7=\left(\sqrt{2}\right)^n$$
$$n=7$$

**15.** Use $t_n=ar^{n-1}$, where
$a=60\ 000$, $r=1.10$, and $t_n=500\ 000$.

$$60\ 000(1.10)^{n-1}>500\ 000$$
$$(1.10)^{n-1}>\frac{25}{3}$$
$$n-1>\frac{\log\left(\frac{25}{3}\right)}{\log(1.10)}$$
$$n>23.25$$

By the 24th year, the doctor's salary will be over \$500 000.

**16.** **a)** Enter into $L_1$: year (0,1,2,3,4,5,6)
Enter into $L_2$: numbers of cell phones
Exp. Reg. gives $y = (459.81)(2.64)^x$

**b)** In the year 2008, $x = 8$.
You are looking for y.

$$y = (459.81)(2.64)^8$$
$$y = 1\ 084\ 950$$

By year 2008 there will be approximately 1 084 950 cell phones.

**17.**

$$\sum_{n=k+1}^{3k} (-4)^n \qquad n = 38$$

$n$ = upper limit – lower limit + 1

$$38 = 3k - (k+1) + 1$$
$$38 = 2k$$
$$k = 19$$

**18.** Notice that the series is geometric, and $r = -\frac{1}{2}$.

$$\sum_{n=1}^{5} 48\left(-\frac{1}{2}\right)^{n-1}$$

Number of terms, $n = 5 - 1 + 1 = 5$

$$a = 48\left(-\frac{1}{2}\right)^{1-1} = 48$$

Use $S_n = \dfrac{a(r^n - 1)}{r-1}$.

$$S_5 = \frac{48\left(\left(-\frac{1}{2}\right)^5 - 1\right)}{\left(-\frac{1}{2}\right) - 1}$$

$$S_5 = 33$$

The sum of the series is 33.

**19.** The series is $12 + 8 + 8 + \frac{16}{3} + \frac{16}{3} + \ldots$

Total Distance Falling:

$$t_1 = 12$$
$$t_2 = 8$$
$$t_3 = \frac{16}{3}$$

Thus, $a = 12$ and $r = \frac{2}{3}$.

Since the ball falls 8 times, $n = 8$.

Use $S_n = \dfrac{a(r^n - 1)}{r-1}$

$$S_8 = \frac{12\left(\left(\frac{2}{3}\right)^8 - 1\right)}{\left(\frac{2}{3}\right) - 1} \doteq 34.60$$

$$S_8 \doteq 34.60$$

Total Distance Rising:

$$t_1 = 8$$
$$t_2 = \frac{16}{3}$$
$$t_3 = \frac{32}{9}$$

Thus, $a = 8$ and $r = \frac{2}{3}$.

Since the ball falls 7 times, $n = 7$.

Use $S_n = \dfrac{a(r^n - 1)}{r-1}$, $n = 7$.

$$S_7 = \frac{8\left(\left(\frac{2}{3}\right)^7 - 1\right)}{\left(\frac{2}{3}\right) - 1} \doteq 22.60$$

Total distance = total distance rising + total distance falling.

Thus, total distance $\doteq$ 22.60 + 34.60 = 57.20.

The ball has travelled is approximately 57.2 m when it hits the ground for the 8th time.

 

**20.** **A**

$$\sum_{n=1}^{1\,000} 2(-1)^n$$

$n = 1\,000 - 1 + 1 = 1\,000$

$t_1 = -2 \quad t_2 = 2$

Thus, $a = -2$ and $r = -1$.

Use $S_n = \dfrac{a(r^n - 1)}{r-1}$:

$$S_{1\,000} = \frac{-2\left((-1)^{1\,000} - 1\right)}{(-1)-1}$$

$$S_{1\,000} = 0$$

Or, this is clearly the answer if the series is written out: $-2 + 2 - 2 + 2 \ldots - 2 + 2$

**21.** For milk $\left[H^+\right] = \dfrac{1}{10^{5.8}}$

For pure water $\left[H^+\right] = \dfrac{1}{10^7}$

The ratio of their acidities is:

$$\frac{\frac{1}{10^{5.8}}}{\frac{1}{10^7}} = \frac{10^7}{10^{5.8}} = 10^{1.2} \doteq 15.85$$

Thus, milk is approximately 16 times more acidic than pure water.

| | | | |
|---|---|---|---|
| 1 | 6 | | |

# TRIGONOMETRY

## *Lesson 1 Review, Reciprocal Ratios, Special Angles*

### CLASS EXERCISES ANSWERS AND SOLUTIONS

**1.** On the home screen, press [COS] [7] [9] [)] [ENTER].

cos 79° $\doteq$ 0.191.

**2.** Press [2nd] [SIN] [0] [.] [2] [3] [4] [ENTER].

$\theta \doteq 13.5°$

**3.** We are given the length of the hypotenuse and we are asked to find the length of side $x$—the side opposite the angle with a measure of 28°.

Which of the three trig ratios relates to the opposite side and the hypotenuse? It is sine.

$$\sin\theta = \frac{\text{opposite}}{\text{hypotenuse}}$$

$$\sin 28° = \frac{x}{14}$$

- Substitute in $\theta = 28°$ and hypotenuse = 14
- Use a calculator to find sin 28°

$$0.469 \doteq \frac{x}{14}$$

$$x \doteq 0.469 \times 14$$

$$x \doteq 6.6$$

The length of side $x$ is approximately 6.6 cm.

**4.** With respect to the 38° angle, we are given the adjacent side as 13 cm. We need to find the length of the hypotenuse, *m*.

Which of the three trig ratios relates to the adjacent side and the hypotenuse? It is cosine.

$$\cos\theta = \frac{\text{adjacent}}{\text{hypotenuse}}$$

$$\cos 38° = \frac{13}{m}$$

- Substitute in $\theta = 38°$ and adjacent = 13

$$0.788 \doteq \frac{13}{m}$$

- Use a calculator to find $\cos 38°$

$$0.788m \doteq 13$$
$$m \doteq 16.5$$

The length of side *m* is approximately 16.5 cm.

**5.** With respect to angle $\theta$, we are given the opposite and the adjacent sides.

Which of the three trig ratios relates to the opposite side and the adjacent side? It is tangent.

$$\tan\theta = \frac{\text{opposite}}{\text{adjacent}}$$

$$\tan\theta = \frac{12}{21}$$

- Substitute in opposite = 21 and adjacent = 12

$$\tan\theta \doteq 0.571$$
$$\theta \doteq 29.7°$$

- Use a calculator to find $\tan^{-1} 0.571$

The measure of angle $\theta$ is approximately 29.7°.

**6.** Since $\sec\theta = \frac{1}{\cos\theta}$, we know that

$$\sec 72° = \frac{1}{\cos 72°}$$
$$\doteq 3.236$$

**7.** $\cot\theta = 2.894$

$$\tan\theta = \frac{1}{2.894}$$

- Since $\cot\theta$ is the reciprocal of $\tan\theta$

$$\tan\theta \doteq 0.346$$
$$\theta \doteq 19.1°$$

- Use a calculator to find $\tan^{-1} 0.346$

**8.** To find the exact value, use the 45-45-90 triangle.

$$\sin 45° = \frac{\text{opposite}}{\text{hypotenuse}} = \frac{1}{\sqrt{2}} \text{ or } \frac{\sqrt{2}}{2}$$

## PRACTICE EXERCISES ANSWERS AND SOLUTIONS

**1.** With respect to $\theta$, you are given the adjacent side and the hypotenuse. Thus, the trig ratio to use is cosine.

$$\cos\theta = \frac{\text{adjacent}}{\text{hypotenuse}}$$
$$\cos\theta = \frac{24}{37}$$
$$\cos\theta \doteq 0.648\ 6$$
$$\theta \doteq \cos^{-1}(0.684\ 6)$$
$$\theta \doteq 49.6°$$

**2.**

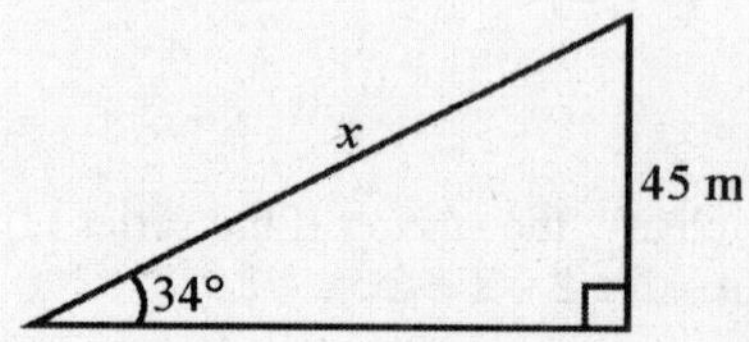


Let $x$ = the length of the kite's string with respect to the 34° angle of elevation.

You are given the opposite side, so you need to determine the length of the hypotenuse. Thus, the trig ratio to use is sine.

$$\sin\theta = \frac{\text{opposite}}{\text{hypotenuse}}$$
$$\sin 34° = \frac{45}{x}$$
$$0.559\ 2x \doteq \frac{45}{x}$$
$$0.559\ 2x \doteq 45$$
$$x \doteq \frac{45}{0.559\ 2}$$
$$x \doteq 80.5 \text{ m}$$

 

**3.** If $\sin\theta = \frac{3}{5}$, then in relation to $\theta$ the opposite side is 3 and the hypotenuse is 5.

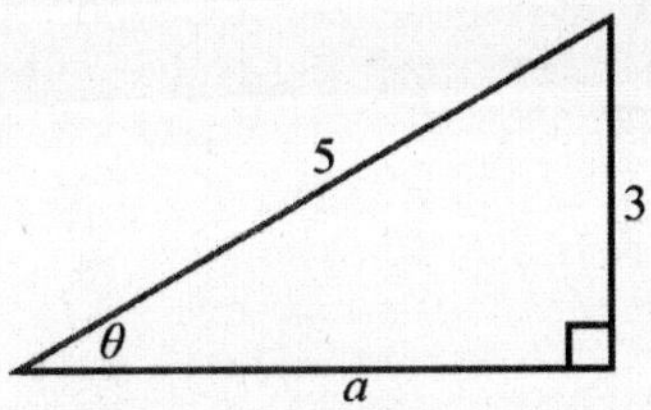


Use the Pythagorean Theorem to determine the length of the adjacent side, *a*.

$$a^2 + 3^2 = 5^2$$
$$a^2 + 9 = 25$$
$$a^2 = 16$$
$$a = 4$$
$$\tan\theta = \frac{\text{opposite}}{\text{adjacent}}$$
$$\tan\theta = \frac{3}{4}$$

**4.** If $\tan\theta = \frac{12}{13}$, then in relation to $\theta$ the opposite side is 12 and the adjacent side is 5.

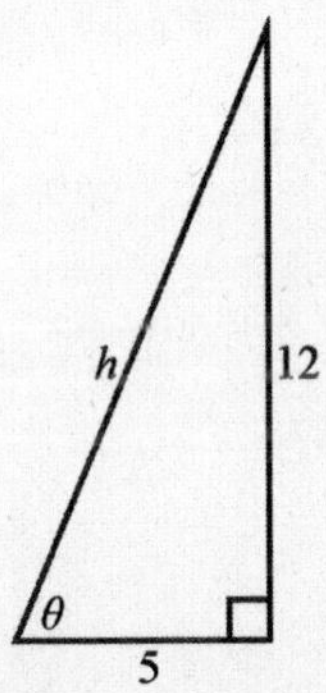


Use the Pythagorean Theorem to determine the length of the hypotenuse, *h*.

$$h^2 = 12^2 + 5^2$$
$$h^2 = 144 + 25$$
$$h^2 = 169$$
$$h = \sqrt{169} = 13$$
$$\cos\theta = \frac{\text{adjacent}}{\text{hypotenuse}}$$
$$\cos\theta = \frac{5}{13}$$

**5.** With respect to the 52° angle, you are given the hypotenuse. You need to determine the length of the opposite side. Thus, the trig ratio to use is sine.

$$\sin\theta = \frac{\text{opposite}}{\text{hypotenuse}}$$
$$\sin 52^\circ = \frac{x}{15}$$
$$0.788\,0 \doteq \frac{x}{15}$$
$$x \doteq 0.788\,0 \times 15$$
$$x \doteq 11.8$$

**6.** With respect to $\theta$, you are given the adjacent side and the opposite side. Thus, the trig ratio to use is tangent.

$$\tan\theta = \frac{\text{opposite}}{\text{adjacent}}$$
$$\tan\theta = \frac{9\text{ cm}}{5\text{ cm}}$$
$$\tan\theta = 1.8$$
$$\theta = \tan^{-1}(1.8)$$
$$\theta \doteq 60.9^\circ$$

**7.** If $\tan\theta = \frac{4}{3}$, then in relation to $\theta$ the opposite side is 4 and the adjacent side is 3.

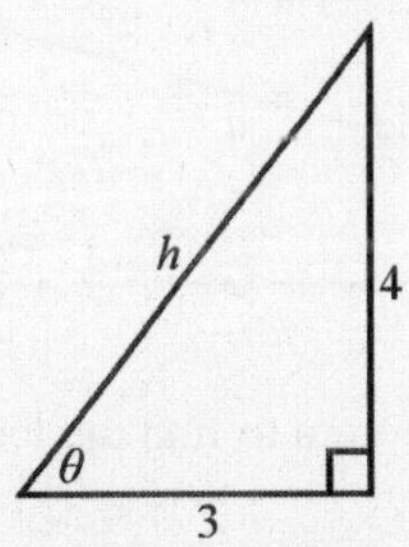


Use the Pythagorean Theorem to determine the length the hypotenuse, *h*.

$$h^2 = 4^2 + 3^2$$
$$h^2 = 16 + 9$$
$$h^2 = 25$$
$$h = 5$$

Thus,

$$\sin\theta = \frac{\text{opposite}}{\text{hypotenuse}}$$
$$\sin\theta = \frac{4}{5}$$

**8.** With respect to $\theta$, you are given the adjacent side and the hypotenuse. The trig ratio to use is cosine.

$$\cos\theta = \frac{\text{adjacent}}{\text{hypotenuse}}$$
$$\cos\theta = \frac{6}{8}$$
$$\cos\theta = 0.75$$
$$\theta = \cos^{-1}(0.75)$$
$$\theta \doteq 41.4°$$

**9.** With respect to the 72° angle, you are given the opposite side. You need to find the length of the hypotenuse. Thus, the trig ratio to use is sine.

$$\sin\theta = \frac{\text{opposite}}{\text{hypotenuse}}$$
$$\sin 72° = \frac{8 \text{ cm}}{x}$$
$$0.9511 \doteq \frac{8}{x}$$
$$0.9511x \doteq 8$$
$$x \doteq 8.4 \text{ cm}$$

**10.** If $\sin\theta = \frac{7}{25}$, then in relation to $\theta$, the opposite side is 7 and the hypotenuse is 25.

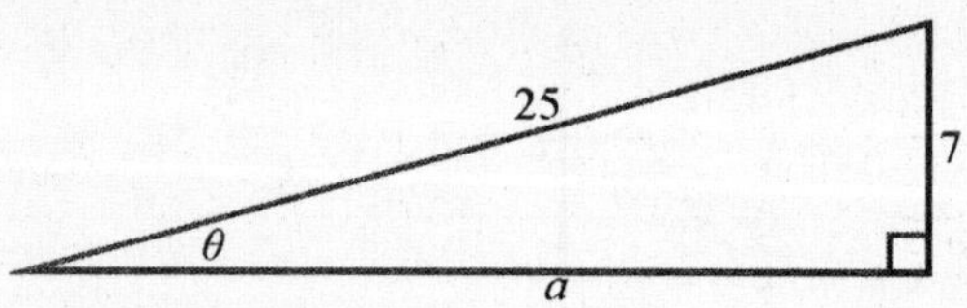


Use the Pythagorean Theorem to find the length of the adjacent side, $a$.

$$a^2 + 7^2 = 25^2$$
$$a^2 + 49 = 625$$
$$a^2 = 576$$
$$a = 24$$

Thus,

$$\cos\theta = \frac{\text{adjacent}}{\text{hypotenuse}}$$
$$\cos\theta = \frac{24}{25}$$

**11.**
$$\cot\theta = \frac{1}{\tan\theta}$$
$$\cot 73° = \frac{1}{\tan 73°}$$
$$\cot 73° \doteq \frac{1}{3.27}$$
$$\cot 73° \doteq 0.31$$

**12.**
$$\sec\theta = \frac{1}{\cos\theta}$$
$$\sec 52° = \frac{1}{\cos 52°}$$
$$\sec 52° \doteq \frac{1}{0.616}$$
$$\sec 52° \doteq 1.62$$

**13.**
$$\csc\theta = \frac{1}{\sin\theta}$$
$$\csc 10° = \frac{1}{\sin 10°}$$
$$\csc 10° \doteq \frac{1}{0.174}$$
$$\csc 10° \doteq 5.76$$

**14.**
$$\sec\theta = \frac{1}{\cos\theta}$$
$$\sec 70° = \frac{1}{\cos 70°}$$
$$\sec 70° \doteq \frac{1}{0.342}$$
$$\sec 70° \doteq 2.92$$

**15.**
$$\cot\theta = \frac{1}{\tan\theta}$$
$$\cot 12° = \frac{1}{\tan 12°}$$
$$\cot 12° \doteq \frac{1}{0.213}$$
$$\cot 12° \doteq 4.70$$

 

**16.**

$$\csc\theta = \frac{1}{\sin\theta}$$
$$\csc 41° = \frac{1}{\sin 41°}$$
$$\csc 41° \doteq \frac{1}{0.656}$$
$$\csc 41° \doteq 1.52$$

**17.**

$$\sec\theta = 2.317$$
$$\cos\theta = \frac{1}{\sec\theta}$$
$$\cos\theta = \frac{1}{2.317}$$
$$\cos\theta \doteq 0.432$$
$$\theta \doteq \cos^{-1}(0.432)$$
$$\theta \doteq 64.4°$$

**18.**

$$\cot\theta = 0.456$$
$$\tan\theta = \frac{1}{\cot\theta}$$
$$\tan\theta = \frac{1}{0.456}$$
$$\tan\theta \doteq 2.19$$
$$\theta \doteq \tan^{-1}(2.19)$$
$$\theta \doteq 65.5$$

**19.**

$$\csc\theta = 2$$
$$\sin\theta = \frac{1}{\csc\theta}$$
$$\sin\theta = \frac{1}{2}$$
$$\sin\theta = 0.5$$
$$\theta = \sin^{-1}(0.5)$$
$$\theta = 30°$$

**20.**

$$\csc\theta = 1.693$$
$$\sin\theta = \frac{1}{\csc\theta}$$
$$\sin\theta \doteq \frac{1}{1.693}$$
$$\sin\theta \doteq 0.591$$
$$\theta \doteq \sin^{-1}(0.591)$$
$$\theta \doteq 36.2°$$

**21.**

$$\sec\theta = 7.011$$
$$\cos\theta = \frac{1}{\sec\theta}$$
$$\cos\theta = \frac{1}{7.011}$$
$$\cos\theta \doteq 0.143$$
$$\theta \doteq \cos^{-1}(0.143)$$
$$\theta \doteq 81.8°$$

**22.**

$$\cot\theta = 2.475$$
$$\tan\theta = \frac{1}{\cot\theta}$$
$$\tan\theta = \frac{1}{2.475}$$
$$\tan\theta \doteq 0.404$$
$$\theta \doteq \tan^{-1}(0.404)$$
$$\theta \doteq 22.0°$$

**23.** Use the 45-45-90 and the 30-60-90 triangles and:

$$\sin\theta = \frac{\text{opposite}}{\text{hypotenuse}}$$
$$\cos\theta = \frac{\text{adjacent}}{\text{hypotenuse}}$$
$$\tan\theta = \frac{\text{opposite}}{\text{adjacent}}$$

These triangles will give you the exact value for each of these trigonometric ratios

| | $\sin\theta$ | $\cos\theta$ | $\tan\theta$ |
|---|---|---|---|
| $\theta = 30°$ | $\frac{1}{2}$ | $\frac{\sqrt{3}}{2}$ | $\frac{1}{\sqrt{3}}$ |
| $\theta = 45°$ | $\frac{1}{\sqrt{2}} = \frac{\sqrt{2}}{2}$ | $\frac{1}{\sqrt{2}} = \frac{\sqrt{2}}{2}$ | 1 |
| $\theta = 60°$ | $\frac{\sqrt{3}}{2}$ | $\frac{1}{2}$ | $\sqrt{3}$ |

**24.**

$$\cot 45° = \frac{1}{\tan 45°}$$

(use the exact value for tan 45°)

$$\cot 45° = \frac{1}{1}$$
$$\cot 45° = 1$$

**25.**
$$\csc 30° = \frac{1}{\sin 30°}$$
$$\csc 30° = \frac{1}{\frac{1}{2}} = 2$$ (use the exact value for sin 30°)

**26.**
$$\cot 60° = \frac{1}{\tan 60°}$$ (use the exact value for tan 60°)
$$\cot 60° = \frac{1}{\sqrt{3}} = \frac{\sqrt{3}}{3}$$

**27.**
$$\sec 45° = \frac{1}{\cos 45°}$$
$$\sec 45° = \frac{1}{\frac{1}{\sqrt{2}}}$$ (use the exact value for cos 45°)
$$\sec 45° = \sqrt{2}$$

**28.**
$$\csc 45° = \frac{1}{\sin 45°}$$
$$\csc 45° = \frac{1}{\frac{1}{\sqrt{2}}}$$ (use the exact value for sin 45°)
$$\csc 45° = \sqrt{2}$$

**29.**
$$\sec 30° = \frac{1}{\cos 30°}$$
$$\sec 30° = \frac{1}{\frac{\sqrt{3}}{2}}$$ (use the exact value for cos 30°)
$$\sec 30° = 1 \div \frac{\sqrt{3}}{2}$$
$$\sec 30° = 1 \times \frac{2}{\sqrt{3}}$$
$$\sec 30° = \frac{2}{\sqrt{3}} = \frac{2\sqrt{3}}{3}$$

## *Lesson 2 Angles in Standard Position*

### CLASS EXERCISES ANSWERS AND SOLUTIONS

**1.** A negative angle means that the rotating arm rotates clockwise.

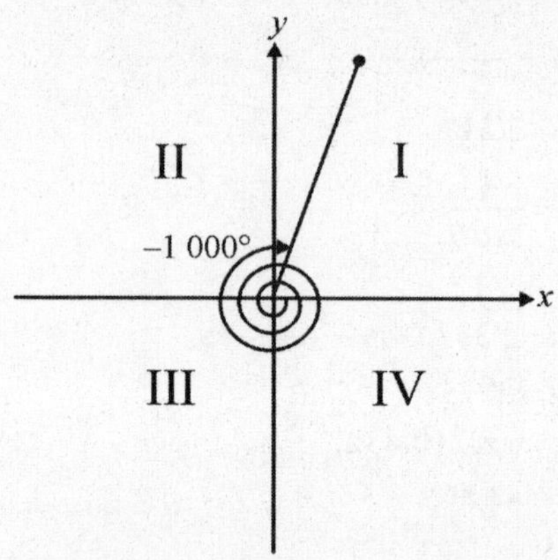


Rotating once around is equal to –360°. Rotating twice around is equal to –720°. After rotating twice around, we still have another 280° to go, which puts the terminal arm in quadrant I.

**2.** To draw an angle of 397°, we need to make one complete revolution (360°) and then rotate another 37°.

So, 37° is the reference angle of 397°.

**3.** We use $\sin\theta = \frac{y}{r}$.

From the coordinates of *P*, we know that $y = -3$, so we only need to determine the length of the radius, *r*.

Use the formula $r = \sqrt{x^2 + y^2}$.

The coordinates of *P* tell us that $x = -4$ and $y = -3$

Substituting into the formula gives:
$$r = \sqrt{(-4)^2 + (-3)^2}$$
$$= \sqrt{16+9}$$
$$= \sqrt{25}$$
$$= 5$$

So, $\sin\theta = \frac{y}{r} = \frac{-3}{5} = -\frac{3}{5}$.

4. Begin by sketching an angle with the given information.

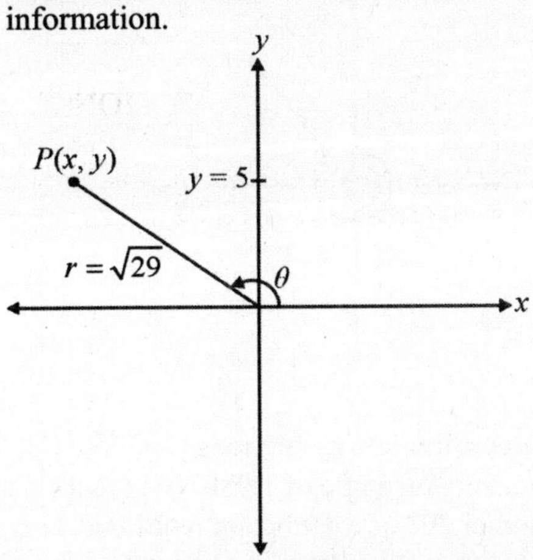


We have an angle terminating in the 2nd quadrant with $r = \sqrt{29}$ and $y = 5$.

The cotangent ratio is $\cot\theta = \frac{x}{y}$. We know the value of $y$, so all we need is $x$. How do we find $x$?

We know that $y = 5$ and that $r = \sqrt{29}$.

Substituting these values into $r = \sqrt{x^2 + y^2}$ gives:

$$\sqrt{29} = \sqrt{x^2 + 5^2}$$
$$\sqrt{29} = \sqrt{x^2 + 25}$$
$$29 = x^2 + 25 \quad \text{• Square both sides}$$
$$x^2 = 4$$
$$x = \pm 2$$

How do we know which value to use for $x$? $x = 2$ or $-2$?

Since $\theta$ is an angle in quadrant II, we know that $x$ must be negative since $x$ is negative for *all* points in quadrant II.

So, using $x = -2$, we get $\cot\theta = \frac{x}{y} = \frac{-2}{5} = -\frac{2}{5}$.

5. As seen in the diagram, the angle 570° terminates in quadrant III.

We need to use a relationship that involves the measure of the angle and the values for $y$ and $r$.

This is the sine ratio, $\sin\theta = \frac{y}{r}$.

So, $\sin 570° = \frac{-2}{r}$.

Analyzing the angle, it can be seen that the reference angle is 30° because $570° = 360° + 180° + 30°$.

Since the sine ratio is negative in quadrant III, $\sin 570° = -\sin 30° = -0.5$.

Thus,

$$-0.5 = \frac{-2}{r}$$
$$-0.5r = -2$$
$$r = \frac{-2}{-0.5}$$
$$r = 4$$

## PRACTICE EXERCISES ANSWERS AND SOLUTIONS

1.

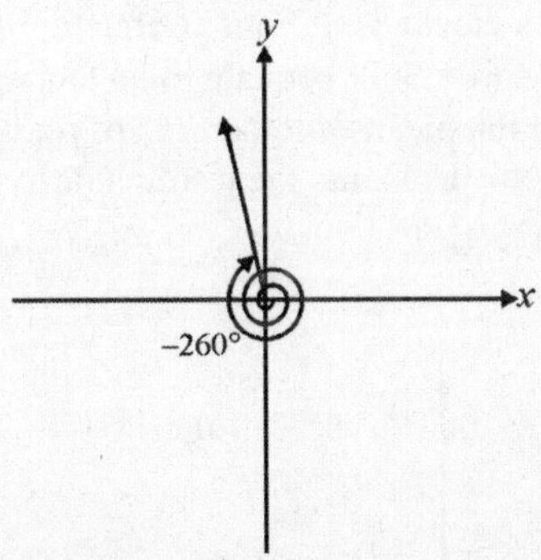


A negative angle means that the rotating arm will rotate in a clockwise direction. The angle is between −180° and −270°. This puts the terminal arm in quadrant II.

**2.**

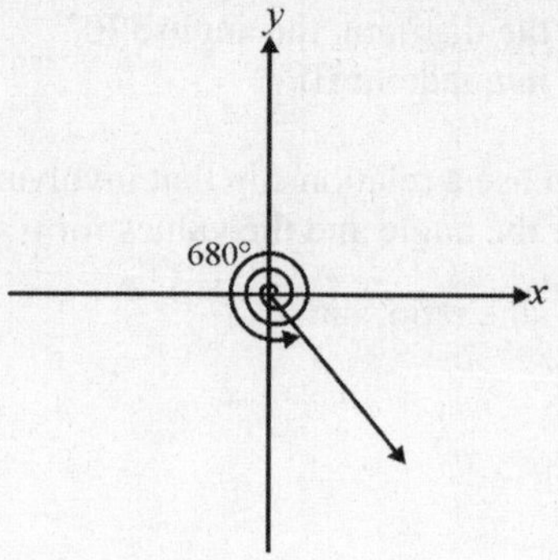


One complete revolution equates to an angle of 360°, which means another 320° of rotation are required. This puts the terminal arm in quadrant IV.

**3.**

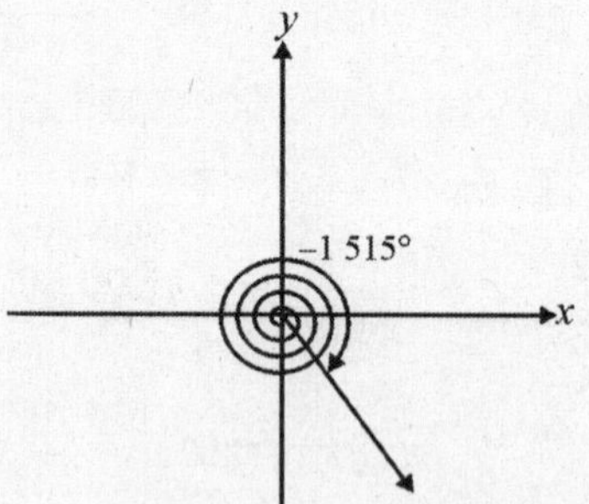


A negative angle means that the rotating arm rotates clockwise. Four complete revolutions (4 × 360°) equate to an angle of 1 440°, which means another 75° of rotation are required. This puts the terminal arm in quadrant IV.

**4.**

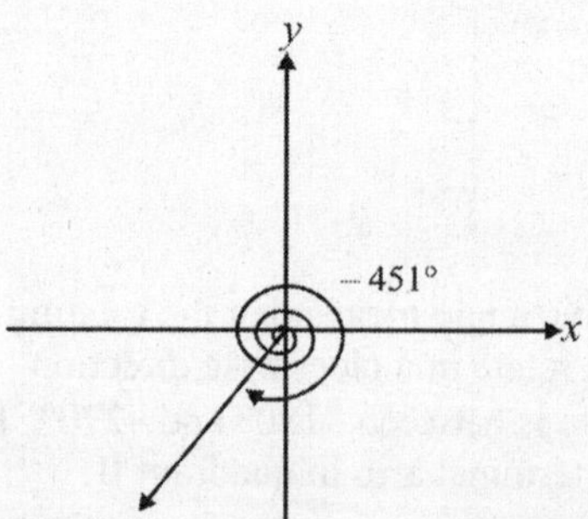


One complete revolution equates to an angle of 360°, which means another 91° of rotation are required in a clockwise direction. This puts the terminal arm in quadrant III.

**5.**

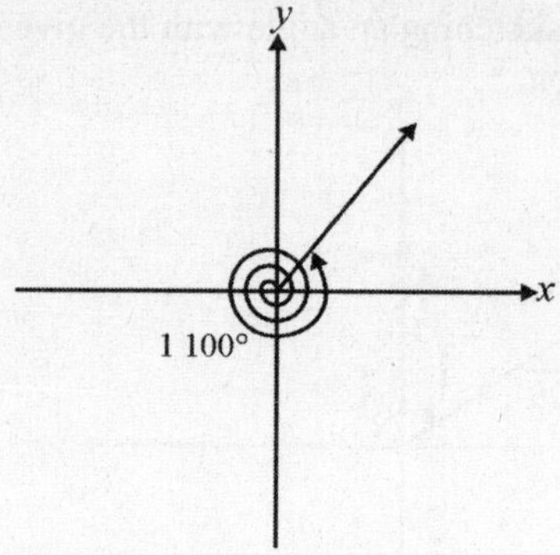


Three complete revolutions (3 × 360°) equate to an angle of 1 080°, which means another 20° of rotation are required. This puts the terminal arm in quadrant I.

**6.**

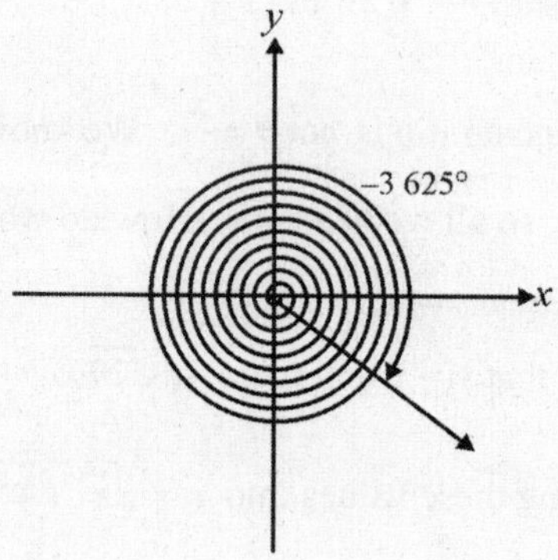


Ten revolutions (10 × 360°) equate to an angle of 3 600°, which means another 25° of rotation are required in a clockwise direction. This puts the terminal arm in quadrant IV.

**7.**

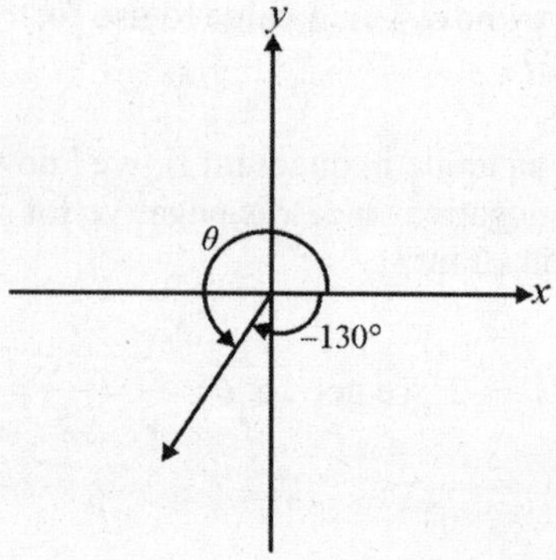


Let $\theta$ represent the principle angle.

$\theta = -130° + 360°$

$\theta = 230°$

 

**8.**

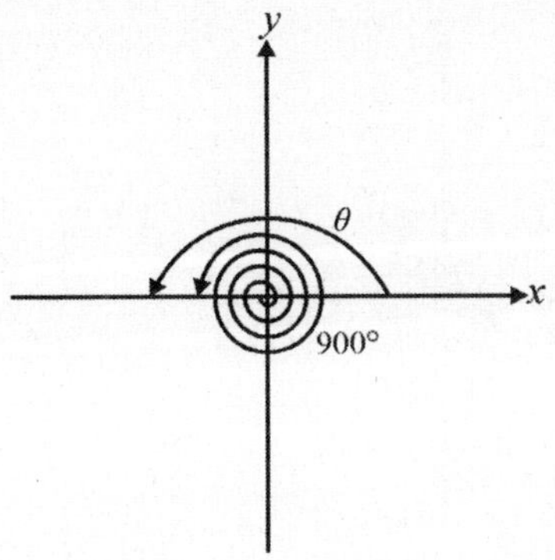


Let $\theta$ represent the principle angle.

Two complete revolutions ($2 \times 360°$) equate to an angle of 720°, which means another 180° of rotation are required. Thus, the principle angle is 180°.

**9.**

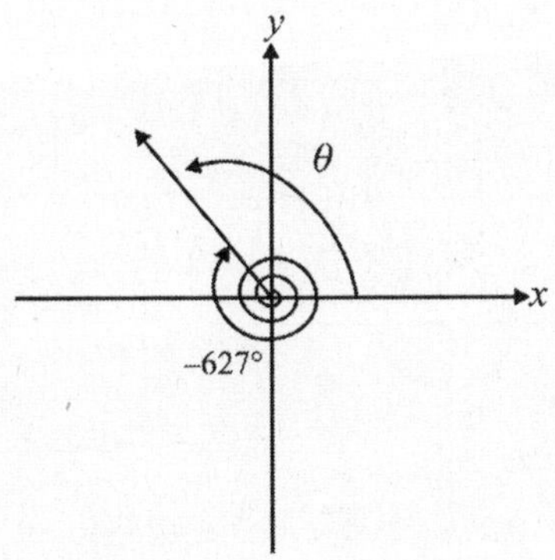


Let $\theta$ represent the principle angle. (NOTE: a negative angle means that the rotating arm rotates clockwise.)

One complete revolution equates to an angle of 360°, which means another 267° of rotation in a clockwise direction are required.

Since the angle is negative,

$\theta = -267° + 360°$

$\theta = 93°$

**10.**

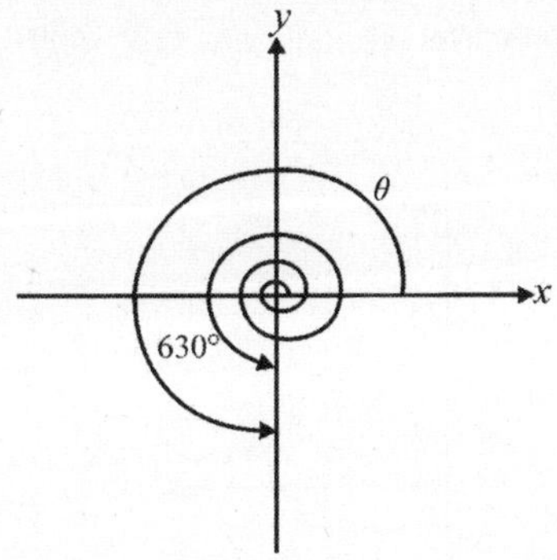


Let $\theta$ represent the principle angle. One complete revolution equates to an angle of 360°, which means another 270° of rotation are required.

Thus, the principle angle is 270°.

**11.**

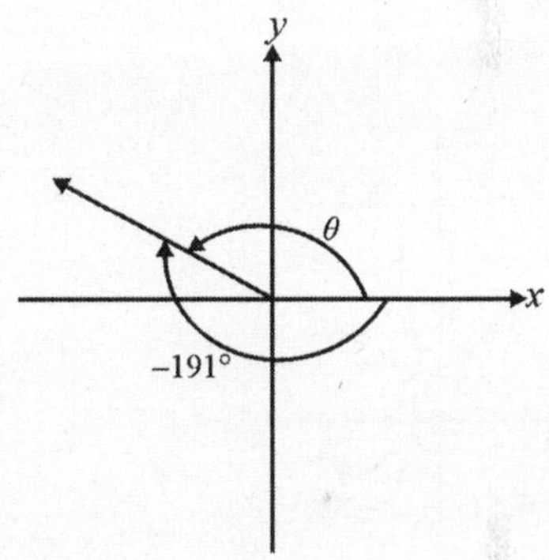


Let $\theta$ represent the principle angle

$\theta = -191° + 360°$

$\theta = 169°$

**12.**

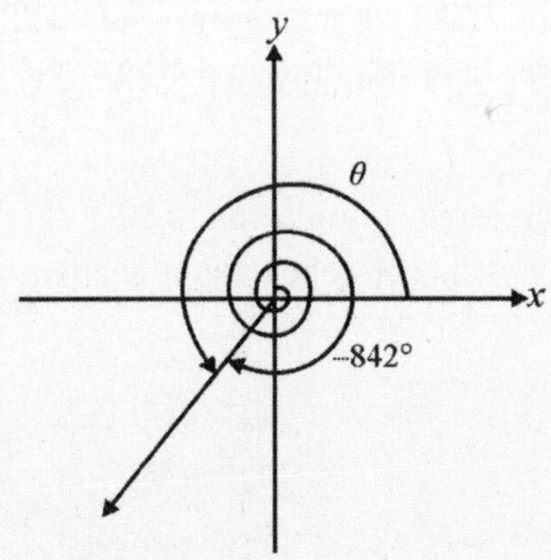


Let $\theta$ represent the principle angle. Two complete revolutions ($2 \times 360°$) equate to an angle of 720°, which means another 122° of rotation in a clockwise direction are required.

$\theta = -122° + 360°$

$\theta = 238°$

**13.** Angles coterminal with 13° are $(13+360n)°$, where n∈$I$.

**14.**

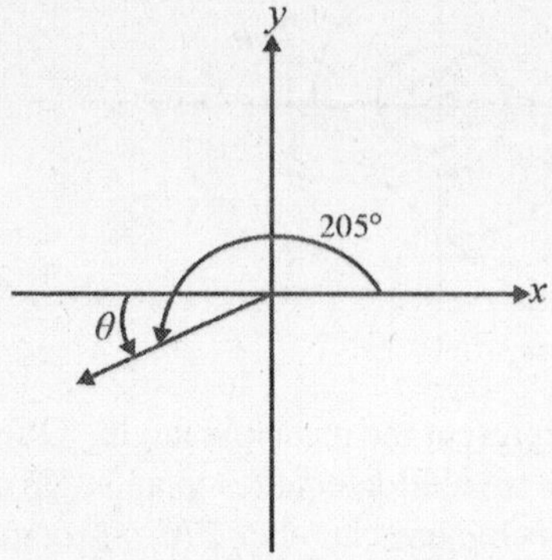


Let $\theta$ represent the reference angle. Since the $x$-axis represents an angle of 180°,

$\theta + 180° = 205°$

$\theta = 25°$

**15.**

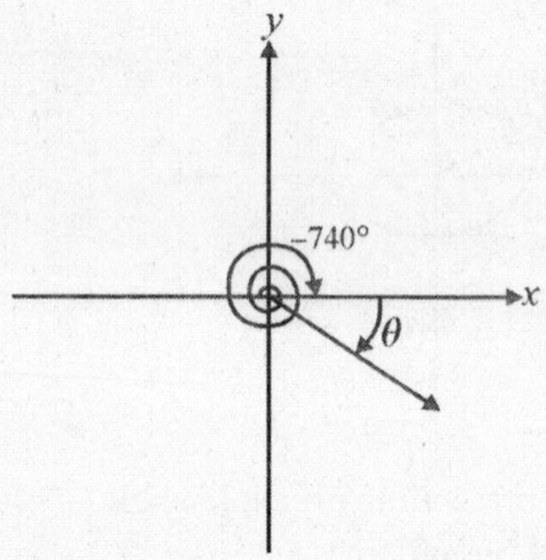


Let $\theta$ represent the reference angle. Two complete revolutions (2 × 360°) equate to an angle of 720°, which means another 20° of revolution are required in a clockwise direction.

Thus, the reference angle, $\theta$, is 20°. (Remember: the reference angle is always positive.)

**16.**

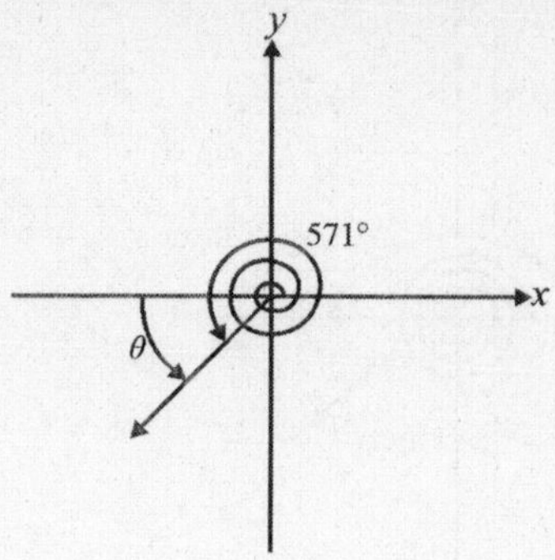


Let $\theta$ represent the reference angle. One complete revolution equates to an angle of 360°, which means another 211° of revolution are required.

Since the $x$-axis represents an angle of 180°,

$180° + \theta = 211°$

$\theta = 31°$

**17.**

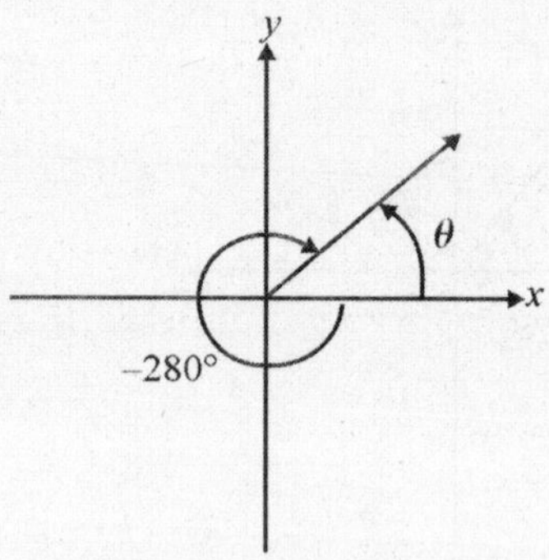


Let $\theta$ represent the reference angle. Since the $x$-axis represents a 360° angle.

$\theta = -280° + 360°$

$\theta = 80°$

**18.**

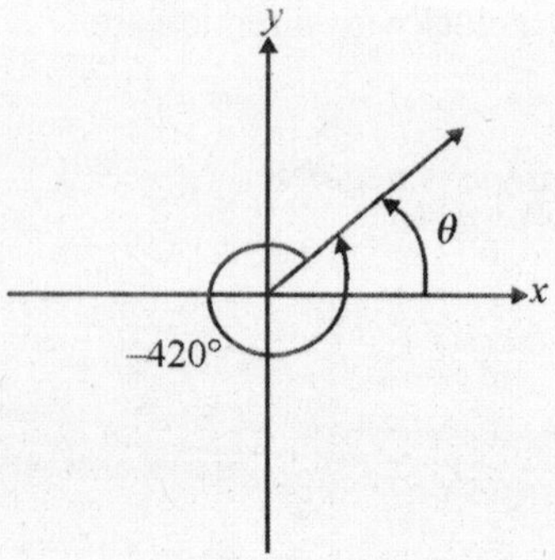


Let $\theta$ represent the reference angle. One complete revolution equates to an angle of 360°, which means another 60° of rotation in a clockwise direction are required.

$\theta = 60°$

 

**19.**

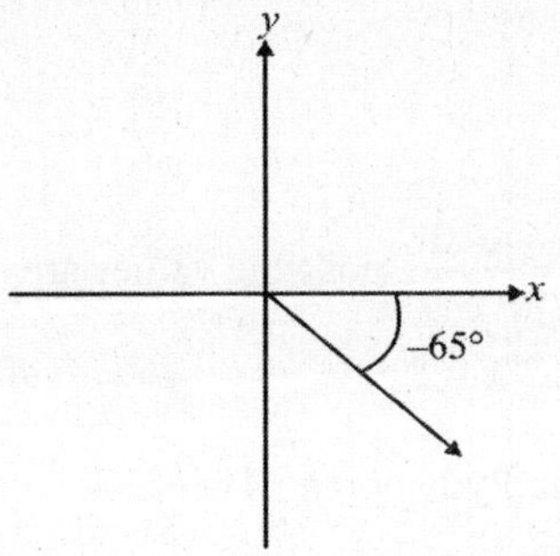


The reference angle is 65°. (Remember the reference angle is always positive).

**20.**

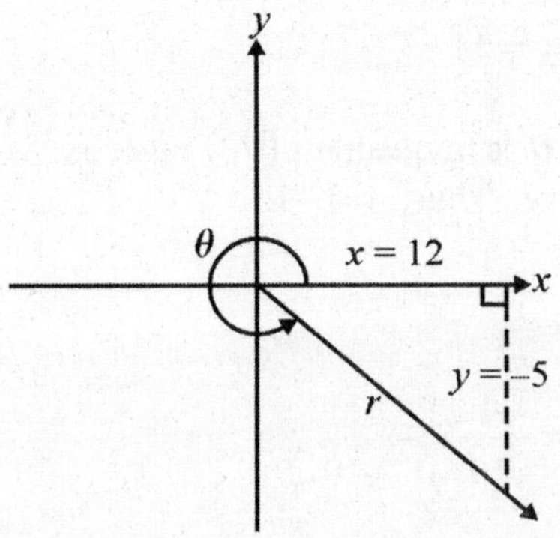


Applying the Pythagorean Theorem.

$r=\sqrt{x^2+y^2}$

$r=\sqrt{12^2+(-5)^2}$

$r=\sqrt{144+25}$

$r=\sqrt{169}$

$r=13$

(Remember, $r$ is always positive.)

$\sin\theta=\frac{y}{r}$ $\quad\cos\theta=\frac{x}{r}$ $\quad\tan\theta=\frac{y}{x}$

$\sin\theta=\frac{-5}{13}=-\frac{5}{13}$ $\quad\cos\theta=\frac{12}{13}$ $\quad\tan\theta=\frac{-5}{12}=-\frac{5}{12}$

$\csc\theta=\frac{r}{y}$ $\quad\sec\theta=\frac{r}{x}$ $\quad\cot\theta=\frac{x}{y}$

$\csc\theta=\frac{13}{-5}=-\frac{13}{5}$ $\quad\sec\theta=\frac{13}{12}$ $\quad\cot\theta=\frac{12}{-5}=-\frac{12}{5}$

**21.**

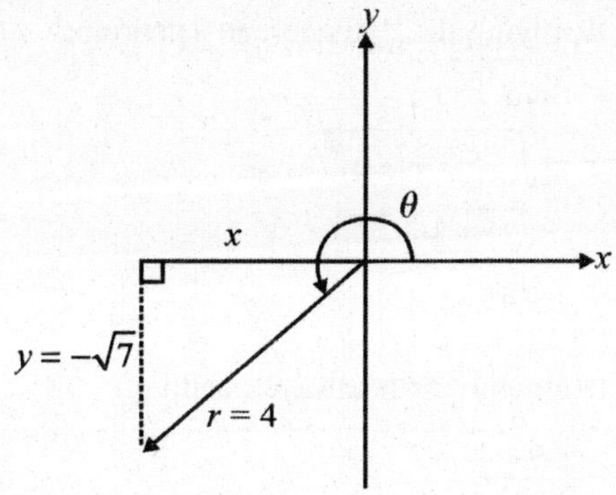


$\sin\theta=\frac{y}{r}$

Since $\sin\theta=\frac{-\sqrt{7}}{4}$, possible values are $y=-\sqrt{7}$ and $r=4$.

Applying the Pythagorean Theorem:

$x^2+y^2=r^2$

$x^2+\left(-\sqrt{7}\right)=4^2$

$x^2+7=16$

$x^2=9$

$x=\pm 3$

Because $\theta$ is in quadrant III, $x$ must be negative. Thus, $x=-3$.

$\sec\theta=\frac{r}{x}$

$\sec\theta=\frac{4}{-3}=-\frac{4}{3}$

**22.** Angles coterminal with –65° are $(-65+360n)°$, where $n\in I$.

**23.**

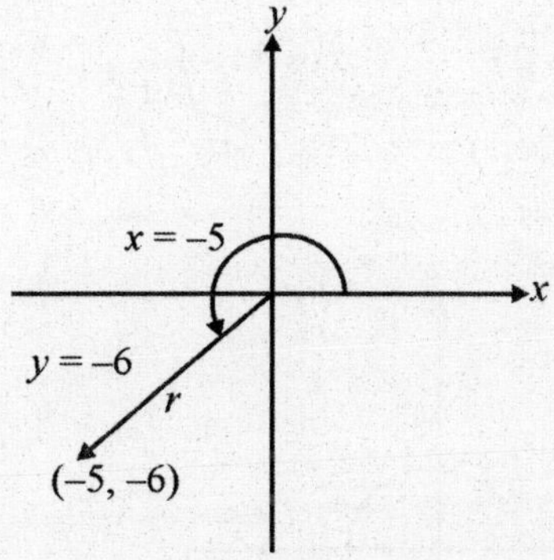

Applying the Pythagorean Theorem:

$r = \sqrt{x^2 + y^2}$

$r = \sqrt{(-5)^2 + (-6)^2}$

$r = \sqrt{25 + 36}$

$r = \sqrt{61}$

(Remember $r$ is always positive.)

$\sec\theta = \dfrac{r}{x}$

$\sec\theta = \dfrac{\sqrt{61}}{-5} = -\dfrac{\sqrt{61}}{5}$

**24.**

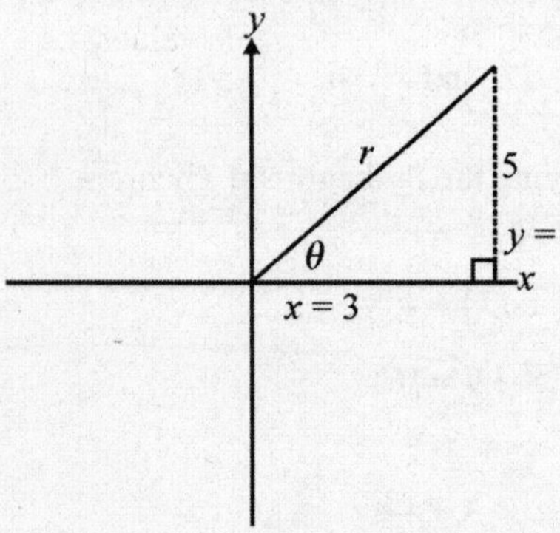


$\cot\theta = \dfrac{x}{y}$

Since $\cot\theta = \dfrac{3}{5}$, possible values are $x = 3$ and $y = 5$.

Applying the Pythagorean Theorem:

$r = \sqrt{x^2 + y^2}$

$r = \sqrt{(3)^2 + (5)^2}$

$r = \sqrt{9 + 25}$

$r = \sqrt{34}$

$\cos\theta = \dfrac{x}{r}$

$\cos\theta = \dfrac{3}{\sqrt{34}}$

**25.**

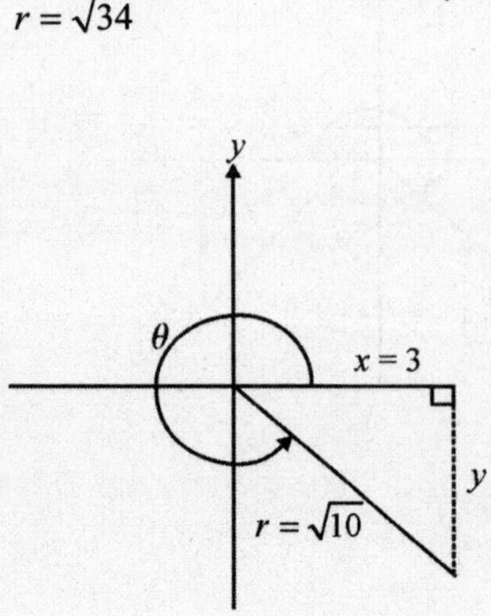


$\sec\theta = \dfrac{r}{x}$

Since $\sec\theta = \dfrac{\sqrt{10}}{3}$, possible values are $r = \sqrt{10}$ and $x = 3$.

Applying the Pythagorean Theorem:

$x^2 + y^2 = r^2$

$3^2 + y^2 = \left(\sqrt{10}\right)^2$

$9 + y^2 = 10$

$y^2 + 1$

$y = \pm 1$

Since $\theta$ is in quadrant IV, $y$ must be negative. Thus, $y = -1$.

$\sin\theta = \dfrac{y}{r}$

$\sin\theta = \dfrac{-1}{\sqrt{10}} = -\dfrac{1}{\sqrt{10}}$

**26.** Angles coterminal with 321° are $(321 + 360n)°$, $n \in I$.

**27.**

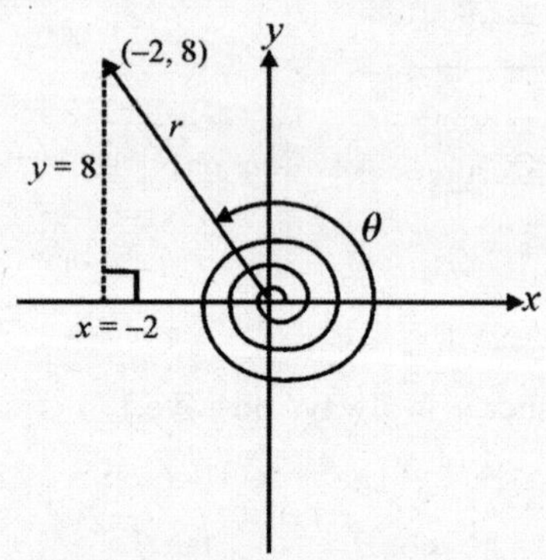


Applying the Pythagorean Theorem:

$r = \sqrt{x^2 + y^2}$

$r = \sqrt{(-2)^2 + 8^2}$

$r = \sqrt{4 + 64}$

$r = \sqrt{68}$

$r = 2\sqrt{17}$

(Remember, $r$ is always positive.)

 

$\sin\theta = \dfrac{y}{r}$

$\sin\theta = \dfrac{8}{2\sqrt{17}}$

$= \dfrac{4}{\sqrt{17}} = \dfrac{4\sqrt{17}}{17}$

$\cos\theta = \dfrac{x}{r}$

$\cos\theta = \dfrac{-2}{2\sqrt{17}} = \dfrac{-1}{\sqrt{17}}$

$= -\dfrac{\sqrt{17}}{17}$

$\tan\theta = \dfrac{y}{x}$

$\tan\theta = \dfrac{8}{-2}$

$= -4$

$\csc\theta = \dfrac{r}{y}$

$\csc\theta = \dfrac{2\sqrt{17}}{8}$

$= \dfrac{\sqrt{17}}{4}$

$\sec\theta = \dfrac{r}{x}$

$\sec\theta = \dfrac{2\sqrt{17}}{-2}$

$\sec\theta = -\sqrt{17}$

$\cot\theta = \dfrac{x}{y}$

$\cot\theta = \dfrac{-2}{8}$

$= -\dfrac{1}{4}$

**28.**

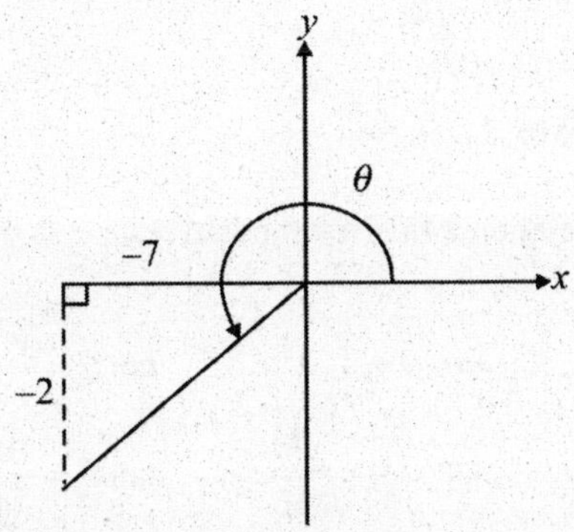


$\tan\theta = \dfrac{y}{x}$

Since $\tan\theta = \dfrac{2}{7}$, possible values are $y = -2$ and $x = -7$.

(Remember, in quadrant III, $x$ and $y$ are negative.)

Applying the Pythagorean Theorem:

$r = \sqrt{x^2 + y^2}$

$r = \sqrt{(-7)^2 + (-2)^2}$

$r = \sqrt{49 + 4}$

$r = \sqrt{53}$

$\sin\theta = \dfrac{y}{r}$

$\sin\theta = \dfrac{-2}{\sqrt{53}} = -\dfrac{2\sqrt{53}}{53}$

**29.**

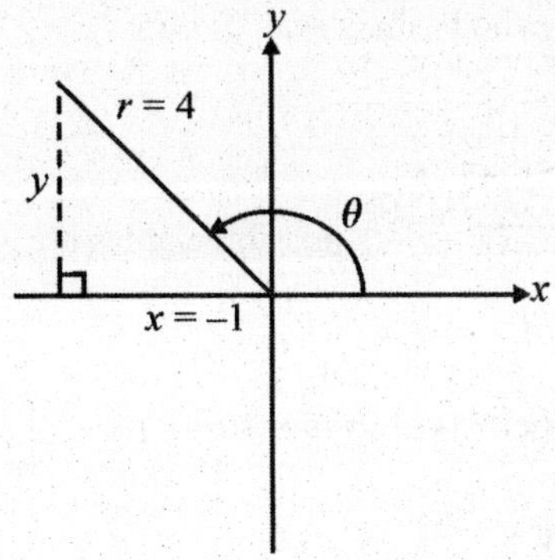


$\cos\theta = \dfrac{x}{r}$

Since $\cos\theta = -\dfrac{1}{4}$, possible values are $x = -1$ and $r = 4$.

Applying the Pythagorean Theorem:

$x^2 + y^2 = r^2$

$(-1)^2 + y^2 = 4^2$

$1 + y^2 = 16$

$y^2 = 15$

$y = \pm\sqrt{15}$

Since $\theta$ is in quadrant II, $y$ must be positive. Thus, $y = \sqrt{15}$.

$\csc\theta = \dfrac{r}{y}$

$\csc\theta = \dfrac{4}{\sqrt{15}} = \dfrac{4\sqrt{15}}{15}$

**30.**

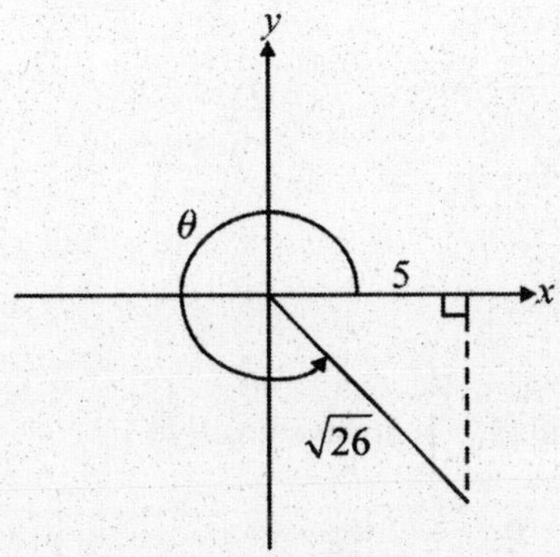

$\sec\theta = \dfrac{r}{x}$

Since $\sec\theta = \dfrac{\sqrt{26}}{5}$, possible values are $x = 5$ and $r = \sqrt{26}$.

$$x^2 + y^2 = r^2$$
$$5^2 + y^2 = \left(\sqrt{26}\right)^2$$
$$25 + y^2 = 26$$
$$y^2 = 1$$
$$y \pm 1$$

Since $\theta$ is in quadrant IV, $y$ must be negative. Thus, $y = -1$.

$$\tan\theta = \frac{y}{x}$$
$$\tan\theta = \frac{-1}{5} = -\frac{1}{5}$$

**31.**

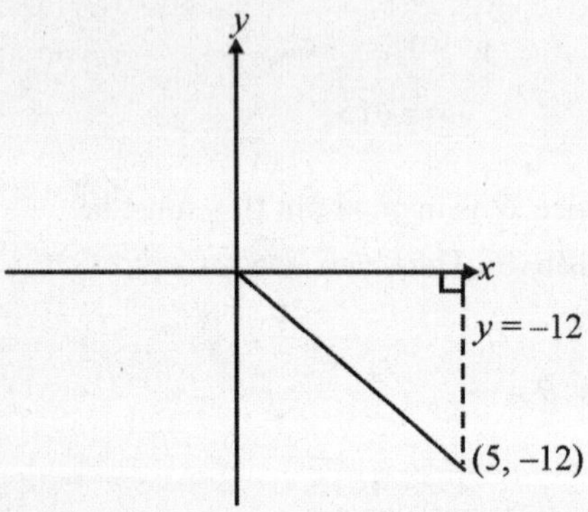


Applying the Pythagorean Theorem:

$$r = \sqrt{x^2 + y^2}$$
$$r = \sqrt{5^2 + (-12)^2}$$
$$r = \sqrt{25 + 144}$$
$$r = \sqrt{169}$$
$$r = 13$$

(Remember, $r$ is always positive.)

$\sin\theta = \dfrac{y}{r}$ $\qquad$ $\cos\theta = \dfrac{x}{r}$ $\qquad$ $\tan\theta = \dfrac{y}{x}$

$\sin\theta = \dfrac{-12}{13}$ $\qquad$ $\cos\theta = \dfrac{5}{13}$ $\qquad$ $\tan\theta = \dfrac{-12}{5}$

**32.**

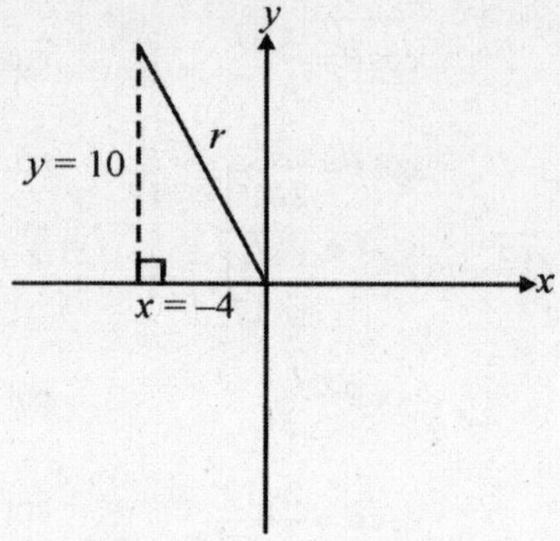


Applying the Pythagorean Theorem:

$$r = \sqrt{x^2 + y^2}$$
$$r = \sqrt{(-4)^2 + 10^2}$$
$$r = \sqrt{16 + 100}$$
$$r = \sqrt{116}$$

(Remember, $r$ is always positive.)

$\sin\theta = \dfrac{y}{r}$ $\qquad$ $\cos\theta = \dfrac{x}{r}$ $\qquad$ $\tan\theta = \dfrac{y}{x}$

$\sin\theta = \dfrac{10}{\sqrt{116}}$ $\qquad$ $\cos\theta = \dfrac{-4}{\sqrt{116}}$ $\qquad$ $\tan\theta = \dfrac{10}{-4}$

$= \dfrac{10}{2\sqrt{29}} = \dfrac{5}{29}$ $\qquad$ $= \dfrac{-4}{2\sqrt{29}} = \dfrac{-2}{\sqrt{29}}$ $\qquad$ $= -\dfrac{5}{2}$

$= \dfrac{5\sqrt{29}}{29}$ $\qquad$ $= \dfrac{-2\sqrt{29}}{29}$

**33.**

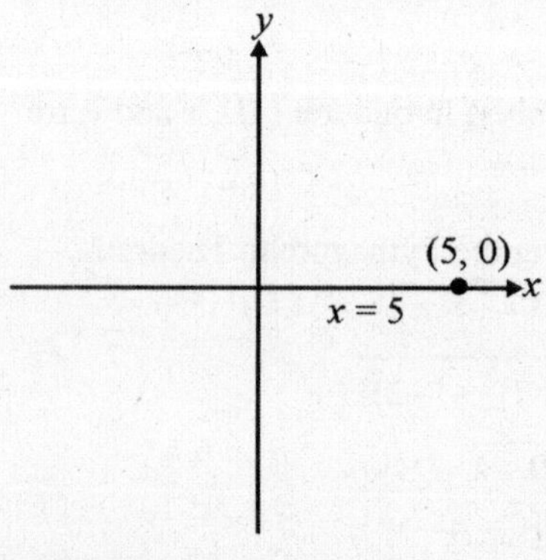


Applying the Pythagorean Theorem:

$$r = \sqrt{x^2 + y^2}$$
$$r = \sqrt{5^2 + 0^2}$$
$$r = \sqrt{25 + 0}$$
$$r = 5$$

(Remember, $r$ is always positive.)

$\sin\theta = \frac{y}{r}$ $\quad$ $\cos\theta = \frac{x}{r}$ $\quad$ $\tan\theta = \frac{y}{x}$

$\sin\theta = \frac{0}{5}$ $\quad$ $\cos\theta = \frac{5}{5}$ $\quad$ $\tan\theta = \frac{0}{5}$

$\sin\theta = 0$ $\quad$ $\cos\theta = 1$ $\quad$ $\tan\theta = 0$

**34. C**

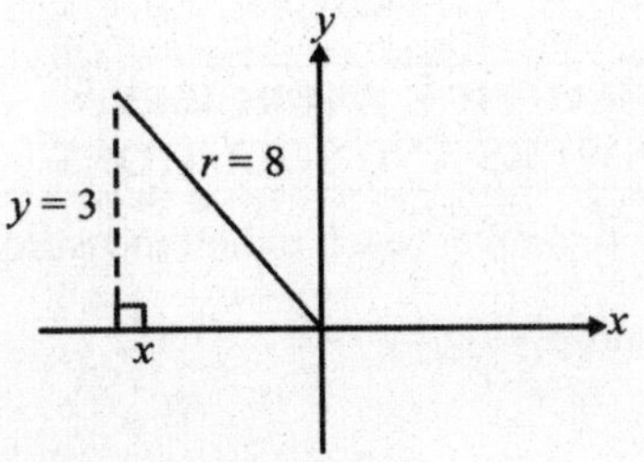


$\theta$ is in quadrant II.

$\sin\theta = \frac{y}{r}$

Since $\sin\theta = \frac{3}{8}$, possible values are $y = 3$ and $r = 8$.

Applying the Pythagorean Theorem:

$$x^2 + y^2 = r^2$$
$$x^2 + 3^2 = 8^2$$
$$x^2 + 9 = 64$$
$$x^2 = 55$$
$$x = \pm\sqrt{55}$$

Since $\theta$ is in quadrant II, the value $x$ is negative. Thus, $x = -\sqrt{55}$.

$$\sec\theta + \tan\theta = \frac{r}{x} + \frac{y}{x}$$

$$\sec\theta + \tan\theta = \frac{8}{-\sqrt{55}} + \frac{3}{-\sqrt{55}}$$

$$\sec\theta + \tan\theta = \frac{-11}{\sqrt{55}} = \frac{-11\sqrt{55}}{55} = \frac{-\sqrt{55}}{5}$$

**35.** $\theta$ is in quadrant III.

If $\sec\theta = \csc\theta$, then $\frac{x}{r} = \frac{y}{r}$. Thus, $x = y$.

If $x = y$, then the reference angle must be 45°.

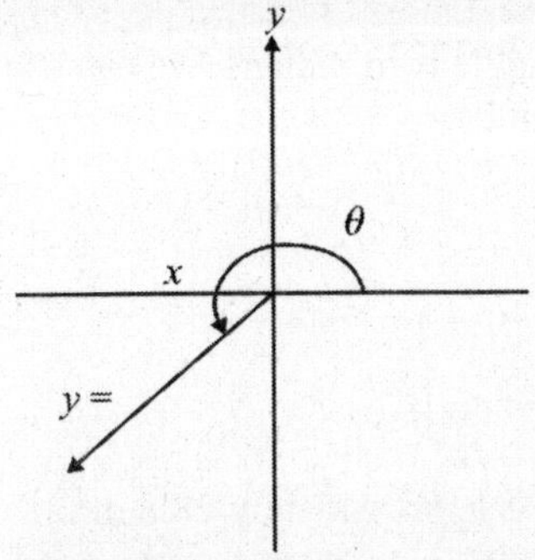


The value of $\theta$ (the principle angle):

$\theta = 180° + 45°$

$\theta = 225°$

| 2 | 2 | 5 | |
|---|---|---|---|

## *Lesson 3 Radian Measure*

### CLASS EXERCISES ANSWERS AND SOLUTIONS

**1.** $1 \text{ radian} = 1 \times \frac{180}{\pi} \doteq 57.3°$

So, 1 radian is approximately 57.3°.

**2.** $\frac{13\pi}{12} \text{ rad} = \frac{13\pi}{12} \times \frac{180}{\pi} = 195°$

**3.** To convert to radians, multiply 135 by $\frac{\pi}{180}$.

$$135° = 135 \times \frac{\pi}{180} \text{ rad}$$
$$= \frac{135\pi}{180} \text{ rad}$$
$$= \frac{3\pi}{4} \text{ rad}$$

**4.** To convert to radians we multiply 89 by $\frac{\pi}{180}$.

$$89° = 89 \times \frac{\pi}{180} \text{ rad} = \frac{89\pi}{180} \text{ rad}$$

The exact value is $\frac{89\pi}{180}$. When converted to a decimal, this is approximately 1.55.

$$89° = \frac{89\pi}{180} \text{ rad} \doteq 1.55 \text{ rad}$$

**5.** Since the angle is given in degrees, we must first convert the angle to radians:

$150° = 150 \times \frac{\pi}{180} = \frac{150\pi}{180} = \frac{5\pi}{6}$ rad

Now that our angle is in radians, we can use the arc length formula:

$$a = r\theta$$
$$= (11)\left(\frac{5\pi}{6}\right)$$
$$\doteq 28.8$$

The arc length is approximately 28.8 mm.

**6.** Once again, we will use the formula: $a = r\theta$

This time we know that $r = 18.5$ and $a = 88.8$.

Substitute these values into the formula:
$88.8 = (18.5)\ \theta$
$88.8 \div 18.5 = \theta$
$4.8 \doteq \theta$

Therefore, $\theta \doteq 4.8$ rad.

**7.** Here, we are subtracting, $1\pi - \frac{3}{10}\pi$. You can think of this expression as $1 - \frac{3}{10}$ and then add the $\pi$ once you have performed the necessary calculations.

To do the subtraction, $1 - \frac{3}{10}$, we will rewrite 1 as $\frac{10}{10}$ to get:

$$\frac{10}{10} - \frac{3}{10} = \frac{7}{10}$$

Now, put the $\pi$ back and get $\frac{7}{10}\pi$ or $\frac{7\pi}{10}$.

**8.** We can temporarily remove the $\pi$s to get: $\frac{2}{3} + \frac{1}{6}$

Rewrite with a common denominator of 6 and add: $\frac{4}{6} + \frac{1}{6} = \frac{5}{6}$

Replace the $\pi$ to get $\frac{5}{6}\pi$ or $\frac{5\pi}{6}$.

**9.** Remove the $\pi$s to get: $2 - \frac{7}{6}$

Rewrite with common denominators:
$\frac{12}{6} - \frac{7}{6} = \frac{5}{6}$

Replace the $\pi$ to get $\frac{5}{6}\pi$ or $\frac{5\pi}{6}$.

## PRACTICE EXERCISES ANSWERS AND SOLUTIONS

For questions 1–6, convert each radian measure to degrees by multiplying by $\frac{180}{\pi}$.

**1.** $\frac{2\pi}{3} \times \frac{180}{\pi} = 120°$

**2.** $2.7 \times \frac{180}{\pi} \doteq 154.7°$

**3.** $\frac{17\pi}{12} \times \frac{180}{\pi} = 255°$

**4.** $0.85 \times \frac{180}{\pi} \doteq 48.7°$

**5.** $\frac{7\pi}{4} \times \frac{180}{\pi} = 315°$

**6.** $\frac{\pi}{8} \times \frac{180}{\pi} = 22.5°$

**7.** $3\pi \times \frac{180}{\pi} = 540.0°$

**8.** $5.6 \times \frac{180}{\pi} \doteq 320.9°$

**9.** $\frac{11\pi}{3} \times \frac{180}{\pi} = 660.0°$

 

**10.** $4.38\times\frac{180}{\pi}\doteq 251.0°$

**11.** $\frac{9\pi}{2}\times\frac{180}{\pi}=810.0°$

**12.** $\frac{\pi}{12}\times\frac{180}{\pi}=15.0°$

For questions 13–24, convert each degree measure to radians by multiplying by $\frac{\pi}{180}$.

**13.** $25\times\frac{\pi}{180}=\frac{25\pi}{180}\text{ rad}=\frac{5\pi}{36}\text{ rad}$

**14.** $720\times\frac{\pi}{180}=4\pi\text{ rad}$

**15.** $53\times\frac{\pi}{180}=\frac{53\pi}{180}\text{ rad}$

**16.** $280\times\frac{\pi}{180}=\frac{14\pi}{9}\text{ rad}$

**17.** $1\times\frac{\pi}{180}=\frac{\pi}{180}\text{ rad}$

**18.** $25\times\frac{\pi}{180}=\frac{5\pi}{36}\text{ rad}$

**19.** $40\times\frac{\pi}{180}=\frac{40\pi}{180}=\frac{2\pi}{9}\text{ rad}$

**20.** $110\times\frac{\pi}{180}=\frac{11\pi}{18}\text{ rad}$

**21.** $203\times\frac{\pi}{180}=\frac{203\pi}{180}\text{ rad}$

**22.** $420\times\frac{\pi}{180}=\frac{7\pi}{3}\text{ rad}$

**23.** $90\times\frac{\pi}{180}=\frac{90\pi}{180}=\frac{\pi}{2}\text{ rad}$

**24.** $0\times\frac{\pi}{180}=0\text{ rad}$

**25.**
$$a=r\theta$$
$$16.7=r(5.4)$$
$$3.1\doteq r$$
Thus, the radius is approximately 3.1 cm.

**26.**
$$a=r\theta$$
$$90=30\theta$$
$$3=\theta$$
Thus, the measure of the angle is 3 rad.

**27.**
$$a=r\theta$$
$$a=(19)\left(\frac{\pi}{6}\right)$$
$$a=\frac{19\pi}{6}\text{ cm}\doteq 9.9\text{ cm}$$
(Using a decimal approximation for $\pi$)

**28.** First, convert 37° to rad:

$$37\times\frac{\pi}{180}=\frac{37\pi}{180}\text{rad}$$
$$a=r\theta$$
$$24=r\left(\frac{37\pi}{180}\right)$$
$$r=\frac{24}{\frac{37\pi}{180}}$$
$$r\doteq 37.2\text{ mm}$$

(Using a decimal approximation for $\pi$)

**29.** First convert 160° to rad:

$$160\times\frac{\pi}{180}=\frac{160\pi}{180}=\frac{8\pi}{9}\text{ rad}$$

$$a=r\theta$$

$$a=(80)\left(\frac{8\pi}{9}\right)$$

$$a=\frac{690\pi}{9}\doteq 223.4\text{ cm}$$

(Using decimal approximation for $\pi$)

**30.**

$$a=r\theta$$

$$11=24.5\theta$$

$$0.4\text{ rad}\doteq\theta$$

**31.**

$$a=r\theta$$

$$40.1=r\left(\frac{3\pi}{4}\right)$$

$$\frac{40.1}{\frac{3\pi}{4}}=r$$

$$17.0\text{ m}\doteq r$$

**32.**

$$a=r\theta$$

$$24=r(2.05)$$

$$11.7\text{ mm}\doteq r$$

**33.** Since the length of the diameter is twice the radius, 8 times the diameter is 16 times the radius.

Thus, the arc length is 16 times the length of the radius.

$$a=r\theta$$

$$16r=r\theta$$

$$\frac{16r}{r}=\theta$$

$$16\text{ rad}=\theta$$

Now, convert the radian measure to degrees.

$$16\times\frac{180}{\pi}\doteq 916.7°$$

**34.** $\pi-\frac{3\pi}{4}=\frac{4\pi}{4}-\frac{3\pi}{4}=\frac{\pi}{4}$

**35.** $2\pi-\frac{5\pi}{6}=\frac{12\pi}{6}-\frac{5\pi}{6}=\frac{7\pi}{6}$

**36.** $\frac{2\pi}{3}+\frac{\pi}{4}=\frac{8\pi}{12}+\frac{3\pi}{12}=\frac{11\pi}{12}$

**37.** $\pi+\frac{\pi}{7}=\frac{7\pi}{7}\times\frac{\pi}{7}=\frac{8\pi}{7}$

**38.** $\frac{\pi}{4}+\frac{5\pi}{8}=\frac{2\pi}{8}+\frac{5\pi}{8}=\frac{7\pi}{8}$

**39.** $\frac{7\pi}{6}+\frac{3\pi}{4}=\frac{14\pi}{12}+\frac{9\pi}{12}=\frac{23\pi}{12}$

**40.** $\pi-\frac{2\pi}{5}=\frac{5\pi}{5}-\frac{2\pi}{5}=\frac{3\pi}{5}$

**41.** $2\pi-\frac{5\pi}{4}=\frac{8\pi}{4}-\frac{5\pi}{4}=\frac{3\pi}{4}$

**42.** $\frac{\pi}{3}+\frac{5\pi}{6}=\frac{2\pi}{6}+\frac{5\pi}{6}=\frac{7\pi}{6}$

**43.** $\pi+\frac{\pi}{9}=\frac{9\pi}{9}+\frac{\pi}{9}=\frac{10\pi}{9}$

**44.** $2\pi-\frac{3\pi}{10}=\frac{20\pi}{10}-\frac{3\pi}{10}=\frac{17\pi}{10}$

**45.** $\frac{4\pi}{3}+\frac{3\pi}{4}=\frac{16\pi}{12}+\frac{9\pi}{12}=\frac{25\pi}{12}$

**46. D**
The diameter of a circle is 2 times the radius. Since the diameter is 70 cm, then the radius is 35 cm.

$a = r\theta$
$a = 35 \times 16$
$a = 560$

The bicycle will travel 560 cm, or 5.6 m.

**47. B**
Since triangle $CAO$ is isosceles, then angle $ACO$ is 30° and angle $COA$ is $180 - 2(30) = 120°$.

$$120° = 120 \times \frac{\pi}{180} \text{rad}$$
$$= \frac{2\pi}{3} \text{rad}$$
$$a = r\theta$$
$$la = r\left(\frac{2\pi}{3}\right)$$
$$r = \frac{30}{2\pi} \doteq 4.8 \text{ cm}$$

## Lesson 4 Using Radians With Angles in Standard Position

### CLASS EXERCISES ANSWERS AND SOLUTIONS

1. The angle is negative, which means the terminal arm is rotating clockwise. In the last question, we saw that $\pi$ radians generates half a revolution. So, $\frac{\pi}{3}$ radians generates $\frac{1}{3}$ of half a revolution.

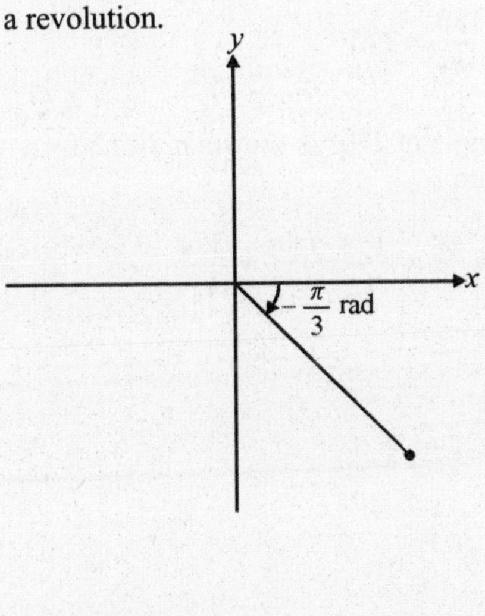


2. A positive angle means a counter-clockwise rotation. If $\pi$ radians generates half a revolution, then $\frac{5\pi}{4}$ radians generates more than half a revolution. In fact, we can rewrite $\frac{5\pi}{4}$ as $\pi + \frac{\pi}{4}$, which results in half a revolution, and then $\frac{\pi}{4}$ of half a revolution more.

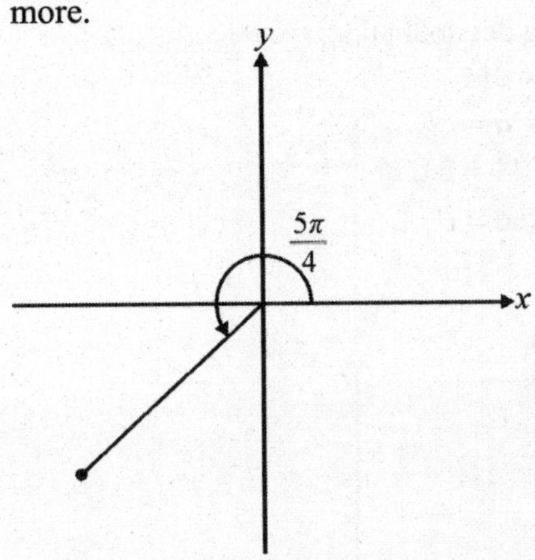


3. A negative angle means that the terminal arm rotates clockwise. If $2\pi$ rad equals one revolution then we can also say that $\frac{8\pi}{4}$ is one revolution. This means that $\frac{16\pi}{4}$ is two revolutions and $\frac{24\pi}{4}$ is three revolutions.

After rotating $\frac{24\pi}{4}$ radians (3 revolutions), another $\frac{3\pi}{4}$ radians is required.

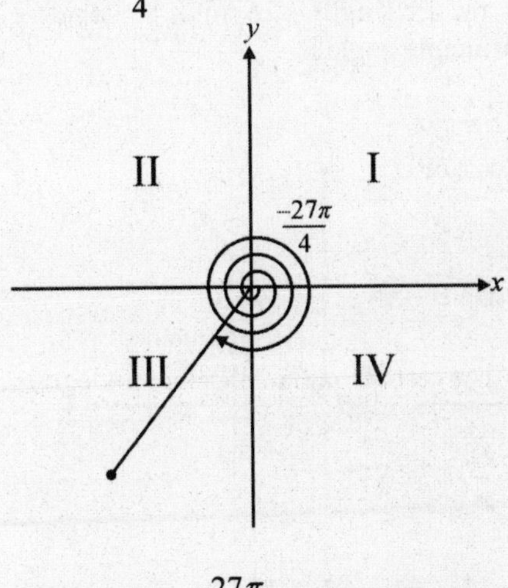


The angle of $\frac{-27\pi}{4}$ radians terminates in quadrant III.

**4.** For this question, it is likely easiest to convert 8 rad to degrees and then sketch the angle.

To convert to degrees, multiply 8 by $\frac{180}{\pi}$

$8\times\frac{180}{\pi} \doteq 458.4$

The angle is approximately 458°.

We can easily sketch 458° because it is 360° + 98°.

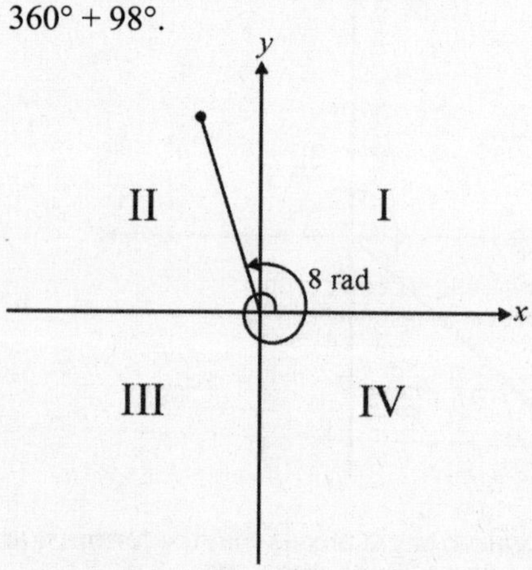


So, an angle of 8 radians terminates in quadrant II.

**5.** Since $2\pi = \frac{6\pi}{3}$, multiples of $2\pi$ are

$\frac{6\pi}{3}, \frac{12\pi}{3}, \frac{18\pi}{3}, \ldots$

The largest multiple of $2\pi$ that can be subtracted from $\frac{17\pi}{3}$ and leave a positive result is $\frac{12\pi}{3}$ so, the principle angle is $\frac{17\pi}{3} - \frac{12\pi}{3} = \frac{5\pi}{3}$.

**6.**

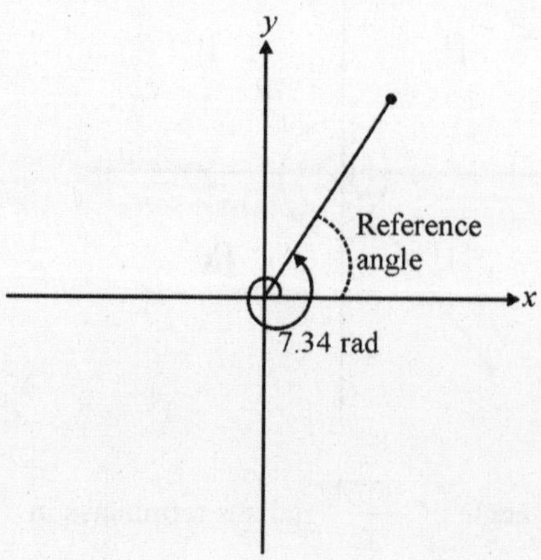


An angle of 7.34 radians consists of one complete revolution of $2\pi$ radians (approximately 6.28 rad) and then a bit more.

How much more?

To determine this, we subtract: $7.34 - 2\pi \doteq 7.34 - 6.28 = 1.06$ rad

So, an angle of approximately 1.06 radians is the reference angle for one of 7.34 radians.

## PRACTICE EXERCISES ANSWERS AND SOLUTIONS

**1.** Notice that $\frac{3\pi}{2} = \pi + \frac{\pi}{2}$.

Recall that an angle of $\pi$ radians generates half of a complete revolution, so an angle of $\frac{3\pi}{2}$ rad generates half of a revolution plus a quarter of a revolution.

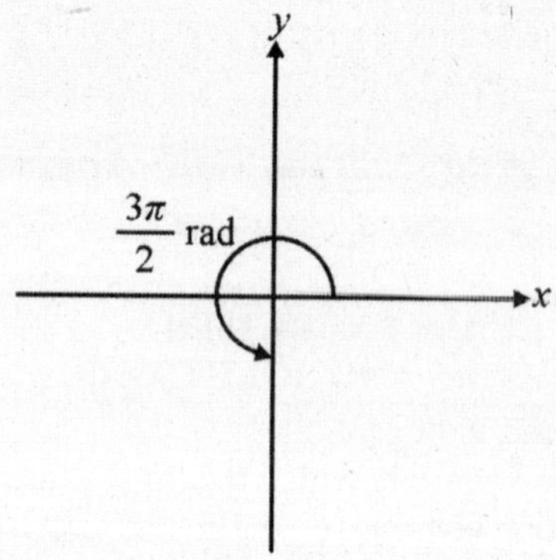


or convert $\frac{3\pi}{2}$ rad to degrees:

$\frac{3\pi}{2}\times\frac{180}{\pi} = 270°$

The angle of 270° is shown in the above diagram.

 

**2.** $\dfrac{-11\pi}{6} = -\pi + \dfrac{-5\pi}{6}$

The angle is negative, which means the terminal arm is rotating in a clockwise direction. Since $\pi$ radians generates half of a revolution, $\dfrac{-11\pi}{6}$ rad generates half of a revolution plus $\dfrac{5}{6}$ of a half-revolution in a clockwise direction.

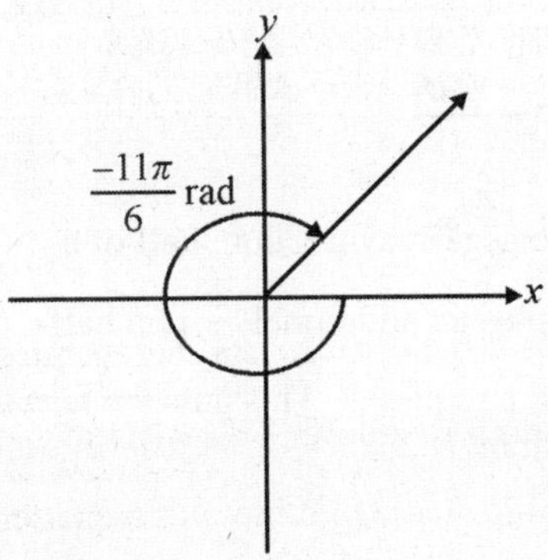


or convert $\dfrac{-11\pi}{6}$ rad to degrees:

$$\frac{-11\pi}{6} \times \frac{180}{\pi} = -330°$$

The angle of –330° is shown in the above diagram.

**3.** Convert 1 rad to degrees:

$$1 \times \frac{180}{\pi} \doteq 57.3°$$

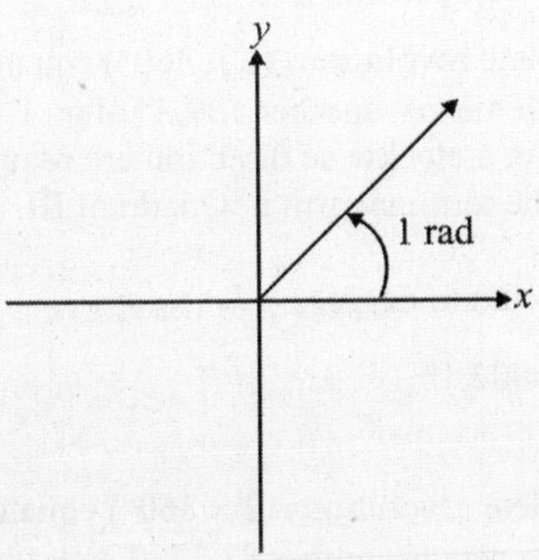


**4.** An angle of $\pi$ rad generates half of a revolution, so an angle of $\dfrac{4\pi}{5}$ rad results in $\dfrac{4}{5}$ of half of a revolution.

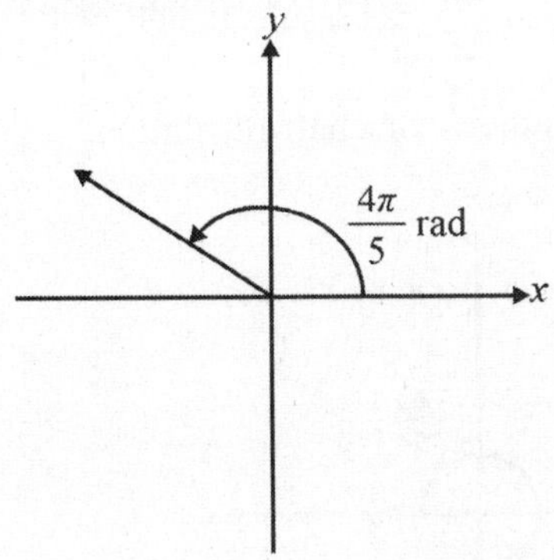


or convert $\dfrac{4\pi}{5}$ rad to degrees:

$$\frac{4\pi}{5} \times \frac{180}{\pi} = 144°$$

An angle of 144° is shown in the above diagram.

**5.** The angle is negative, which means the terminal arm is rotating in a clockwise direction.

Convert –3 rad to degrees:

$$-3 \times \frac{180}{\pi} \doteq -171.9°$$

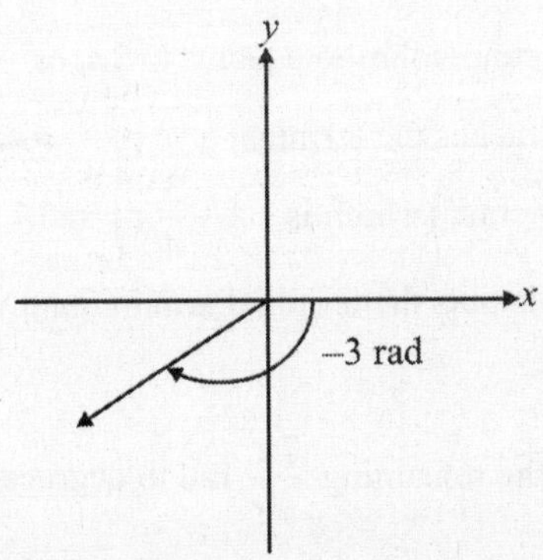

**6.** $\dfrac{4\pi}{3} = \pi + \dfrac{\pi}{3}$

An angle of π radians generates half of a revolution, so $\dfrac{4\pi}{3}$ rad results in half of a revolution plus $\dfrac{1}{3}$ of a half-revolution.

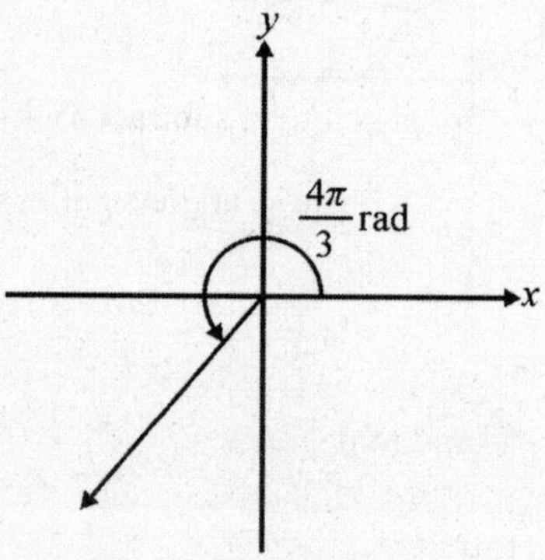


or convert $\dfrac{4\pi}{3}$ rad to degrees:

$\dfrac{4\pi}{3} \times \dfrac{180°}{\pi} = 240°$

An angle of 240° is shown in the above diagram.

**7.** $\dfrac{15\pi}{4} = \dfrac{8\pi}{4} + \dfrac{7\pi}{4}$

One complete revolution equates to $2\pi$, or $\dfrac{8\pi}{4}$ rad, which means the terminal arm must rotate another $\dfrac{7\pi}{4}$ rad, (which is $\left(\pi + \dfrac{3}{4}\pi\right)$ rad) farther. This puts the terminal arm in quadrant IV.

or convert the remaining $\dfrac{7\pi}{4}$ rad to degrees:

$\dfrac{7\pi}{4} \times \dfrac{180}{\pi} = 315°$

An angle of 315° terminates in quadrant IV.

**8.** The angle is negative, which means the terminal arm is rotating in a clockwise direction.

$\dfrac{-23\pi}{3} = \dfrac{-18\pi}{3} + \dfrac{-5\pi}{3}$

Three complete revolutions $(3 \times 2\pi)$ equates to $6\pi$, or $\dfrac{18\pi}{3}$ rad. The terminal arm must rotate $\dfrac{-5\pi}{3}$ rad further in a clockwise direction.

$\dfrac{-5\pi}{3} = -\pi + \dfrac{-2\pi}{3}$

Thus, in the negative direction, half of a revolution plus an additional $\dfrac{2}{3}$ of a half-revolution are required. This puts the terminal arm in quadrant I.

or convert the remaining $\dfrac{-5\pi}{3}$ rad to degrees

$\dfrac{-5\pi}{3} \times \dfrac{180}{\pi} = -300°$

An angle of –300° terminates in quadrant I.

**9.** The angle is negative, which means the terminal arm is rotating in a clockwise direction.

Convert –15 rad to degrees:

$-15 \times \dfrac{180}{\pi} \doteq 859.4°$

Two complete revolutions $(2 \times 360°)$ equate to 720°, which means, another 139.4° of a revolution in a clockwise direction are required. This puts the terminal arm in Quadrant III.

**10.** Convert 14 rad to degrees:

$14 \times \dfrac{180}{\pi} \doteq 802.1°$

Two complete revolutions $(2 \times 360°)$ equate to 720°, which means another 82.1° of a revolution are required. This puts the terminal arm in quadrant I.

 

**11.** The angle is negative, which means the terminal arm is rotating in a clockwise direction.

$$\frac{-31\pi}{6}=\frac{-24\pi}{6}+\frac{-7\pi}{6}$$

Two complete revolutions $(2\times 2\pi)$ equate to $4\pi$, or $\frac{24\pi}{6}$ rad, which means the terminal arm must rotate another $\frac{7\pi}{6}$ rad in a clockwise direction.

$$\frac{-7\pi}{6}=-\pi+\frac{-\pi}{6}$$

Thus, in the negative direction, one half of a revolution plus an additional $\frac{1}{6}$ of a half-revolution are required. This puts the terminal arm in quadrant II.

Or convert the remaining $\frac{-7\pi}{6}$ rad to degrees:

$$\frac{-7\pi}{6}\times\frac{180}{\pi}=-210°$$

An angle of 210° terminates in quadrant II.

**12.** $\frac{7\pi}{3}=\frac{6\pi}{3}+\frac{\pi}{3}$

One complete revolution equates to $2\pi$ or $\frac{6\pi}{3}$ rad, which means the terminal arm must rotate another $\frac{\pi}{3}$. This puts the terminal arm in quadrant I.

or convert the remaining $\frac{\pi}{3}$ rad to degrees:

$$\frac{\pi}{3}\times\frac{180}{\pi}=60°$$

An angle of 60° terminates in quadrant I.

**13.** Because $\frac{-\pi}{3}$ rad is negative, add multiples of $2\pi$ until you get a positive angle smaller than $2\pi$.

$$\frac{-\pi}{3}+2\pi=\frac{-\pi}{3}+\frac{6\pi}{3}=\frac{5\pi}{3}$$

The principal angle is $\frac{5\pi}{3}$ rad.

**14.** Because $\frac{35\pi}{6}$ rad is positive, subtract multiples of $2\pi$ until you get a positive angle smaller than $2\pi$.

$$\frac{35\pi}{6}-2(2\pi)=\frac{35\pi}{5}-4\pi=\frac{35\pi}{6}-\frac{24\pi}{6}=\frac{11\pi}{6}$$

The principle angle is $\frac{11\pi}{6}$ rad.

**15.** Because 7 rad is positive, subtract multiples of $2\pi$ until you get a positive angle smaller than $2\pi$.

The principal angle is $7-2\pi$ or approximately 0.72 rad.

**16.** Since $\frac{21\pi}{5}$ is positive, subtract multiples of $2\pi$ until you get a positive angle smaller than $2\pi$.

$$\frac{21\pi}{5}-2(2\pi)=\frac{21\pi}{5}-4\pi=\frac{21\pi}{5}-\frac{20\pi}{5}=\frac{\pi}{5}$$

The principal angle is $\frac{\pi}{5}$ rad.

**17.** Because $\frac{-11\pi}{2}$ rad is negative, add multiples of $2\pi$ until you get a positive angle smaller than $2\pi$ rad.

$$\frac{-11\pi}{2}+3(2\pi)=\frac{-11\pi}{2}+6\pi=\frac{-11\pi}{2}+\frac{12\pi}{2}=\frac{\pi}{2}$$

The principle angle is $\frac{\pi}{2}$ rad.

**18.** Since –8 rad is negative, add multiples of $2\pi$ until you get a positive angle smaller than $2\pi$.

$$-8+2(2\pi)=-8+4\pi$$
$$\doteq 8+12.56 \doteq 4.56$$

The principle angle is $(4\pi - 8)$ rad (approximately 4.56 rad).

**19.** Angles coterminal with $\frac{2\pi}{11}$ rad are $\left(\frac{2\pi}{11}+2n\pi\right)$ rad, $n \in I$.

**20.** Angles coterminal with 2.4 rad are $(2.4+2n\pi)$ rad, $n \in I$.

**21.** Angles coterminal with $\frac{\pi}{8}$ rad are $\left(\frac{\pi}{8}+2n\pi\right)$ rad, $n \in I$.

**22.** $\pi +$ reference angle $=\frac{7\pi}{6}$

reference angle $=\frac{7\pi}{6}-\pi=\frac{7\pi}{6}-\frac{6\pi}{6}$

reference angle $=\frac{\pi}{6}$ rad

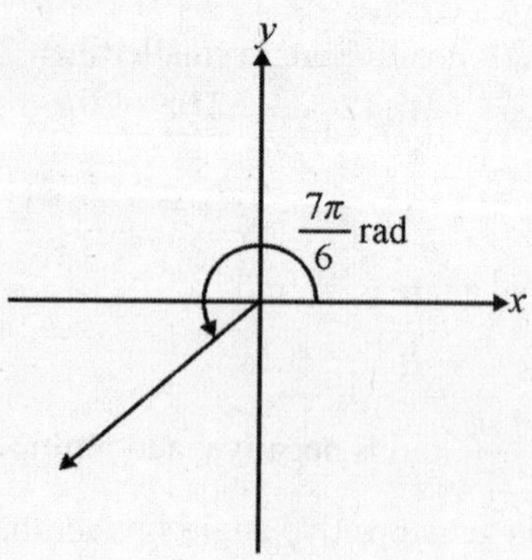


**23.** $\pi +$ reference angle $=\frac{17\pi}{12}$

reference angle $=\frac{17\pi}{12}-\pi=\frac{17\pi}{12}-\frac{12\pi}{12}$

reference angle $=\frac{5\pi}{6}$ rad

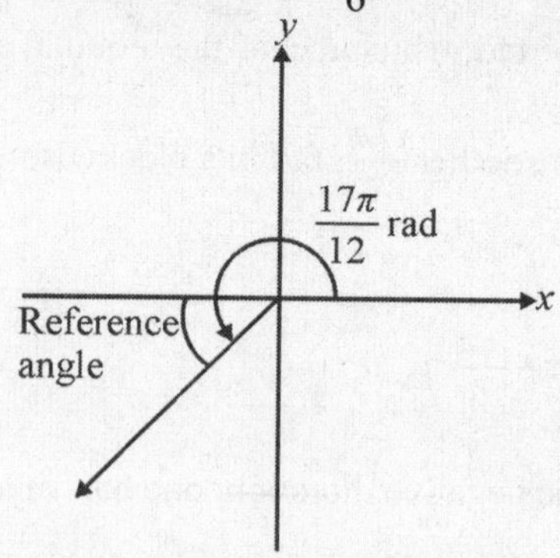


**24.** $-\pi-$ reference angle $=-4$
$\pi+$ reference angle $=4$ (multiply both sides by –1)

reference angle $=4-\pi$
reference angle $\doteq .86$ rad

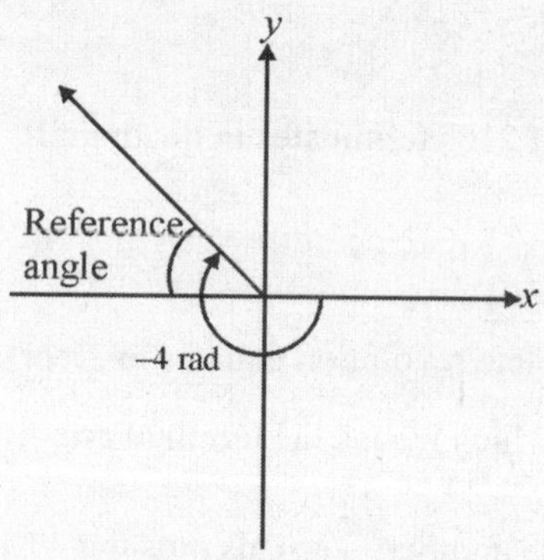


**25.** reference angle $=2\pi-5.24$
reference angle $\doteq$ 1.04 rad

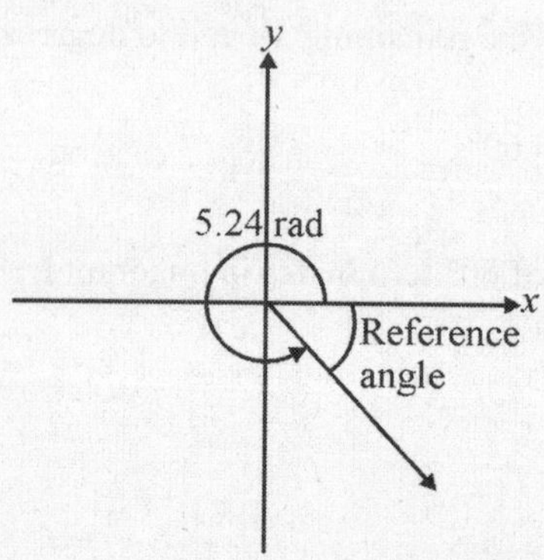

**26.** $-\pi - \text{reference angle} = \frac{-5\pi}{4}$

$\pi + \text{reference angle} = \frac{5\pi}{4}$ (multiply both sides by $-1$)

$\text{reference angle} = \frac{5\pi}{4} - \pi = \frac{5\pi}{4} - \frac{4\pi}{4}$

$\text{reference angle} = \frac{\pi}{4}$ rad

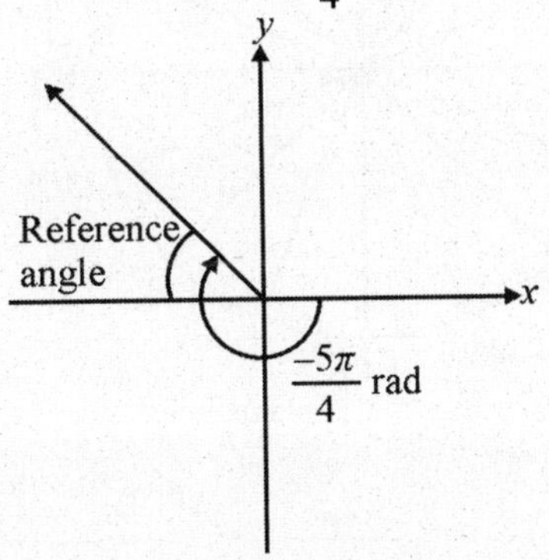


**27.** $\text{reference angle} = 2\pi - \frac{5\pi}{3}$

$\text{reference angle} = \frac{6\pi}{3} - \frac{5\pi}{3}$

$\text{reference angle} = \frac{\pi}{3}$ rad

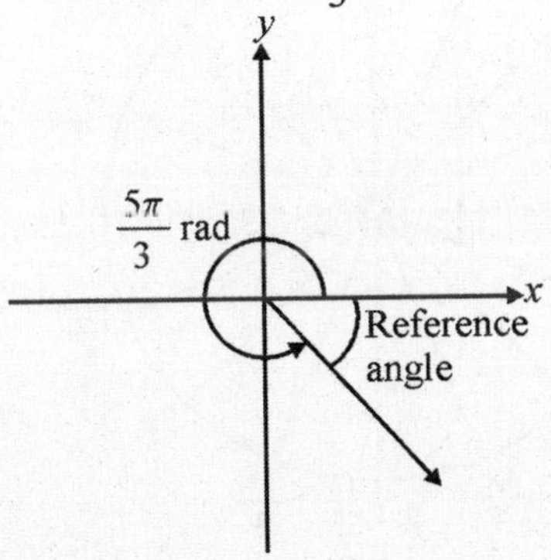


**28.** $\theta$ is in quadrant II.

$\sin\theta = \frac{y}{r}$

Since $\sin\theta = \frac{3}{8}$, a representative triangle has $y = 3$ and $r = 8$.

Applying the Pythagorean Theorem:

$$x^2 + y^2 = r^2$$
$$x^2 + 3 = 8^2$$
$$x^2 + 9 = 64$$
$$x^2 = 55$$
$$x = \pm\sqrt{55}$$

In quadrant II, the value of $x$ is negative. Thus, $x = -\sqrt{55}$.

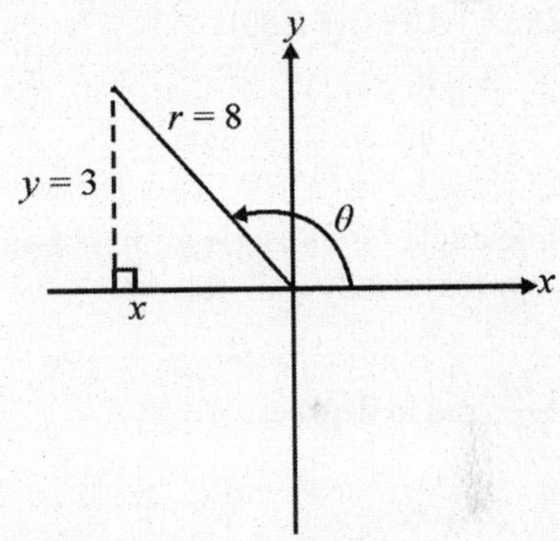


$\cos\theta = \frac{x}{r}$

$\cos\theta = \frac{-\sqrt{55}}{8}$

$\tan\theta = \frac{y}{x}$

$\tan\theta = \frac{3}{-\sqrt{55}} = \frac{-3\sqrt{55}}{55}$

**29.** $\theta$ is in quadrant II.

$\tan\theta = \frac{y}{x}$

Since $\tan\theta = \frac{-5}{2}$, a representative triangle has $x = -2$ and $y = 5$.

Applying the Pythagorean Theorem:

$$r = \sqrt{x^2 + y^2}$$
$$r = \sqrt{(-2)^2 + 5^2}$$
$$r = \sqrt{4 + 25}$$
$$r = \sqrt{29}$$

(Remember, $r$ is always positive.)

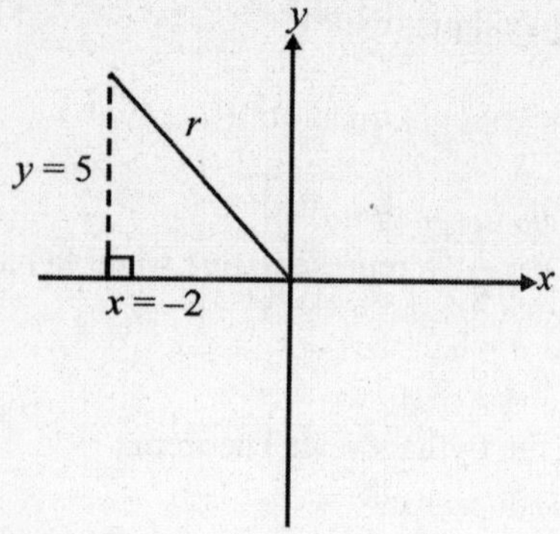


$\sin\theta = \dfrac{y}{r}$ $\cos\theta = \dfrac{x}{r}$

$\sin\theta = \dfrac{5}{\sqrt{29}} = \dfrac{5\sqrt{29}}{29}$ $\cos\theta = \dfrac{-2}{\sqrt{29}} = \dfrac{-2\sqrt{29}}{29}$

**30.** Since −10 rad is negative, add multiples of $2\pi$ until you get a positive angle smaller than $2\pi$.

$$\begin{aligned} -10+2(2\pi) &\doteq -10+2(6.283) \\ &\doteq -10+12.57 \\ &\doteq 2.57 \end{aligned}$$

The principle angle for an angle with measure −10 rad is approximately 2.57 rad.

**31.** Convert $\dfrac{-7\pi}{12}$ rad to degrees:

$$\frac{-7\pi}{12}\times\frac{180}{\pi} = -105°$$

Add multiples of 360° until you get a positive angle with a measure between 360° and 720°.

$$\begin{aligned} -105+2(360) &= -105+720° \\ &= 615° \end{aligned}$$

The measure of angle $B$ is 615°.

## *Lesson 5 Unit Circle—Part I*

### PRACTICE EXERCISES ANSWERS AND SOLUTIONS

**1.**

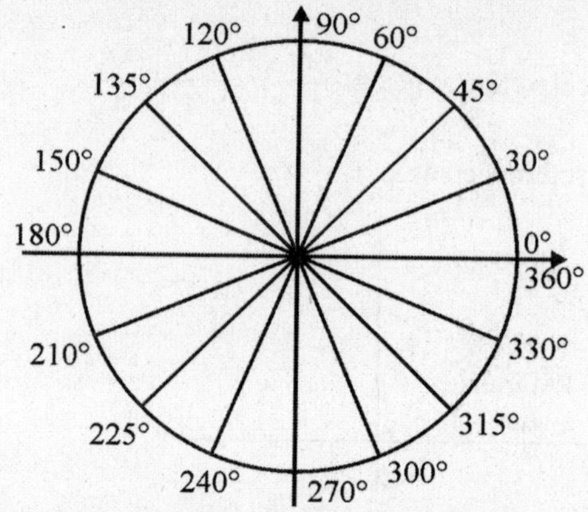


**2.**

- 30°
- 135°
- 210°
- 240°
- 270°
- 315°
- 330°

**3.**

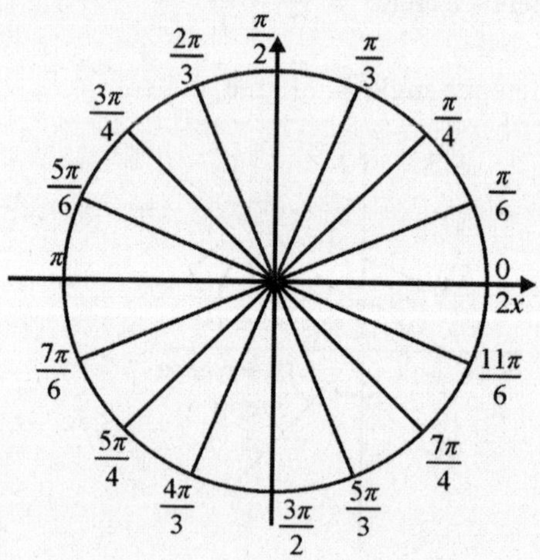


 

**4.**

- $\frac{\pi}{6}$ rad
- $\frac{3\pi}{2}$ rad
- $\frac{3\pi}{4}$ rad
- $\frac{7\pi}{4}$ rad
- $\frac{7\pi}{6}$ rad
- $\frac{11\pi}{6}$ rad
- $\frac{4\pi}{3}$ rad

**5.**

- 60°
- 225°
- 150°
- 300°
- 180°

**6.**

- $\frac{\pi}{3}$ rad
- $\frac{5\pi}{4}$ rad
- $\frac{5\pi}{6}$ rad
- $\frac{5\pi}{3}$ rad
- $\pi$ rad

**7.**

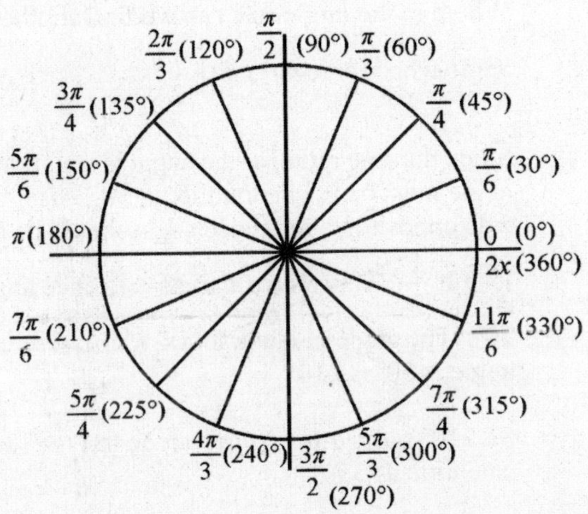


**8.**

- 30°, $\frac{\pi}{6}$ rad
- 150°, $\frac{3\pi}{4}$ rad
- 60°, $\frac{\pi}{3}$ rad
- 225°, $\frac{5\pi}{4}$ rad
- 90°, $\frac{\pi}{2}$ rad
- 270°, $\frac{3\pi}{2}$ rad
- 120°, $\frac{2\pi}{3}$ rad
- 300°, $\frac{5\pi}{3}$ rad
- 135°, $\frac{3\pi}{4}$ rad
- 330°, $\frac{11\pi}{6}$ rad

**9.** Tall triangle in quadrant II

**10.** Medium triangle in quadrant IV

**11.** Tall triangle in quadrant III

**12.** Short triangle in quadrant IV

**13.** Bottom corner of the circle

**14.** Short triangle in quadrant I

**15.** Short triangle in quadrant II

**16.** Medium triangle in quadrant III

**17.** Tall triangle in quadrant III

**18.** Medium triangle in quadrant IV

**19.** Tall triangle in quadrant I

**20.** Short triangle in quadrant IV

**21.** Top corner of the circle

**22.** Short triangle in quadrant III

## Lesson 6 Unit Circle—Part II

### CLASS EXERCISES ANSWERS AND SOLUTIONS

**1.** Consider the special angle 315° in the unit circle. It is an angle in standard position, and corresponds to a medium triangle, with a reference angle of 45°, in quadrant IV.

We label the sides, $x$, $y$, and $r$, and we know that $r = 1$.

To find sin 315°, we use the reference angle, 45°.

$$\sin 315° = \frac{\text{opposite}}{\text{hypotenuse}} = \frac{y}{r} = \frac{y}{1} = y.$$

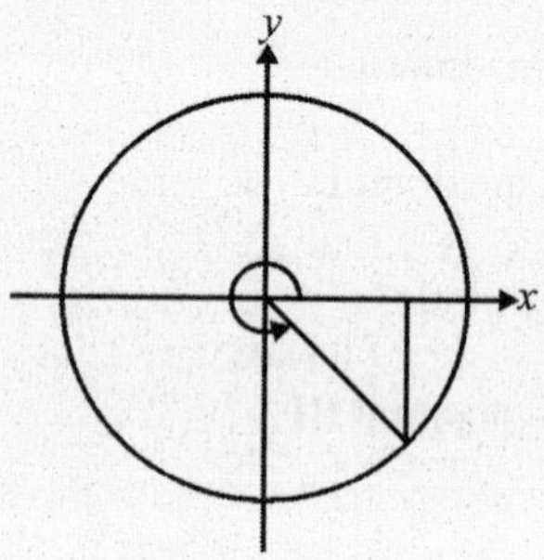


Again, the sine ratio for the angle is simply $y$. So, what is $y$?

Because we are dealing with a 45° triangle, the side opposite the 45° angle is of length $\frac{\sqrt{2}}{2}$, and the side adjacent to the 45° angle is also of length $\frac{\sqrt{2}}{2}$. Because point $(x, y)$, where the rotating arm intersects the circle, is in quadrant III, its coordinates are $\left(-\frac{\sqrt{2}}{2}, -\frac{\sqrt{2}}{2}\right)$. Since $y = -\frac{\sqrt{2}}{2}$, $\sin 315° = -\frac{\sqrt{2}}{2}$.

**2.** Consider the special angle $\frac{4\pi}{3}$ rad in the unit circle. It corresponds to a tall triangle with a reference angle of $\frac{\pi}{3}$ rad (60°) in quadrant III.

To find the sine ratio for the angle, we examine the side opposite the reference angle. It is of length $\frac{\sqrt{3}}{2}$. Because the angle is in quadrant III, $\sin\frac{4\pi}{3}\text{rad} = -\frac{\sqrt{3}}{2}$.

**3.** Consider the special angle 330° in the unit circle. It corresponds to a short triangle with a reference angle of 30° in quadrant IV.

To find sin 330°, we examine the side opposite the reference angle. It is of length $\frac{1}{2}$. Because the angle is in quadrant IV, $\sin 330° = -\frac{1}{2}$.

Where in the unit circle can we find another angle whose sine ratio is $-\frac{1}{2}$?

Since the sine ratio for the angle is negative, the angle must be in quadrant III or IV. Also, the side opposite its reference angle should be of length $\frac{1}{2}$. This means that its reference angle is 30°. The angle in quadrant III whose reference angle is 30° is 210°.

So 210° is the other angle, which has the same sine ratio as 330°.

**4.** The angle $\frac{3\pi}{2}$ rad is at the bottom corner of the unit circle. Its coordinates are (0, –1). Clearly, $y = -1$. So, $\sin\frac{3\pi}{2} = -1$.

 

5. Consider the special angle $\frac{11\pi}{6}$ rad in the unit circle. It corresponds to a short triangle with a reference angle of $\frac{\pi}{6}$ rad (30°) in quadrant IV.

To find the cosine ratio for the angle, we examine the side adjacent to the reference angle. It is a long side with length $\frac{\sqrt{3}}{2}$. Because the angle is in quadrant IV, $\cos\frac{11\pi}{6}=\frac{\sqrt{3}}{2}$.

6. Consider the special angle 210° in the unit circle. It corresponds to a short triangle with a reference angle of 30° in quadrant III.

To find cos 210°, we examine the side adjacent to the reference angle. It is of length $\frac{\sqrt{3}}{2}$. Because the angle is in quadrant III, $\cos 210° = -\frac{\sqrt{3}}{2}$.

Where in the unit circle can we find another angle whose cosine ratio is $-\frac{\sqrt{3}}{2}$?

Since the cosine ratio for the angle is negative, the angle must be in quadrant II or III. Also, the side adjacent to its reference angle should be of length $\frac{\sqrt{3}}{2}$. This means that its reference angle is also 30°. The angle in quadrant II whose reference angle is 30° is 150°.

So, 150° is the other angle, which has the same cosine ratio as 210°.

7. The angle 270° is at the bottom corner of the unit circle. Its coordinates are (0, –1). Clearly, $x = 0$. So, cos 270° = 0.

8. One complete revolution is $2\pi$ radians. $\frac{14\pi}{3}$ is $\frac{12\pi}{3}+\frac{2\pi}{3}$, which is $4\pi+\frac{2\pi}{3}$. Since $4\pi$ radians is 2 complete revolutions, the angle $\frac{14\pi}{3}$ rad will have the same terminal side as the angle $\frac{2\pi}{3}$ rad.

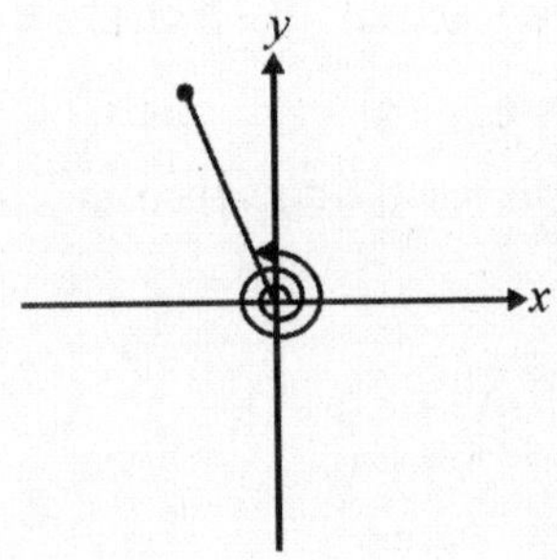


The reference angle for $\frac{2\pi}{3}$ rad is $\frac{\pi}{3}$ rad. Since the angle terminates in quadrant II, the sine ratio is positive, $+\sin\left(\frac{\pi}{3}\right)$, which is $\frac{\sqrt{3}}{2}$.

Thus, $\sin\frac{14\pi}{3}=\frac{\sqrt{3}}{2}$.

9. An angle of –150° terminates in quadrant III. It has the same terminal side as $-150° + 360° = 210°$.

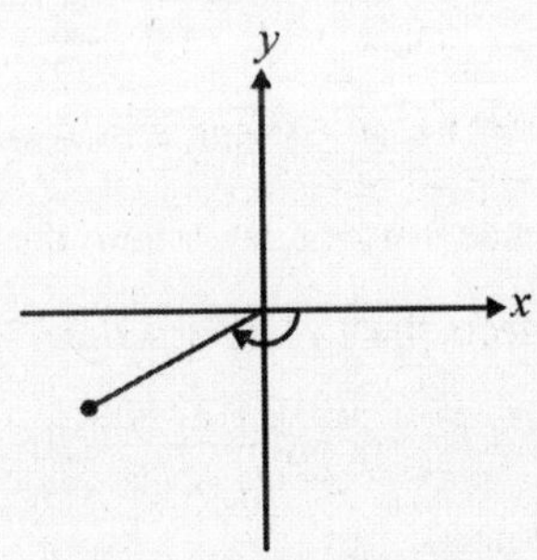


The reference angle is 30°. Since the angle terminates in quadrant III, the sine ratio is negative, –sin 30°, which is $-\frac{1}{2}$.

Thus, $\sin(-150°) = -\frac{1}{2}$.

## PRACTICE EXERCISES ANSWERS AND SOLUTIONS

1. $\sin 45° = \frac{\sqrt{2}}{2}$

2. $\sin 330° = -\frac{1}{2}$

3. $\sin 270° = -1$

4. $\sin 240° = \frac{-\sqrt{3}}{2}$

5. $\sin 210° = -\frac{1}{2}$

6. $\sin 315° = \frac{-\sqrt{2}}{2}$

7. $\sin \frac{5\pi}{4} = \frac{-\sqrt{2}}{2}$

8. $\sin \frac{\pi}{2} = 1$

9. $\sin \frac{7\pi}{6} = \frac{-1}{2}$

10. $\sin \frac{11\pi}{6} = -\frac{1}{2}$

11. $\sin \pi = 0$

12. $\sin \frac{5\pi}{3} = -\frac{\sqrt{3}}{2}$

13. $\sin \frac{3\pi}{4} = \frac{\sqrt{3}}{2}$

14. $\sin \frac{5\pi}{4} = \frac{-\sqrt{2}}{2}$

15. $\sin \frac{\pi}{3} = \frac{\sqrt{3}}{2}$

16. $\sin \frac{2\pi}{3} = \frac{\sqrt{3}}{2}$

17. $\cos 315° = \frac{\sqrt{2}}{2}$

18. $\sin 135° = \frac{\sqrt{2}}{2}$

19. $\sin 150° = \frac{1}{2}$

20. $\sin 225° = \frac{-\sqrt{2}}{2}$

21. $\cos 135° = \frac{-\sqrt{2}}{2}$

22. $\cos 150° = \frac{-\sqrt{3}}{2}$

23. $\sin 315° = \frac{-\sqrt{2}}{2}$

24. $\sin 300° = \frac{-\sqrt{3}}{2}$

25. $\cos 45° = \frac{\sqrt{2}}{2}$

26. $\cos 330° = \frac{\sqrt{3}}{2}$

27. $\cos 270° = 0$

28. $\cos 240° = -\frac{1}{2}$

29. $\cos 210° = \frac{-\sqrt{2}}{2}$

30. $\cos 315° = \frac{\sqrt{2}}{2}$

31. $\cos \frac{5\pi}{4} = \frac{-\sqrt{2}}{2}$

32. $\cos \frac{\pi}{2} = 0$

33. $\cos \frac{7\pi}{6} = \frac{-\sqrt{3}}{2}$

 

**34.** $\cos\frac{11\pi}{6}=\frac{\sqrt{3}}{2}$

**35.** $\cos\pi=-1$

**36.** $\cos\frac{5\pi}{3}=\frac{1}{2}$

**37.** $\cos 45°=\frac{\sqrt{2}}{2}$

**38.** $\cos 90°=0$

**39.** $\cos 180°=-1$

**40.** $\sin 120°=\frac{\sqrt{3}}{2}$

**41.** $\cos 225°=\frac{-\sqrt{2}}{2}$

**42.** $\cos 60°=\frac{1}{2}$

**43.** $\cos(-60°)=\frac{1}{2}$

**44.** $\sin(-90°)=-1$

**45.** $\frac{-5\pi}{6}=-2\pi+\frac{7\pi}{6}$

Thus, $\cos\left(\frac{-5\pi}{6}\right)=\cos\left(\frac{7\pi}{6}\right)=\frac{-\sqrt{3}}{2}$.

**46.** $\cos(-135°)=-\frac{\sqrt{2}}{2}$

**47.** $\cos(-300°)=\frac{1}{2}$

**48.** $420°=360°+60°$
An angle of 420° has a principle angle of 60°.
Thus, $\sin 420°=\sin 60°=\frac{\sqrt{3}}{2}$.

**49.** $\sin(-120°)=\frac{-\sqrt{3}}{2}$

**50.** $570°=360°+210°$
An angle of 570° has a corresponding principle angle of 210°. Thus,
$\cos 570°=\cos 210°=\frac{-\sqrt{3}}{2}$.

**51.** $\sin(-300°)=\frac{\sqrt{3}}{2}$

**52.** $780°=2(360°)+60°$
An angle of 780° has a corresponding principle angle of 60°. Thus, $\cos 780°=\cos 60°=\frac{1}{2}$.

**53.** $585°=360°+225°$
An angle of 585° has a corresponding principle angle of 225°. Thus, $\sin 585°=\sin 225°=\frac{\sqrt{2}}{2}$.

**54.** $390°=360°+30°$
An angle of 390° has a corresponding principle of 30°. Thus, $\cos 390°=\cos 30°=\frac{\sqrt{3}}{2}$.

**55.** $-\frac{5\pi}{3}=-2\pi+\frac{\pi}{3}$

An angle of $-\frac{5\pi}{3}$ has a corresponding principle angle of $\frac{\pi}{3}$. Thus, $\cos\left(\frac{-5\pi}{3}\right)=\cos\left(\frac{\pi}{3}\right)=\frac{1}{2}$.

**56.** $-\frac{17\pi}{6}=2(-2\pi)+\frac{7\pi}{6}$

An angle of $-\frac{17\pi}{6}$ has a corresponding principle angle of $\frac{7\pi}{6}$. Thus,

$\sin\left(-\frac{17\pi}{6}\right)=\sin\left(\frac{7\pi}{6}\right)=-\frac{1}{2}$.

**57.** $\frac{15\pi}{6}=2\pi+\frac{3\pi}{6}=2\pi+\frac{\pi}{2}$

Thus, $\cos\left(\frac{15\pi}{6}\right)=\cos\left(\frac{\pi}{2}\right)=0$.

**58.** $\frac{10\pi}{3} = 2\pi + \frac{4\pi}{3}$

Thus, $\sin\left(\frac{10\pi}{3}\right) = \sin\left(\frac{4\pi}{3}\right) = -\frac{\sqrt{3}}{2}$.

**59.** $\frac{15\pi}{3} = 2(2\pi) + \frac{3\pi}{3} = 2(2\pi) + \pi$

Thus, $\cos\left(\frac{15\pi}{3}\right) = \cos(\pi) = -1$.

**60.** $\frac{-19\pi}{4} = 3(-2\pi) + \frac{5\pi}{4}$

Thus, $\cos\left(\frac{-19\pi}{4}\right) = \cos\left(\frac{5\pi}{4}\right) = -\frac{\sqrt{2}}{2}$.

**61.** $\sin 330° = -\frac{1}{2}$

An angle of 330° terminates in quadrant IV. The reference angle is 30°. Since the sine ratio is negative, the second angle must have a reference angle of 30° and be in quadrant III.

The angle must be 180° + 30° = 210°.

**62.** $\sin 45° = \frac{\sqrt{2}}{2}$

An angle of 45° terminates in quadrant I. The reference angle is 45°. Since the sine ratio is positive, the second angle must have a reference angle of 45° and be in quadrant II.

The angle must be 180° – 45° = 135°.

**63.** $\sin 60° = \frac{\sqrt{3}}{2}$

An angle of 60° terminates in quadrant I. The reference angle is 60°. Since the sine ratio is positive, the second angle must have a reference angle of 60° and be in quadrant II.

**64.** $\sin 135° = \frac{\sqrt{2}}{2}$

An angle of 135° terminates in quadrant II. The reference angle is 45°. Since the sine ratio is positive, the second angle must have a reference angle of 45° and be in quadrant I.

The angle must be 45°.

**65.** $\sin 210° = -\frac{1}{2}$

An angle of 210° terminates in quadrant III. The reference angle is 30°. Since the sine ratio is negative, the second angle must have a reference angle of 30° and be in quadrant IV.

The angle must be 360° – 30° = 330°.

**66.** $\sin 300° = -\frac{\sqrt{3}}{2}$

An angle of 300° terminates in quadrant IV. The reference angle is 60°. Since the sine ratio is negative, the second angle must have a reference angle of 60° and be in quadrant III.

The angle must be 180° + 60° = 240°.

**67.** $\cos 330° = \frac{\sqrt{3}}{2}$

An angle of 330° terminates in quadrant IV. The reference angle is 30°. Since the cosine ratio is positive, the second angle must have a reference angle of 30° and be in quadrant I.

The angle must be 30°.

**68.** $\cos 45° = \frac{\sqrt{2}}{2}$

An angle of 45° terminates in quadrant I. The reference angle is 45°. Since the cosine ratio is positive, the second angle must have a reference angle of 45° and be in quadrant IV.

The angle must be 360° – 45° = 315°.

**69.** $\cos 60° = \frac{1}{2}$

An angle of 60° terminates in quadrant I. The reference angle is 60°. Since the cosine ratio is positive, the second angle must have a reference angle of 60° and be in quadrant IV.

The angle must be 360° – 60° = 300°.

 

**70.** $\cos 135° = \frac{-\sqrt{2}}{2}$

An angle of 135° terminates in quadrant II. The reference angle is 45° . Since the cosine ratio is negative, the second angle must have a reference angle of 45° and be in quadrant III.

The angle must be 180° – 45° = 225°.

**71.** $\cos 210° = \frac{-\sqrt{3}}{2}$

An angle of 210° terminates in quadrant III. The reference angle is 30°. Since the cosine ratio is negative, the second angle must have a reference angle of 30° and be in quadrant II.

The angle must be 180° – 30° = 150°.

**72.** $\cos 300° = \frac{1}{2}$

An angle of 300° terminates in quadrant IV. The reference angle is 60°. Since the cosine ratio is positive, the second angle must have a reference angle of 60° and be in quadrant I.

The angle must be 60°.

**73.** $\sin\frac{5\pi}{6} = \frac{1}{2}$

An angle of $\frac{5\pi}{6}$ terminates in quadrant II. The reference angle is $\frac{\pi}{6}$. Since the sine ratio is positive, the second angle must have a reference angle of $\frac{\pi}{6}$ and be in quadrant I.

The angle must be $\frac{\pi}{6}$.

**74.** $\sin\frac{\pi}{4} = \frac{\sqrt{2}}{2}$

An angle of $\frac{\pi}{4}$ terminates in quadrant I. The reference angle is $\frac{\pi}{4}$. Since the sine ratio is positive, the second angle must have a reference angle of $\frac{\pi}{4}$ and be in quadrant II.

The angle must be $\pi - \frac{\pi}{4} = \frac{3\pi}{4}$.

**75.** $\sin\frac{4\pi}{3} = \frac{-\sqrt{3}}{2}$

An angle of $\frac{4\pi}{3}$ terminates in quadrant III. The reference angle is $\frac{\pi}{3}$. Since the sine ratio is negative, the second angle must have a reference angle of $\frac{\pi}{3}$ and be in quadrant IV.

The angle must be $2\pi - \frac{\pi}{3} = \frac{5\pi}{3}$.

**76.** $\sin\frac{\pi}{6} = \frac{1}{2}$

An angle of $\frac{\pi}{6}$ terminates in quadrant I. The reference angle is $\frac{\pi}{6}$. Since the sine ratio is positive, the second angle must have a reference angle of $\frac{\pi}{6}$ and be in quadrant II.

The angle must be $\pi - \frac{\pi}{6} = \frac{5\pi}{6}$.

**77.** $\sin\frac{5\pi}{4}=\frac{-\sqrt{2}}{2}$

An angle of $\frac{5\pi}{4}$ terminates in quadrant III. The reference angle is $\frac{\pi}{4}$. Since the sine ratio is negative The second angle must have a reference angle of $\frac{\pi}{4}$ and be in quadrant IV.

The angle must be $2\pi-\frac{\pi}{4}=\frac{7\pi}{4}$.

**78.** $\sin\frac{5\pi}{3}=\frac{-\sqrt{3}}{2}$

An angle of $\frac{5\pi}{3}$ terminates in quadrant IV. The reference angle is $\frac{\pi}{3}$. Since the sine ratio is negative, the second angle must have a reference angle of $\frac{\pi}{3}$ and be in quadrant III.

The angle must be $\pi+\frac{\pi}{3}=\frac{4\pi}{3}$.

**79.** $\cos\frac{7\pi}{6}=\frac{-\sqrt{3}}{2}$

An angle of $\frac{7\pi}{6}$ terminates in quadrant III. The reference angle is $\frac{\pi}{6}$. Since the cosine ratio is negative, the second angle must have a reference angle of $\frac{\pi}{6}$ and be in quadrant II.

The angle must be $\pi-\frac{\pi}{6}=\frac{5\pi}{6}$.

**80.** $\cos\frac{5\pi}{4}=\frac{-\sqrt{2}}{2}$

An angle of $\frac{5\pi}{4}$ terminates in quadrant III. The reference angle is $\frac{\pi}{4}$. Since the cosine ratio is negative, the second angle must have a reference angle of $\frac{\pi}{4}$ and be in quadrant II.

The angle must be $\pi-\frac{\pi}{4}=\frac{3\pi}{4}$.

**81.** $\cos\frac{\pi}{3}=\frac{1}{2}$

An angle of $\frac{\pi}{3}$ terminates in quadrant I. The reference angle is $\frac{\pi}{3}$. Since the cosine ratio is positive, the second angle must have a reference angle of $\frac{\pi}{3}$ and be in quadrant IV.

The angle must be $2\pi-\frac{\pi}{3}=\frac{5\pi}{3}$.

**82.** $\cos\frac{5\pi}{6}=\frac{-\sqrt{3}}{2}$

An angle of $\frac{5\pi}{6}$ terminates in quadrant II. The reference angle is $\frac{\pi}{6}$. Since the cosine ratio is negative, the second angle must have a reference angle of $\frac{\pi}{6}$ and be in quadrant III.

The angle must be $\pi+\frac{\pi}{6}=\frac{7\pi}{6}$.

**83.** $\cos\frac{\pi}{4}=\frac{\sqrt{2}}{2}$

An angle of $\frac{\pi}{4}$ terminates in quadrant I. The reference angle is $\frac{\pi}{4}$. Since the cosine ratio is positive, the second angle must have a reference angle of $\frac{\pi}{4}$ and be in quadrant IV.

The angle must be $2\pi-\frac{\pi}{4}=\frac{7\pi}{4}$.

**84.** $\cos\frac{4\pi}{3}=\frac{-1}{2}$

An angle of $\frac{4\pi}{4}$ terminates in quadrant III. The reference angle is $\frac{\pi}{3}$. Since the cosine ratio is negative, the second angle must have a reference angle of $\frac{\pi}{3}$ and be in quadrant II.

The angle must be $\pi-\frac{\pi}{3}=\frac{2\pi}{3}$.

 

**85.** **A**

$$\sin^2\left(\frac{7\pi}{6}\right)+\cos^2\left(\frac{7\pi}{6}\right)$$
$$=\left(\frac{1}{2}\right)^2+\left(\frac{\sqrt{3}}{2}\right)^2$$
$$=\frac{1}{4}+\frac{3}{4}$$
$$=1$$

$$10\log\left(\sin^2\frac{7\pi}{6}+\cos^2\frac{7\pi}{6}\right)$$
$$=10\log 1 \qquad (\log 1=0)$$
$$=10\times 0$$
$$=0$$

## *Lesson 7 Unit Circle—Part III—Tangent Ratio*

### CLASS EXERCISES ANSWERS AND SOLUTIONS

**1.** Using the identity $\tan\theta=\dfrac{\sin\theta}{\cos\theta}$ gives

$$\tan\frac{7\pi}{4}=\frac{\sin\frac{7\pi}{4}}{\cos\frac{7\pi}{4}}$$

We can determine that $\sin\dfrac{7\pi}{4}=-\dfrac{\sqrt{2}}{2}$ and that $\cos\dfrac{7\pi}{4}=\dfrac{\sqrt{2}}{2}$.

Substitute these values into the identity:

$$\tan\frac{7\pi}{4}=\frac{-\frac{\sqrt{2}}{2}}{\frac{\sqrt{2}}{2}}$$
$$=-\frac{\sqrt{2}}{2}\times\frac{2}{\sqrt{2}}$$
$$=-1$$

So, $\tan\dfrac{7\pi}{4}=-1$.

**2.** Using the unit circle, the point on the terminal side of an angle of $\pi$ radians is (–1, 0).

Substituting these coordinates into $\tan\theta=\dfrac{y}{x}$ gives:

$$\tan\theta=\frac{y}{x}\rightarrow\tan\pi=\frac{0}{-1}$$
$$=0$$

So, $\tan\pi=0$.

**3.** Using the unit circle, the point on the terminal side of an angle of 90° is (0, 1).

Substituting these coordinates into $\tan\theta=\dfrac{y}{x}$ gives:

$$\tan\theta=\frac{y}{x}\rightarrow\tan 90°=\frac{1}{0}$$

But $\dfrac{1}{0}$ is undefined, so tan 90° is undefined as well.

**4.**

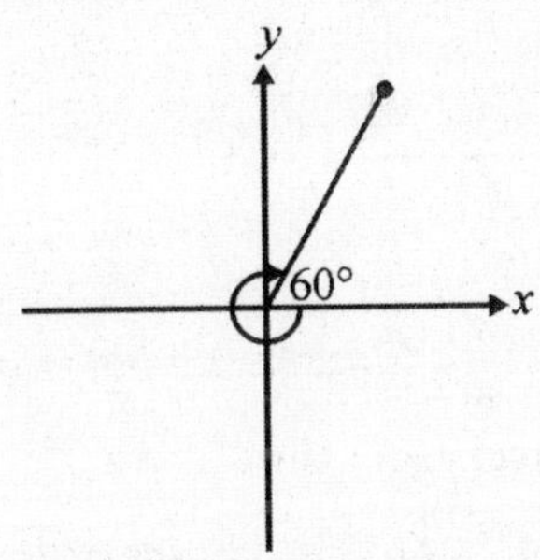


The reference angle is –300° + 360° = 60°.

Thus, tan(–300°) = tan 60° = $\sqrt{3}$ (see Example 1).

### PRACTICE EXERCISES ANSWERS AND SOLUTIONS

**1.** $\tan 30°=\dfrac{\frac{1}{2}}{\frac{\sqrt{3}}{2}}=\dfrac{1}{2}\times\dfrac{2}{\sqrt{3}}=\dfrac{1}{\sqrt{3}}$ or $\dfrac{\sqrt{3}}{3}$

**2.** $\tan\dfrac{7\pi}{6}=\dfrac{\frac{-1}{2}}{\frac{-\sqrt{3}}{2}}=\dfrac{-1}{2}\times\dfrac{-2}{\sqrt{3}}=\dfrac{1}{\sqrt{3}}$ or $\dfrac{\sqrt{3}}{3}$

**3.** $\tan\frac{3\pi}{4}=\frac{\frac{\sqrt{2}}{2}}{\frac{-\sqrt{2}}{2}}=\frac{\sqrt{2}}{2}\times\frac{-2}{\sqrt{2}}=-1$

**4.** $\tan 0°=\frac{0}{1}=0$

**5.** $\tan\frac{5\pi}{3}=\frac{\frac{-\sqrt{3}}{2}}{\frac{1}{2}}=\frac{-\sqrt{3}}{2}\times\frac{2}{1}=-\sqrt{3}$

**6.** $\tan 330°=\frac{-\frac{1}{2}}{\frac{\sqrt{3}}{2}}=\frac{-1}{2}\times\frac{2}{\sqrt{3}}=\frac{-1}{\sqrt{3}}$ *or* $\frac{-\sqrt{3}}{3}$

**7.** $\tan 225°=\frac{\frac{-\sqrt{2}}{2}}{\frac{-\sqrt{2}}{2}}=\frac{-\sqrt{2}}{2}\times\frac{-2}{\sqrt{2}}=1$

**8.** $\tan\frac{3\pi}{2}=\frac{-1}{0}$, which is undefined

**9.** $\tan 120°=\frac{\frac{\sqrt{3}}{2}}{\frac{-1}{2}}=\frac{\sqrt{3}}{2}\times\frac{-2}{1}=-\sqrt{3}$

**10.** $\tan\pi=\frac{0}{-1}=0$

**11.** $\tan\frac{5\pi}{6}=\frac{\frac{1}{2}}{\frac{-\sqrt{3}}{2}}=\frac{1}{2}\times\frac{-2}{\sqrt{3}}=\frac{-1}{\sqrt{3}}$ or $\frac{-\sqrt{3}}{3}$

**12.** $\tan 240°=\frac{\frac{-\sqrt{3}}{2}}{\frac{-1}{2}}=\frac{-\sqrt{3}}{2}\times\frac{-2}{1}=\sqrt{3}$

**13.** $-120°=-360°+240°$

An angle of $-120°$ has a corresponding principle angle of 240°, thus,

$\tan(-120°)=\tan 240°-\frac{\frac{-\sqrt{3}}{2}}{\frac{-1}{2}}=\frac{-\sqrt{3}}{2}\times\frac{-2}{1}=\sqrt{3}.$

**14.** $390°=360°+30°$

An angle of 390° has a corresponding principle angle of 30°, thus,

$\tan 390°=\tan 30°=\frac{\frac{1}{2}}{\frac{\sqrt{3}}{2}}=\frac{1}{2}\times\frac{2}{\sqrt{3}}=\frac{1}{\sqrt{3}}$ or $\frac{\sqrt{3}}{3}$.

**15.** $\frac{11\pi}{4}=2\pi+\frac{3\pi}{4}$

An angle of $\frac{11\pi}{4}$ has a corresponding principle angle of $\frac{3\pi}{4}$, thus,

$\tan\frac{11\pi}{4}=\tan\frac{3\pi}{4}=\frac{\frac{\sqrt{2}}{2}}{\frac{-\sqrt{2}}{2}}=\frac{\sqrt{2}}{2}\times\frac{-2}{\sqrt{2}}=-1.$

**16.** $\frac{-7\pi}{6}=-2\pi+\frac{5\pi}{6}$

An angle of $\frac{-7\pi}{6}$ has a corresponding principle angle of $\frac{5\pi}{6}$, thus,

$\tan\left(\frac{-7\pi}{6}\right)=\tan\left(\frac{5\pi}{6}\right)=\frac{\frac{1}{2}}{\frac{-\sqrt{3}}{2}}=\frac{1}{2}\times\frac{-2}{\sqrt{3}}=\frac{-1}{\sqrt{3}}=\frac{-\sqrt{3}}{3}.$

**17.** $540°=360°+180°$

An angle of 540° has a corresponding principle angle of 180° thus,

$\tan 540°=\tan 180°=\frac{0}{-1}=0.$

**18.** $\frac{-8\pi}{3}=2(-2\pi)+\frac{4\pi}{3}$

An angle of $\frac{-8\pi}{3}$ has a corresponding principle angle of $\frac{4\pi}{3}$ thus,

$\tan\left(\frac{-8\pi}{3}\right)=\tan\frac{4\pi}{3}=\frac{\frac{-\sqrt{3}}{2}}{\frac{-1}{2}}=\frac{-\sqrt{3}}{2}\times\frac{-2}{1}=\sqrt{3}.$

 

**19.** $-135° = -360° + 225°$

$$\tan(-135°) = \tan 225° = \frac{\frac{-\sqrt{2}}{2}}{\frac{-\sqrt{2}}{2}} = \frac{-\sqrt{2}}{2} \times \frac{-2}{\sqrt{2}} = 1$$

**20.** $\frac{-7\pi}{4} = -2\pi + \frac{\pi}{4}$

$$\tan\left(\frac{-7\pi}{4}\right) = \tan\frac{\pi}{4} = \frac{\frac{\sqrt{2}}{2}}{\frac{\sqrt{2}}{2}} = \frac{\sqrt{2}}{2} \times \frac{2}{\sqrt{2}} = 1$$

**21.** $\frac{23\pi}{6} = 2\pi + \frac{11\pi}{6}$

$$\tan\frac{23\pi}{6} = \tan\frac{11\pi}{6} = \frac{\frac{-1}{2}}{\frac{\sqrt{3}}{2}} = \frac{-1}{2} \times \frac{2}{\sqrt{3}} = \frac{-1}{\sqrt{3}} \text{ or } \frac{-\sqrt{3}}{3}$$

**22.** $\tan 0 = \frac{0}{1} = 0$

$$\tan 30° = \tan\frac{\pi}{6} = \frac{\frac{1}{2}}{\frac{\sqrt{3}}{2}} = \frac{1}{2} \times \frac{2}{\sqrt{3}} = \frac{1}{\sqrt{3}} \text{ or } \frac{\sqrt{3}}{3}$$

$$\tan 45° = \tan\frac{\pi}{4} = \frac{\frac{\sqrt{2}}{2}}{\frac{\sqrt{2}}{2}} = \frac{\sqrt{2}}{2} \times \frac{2}{\sqrt{2}} = 1$$

$$\tan 60° = \tan\frac{\pi}{3} = \frac{\frac{\sqrt{3}}{2}}{\frac{1}{2}} = \frac{\sqrt{3}}{2} \times \frac{2}{1} = \sqrt{3}$$

$$\tan 90° = \tan\frac{\pi}{2} = \frac{1}{0}, \text{ which is undefined}$$

$$\tan 120° = \tan\frac{2\pi}{3} = \frac{\frac{\sqrt{3}}{2}}{\frac{-1}{2}} = \frac{\sqrt{3}}{2} \times \frac{-2}{1} = -\sqrt{3}$$

$$\tan 135° = \tan\frac{3\pi}{4} = \frac{\frac{\sqrt{2}}{2}}{\frac{-\sqrt{2}}{2}} = \frac{\sqrt{2}}{2} \times \frac{-2}{\sqrt{2}} = -1$$

$$\tan 150° = \tan\frac{5\pi}{6} = \frac{\frac{1}{2}}{\frac{-\sqrt{3}}{2}} = \frac{1}{2} \times \frac{-2}{\sqrt{3}} = \frac{-1}{\sqrt{3}} \text{ or } \frac{-\sqrt{3}}{3}$$

$$\tan 180° = \tan\pi = \frac{0}{1} = 0$$

$$\tan 210° = \tan\frac{7\pi}{6} = \frac{\frac{-1}{2}}{\frac{-\sqrt{3}}{2}} = \frac{-1}{2} \times \frac{-2}{\sqrt{3}} = \frac{1}{\sqrt{3}} \text{ or } \frac{\sqrt{3}}{3}$$

$$\tan 225° = \tan\frac{5\pi}{4} = \frac{\frac{-\sqrt{2}}{2}}{\frac{-\sqrt{2}}{2}} = \frac{-\sqrt{2}}{2} \times \frac{-2}{\sqrt{2}} = 1$$

$$\tan 240° = \tan\frac{4\pi}{3} = \frac{\frac{-\sqrt{2}}{2}}{\frac{-1}{2}} = \frac{-\sqrt{3}}{2} \times \frac{-2}{1} = \sqrt{3}$$

$$\tan 270° = \tan\frac{3\pi}{2} = \frac{-1}{0}, \text{ which is undefined}$$

$$\tan 300° = \tan\frac{5\pi}{3} = \frac{\frac{-\sqrt{3}}{2}}{\frac{1}{2}} = \frac{-\sqrt{3}}{2} \times \frac{2}{1} = -\sqrt{3}$$

$$\tan 315° = \tan\frac{7\pi}{4} = \frac{\frac{-\sqrt{2}}{2}}{\frac{\sqrt{2}}{2}} = \frac{-\sqrt{2}}{2} \times \frac{2}{\sqrt{2}} = -1$$

$$\tan 330° = \tan\frac{11\pi}{6} = \frac{\frac{-1}{2}}{\frac{\sqrt{3}}{2}} = \frac{-1}{2} \times \frac{2}{\sqrt{3}} = \frac{-1}{\sqrt{3}} \text{ or } \frac{-\sqrt{3}}{3}$$

$$\tan 360° = \tan 2\pi = \frac{0}{1} = 0$$

**23.**

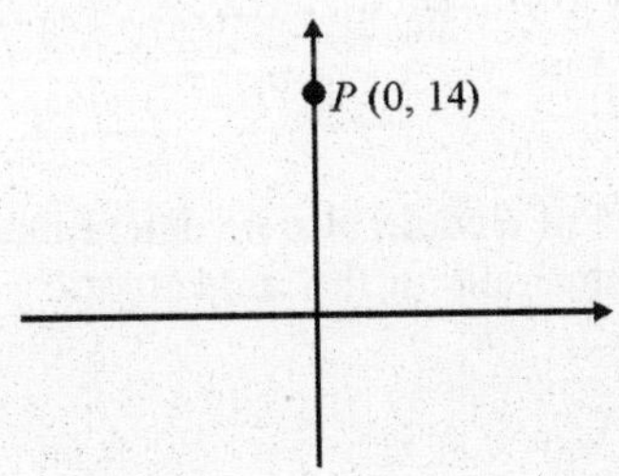


$x = 0$
$y = 14$
$r = 14$

$\sin\theta = \frac{y}{r}$ $\cos\theta = \frac{x}{r}$

$\sin\theta = \frac{14}{14} = 1$ $\cos\theta = \frac{0}{10} = 0$

$\tan\theta = \frac{y}{x}$

$\tan\theta = \frac{14}{0}$

There is no exact value for tan $\theta$ because $\frac{14}{0}$ is undefined. Notice that $\theta = 90°$.

**24.**

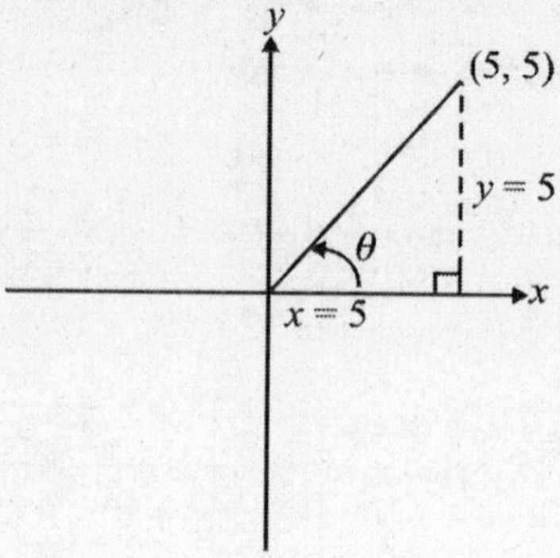


Applying the Pythagorean Theorem:

$$r = \sqrt{x^2 + y^2}$$
$$r = \sqrt{5^2 + 5^2}$$
$$r = \sqrt{25 + 25}$$
$$r^2 = \sqrt{50}$$
$$r = 5\sqrt{2}$$

(Remember, $r$ is always positive.)

$\sin\theta = \frac{y}{r}$ $\cos\theta = \frac{x}{r}$ $\tan\theta = \frac{y}{x}$

$\sin\theta = \frac{5}{5\sqrt{2}}$ $\cos\theta = \frac{5}{5\sqrt{2}}$ $\tan\theta = \frac{5}{5} = 1$

or $\frac{\sqrt{2}}{2}$ or $\frac{\sqrt{2}}{2}$

$\theta$ is in quadrant I.

Since $\sin\theta = \frac{\sqrt{2}}{2}$, then $\theta$ must have a measure of $45^\circ$ or $\frac{\pi}{4}$ rad.

(The measure of $\theta$ could also be determined using the cosine ratio or the tangent ratio.)

## *Lesson 8 Reciprocal Trigonometric Ratios for Special Angles*

### CLASS EXERCISES ANSWERS AND SOLUTIONS

**1.** Since $\csc\theta = \frac{1}{\sin\theta}$, we have

$$\csc(-225^\circ) = \frac{1}{\sin(-225^\circ)}.$$

First, find $\sin(-225^\circ)$.

$-225^\circ$ has a corresponding principal angle of $135^\circ$.

We know that $\sin(135^\circ) = \frac{\sqrt{2}}{2}$, thus,

$$\sin(-225^\circ) = \frac{\sqrt{2}}{2}.$$

$$\therefore \csc(-225^\circ) = \frac{1}{\frac{\sqrt{2}}{2}}$$
$$= \frac{2}{\sqrt{2}}$$
$$= \sqrt{2}$$

(rationalize the denominator)

**2.** Since $\cot\theta = \frac{1}{\tan\theta}$, we have

$$\cot 240^\circ = \frac{1}{\tan 240^\circ}.$$

We know that $\tan 240^\circ = \sqrt{3}$.

$$\therefore \cot 240^\circ = \frac{1}{\sqrt{3}}$$
$$= \frac{\sqrt{3}}{3}$$

(rationalize the denominator)

**3.** $\sec(-3\pi) = \frac{1}{\cos(-3\pi)} = \frac{1}{-1} = -1$

 

**4.** A representation is:

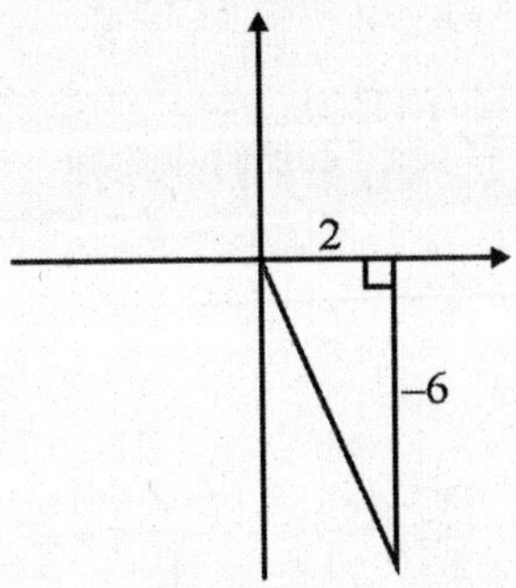


$r=\sqrt{x^2+y^2}=\sqrt{2^2+(-6)^2}$

$r=\sqrt{4+36}=\sqrt{40}=2\sqrt{10}$

$\csc\theta=\frac{r}{y}=\frac{2\sqrt{10}}{-6}=-\frac{\sqrt{10}}{3}$

$\sec\theta=\frac{r}{x}=\frac{2\sqrt{10}}{2}=\sqrt{10}$

$\cot\theta=\frac{x}{y}=\frac{2}{-6}=-\frac{1}{3}$

## PRACTICE EXERCISES ANSWERS AND SOLUTIONS

**1.** $\sec 30° = \frac{1}{\frac{\sqrt{3}}{2}} = 1\times\frac{2}{\sqrt{3}} = \frac{2}{\sqrt{3}} = \frac{2\sqrt{3}}{3}$

**2.** $\cot\left(\frac{7\pi}{6}\right) = \frac{1}{\frac{1}{\sqrt{3}}} = 1\times\frac{\sqrt{3}}{1} = \sqrt{3}$

**3.** $\sec\left(\frac{3\pi}{4}\right) = \frac{1}{\frac{-\sqrt{2}}{2}} = 1\times\frac{-2}{\sqrt{2}}$ or $-\sqrt{2}$

**4.** $\csc 0° = \frac{1}{0}$, which is undefined

**5.** $\tan\left(\frac{5\pi}{3}\right) = \frac{1}{\frac{-1}{\sqrt{3}}} = 1\times\frac{-\sqrt{3}}{1} = -\sqrt{3}$

**6.** $\cot 120° = \frac{1}{\frac{-1}{\sqrt{3}}} = 1\times\frac{-\sqrt{3}}{1} = -\sqrt{3}$

**7.** $\csc 225° = \frac{1}{\frac{-\sqrt{2}}{2}} = 1\times\frac{-2}{\sqrt{2}} = \frac{-2}{\sqrt{2}}$ or $-\sqrt{2}$

**8.** $\cot\left(\frac{3\pi}{2}\right) = \frac{0}{-1} = 0$

**9.** $\sec 120° = \frac{1}{-\frac{1}{2}} = 1\times\frac{-2}{1} = -2$

**10.** $\cot\pi = \frac{-1}{0}$, which is undefined

**11.** $\frac{-13\pi}{6} = -2(2\pi)+\frac{11\pi}{6}$

An angle of $\frac{-13\pi}{6}$ has a corresponding principle angle of $\frac{11\pi}{6}$. Thus,

$\cot\left(\frac{-13\pi}{6}\right)=\cot\left(\frac{11\pi}{6}\right)=\frac{1}{\frac{-1}{\sqrt{3}}}=1\times\frac{-\sqrt{3}}{1}=-\sqrt{3}.$

**12.** $420° = 360° + 60°$

An angle of 420° has a corresponding principle angle of 60°.

Thus, $\csc 420° = \csc 60° =$

$\frac{1}{\frac{\sqrt{3}}{2}} = 1\times\frac{2}{\sqrt{3}} = \frac{2}{\sqrt{3}}$ or $\frac{2\sqrt{3}}{3}$.

**13.** $-330° = -360° + 30°$

An angle of –330° has a corresponding principle angle of 30°.

Thus, $\sec(-330°) = \sec 30° =$

$\frac{1}{\frac{\sqrt{3}}{2}} = 1\times\frac{2}{\sqrt{3}} = \frac{2}{\sqrt{3}}$ or $\frac{2\sqrt{3}}{3}$.

**14.** $-135° = -360° + 225°$

An angle of $-135°$ has a corresponding principle angle of 225°.

Thus, $\csc(-135°) = \csc 225° =$
$$\frac{1}{\frac{-\sqrt{2}}{2}} = 1\times\frac{-2}{\sqrt{2}} = \frac{-2}{\sqrt{2}} \text{ or } -\sqrt{2}.$$

**15.** $\frac{-5\pi}{6} = -2\pi + \frac{7\pi}{6}$

$\frac{-5\pi}{6}$ has a corresponding principle angle of $\frac{7\pi}{6}$. Thus,
$$\sec\left(-\frac{5\pi}{6}\right) = \sec\frac{7\pi}{6} = \frac{1}{\frac{-\sqrt{3}}{2}} = 1\times\frac{-2}{\sqrt{3}} \text{ or } \frac{-2\sqrt{3}}{3}.$$

**16.** $480° = 360° + 120°$

An angle of 480° has a corresponding principle angle of 120°.

Thus, $\cot 480° = \cot 120° =$
$$\frac{\frac{-1}{2}}{\frac{\sqrt{3}}{2}} = \frac{-1}{2}\times\frac{2}{\sqrt{3}} = \frac{-1}{\sqrt{3}} \text{ or } \frac{-\sqrt{3}}{3}.$$

**17.** $750° = 2(360°) + 30°$

An angle of 750° has a corresponding principle angle of 30°.

Thus, $\csc 750° = \csc 30° = \frac{1}{\frac{1}{2}} = 1\times\frac{2}{1} = 2.$

**18.** $-\frac{20\pi}{3} = -4(2\pi) + \frac{4\pi}{3}$

An angle of $\frac{-20\pi}{3}$ has a corresponding principle angle of $\frac{4\pi}{3}$.

Thus,
$$\sec\left(\frac{-20\pi}{3}\right) = \sec\frac{4\pi}{3} = \frac{1}{\frac{-1}{2}} = 1\times\frac{-2}{1} = -2.$$

**19.** $\cot 0 = \frac{1}{0}$, which is undefined

$$\cot 30° = \cot\frac{\pi}{6} = \frac{\frac{\sqrt{3}}{2}}{\frac{1}{2}} = \frac{\sqrt{3}}{2}\times\frac{2}{1} = \sqrt{3}$$

$$\cot 45° = \cot\frac{\pi}{4} = \frac{\frac{\sqrt{2}}{2}}{\frac{\sqrt{2}}{2}} = \frac{\sqrt{2}}{2}\times\frac{2}{\sqrt{2}} = 1$$

$$\cot 60° = \cot\frac{\pi}{3} = \frac{\frac{1}{2}}{\frac{\sqrt{3}}{2}} = \frac{1}{2}\times\frac{2}{\sqrt{3}} = \frac{1}{\sqrt{3}} \text{ or } \frac{\sqrt{3}}{3}$$

$$\cot 90° = \cot\frac{\pi}{2} = \frac{0}{1} = 0$$

$$\cot 120° = \cot\frac{2\pi}{3} = \frac{\frac{-1}{2}}{\frac{\sqrt{3}}{2}} = \frac{-1}{2}\times\frac{2}{\sqrt{3}} = \frac{-1}{\sqrt{3}} \text{ or } \frac{-\sqrt{3}}{3}$$

$$\cot 135° = \cot\frac{3\pi}{4} = \frac{\frac{-\sqrt{2}}{2}}{\frac{\sqrt{2}}{2}} = \frac{-\sqrt{2}}{2}\times\frac{2}{\sqrt{2}} = -1$$

$$\cot 150° = \cot\frac{5\pi}{6} = \frac{\frac{-\sqrt{3}}{2}}{\frac{1}{2}} = \frac{-\sqrt{3}}{2}\times\frac{2}{1} = -\sqrt{3}$$

$\cot 180° = \cot\pi = \frac{1}{0}$, which is undefined.

$$\cot 210° = \cot\frac{7\pi}{6} = \frac{\frac{-\sqrt{3}}{2}}{\frac{-1}{2}} = \frac{-\sqrt{3}}{2}\times\frac{-2}{1} = \sqrt{3}$$

 

$$\cot 225° = \cot\frac{5\pi}{4} = \frac{\frac{-\sqrt{3}}{2}}{\frac{-\sqrt{2}}{2}} = \frac{-\sqrt{2}}{2}\times\frac{-2}{\sqrt{2}} = 1$$

$$\cot 240° = \cot\frac{4\pi}{2} = \frac{\frac{-1}{2}}{\frac{-\sqrt{3}}{2}} = \frac{-1}{2}\times\frac{-2}{\sqrt{3}} = \frac{1}{\sqrt{3}} \text{ or } \frac{\sqrt{3}}{3}$$

$$\cot 270° = \cot\frac{3\pi}{2} = \frac{0}{-1} = 0$$

$$\cot 300° = \cot\frac{5\pi}{3} = \frac{\frac{1}{2}}{\frac{-\sqrt{3}}{2}} = \frac{1}{2}\times\frac{-2}{\sqrt{3}} = \frac{-1}{\sqrt{3}} \text{ or } \frac{-\sqrt{3}}{3}$$

$$\cot 315° \cot\frac{7\pi}{4} = \frac{\frac{\sqrt{2}}{2}}{\frac{-\sqrt{2}}{2}} = \frac{\sqrt{2}}{2}\times\frac{-2}{\sqrt{2}} = -1$$

$$\cot 330° \cot\frac{11\pi}{6} = \frac{\frac{\sqrt{2}}{2}}{\frac{-1}{2}} = \frac{\sqrt{3}}{2}\times\frac{-2}{1} = -\sqrt{3}$$

$$\cot 360° = \cot 2\pi = \frac{1}{0}, \text{ which is undefined}$$

**20.**

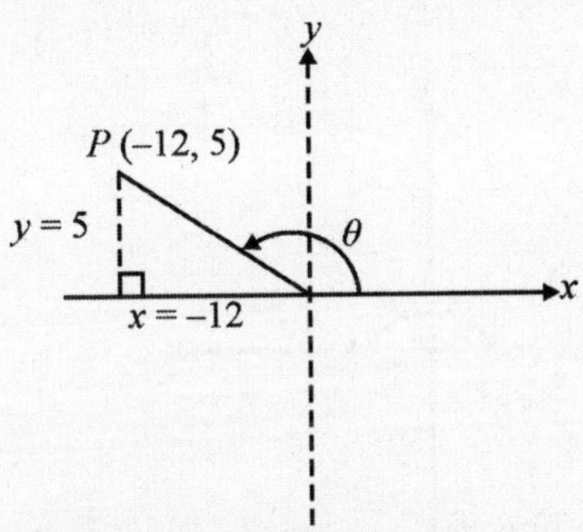


Applying the Pythagorean Theorem:

$r = \sqrt{x^2 + y^2}$

$r = \sqrt{(-12)^2 + 5^2}$

$r = \sqrt{144 + 25}$

$r = \sqrt{169}$

$r = 13$

(Remember, $r$ is always positive.)

$\csc\theta = \dfrac{r}{y}$ $\quad \sec\theta = \dfrac{r}{x}$ $\quad \cot\theta = \dfrac{x}{y}$

$\csc\theta = \dfrac{13}{5}$ $\quad \sec\theta = \dfrac{-13}{12}$ $\quad \cot\theta = \dfrac{-12}{5}$

**21.**

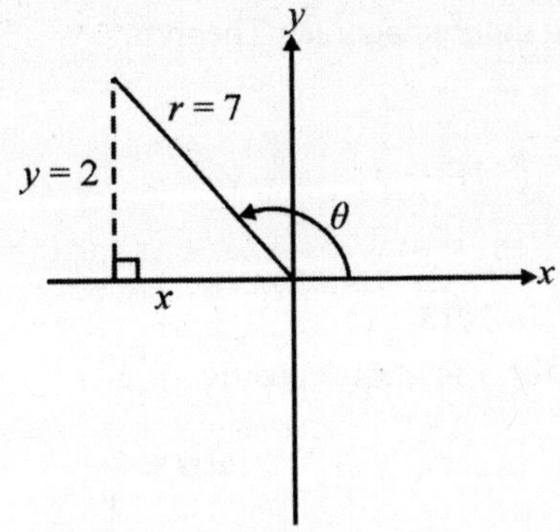


$\theta$ is in quadrant II.

$\csc\theta = \dfrac{r}{y}$

Since $\csc\theta = \dfrac{7}{2}$, a representative triangle has $r = 7$ and $y = 2$.

Applying the Pythagorean Theorem:

$x^2 + y^2 = r^2$

$x^2 + 2^2 = 72$

$x^2 + 4 = 49$

$x = \pm\sqrt{45} \text{ or } \pm 3\sqrt{5}$

Since $\theta$ is in quadrant II, the value of $x$ is negative. Thus, $x = -3\sqrt{5}$.

$\sec\theta = \dfrac{r}{x}$ $\quad \cot\theta = \dfrac{x}{y}$

$\sec\theta = \dfrac{7}{-3\sqrt{5}} \text{ or } \dfrac{-7\sqrt{5}}{15}$ $\quad \cot\theta \ \dfrac{-3\sqrt{5}}{2}$

**22.**

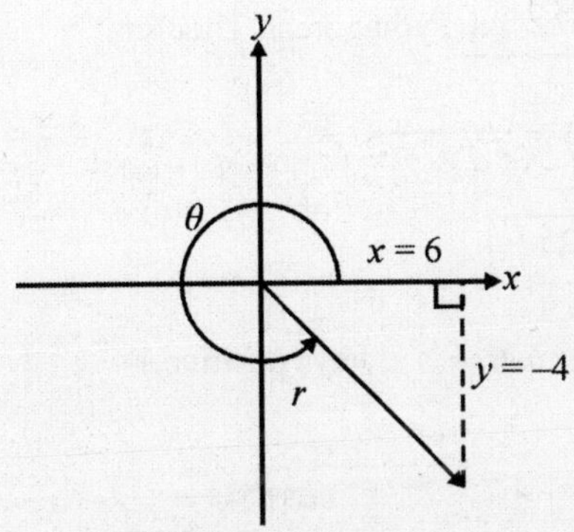

Applying the Pythagorean Theorem:

$r = \sqrt{x^2 + y^2}$

$r = \sqrt{6^2 + (-4)^2}$

$r = \sqrt{36 + 16}$

$r = \sqrt{52}$ or $2\sqrt{13}$

(Remember, $r$ is always positive.)

$\csc\theta = \frac{r}{y}$ $\quad \sec\theta = \frac{r}{x}$ $\quad \cot\theta = \frac{x}{y}$

$\csc\theta = \frac{2\sqrt{13}}{-4}$ $\quad \sec\theta \frac{2\sqrt{3}}{6}$ $\quad \cot\theta = \frac{6}{-4}$

$= \frac{-\sqrt{13}}{2}$ $\quad = \frac{\sqrt{13}}{3}$ $\quad = \frac{-3}{2}$

**23.**

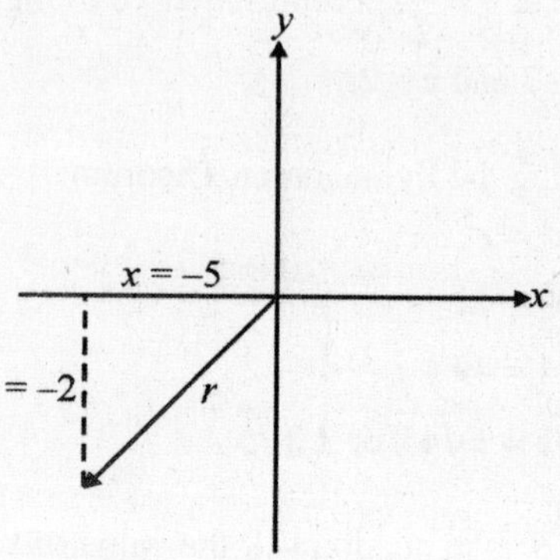


$\theta$ is in quadrant III.

$\cot\theta = \frac{x}{y}$

Since $\cot\theta = \frac{5}{2}$, a representative triangle has $x = -5$ and $y = -2$.

Applying the Pythagorean Theorem:

$r = \sqrt{x^2 + y^2}$

$r = \sqrt{(-5)^2 + (-2)^2}$

$r = \sqrt{25 + 4}$

$r = \sqrt{29}$

(Remember, $r$ is always positive.)

$\sin\theta = \frac{y}{r}$ $\quad \cos\theta = \frac{x}{y}$

$\sin\theta = \frac{-2}{\sqrt{29}}$ or $\frac{-2\sqrt{29}}{29}$ $\quad \cos\theta = \frac{-5}{\sqrt{29}}$ or $\frac{5\sqrt{29}}{29}$

**24.**

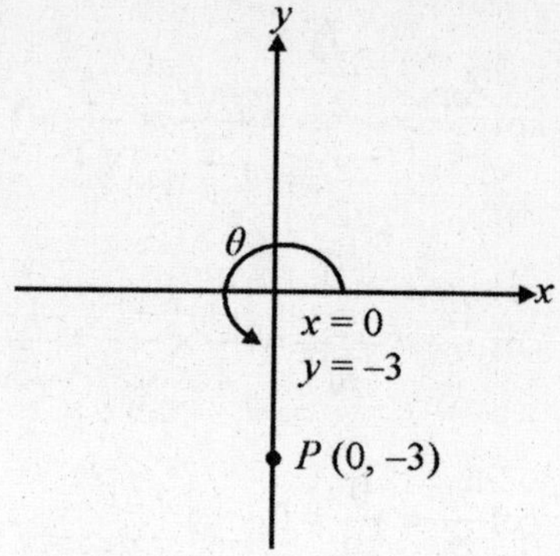


Applying the Pythagorean Theorem:

$r = \sqrt{x^2 + y^2}$

$r = \sqrt{0 + (-3)^2}$

$r = \sqrt{0 + 9}$

$r = 3$

(Remember, $r$ is always positive.)

$\csc\theta = \frac{r}{y}$ $\quad \sec\theta = \frac{r}{x}$ $\quad \cot\theta = \frac{x}{y}$

$\csc\theta = \frac{3}{-3} = -1$ $\quad \sec\theta = \frac{3}{0}$ $\quad \cot\theta = \frac{0}{-3}$

$\sec\theta$ is undefined $\quad \cot\theta = 0$

**25.**

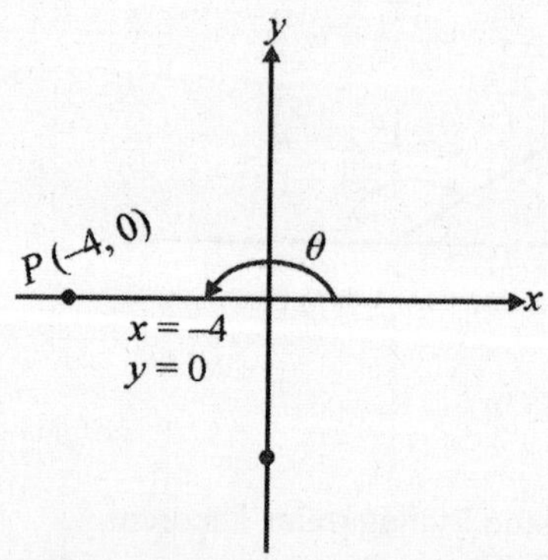


Applying the Pythagorean Theorem:

$r = \sqrt{x^2 + y^2}$

$r = \sqrt{(-4)^2 + 0}$

$r = \sqrt{16 + 0}$

$r = 4$

(Remember, $r$ is always positive.)

 

| | | |
|---|---|---|
| $\sin\theta = \dfrac{y}{r}$ | $\cos\theta = \dfrac{x}{r}$ | $\tan\theta = \dfrac{y}{x}$ |
| $\sin\theta = \dfrac{0}{4} = 0$ | $\cos\theta = \dfrac{-4}{4} = -1$ | $\tan\theta = \dfrac{0}{-4} = 0$ |
| $\csc\theta = \dfrac{r}{y}$ | $\sec\theta = \dfrac{r}{x}$ | $\cot\theta = \dfrac{x}{y}$ |
| $\csc\theta = \dfrac{-4}{0}$ | $\sec\theta = \dfrac{4}{-4} = -1$ | $\cot\theta = \dfrac{-4}{0}$ |
| $\csc\theta$ is undefined | | $\cot\theta$ is undefined |

If $\cos\theta = -1$, the measure of $\theta$ must be 180°, or $\pi$ rad.

**26.**

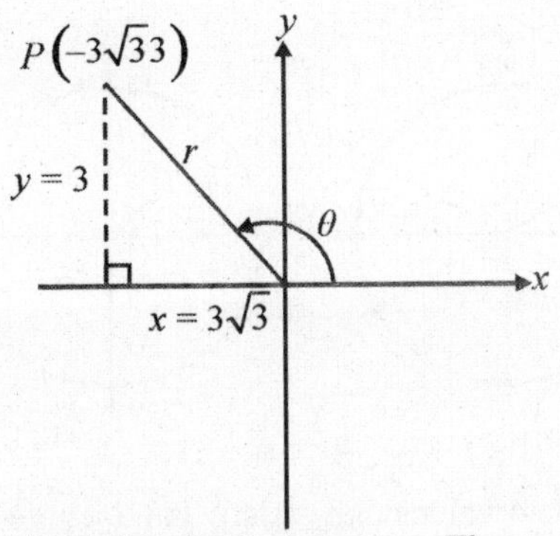


Applying the Pythagorean Theorem:

$r = \sqrt{x^2 + y^2}$

$r = \sqrt{\left(-3\sqrt{3}\right)^2 + 3^2}$

$r = \sqrt{27 + 9}$

$r = \sqrt{36}$

$r = 6$

(Remember, $r$ is always positive.)

| | | |
|---|---|---|
| $\csc\theta = \dfrac{r}{y}$ | $\sec\theta = \dfrac{r}{x}$ | $\cot\theta = \dfrac{x}{y}$ |
| $\csc\theta = \dfrac{6}{3} = 2$ | $\sec\theta = \dfrac{6}{-3\sqrt{3}} = \dfrac{-2}{\sqrt{3}} = \dfrac{-2\sqrt{3}}{3}$ | $\cot\theta = \dfrac{-3\sqrt{3}}{3} = -\sqrt{3}$ |

If $\csc\theta = 2$, then $\sin\theta = \dfrac{1}{2}$.

$\theta$ is in quadrant II and the reference angle is 30°. Thus, $\theta = 180° - 30° = 150°$.

## *Lesson 9 Solving Trigonometric Equations Involving Special Angles*

### CLASS EXERCISES ANSWERS AND SOLUTIONS

**1.**

$\csc^2\theta = \dfrac{4}{3}$
- Divide both sides by 3

$\csc\theta = \pm\dfrac{2}{\sqrt{3}}$
- Take the square root of both sides

$\sin\theta = \pm\dfrac{\sqrt{3}}{2}$
- Take the reciprocal of both sides to get primary trig ratios

On the unit circle, we need to find locations where $\sin\theta = \pm\dfrac{\sqrt{3}}{2}$, which is equivalent to locations where $y = \pm\dfrac{\sqrt{3}}{2}$.

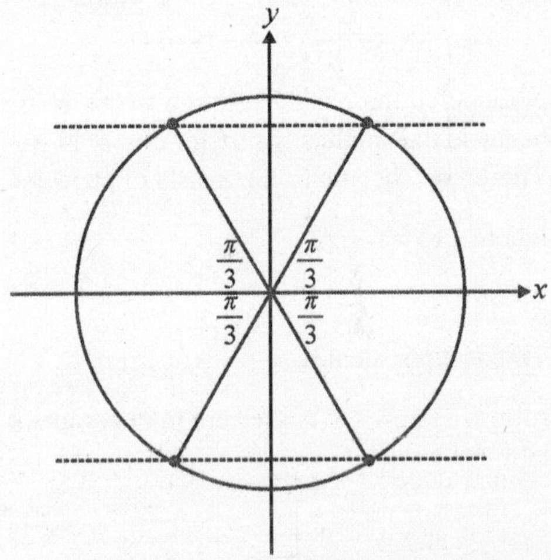


This time there are 4 angles that satisfy the equation. Again, it should be clear that the reference angle for each of them is 60°, or $\dfrac{\pi}{3}$ rad. We know the answers are to be given in radians because potential values for $\theta$ are given as $0 \le \theta < 2\pi$.

The solution consists of 4 answers:

$$\theta=\frac{\pi}{3}$$

$$\pi-\frac{\pi}{3}=\frac{2\pi}{3}$$

$$\pi+\frac{\pi}{3}=\frac{4\pi}{3}$$

$$2\pi-\frac{\pi}{3}=\frac{5\pi}{3}$$

Alternatively, by counting one rotation in terms of $\frac{\pi}{3}$ increments $(\frac{\pi}{3},\frac{2\pi}{3},\pi,\frac{4\pi}{3},\frac{5\pi}{3},2\pi)$ you should see that the answers are $\frac{\pi}{3},\frac{2\pi}{3},\frac{4\pi}{3},\frac{5\pi}{3}$.

**2.**

$(2\cos\theta-1)(\cos\theta+2)=0$ • Factor

$\cos\theta=\frac{1}{2}$ or $\cos\theta=-2$ • Solve

Notice that there is no solution to cos $\theta=-2$ because the smallest value for cos $\theta$ is –1. Therefore, the solution consists of answers where $\cos\theta=\frac{1}{2}$.

This is equivalent to $x=\frac{1}{2}$ (or from the CAST rule, we know the angles are in quadrants I and IV).

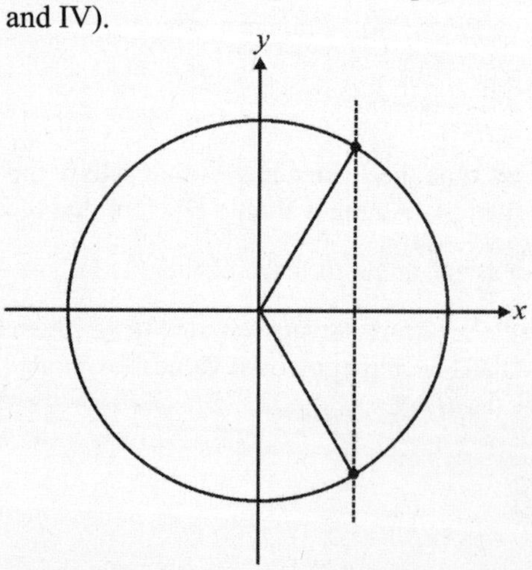


The solution consists of 2 answers as indicated on the graph.

$\theta=\frac{\pi}{3},\frac{5\pi}{3}$

**3.**

$3\tan^2\theta(\tan\theta+1)-1(\tan\theta+1)=0$ • Group and factor to get a common factor

$(3\tan^2\theta-1)(\tan\theta+1)=0$ • Common factor of $(\tan\theta+1)$

$\tan\theta=\pm\frac{1}{\sqrt{3}}$ or $\tan\theta=-1$ • Solve

Now, we examine the unit circle to determine where the ratio $\frac{y}{x}$ is equal to $\pm\frac{1}{\sqrt{3}}$ or –1.

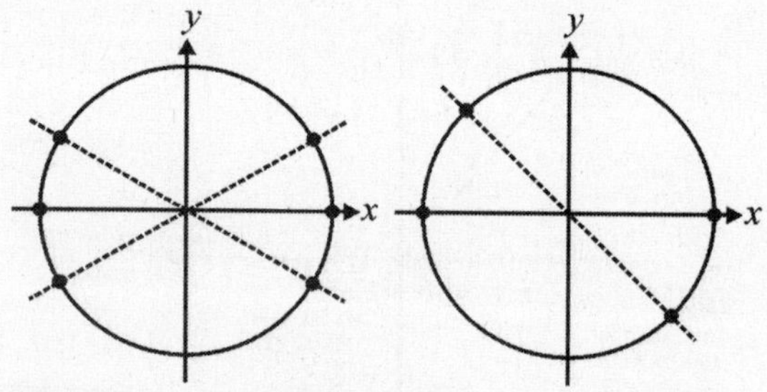


There are 4 values that satisfy $\tan\theta=\pm\frac{1}{\sqrt{3}}$:

30°, 150°, 210°, 330°

There are 2 values that satisfy tan $\theta=-1$:
135°, 315°

Therefore, the solution consists of 6 values.
$\theta=30°, 135°, 150°, 210°, 315°, 330°$

**4.**

$2\tan\theta\cos\theta+\tan\theta-2\cos\theta-1=0$ • Subtract $2\cos\theta+1$ from both sides

$\tan\theta(2\cos\theta+1)-1(2\cos\theta+1)=0$ • Group and factor to get a common factor

$(\tan\theta-1)(2\cos\theta+1)=0$ • Factor out the common factor, $(2\cos\theta+1)$

$\tan\theta=1$ or $\cos\theta=-\frac{1}{2}$ • Solve

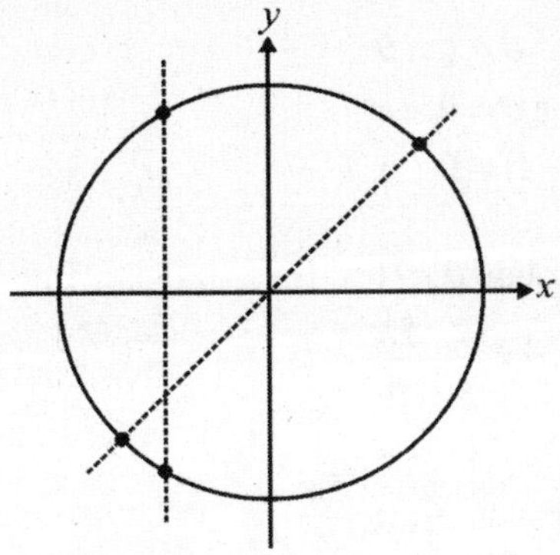


- Determine the values in the unit circle that satisfy each part of the equation

If $\tan\theta = 1$, $\theta = \ldots 45°, 225°\ldots$

If $\cos\theta = -\frac{1}{2}$, then $\theta = 120°, 240°, 480°, 600°\ldots$

Therefore, $\theta = (45 + 180n)°$, $(120 + 360n)°$, $(240 + 360n)°$, $n \in I$.

- Write out the general solution

**5.**

If $0° \le x < 360°$, then

$0° \le \frac{1}{2}x < 180°$

- Determine possible values for $\frac{1}{2}x$

$\tan\left(\frac{1}{2}x\right)\left[\tan\left(\frac{1}{2}x\right) + 1\right] = 0$

- Factor

$\tan\left(\frac{1}{2}x\right) = 0$ or $\tan\left(\frac{1}{2}x\right) = -1$

- Solve for $\tan\left(\frac{1}{2}x\right)$

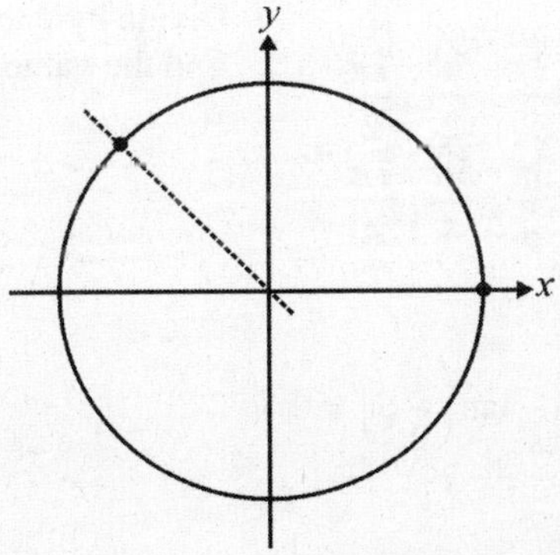


$\frac{1}{2}x = 0°, 135°$

- From the unit circle, for $0° \le \frac{1}{2}x < 180°$

$x = 0°, 270°$

- Multiply by 2 to get solutions for $0° \le x < 360°$

## PRACTICE EXERCISES ANSWERS AND SOLUTIONS

**1.** $2\cos\theta + 1 = 0$

$2\cos\theta = -1$

$\cos\theta = -\frac{1}{2}$

$\cos\theta = -\frac{1}{2}$

$\theta = 120°, 240°$

**2.** $2\sin\theta = \sin\theta$

$2\sin\theta - \sin\theta = 0$

$\sin\theta = 0$

$\theta = 0°, 180°$

**3.** $\tan\theta = \sqrt{3}$

$\theta = 60°, 240°$

**4.** $2\cos^2\theta = 1$

$\cos^2\theta = \frac{1}{2}$

$\cos\theta = \pm\frac{1}{\sqrt{2}}$ or $\pm\frac{\sqrt{2}}{2}$

$\theta = 45°, 135°, 225°, 315°$

**5.** $\csc\theta = \sin\theta$

$\frac{1}{\sin\theta} = \sin\theta$

$\sin^2\theta = 1$

$\sin\theta = \pm 1$

$\theta = 90°, 270°$

**6.** $\sin(2\theta)\cos(2\theta) = 0$

Note that if $0° \le \theta < 360°$, then $0° \le 2\theta < 720°$. Thus, the values for $2\theta$ can be any degree measure from 0° up to, but not including, 720°.

$\sin(2\theta)\cos(2\theta) = 0$

$\sin(2\theta) = 0$ or $\cos(2\theta) = 0$

$2\theta = 0°, 0 + 360° = 360°$ or
$2\theta = 90°, 90° + 360° = 450°$

$180°, 180° + 360° = 590°$ or
$270°, 270° + 360° = 630°$

To find the values for $\theta$, divide all values by 2.
$\theta = 0°, 180°$ or $\theta = 45°, 225°$
$90°, 270°$ $\quad 135°, 315°$

Therefore, $\theta = 0°, 45°, 90°, 135°, 180°, 225°, 270°$, and $315°$.

**7.** $4\sin\theta + 2 = 0$

$$4\sin\theta = -2$$
$$\sin\theta = \frac{-2}{4} = \frac{-1}{2}$$
$$\theta = \frac{7\pi}{6}, \frac{11\pi}{6}$$

**8.** $\cot^2\theta = 1$

$\cot\theta = \pm 1$

$\tan\theta = \pm 1$

Take the reciprocals of both sides of the equation to get primary trig ratios.

$$\theta = \frac{\pi}{4}, \frac{3\pi}{4}, \frac{5\pi}{4}, \frac{7\pi}{4}$$

**9.** $\sin\theta = \cos\theta$

- Divide both sides of the equation by $\cos\theta$

$$\frac{\sin\theta}{\cos\theta} = \frac{\cos\theta}{\cos\theta}$$
$$\tan\theta = 1$$
$$\theta = \frac{\pi}{4}, \frac{5\pi}{4}$$

**10.**

$$\tan^2\theta = \tan\theta$$
$$\tan^2\theta - \tan\theta = 0$$
$$\tan\theta(\tan\theta - 1) = 0$$

$\tan\theta = 0$ or $\tan\theta = 1$

$\theta = 0, \pi$ or $\theta = \frac{\pi}{4}, \frac{5\pi}{4}$

Therefore, $\theta = 0, \frac{\pi}{4}, \pi, \frac{5\pi}{4}$.

**11.** $\cos^2(2\theta) = 4\cos(2\theta)$

Note that if $0 \le \theta < 2\pi$, then $0 \le 2\theta < 4\pi$.

$$\cos^2(2\theta) - 4\cos(2\theta) = 0$$
$$\cos(2\theta)(\cos(2\theta) - 4) = 0$$

$\cos(2\theta) = 0$ or $\cos(2\theta) = 4$

The ratio $\cos 2\theta = 4$ is not possible because the cosine ratio is never greater than 1.

$$2\theta = \frac{\pi}{2}, \frac{\pi}{2} + 2\pi = \frac{5\pi}{2}, \text{ or}$$
$$2\theta = \frac{3\pi}{2}, \frac{3\pi}{2} + 2\pi = \frac{7\pi}{2}$$

Thus,

- Divide by 2 to find the value of $\theta$

$$2\theta = \frac{\pi}{2}, \frac{3\pi}{2}, \frac{5\pi}{2}, \frac{7\pi}{2}$$
$$\theta = \frac{\pi}{4}, \frac{3\pi}{4}, \frac{5\pi}{4}, \frac{7\pi}{4}$$

**12.** $\cot\left(\frac{1}{2}\theta\right) - \tan\left(\frac{1}{2}\theta\right) = 0$

If $0 \le \theta < 2\pi$, then $0 \le \frac{1}{2}\theta < \pi$.

$$\cot\left(\frac{1}{2}\theta\right) - \tan\left(\frac{1}{2}\theta\right) = 0$$
$$\frac{1}{\tan\left(\frac{1}{2}\theta\right)} - \tan\left(\frac{1}{2}\theta\right) = 0$$
$$1 - \tan^2\left(\frac{1}{2}\theta\right) = 0$$

Multiply all parts by $\tan\left(\frac{1}{2}\theta\right)$:

$$\tan^2\left(\frac{1}{2}\theta\right) = 1$$
$$\tan\left(\frac{1}{2}\theta\right) = \pm 1$$
$$\frac{1}{2}\theta = \frac{\pi}{4}, \frac{3\pi}{4}$$

Note that these are the values for $\frac{1}{2}\theta$, $0 \le \frac{1}{2}\theta < \pi$. Multiply by 2 to find the values for $\theta$:

$$\theta = \frac{\pi}{2}, \frac{3\pi}{2}$$

**13.**

$$\sin^3\theta - \sin\theta = 0$$
$$\sin\theta\left(\sin^2\theta - 1\right) = 0$$
$$\sin\theta(\sin\theta - 1)(\sin\theta + 1) = 0$$

$\sin\theta = 0$ or $\sin\theta = 1$ or $\sin\theta = -1$
$\theta = 0°, 180°$ or $\theta = 90°$ or $\theta = 270°$

$\theta = 0°, 90°, 180°, 270°$

**14.**

$$\cos\theta = \cos^4\theta$$
$$\cos\theta - \cos^4\theta = 0$$
$$\cos\theta\left(1 - \cos^3\theta\right) = 0$$

$\cos\theta = 0$ or
$\cos^3\theta = 1$
$\cos\theta = 1$
$\theta = 90°, 270°$ or $\theta = 0°$

$\theta = 0°, 90°, 270°$

**15.**

$$\cos^2\theta + 2\cos\theta - 3 = 0$$
$$(\cos\theta - 1)(\cos\theta + 3) = 0$$

$\cos\theta = 1$ or $\cos\theta = -3$

The ratio $\cos\theta = -3$ is not possible, since $\cos\theta$ is never less than $-1$.

$\theta = 0°$

**16.**

$$2\sin^2\theta - 3\sin\theta - 2 = 0$$
$$(2\sin\theta + 1)(\sin\theta - 2) = 0$$

$\sin\theta = -\frac{1}{2}$ or $\sin\theta = 2$

The ratio $\sin\theta = 2$ is not possible since sin $\theta$ is never greater than 1.

$\theta = 210°, 330°$

**17.**

$$2\cos^2\theta - \cos\theta - 1 = 0$$
$$(2\cos\theta + 1)(\cos\theta - 1) = 0$$

$\cos\theta = \frac{-1}{2}$ or $\cos\theta = 1$
$\theta = 120°, 240°$ or $\theta = 0°$

$\theta = 0°, 120°, 240°$

**18.** $2\sin^2(2\theta) - 3\sin(2\theta) + 1 = 0$

If $0° \le \theta < 360°$, then $0° \le 2\theta < 720°$.

$$2\sin^2(2\theta) - 3\sin(2\theta) + 1 = 0$$
$$(2\sin(2\theta) - 1)(\sin(2\theta) - 1) = 0$$

$\sin(2\theta) = \frac{1}{2}$ or $\sin(2\theta) = 1$

$2\theta = 30°, 30° + 360° = 390°$
$150°, 150° + 360° = 510°$ or
$2\theta = 90°, 90° + 360° = 450°$

If $2\theta = 30°, 90°, 150°, 390°, 450°, 510°$, then $\theta = 15°, 45°, 75°, 195°, 225°, 255°$.

**19.**

$$2\cos^2\theta - \sqrt{3}\cos\theta = 0$$
$$\cos\theta\left(2\cos\theta - \sqrt{3} = 0\right)$$

$\cos\theta = 0$ or $\cos\theta = \frac{\sqrt{3}}{2}$

$\theta = \frac{\pi}{2}, \frac{3\pi}{2}$ or $\theta = \frac{\pi}{6}, \frac{11\pi}{6}$

**20.**

$$\tan^2\theta = \sqrt{3}\tan\theta$$
$$\tan^2\theta - \sqrt{3}\tan\theta = 0$$
$$\tan\theta\left(\tan\theta - \sqrt{3}\right) = 0$$

$$\tan\theta = 0 \text{ or } \tan\theta = \sqrt{3}$$
$$\theta = 0, \pi \text{ or } \theta = \frac{\pi}{3}, \frac{4\pi}{3}$$

**21.**

$$2\sin\theta\cos\theta = \cos\theta$$
$$2\sin\theta\cos\theta - \cos\theta = 0$$
$$\cos\theta(2\sin\theta - 1) = 0$$

$$\cos\theta = 0 \text{ or } \sin\theta = \frac{1}{2}$$
$$\theta = \frac{\pi}{2}, \frac{3\pi}{2} \text{ or } \theta = \frac{\pi}{6}, \frac{5\pi}{6}$$

**22.**

$$2\sin^2\theta + 9\sin\theta + 4 = 0$$
$$(2\sin\theta + 1)(\sin\theta + 4) = 0$$

$$\sin\theta = -\frac{1}{2} \text{ or } \sin\theta = -4$$

The ratio $\sin\theta = -4$ is not possible, since $\sin\theta$ is never less than $-1$.

$$\theta = \frac{7\pi}{6}, \frac{11\pi}{6}$$

**23.**

$$\sec^2\theta - \sec\theta - 2 = 0$$
$$(\sec\theta + 1)(\sec\theta - 2) = 0$$

$$\sec\theta = -1 \text{ or } \sec\theta = 2$$

Take the reciprocals of both sides to get primary trig ratios.

$$\cos\theta = -1 \text{ or } \cos\theta = \frac{1}{2}$$
$$\theta = \pi \text{ or } \theta = \frac{\pi}{3}, \frac{5\pi}{3}$$

**24.**

$$\csc^2\theta + \csc\theta - 2 = 0$$
$$(\csc\theta - 1)(\csc\theta + 2) = 0$$

$$\csc\theta = 1 \text{ or } \csc\theta = -2$$

Take the reciprocals of both sides to get primary trig ratios.

$$\sin\theta = 1 \text{ or } \sin\theta = -\frac{1}{2}$$
$$\theta = \frac{\pi}{2} \text{ or } \theta = \frac{7\pi}{6}, \frac{11\pi}{6}$$

**25.**

$$\sec^2(2\theta) + 3\sec(2\theta) + 2 = 0$$

If $0 \le \theta < 2\pi$, then $0 \le 2\theta < 4\pi$.

$$\sec^2(2\theta) + 3\sec(2\theta) + 2 = 0$$
$$(\sec(2\theta) + 1)(\sec(2\theta) + 2) = 0$$

$$\sec(2\theta) = -1 \text{ or } \sec(2\theta) = -2$$

Take the reciprocals of both sides to get primary trig ratios.

$$\cos(2\theta) = -1 \text{ or } \cos(2\theta) = -\frac{1}{2}$$
$$2\theta = \pi,\ \pi + 2\pi = 3\pi \text{ or}$$
$$2\theta = \frac{2\pi}{3}, \frac{2\pi}{3} + 2\pi = \frac{8\pi}{3}, \frac{4\pi}{3},$$
$$\frac{4\pi}{3} + 2\pi = \frac{10\pi}{3}$$

If $2\theta = \frac{2\pi}{3}, \pi, \frac{4\pi}{3}, \frac{8\pi}{3}, 3\pi, \frac{10\pi}{3}$,

then $\theta = \frac{\pi}{3}, \frac{\pi}{2}, \frac{2\pi}{3}, \frac{4\pi}{3}, \frac{3\pi}{2}, \frac{5\pi}{3}$.

 

**26.**

$$\cot^3\theta+\cot^2\theta=\cot\theta+1$$
$$\cot^3\theta+\cot^2\theta-\cot\theta-1=0$$
$$\cot^2\theta(\cot\theta+1)-1(\cot\theta+1)=0$$
$$(\cot\theta+1)(\cot^2\theta-1)=0$$

$\cot\theta=-1$ or $\cot^2\theta=1$
$\cot\theta=\pm1$

Take the reciprocal of both sides to get primary trig ratios.

$\tan\theta=-1$ or $\tan\theta=\pm1$

$\theta=\frac{3\pi}{4},\frac{7\pi}{4}$ or $\theta=\frac{\pi}{4},\frac{3\pi}{4},\frac{5\pi}{4},\frac{7\pi}{4}$

$\theta=\frac{\pi}{4},\frac{3\pi}{4},\frac{5\pi}{4},\frac{7\pi}{4}$

**27.** $2\cos\theta-\sqrt{3}=0$

$\cos\theta=\frac{\sqrt{3}}{2}$
$\theta=30°, 330°$

General solution: $\theta=(30+360n)°$ or $\theta=(330+360n)°$, where $n\in I$

**28.** $2\sin^2\theta+2\sin\theta-\sqrt{3}\sin\theta-\sqrt{3}=0$
$$2\sin\theta(\sin\theta+1)-\sqrt{3}(\sin\theta+1)=0$$
$$(\sin\theta+1)(2\sin\theta-\sqrt{3})=0$$

$\sin\theta=-1$ or $\sin\theta=\frac{\sqrt{3}}{2}$
$\theta=270°$ or $\theta=60°, 120°$

General solution:
$\theta=(60+360n)°$
or $\theta=(120+360n)°$
or $\theta=(270+360n)°$, where $n\in I$

**29.**

$$2\cos^2\theta-2\cos\theta=\sqrt{3}-\sqrt{3}\cos\theta$$
$$2\cos^2\theta-2\cos\theta+\sqrt{3}\cos\theta-\sqrt{3}=0$$
$$2\cos\theta(\cos\theta-1)+\sqrt{3}(\cos\theta-1)=0$$
$$(\cos\theta-1)(2\cos\theta+\sqrt{3})=0$$

$\cos\theta=1$ or $\cos\theta=\frac{-\sqrt{3}}{2}$
$\theta=0°$ or $\theta=150°, 210°$

General solution:
$\theta=(360n)°$
or $\theta=(150+360n)°$
or $\theta=(210+360n)°$

**30.** $\tan^2\theta-1=0$
$\tan^2\theta=1$
$\tan\theta=\pm1$

$\theta=45°, 135°, 225°, 315°$

General solution:
Notice that $45°+180°=225°$ and that $135°+150°=315°$
$\theta=(45+180n)°$ or $\theta=(135+180n)°$, where $n\in I$

**31.** $\cos^2\theta-\frac{\cos\theta}{2}=0$

- Multiply all parts by 2

$2\cos^2\theta-\cos\theta=0$

$\cos\theta(2\cos\theta-1)=0$

$\cos\theta=0$ or $\cos\theta=\frac{1}{2}$

$\theta=\frac{\pi}{2},\frac{3\pi}{2}$ or $\theta=\frac{\pi}{3},\frac{5\pi}{3}$

Notice that $\frac{\pi}{2}+\pi=\frac{3\pi}{2}$.

General solution:

$\theta=\frac{\pi}{2}+n\pi$ or $\theta=\frac{\pi}{3}+2n\pi$ or

$\theta=\frac{5\pi}{3}+2n\pi$, where $n\in I$

**32.**

$$\sin^2\theta - 2\sin\theta = 3$$
$$\sin^2 - 2\sin\theta - 3 = 0$$
$$(\sin\theta + 1)(\sin\theta - 3) = 0$$

$\sin\theta = -1$ or $\sin\theta = 3$
The ratio $\sin\theta = 3$ is not possible since sin $\theta$ is never greater than 1.

$$\theta = \frac{3\pi}{2}$$

General solution: $\theta = \frac{3\pi}{2} + 2n\pi$, where $n \in I$

**33.**

$$3\tan\theta = \cot\theta$$
$$3\tan\theta = \frac{1}{\tan\theta}$$
$$3\tan^2\theta = 1$$
$$\tan^2\theta = \frac{1}{3}$$
$$\tan\theta = \pm\frac{1}{\sqrt{3}} \text{ or } \pm\frac{\sqrt{3}}{3}$$
$$\theta = \frac{\pi}{6}, \frac{5\pi}{6}, \frac{7\pi}{6}, \frac{11\pi}{6}$$

Notice that $\frac{\pi}{6} + \pi = \frac{7\pi}{6}$, and that $\frac{5\pi}{6} + \pi = \frac{11\pi}{6}$.

General solution:

$\theta = \frac{\pi}{6} + n\pi$ or $\theta = \frac{5\pi}{6} + n\pi$, where $n \in I$

**34.**

$$4\cos^2\theta + 4\cos\theta + 1 = 0$$
$$(2\cos\theta + 1)(2\cos\theta + 1) = 0$$
$$\cos\theta = -\frac{1}{2}$$
$$\theta = \frac{2\pi}{3}, \frac{4\pi}{3}$$

General solution:

$\theta = \frac{2\pi}{3} + 2n\pi$ or $\theta = \frac{4\pi}{3} + 2\pi$, where $n \in I$

**35. D**

$$\cot^2\theta = \cot\theta$$
$$\cot^2 - \cot\theta = 0$$
$$\cot\theta(\cot\theta - 1) = 0$$

$\cot\theta = 0$ or $\cot\theta = 1$

$\tan\theta = \frac{1}{0}$ or $\tan\theta = 1$

Take the reciprocals of both sides of the equation to get primary trig ratios.

Note that $\tan\theta = \frac{y}{x}$, and tan $\theta$ is undefined when $x = 0$.

$\theta = \frac{\pi}{2}, \frac{3\pi}{2}$ or $\theta = \frac{\pi}{4}, \frac{5\pi}{4}$

There are 4 values that satisfy the equation.

**36. C**

$$\cos\theta\sin\theta = \cos^2\theta$$
$$\cos\theta\sin\theta - \cos^2 = 0$$
$$\cos\theta(\sin\theta - \cos\theta) = 0$$

$\cos\theta = 0$ or $\sin\theta = \cos\theta$

$$\frac{\sin\theta}{\cos\theta} = 1$$
$$\tan\theta = 1$$

$\theta = \frac{\pi}{2}, \frac{3\pi}{2}$ or $\theta = \frac{\pi}{4}, \frac{5\pi}{4}$

Notice that $\frac{\pi}{2} + \pi = \frac{3\pi}{2}$, and that $\frac{\pi}{4} + \pi = \frac{5\pi}{4}$.

General solution:

$\theta = \frac{\pi}{2} + n\pi$ or $\theta = \frac{\pi}{4} + n\pi$, where $n \in I$

 

## *Lesson 10 Solving Trigonometric Equations with a Calculator*

### CLASS EXERCISES ANSWERS AND SOLUTIONS

**1.** To find the first solution of this equation, $\theta_1$, we enter $\tan^{-1}(1.364)$ into a calculator.

When we enter this into a calculator, we find that the first solution, $\theta_1$, is approximately 0.94 rad.

The diagram shows the position of the two solutions:

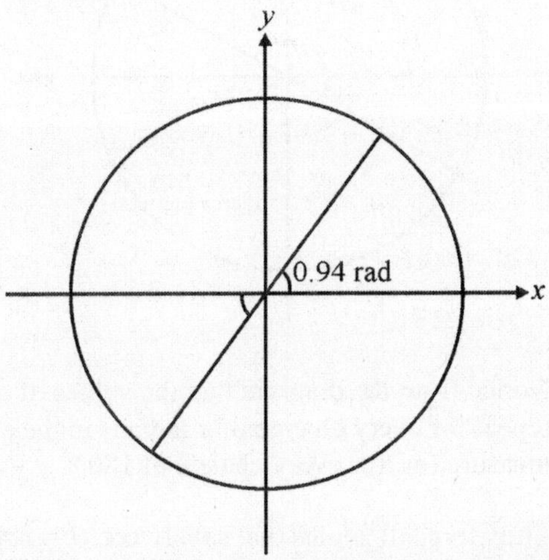


Since the tangent ratio is positive in quadrants I and III, and the reference angles must be equal, it is clear that $\theta_2 = \pi + \theta_1$.

We get:
$\theta_2 \doteq \pi + 0.94$
$\theta_2 \doteq 4.08$

So, the complete solution to the equation $\tan\theta = 1.364$, $0 \leq \theta < 2\pi$ is: $\theta = 0.94$ rad or 4.08 rad.

This may be simply stated as $\theta = 0.94$, 4.08.

**2.** Since a calculator does not have a button for calculating $\sec^{-1}$, we rewrite the equation as a cosine ratio by using the identity $\cos\theta = \frac{1}{\sec\theta}$.

Since $\sec\theta = 3.601$, we get $\cos\theta = \frac{1}{3.601}$.

$\cos\theta \doteq 0.278$

From here, the equation is solved as previous questions have been.

To find the first solution, $\theta_1$, enter $\cos^{-1}(0.278)$ into a calculator. You should get $\theta_1 \doteq 1.29$.

To find the second solution, use the formula for cosine equations, $\theta_2 = 2\pi - \theta_1$, or sketch a diagram. From the CAST rule, we know the other solution is in quadrant IV.

The result is:
$\theta_2 \doteq 2\pi - 1.29$
$\theta_2 \doteq 4.99$

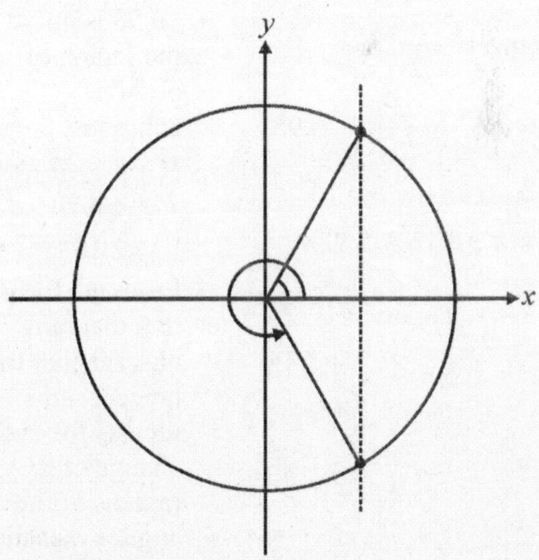


So, the complete solution to the equation, $\sec\theta = 3.601$, $0 \leq \theta < 2\pi$, is $\theta = 1.29$ rad or 4.99 rad.

**3.** $5\tan\theta - 1 = 3$

$5\tan\theta = 4$
- Add 1 to both sides

$\tan\theta = \frac{4}{5}$
- Divide both sides by 5

$\tan^{-1}\left(\frac{4}{5}\right) \doteq 38.7$
- $\theta_1 \doteq 38.7°$, found using a calculator

$\theta_2 = 180° + \theta_1$
$\theta_2 = 180° + 38.7°$
$\theta_2 = 218.7°$
- The second solution is obtained from a diagram or formula

$\theta = 38.7°, 218.7°$
- Complete solution

**4.** $6\tan^2\theta + 7\tan\theta - 3 = 0$

$(3\tan\theta - 1)(2\tan\theta + 3) = 0$
- Factor

$\tan\theta = \frac{1}{3}$ or $\tan\theta = -\frac{3}{2}$
- Solve

If $\tan\theta = \frac{1}{3}$, then
$\tan^{-1}\left(\frac{1}{3}\right) \doteq 0.32.$
- $\theta_1 \doteq 0.32$ rad, found using a calculator

$\theta_2 \doteq \pi + 0.32 \doteq 3.46$
- The second solution is obtained from a diagram or formula

If $\tan\theta = -\frac{3}{2}$, then
$\theta_1 = \tan^{-1}\left(-\frac{3}{2}\right) \doteq -0.98.$
$\theta_1 \doteq -0.98 + \pi \doteq 2.16$
$\theta_2 \doteq \pi + 2.16 \doteq 5.30$
- –0.98 is not in the region of possible solutions, $0 \le \theta < 2\pi$, so $\theta_1 \doteq -0.98 + \pi \doteq 2.16$
- From the formula or a diagram, observe that the tangent ratio repeats for every change of $\pi$ radians of the angle's measure

$\theta = 0.32, 2.16, 3.46, 5.30$
- Complete solution to the nearest hundredth of a radian with answers written in ascending order

**5.** We can rewrite the equation as a tangent instead of a cotangent by using the identity $\tan\theta = \frac{1}{\cot\theta}$.

If $\cot\theta = 4.45$, then $\tan\theta = \frac{1}{4.45}$ and, thus, $\tan\theta \doteq 0.225$.

To find the first solution, $\theta_1$, enter $\tan^{-1}(0.225)$ into a calculator. You should get $\theta_1 \doteq 0.22$ rad.

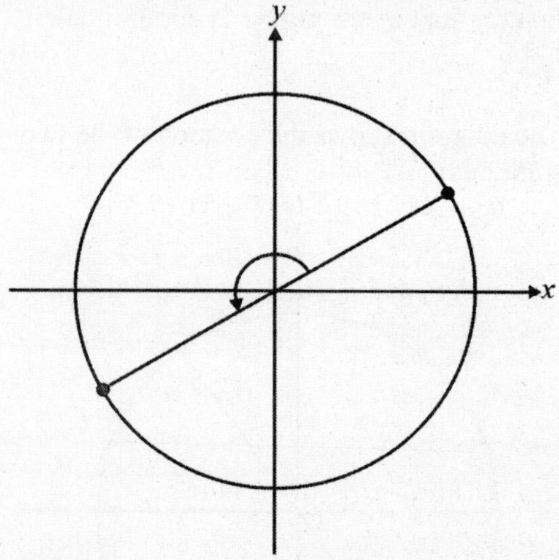


Notice from the diagram that the values of tan $\theta$ repeat for every change of $\pi$ radians in the angle's measure (or, for every change of 180º).

Therefore, all angles that satisfy cot $\theta = 4.45$ can be expressed by adding or subtracting multiples of $\pi$ to 0.22 rad. Therefore, the general solution to cot $\theta = 4.45$ is $\theta = 0.22 + \pi n$, where $n \in I$.

## PRACTICE EXERCISES ANSWERS AND SOLUTIONS

**1.** $\tan\theta = 4.474$

$\theta_1 = \tan^{-1}(4.474) \doteq 77.4°$
$\theta_2 \doteq 180° + \theta_1$
$\doteq 180° + 77.4°$
$\doteq 257.4°$

Thus, $\theta \doteq 77.4°, 257.4°$.

**2.** $\sec\theta = 2.275$

$\cos\theta = \dfrac{1}{2.275}$

$\theta_1 = \cos^{-1}\left(\dfrac{1}{2.275}\right) \doteq 63.9°$

$\theta_2 = 360° - \theta_1$

$\theta_2 \doteq 360° - 63.9°$

$\theta_2 \doteq 296.1°$

Thus, $\theta \doteq 63.9°, 296.1°$.

**3.** $\sin\theta = -0.784$

$\theta_1 = \sin^{-1}(-0.784) \doteq -51.6°$

Find the positive angle coterminal with –51.6°:

$-51.6° + 360° = 308.4°$

$\theta_2 = 180° - \theta_1$

$\theta_2 \doteq 180° - (-51.6°)$

$\theta_2 \doteq 231.6°$

Thus, $\theta \doteq 231.6°, 308.4°$.

**4.** $\cot\theta = -0.771$

$\tan\theta = \dfrac{1}{-0.771}$

$\theta = \tan^{-1}\left(\dfrac{1}{-0.771}\right) \doteq -52.4°$

Find the positive angle coterminal with –52.4°:

$-52.4° + 360° = 307.6°$

$\theta_2 = 180° + \theta_1$

$\theta_2 \doteq 180° + (-52.4°)$

$\theta_2 \doteq 127.6°$

Thus, $\theta \doteq 127.6°, 307.6°$.

**5.** $\cos\theta = -0.266$

$\theta_1 = \cos^{-1}(-0.266) \doteq 105.4°$

$\theta_2 = 360° - \theta_1$

$\theta_2 \doteq 360° - 105.4°$

$\theta_2 \doteq 254.6°$

Thus, $\theta \doteq 105.4°, 254.6°$

**6.** $\csc\theta = 1.890$

$\sin\theta = \dfrac{1}{1.870}$

$\theta_1 = \sin^{-1}\left(\dfrac{1}{1.870}\right) \doteq 31.9°$

$\theta_2 = 180° - \theta_1$

$\theta_2 \doteq 180° - 31.9°$

$\theta_2 \doteq 148.1°$

Thus, $\theta \doteq 31.9°, 148.1°$.

**7.** $\sin\theta = 0.345$

$\theta_1 = \sin^{-1}(0.345) \doteq 20.2°$

$\theta_2 \doteq 180° - \theta_1$

$\theta_2 \doteq 180° - 20.2°$

$\theta_2 \doteq 159.8°$

Thus, $\theta \doteq 20.2°, 159.8°$.

**8.** $\tan\theta = -0.145$

$\theta_1 = \tan^{-1}(-0.145) \doteq -8.3°$

Find the positive angle coterminal with –8.3°:

$-8.3° + 360° = 351.7°$

$\theta_2 = 180° + \theta_1$

$\theta_2 \doteq 180° + (-8.3°)$

$\theta_2 \doteq 171.7°$

Thus, $\theta_2 \doteq 171.7°, 351.7°$.

**9.** $\csc\theta = -2.887$

$$\sin\theta = \frac{1}{-2.887}$$

$$\theta_1 = \sin^{-1}\left(\frac{1}{-2.887}\right) \doteq -20.3°$$

Find the positive angle coterminal with −20.3°:
−20.3° + 360° = 339.7°

$$\theta_2 = 180° - \theta_1$$
$$\theta_2 \doteq 200.3°$$

Thus, $\theta \doteq 200.3°, 339.7°$.

**10.** $\tan\theta = 0.411$

$$\theta_1 = \tan^{-1}(0.411) \doteq 22.3°$$
$$\theta_2 = 180° + \theta_1$$
$$\theta_2 \doteq 180° + 22.3°$$
$$\theta_2 \doteq 202.3°$$

Thus, $\theta \doteq 22.3°, 202.3°$.

**11.** $\csc\theta = -3.345$

$$\sin\theta = \frac{1}{-3.345}$$

$$\theta_1 = \sin^{-1}\left(\frac{1}{-3.345}\right) \doteq -17.4°$$

Find the positive angle coterminal with −17.4°:
−17.4° + 360° = 342.6°.

$$\theta_2 = 180° - \theta_1$$
$$\theta_2 \doteq 180° - (-17.4°)$$
$$\theta_2 \doteq 197.4°$$

Thus, $\theta \doteq 197.4°, 342.6°$

**12.** $\cos\theta = -0.287$

$$\theta_1 = \cos^{-1}(-0.287) \doteq 106.7°$$
$$\theta_2 = 360° - \theta_1$$
$$\theta_2 \doteq 360° - 106.7°$$
$$\theta_2 \doteq 253.3°$$

Thus, $\theta \doteq 106.7°, 253.3°$.

**13.** $\sec\theta = -4.416$

$$\cos\theta = \frac{1}{-4.416}$$

$$\theta_1 = \cos^{-1}\left(\frac{1}{-4.416}\right) \doteq 103.1°$$
$$\theta_2 = 360° - \theta_1$$
$$\theta_2 \doteq 360° - 103.1°$$
$$\theta_2 \doteq 256.9°$$

Thus, $\theta \doteq 103.1°, 256.9°$.

**14.** $\cos\theta = 0.704$

$$\theta_1 = \cos^{-1}(0.704) \doteq 45.3°$$
$$\theta_2 = 360° - \theta_1$$
$$\theta_2 \doteq 360° - 45.3°$$
$$\theta_2 \doteq 314.7°$$

Thus, $\theta \doteq 45.3°, 314.7°$.

**15.** $\csc\theta = 8.45$

$$\sin\theta = \frac{1}{8.45}$$

$$\theta_1 = \sin^{-1}\left(\frac{1}{8.45}\right) \doteq 6.8°$$
$$\theta_2 = 180° - \theta_1$$
$$\theta_2 \doteq 180° - 6.8°$$
$$\theta_2 \doteq 173.2°$$

Thus, $\theta \doteq 6.8°, 173.2°$.

**16.** $\tan(2\theta) = -0.409$

If $0° \le \theta < 360°$, then $0° \le 2\theta < 720°$.

$$\tan 2\theta = -0.409$$
$$2\theta_1 = \tan^{-1}(-0.409) \doteq -22.2°$$

 

Find the positive angle coterminal with –22.2°:
–22.2° + 360° = 337.8°.

$2\theta_2 = 180° + 2\theta_1$
$2\theta_2 \doteq 180° + (-22.2°)$
$2\theta_2 \doteq 157.8°$

$2\theta_2 \doteq 337.8°, 337.8° + 360°, 157.8°, 157.8° + 360°$
$2\theta_2 \doteq 337.8°, 697.8°, 157.8°, 517.8°$

Divide by 2 to solve for $\theta$:
$\theta \doteq$ 168.9°, 348.9°, 70.9°, 258.9°

Thus, $\theta \doteq$ 70.9°, 168.9°, 258.9°, 348.9°.

**17.** $\sin\left(\frac{1}{2}\theta\right) = 0.385$

If $0° \le \theta < 360°$, then $0° \le \frac{1}{2}\theta < 180°$.

$\frac{1}{2}\theta_1 = \sin^{-1}(0.385) \doteq 22.6°$
$\frac{1}{2}\theta_2 \doteq 180° - 22.6° \doteq 157.4°$

Multiply by 2 to solve for $\theta$:
$\theta \doteq$ 45.2°, 314.8°

**18.** $\tan(2\theta) = 4.22$

If $0° \le \theta < 360°$, then $0° \le 2\theta < 720°$.

$\tan 2\theta = 4.22$

$2\theta_1 = \tan^{-1}(4.22) \doteq 76.7°$
$2\theta_2 = 180° + 2\theta_1$
$2\theta_2 \doteq 180 + 76.7$
$2\theta_2 \doteq 256.7°$

$2\theta \doteq 76.7°, 76.7° + 360°, 256.7°, 256.7° + 360°$
$2\theta \doteq 76.7°, 436.7°, 256.7°, 616.7°$

Divide by 2 to solve for $\theta$:
$\theta \doteq$ 38.4°, 218.4°, 128.4°, 308.4°

Thus, $\theta \doteq$ 38.4°, 128.4°, 218.4°, 308.4°.

**19.** $\cos\theta = -0.449$

$\theta_1 = \cos^{-1}(-0.449) \doteq 2.036$
$\theta_2 = 2\pi - \theta_1$
$\theta_2 = 2\pi - 2.036$
$\theta_2 \doteq 4.247$

Thus, $\theta \doteq$ 2.04, 4.25.

**20.** $\cot\theta = -0.575$

$\tan\theta = \frac{1}{-0.575}$

$\theta_1 = \tan^{-1}\left(\frac{1}{-0.575}\right) \doteq -1.049$

Find the positive angle coterminal with –1.049:
$-1.049 + 2\pi \doteq 5.234$

$\theta_2 = \pi + \theta_1$
$\theta_2 \doteq \pi + (-1.049)$
$\theta_2 \doteq 2.093$

Thus, $\theta \doteq$ 2.09, 5.23.

**21.** $\csc\theta = 5.678$

$\sin\theta = \frac{1}{5.678}$

$\theta_1 = \sin^{-1}\left(\frac{1}{5.678}\right) \doteq 0.177$
$\theta_2 = \pi - \theta_1$
$\theta_2 \doteq \pi - 0.177$
$\theta_2 \doteq 2.965$

Thus, $\theta \doteq$ 0.18, 2.97.

**22.** $\sin\theta = -0.124$

$\theta_1 = \sin^{-1}(-0.124) \doteq -0.124$

Find the positive angle coterminal with −0.124:
$-0.124 + 2\pi \doteq 6.159$.

$\theta_2 = \pi - \theta_1$
$\theta_2 \doteq \pi - (0.124)$
$\theta_2 \doteq 3.266$

Thus, $\theta \doteq 3.27, 6.16$.

**23.** $\tan\theta = -1.045$

$\theta_1 = \tan^{-1}(-1045) \doteq -0.807$

Find the positive angle coterminal with −0.807:
$-0.807 + 2\pi \doteq 5.476$.

$\theta_2 = \pi - \theta_1$
$\theta_2 \doteq \pi + (-0.807)$
$\theta_2 \doteq 2.335$

Thus, $\theta \doteq 2.34, 5.48$.

**24.** $\csc\theta = 2.061$

$\sin\theta = \frac{1}{2.061}$

$\theta_1 = \sin^{-1}\left(\frac{1}{2.061}\right) \doteq 0.507$

$\theta_2 = \pi - \theta_1$
$\theta_2 \doteq \pi - 0.507$
$\theta_2 \doteq 2.635$

Thus, $\theta \doteq 0.51, 2.64$

**25.** $\cos\theta = 0.842$

$\theta_1 = \cos^{-1}(0.842) \doteq 0.570$
$\theta_2 = 2\pi - \theta_1$
$\theta_2 \doteq 2\pi - 0.570$
$\theta_2 \doteq 5.713$

Thus, $\theta \doteq 0.57, 5.71$.

**26.** $\sec\theta = -2$

$\cos\theta = \frac{1}{-2}$

$\theta_1 = \cos^{-1}\left(\frac{1}{-2}\right) \doteq 2.094$

$\theta_2 = 2\pi - \theta_1$
$\theta_2 \doteq 2\pi - 2.094$
$\theta_2 \doteq 4.189$

Thus, $\theta \doteq 2.09, 4.19$.

**27.** $\sin\theta = 0.366$

$\theta_1 = \sin^{-1}(0.366) \doteq 0.375$
$\theta_2 = \pi - \theta_1$
$\theta_2 = \pi - 0.375$
$\theta_2 = 2.767$

Thus, $\theta \doteq 0.38, 2.77$.

**28.** $\tan\theta = -3.2$

$\theta_1 = \tan^{-1}(-3.2) \doteq -1.268$

Find the positive angle coterminal with −1.268:
$-1.268 + 2\pi \doteq 5.015$

$\theta_2 = \pi + \theta_1$
$\theta_2 \doteq \pi + (-1.268)$
$\theta_2 \doteq 1.874$

Thus, $\theta \doteq 1.87, 5.02$.

**29.** $\cos\ \theta = -1.55$

Since $\cos\ \theta$ is never less than −1, there is no solution.

 

**30.** $\csc\theta = 2.061$

$\sin\theta = \dfrac{1}{2.061}$

$\theta_1 = \sin^{-1}\left(\dfrac{1}{2.061}\right) \doteq 0.507$

$\theta_2 = \pi - \theta_1$

$\theta_2 \doteq \pi - 0.507$

$\theta_2 \doteq 2.635$

Thus, $\theta \doteq 0.51,\ 2.64$.

**31.** $\cot\theta = -3.451$

$\tan\theta = \dfrac{1}{-3.451}$

$\theta_1 = \tan^{-1}\left(\dfrac{1}{-3.451}\right) \doteq -0.282$

Find the positive angle coterminal with –0.282:
$-0.282 + 2\pi \doteq 6.001$

$\theta_2 = \pi + \theta_1$

$\theta_2 \doteq \pi + (-0.282)$

$\theta_2 = 2.860$

Thus, $\theta \doteq 2.86,\ 6.00$.

**32.** $\csc\theta = -1.882$

$\sin\theta = \dfrac{1}{-1.822}$

$\theta_1 = \sin^{-1}\left(\dfrac{1}{-1.822}\right) \doteq -0.581$

Find the positive angle coterminal with –0.581:
$-0.581 + 2\pi \doteq 5.702$

$\theta_2 = \pi - \theta_1$

$\theta_2 \doteq \pi - (-0.581)$

$\theta_2 \doteq 3.723$

Thus, $\theta \doteq 3.72,\ 5.70$.

**33.** $\cos\theta = -0.595$

$\theta_1 = \cos^{-1}(-0.595) \doteq 2.208$

$\theta_2 = 2\pi - \theta_1$

$\theta_2 \doteq 2\pi - 2.208$

$\theta_2 \doteq 4.075$

Thus, $\theta \doteq 2.21,\ 4.08$.

**34.** $\sec(2\theta) = 9.4$

If $0° \le \theta < 2\pi$, then $0° \le 2\theta < 4\pi$.

$\sec 2\theta = 9.4$

$\cos 2\theta = \dfrac{1}{9.4}$

$2\theta_1 = \cos^{-1}\left(\dfrac{1}{9.4}\right) \doteq 1.464$

$2\theta_2 = 2\pi - 2\theta_1$

$2\theta_2 \doteq 2\pi - (1.464)$

$2\theta_2 \doteq 4.819$

$2\theta \doteq 1.464,\ 1.464 + 2\pi,\ 4.819,\ 4.819 + 2\pi$
$2\theta \doteq 1.464,\ 7.747,\ 4.819,\ 11.102$

Divide by 2 to solve for $\theta$:
$\theta \doteq 0.73,\ 3.87,\ 2.41,\ 5.55$

**35.** $\sin(3\theta) = -0.888$

If $0° \le \theta < 2\pi$, then $0° \le 3\theta < 6\pi$.

$\sin(3\theta) = -0.888$

$3\theta_1 = \sin^{-1}(0.888) \doteq -1.093$

Find the positive angle coterminal with –1.093:
$-1.093 + 2\pi \doteq 5.190$

$3\theta_2 = \pi - 3\theta_1$

$3\theta_2 \doteq \pi - (-1.293)$

$3\theta_2 \doteq 4.235$

$3\theta \doteq 5.190,\ 5.190 + 2\pi,\ 5.190 + 4\pi,\ 4.235,$
$4.235 + 2\pi,\ 4.235 + 4\pi$
$3\theta \doteq 5.190,\ 11.473,\ 17.756,\ 4.235,$
$10.518,\ 16.801$

Divide by 3 to solve for $\theta$:
$\theta \doteq 1.73, 3.82, 5.92, 1.41, 3.51, 5.60$

Thus, $\theta \doteq 1.41, 1.73, 3.51, 3.82, 5.60, 5.92$.

**36.** $\cos\left(\frac{1}{2}\theta\right) = 0.658$

If $0 \le \theta < 2\pi$, then $0 \le \frac{1}{2}\theta < \pi$.

$$\cos\left(\frac{1}{2}\theta\right) = 0.658$$
$$\frac{1}{2}\theta_1 = \cos^{-1}(0.658) \doteq 0.853$$

Note: There is no other value for $\frac{1}{2}\theta$ within the restricted domain, $0 \le \frac{1}{2}\theta < \pi$

Multiply by 2 to solve for $\theta$

Thus, $\theta \doteq 2 \times 0.853 \doteq 1.71$

**37.** $3\cos\theta + 4 = 8$

$$3\cos\theta = 4$$
$$\cos\theta = \frac{4}{3}$$

Since cos $\theta$ is never greater than 1, there is no solution.

**38.** $(4\sin\theta + 3)(5\cos\theta - 2) = 0$

$\sin\theta = \frac{-3}{4}$ or $\cos\theta = \frac{-2}{5}$

$\theta_1 \doteq -48.6°$
$-48.6°$ is coterminal with
$-48.6 + 360 = 311.4°$
$\theta_2 \doteq 180 - (-48.6)$
$\theta_2 \doteq 228.6°$

$\theta_1 \doteq 113.6°$
$\theta_2 \doteq 360 - 113.6$
$\theta_2 \doteq 246.4$

Thus, $\theta \doteq 113.6°, 228.6°, 246.4°, 311.4°$.

**39.** $5 - \sin\theta = 8\sin\theta$

$$5 = 9\sin\theta$$
$$\sin\theta = \frac{5}{9}$$

$\theta_1 = \sin^{-1}\left(\frac{5}{9}\right) \doteq 33.7°$
$\theta_2 \doteq 180 - 37.4$
$\theta_2 \doteq 142.6°$

Thus, $\theta \doteq 37.4°, 147.6°$.

**40.** $\tan^2\theta - 3 = 0$

$$\tan^2\theta = 3$$
$$\tan\theta = \pm\sqrt{3}$$

$\tan\theta = \sqrt{3}$ or $\tan\theta = -\sqrt{3}$

$\theta_1 = 60°$
$\theta_2 = 180 + 60$
$\theta_2 = 240°$

$\theta_1 = 120°$
$\theta_2 = 180 + 120$
$\theta_2 = 300°$

Thus, $\theta = 60°, 120°, 240°, 300°$.

**41.** $7\sin^2(2\theta) = 5\sin(2\theta)$

If $0° \le \theta < 360°$, then $0° \le 2\theta < 720°$.

$$7\sin^2 2\theta = 5\sin 2\theta$$
$$7\sin^2 2\theta - 5\sin 2\theta = 0$$
$$\sin 2\theta(7\sin 2\theta - 5) = 0$$

$\sin 2\theta = 0$ or $\sin 2\theta = \frac{5}{7}$

$2\theta_1 = 0°$
$2\theta_2 = 180° + 0$
$2\theta_2 = 180°$

$2\theta_1 = \sin^{-1}\left(\frac{5}{7}\right) \doteq 45.6°$
$2\theta_2 \doteq 180° - 45.6°$
$2\theta_2 \doteq 134.4°$

$2\theta = 0°, 0° + 360°, 180°, 180° + 360°$
$2\theta = 0°, 360°, 180°, 540°$
$\theta = 0°, 180°, 90°, 270°$

$2\theta \doteq 45.6°, 45.6° + 360°, 134.4°, 134.4° + 360°$
$2\theta \doteq 45.6°, 405.6°, 134.4°, 394.4°$
$\theta \doteq 22.8°, 202.8°, 67.2°, 197.2°$

Thus, $\theta = 0°, 22.8°, 67.2°, 90°, 180°, 197.2°, 202.8°, 270°$.

 

**42.** $3\cos^2\theta + \cos\theta = 6\cos\theta - \cos^2\theta$
$4\cos^2\theta - 5\cos\theta = 0$
$\cos\theta(4\cos\theta - 5) = 0$

$\cos\theta = 0$ or $\cos\theta = \frac{-5}{4}$

$\theta = 90°, 270°$

Since cos $\theta$ is never less than –1, there is no solution.

Thus, $\theta = 90°, 270°$.

**43.** $\tan^2\theta - 5 = 3 + 2\tan\theta$
$\tan^2\theta - 2\tan\theta - 8 = 0$
$(\tan\theta + 2)(\tan\theta - 4) = 0$

$\tan\theta = -2$ or $\tan\theta = 4$

$\theta_1 = \tan^{-1}(-2) \doteq -63.4°$
Find the positive angle coterminal with –63.4°:
$-63.4° + 360° = 296.6°$
$\theta_2 = 180 + \theta_1$
$\theta_2 = 180 + (-63.4)$
$\theta_2 \doteq 116.6°$

$\theta_1 = \tan^{-1}(4) \doteq 76.0°$
$\theta_2 = 180 + \theta_1$
$\theta_2 \doteq 180 + 76.0$
$\theta_2 \doteq 256.0°$

Thus, $\theta \doteq 76.0°, 116.6°, 256.0°, 296.6°$.

**44.** $3\cos^2\left(\frac{1}{2}\theta\right) - \cos\left(\frac{1}{2}\theta\right) - 2 = 0$

$\left(3\cos\left(\frac{1}{2}\theta\right) + 2\right)\left(\cos\left(\frac{1}{2}\theta\right) - 1\right) = 0$

$\cos\left(\frac{1}{2}\theta\right) = \frac{-2}{3}$ or $\cos\left(\frac{1}{2}\theta\right) = 1$

If $0° \le \theta < 360°$, then $0° \le \frac{1}{2}\theta < 180°$.

$\frac{1}{2}\theta = \cos^{-1}\left(\frac{-2}{3}\right) \doteq 131.8°$ or $\frac{1}{2}\theta = \cos^{-1}(1) = 0°$

Note that there are no other values for $\frac{1}{2}\theta$ within the given domain, $0 \le \frac{1}{2}\theta < 180°$.
Multiply by 2 to solve for $\theta$:
$\theta \doteq 2 \times 131.8 \doteq 263.6°$
$2 \times 0° = 0°$
Thus, $\theta \doteq 0°, 263.6°$.

**45.** $5\cos^2\theta = 4\cos\theta$
$5\cos^2\theta - 4\cos\theta = 0$
$\cos\theta(5\cos\theta - 4) = 0$

$\cos\theta = 0$ or $\cos\theta = \frac{4}{5}$

$\theta = 1.571, 4.712$

$\theta_1 = \cos^{-1}\left(\frac{4}{5}\right) = 0.644$
$\theta_2 = 2\pi - \theta_1$
$\theta_2 \doteq 2\pi - 0.644$
$\theta_2 \doteq 5.639$

Thus, $\theta \doteq 0.64, 1.57, 4.71, 5.64$.

**46.** $25\sin^2\theta - 9 = 0$
$(5\sin\theta - 3)(5\sin\theta + 3) = 0$

$\sin\theta = \frac{3}{5}$ or $\sin\theta = \frac{-3}{5}$

$\theta_1 = \sin^{-1}\left(\frac{3}{5}\right)$
$\doteq 0.644$
$\theta_2 = \pi - \theta_1$
$\theta_2 \doteq \pi - 0.644$
$\theta_2 \doteq 2.498$

$\theta_1 = \sin^{-1}\left(\frac{-3}{5}\right)$
$\doteq -0.644$
Find the positive angle coterminal with –0.644:
$-0.644 + 2\pi \doteq 5.640$
$\theta_2 = \pi - \theta_1$
$\theta_2 \doteq \pi - (-0.644)$
$\theta_2 \doteq 3.786$

Thus, $\theta \doteq 0.64, 2.50, 3.79, 5.64$.

**47.** $3(1+\cos 2\theta) = -\cos 2\theta$

$3+3\cos 2\theta = -\cos 2\theta$

$3+4\cos 2\theta = 0$

$4\cos 2\theta = -3$

$\cos 2\theta = \frac{-3}{4}$

If $0 \le \theta < 2\pi$, then $0 \le 2\pi < 4\pi$.

$\cos 2\theta = \frac{-3}{4}$

$2\theta_1 = \cos^{-1}\left(\frac{-3}{4}\right) \doteq 2.419$

$2\theta_2 = 2\pi - 2\theta_1$

$2\theta_2 \doteq 2\pi - 2.419$

$2\theta_2 \doteq 3.864$

$2\theta_2 \doteq 2.419,\ 2.419+2\pi,\ 3.864,\ 3.864+2\pi$

$2\theta \doteq 2.419,\ 8.702,\ 3.864,\ 10.147$

Divide by 2 to solve for $\theta$.

$\theta \doteq 1.21,\ 4.35,\ 1.93,\ 5.07$

**48.** $4\tan^2\theta - \tan\theta = 8\tan\theta$

$4\tan^2\theta - 9\tan\theta = 0$

$\tan\theta(4\tan\theta - 9) = 0$

$\tan\theta = 0$ or $\tan\theta = \frac{9}{4}$

$\theta = 0,\ 3.14$

$\theta_1 = \tan^{-1}\left(\frac{9}{4}\right) \doteq 1.15$

$\theta_2 = \pi + \theta_1$

$\theta_2 \doteq 4.29$

Thus, $\theta \doteq 0,\ 1.15,\ 3.14,\ 4.29$.

**49.** **A**

If $0 \le \theta < 2\pi$, then $0 \le \frac{1}{2}\theta < \pi$.

$5\sin^2\left(\frac{1}{2}\theta\right) - 2\sin\left(\frac{1}{2}\theta\right) - 3 = 0$

$\left(5\sin\left(\frac{1}{2}\theta\right)+3\right)\left(\sin\left(\frac{1}{2}\theta\right)-1\right) = 0$

$\sin\left(\frac{1}{2}\theta\right) = \frac{-3}{5}$ or $\sin\left(\frac{1}{2}\theta\right) = 1$

There is no solution in the domain $0 \le \frac{1}{2}\theta < \pi$.

$\frac{1}{2}\theta = \frac{\pi}{2}$

$\theta = \pi$

Thus, there is one value for $\theta$, $0 \le \theta < 2\pi$.

## *Lesson 11 Solving Trigonometric Equations Graphically*

### PRACTICE EXERCISES ANSWERS AND SOLUTIONS

**1.** $3\sin x = 2\cos(2x-1)$

Graph: $Y_1 = 3\sin x$

$Y_2 = 2\cos(2x-1)$

*x*-values of intersection points $\doteq 0.68,\ 2.98$

Solution: $x \doteq 0.68$ rad, $2.98$ rad

**2.** $\sin x - 2 = 2\cos(2x)+1$

Graph: $Y_1 = 2\cos(2x)+1$

$Y_2 = 2\cos(2x)+1$

*x*-values of intersection points $\doteq 1.57$

Solution: $x \doteq 1.57$ rad

**3.** $\sin(\cos x) = 3\cos(x-2)$

Graph: $Y_1 = \sin(\cos x)$

$Y_2 = 3\cos(x-2)$

*x*-values of intersection points $\doteq 0.67,\ 3.81$

Solution: $x \doteq 0.67$ rad, $3.81$ rad

**4.** $4\sin^3 x = \cos\left(\frac{x}{2}\right)+2$

Graph: $Y_1 = 4(\sin x)^3$

$Y_2 = \cos\left(\frac{x}{2}\right)+2$

*x*-values of intersection points $\doteq$ 1.01, 2.23, 3.84, 5.75

Solution: $x \doteq 1.01$ rad, 2.23 rad, 3.84 rad, 5.75 rad

**5.** $3\sin^2 x + \cos^3 x = \frac{1}{2}x - 1$

Graph: $Y_1 = 3(\sin x)^2 + (\cos x)^3$

$Y_2 = 0.5x - 1$

*x*-values of intersection points $\doteq 2.57,\ 3.88, 5.54$

Solution: $x \doteq 2.57$ rad, 3.88 rad, 5.54 rad

 

**6.** $3^{\cos\left(\frac{x}{3}\right)}+1=4\sin(2x)-1$

Graph: $Y_1=3^{\cos\left(\frac{x}{3}\right)}+1$

$Y_2=4\sin(2x)-1$

$x$-values of intersection points $\doteq$ 3.66, 4.25

Solution: $x\doteq 3.66\text{ rad}, 4.25\text{ rad}$

**7.** $8\sin x\cos^3 x=1$

$8\sin x\cos^3 x-1=0$

Graph: $Y_1=8\sin(x)(\cos x)^3-1$

$x$-intercepts $\doteq$ 0.13, 1.01, 3.27, 4.16

Solution:

$x\doteq 0.13$ rad, 1.01 rad, 3.27 rad, 4.16 rad

**8.** $3^{\sin x}=3\cos x$

$3^{\sin x}-3\cos x=0$

Graph: $Y_1=3^{\sin(x)}-3\cos x$

$x$-intercepts $\doteq$ 0,77, 4.82

Solution: $x\doteq$0.77 rad, 4.82 rad

**9.** $3\sin(3-x)=2\cos(4+x)+1$

$3\sin(3-x)-2\cos(4+x)-1=0$

Graph: $Y_1=3\sin(3-x)-2\cos(4+x)-1$

$x$-intercepts $\doteq$1.81, 5.87

Solution: $\theta\doteq$1.81 rad, 5.87 rad

**10.** $\sin(\theta+30°)=\sin\theta$

Graph: $Y_1=\sin(x+30)$

$Y_2=\sin x$

$x$-values of intersection points = 75, 255

Solution: $\theta=75.0°, 255.0°$ (exact values)

**11.** $\cos^2\theta+4\cos\theta=0$

Graph: $Y_1=(\cos x)^2+4\cos x$

$x$-intercepts = 90, 270

Solution: $\theta=90.0°, 270.0°$ (exact values)

**12.** $\sin^2\theta=2\theta-50$

Graph: $Y_1=(\sin x)^2$

$Y_2=2x-50$

$x$-values of intersection points: $\doteq$ 25.1

Solution: $\theta\doteq 25.1°$

**13.** $\sin(4\theta)=\dfrac{\theta}{60°}$

Graph: $Y_1=\sin(4x)$

$Y_2=\dfrac{x}{60}$

$x$-values of intersection points $\doteq$ 0, 35.8

Solution: $\theta\doteq 0°,\ 35.8°$

**14.** $2\cos\theta=2\sin\theta-\sin^2\theta$

Graph: $Y_1=2\cos(x)$

$Y_2=2\sin(x)-(\sin x)^2$

$x$-values of intersection points $\doteq$ 60.6, 217.5

Solution: $\theta\doteq 60.6°, 217.5°$

**15.** $\sin\theta+\sin(2\theta)+\sin(3\theta)=0$

Graph: $Y_1=\sin x+\sin(2x)+\sin(3x)$

$x$-intercepts

$\doteq -3.14, -2.09, -1.57, 0, 1.57, 2.09$

Solution:

$\theta\doteq -3.14\text{ rad}, -0.29\text{ rad}, -1.57\text{rad}, 0\text{ rad},$

$1.57\text{ rad}, 2.09\text{ rad}$

**16.** $\cos^2\theta-\sin\theta=\cos(\sin\theta)-1$

Graph: $Y_1=(\cos x)^2-\sin x$

$Y_2=\cos(\sin x)-1$

$x$-values of intersection points: $\doteq$ 0.81, 2.33

Solution: $\theta\doteq$0.81 rad, 2.33 rad

## *Practice Quiz*

### ANSWERS AND SOLUTIONS

**1.**

$\sin\theta=\dfrac{3}{8}$

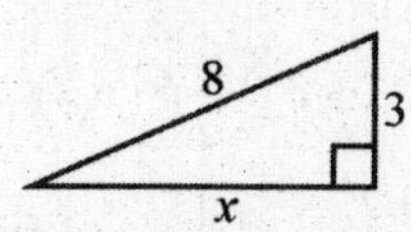


$x^2+3^2=8^2$

$x^2=64-9=55$

$x=\sqrt{55}$

$\tan\theta=\dfrac{3}{\sqrt{55}}=\dfrac{3\sqrt{55}}{55}$

**2.** $\sec\theta=1.640\rightarrow\cos\theta=\dfrac{1}{1.640}$

$\cos^{-1}\left(\dfrac{1}{1.640}\right)\doteq 52.4°$

**3.**

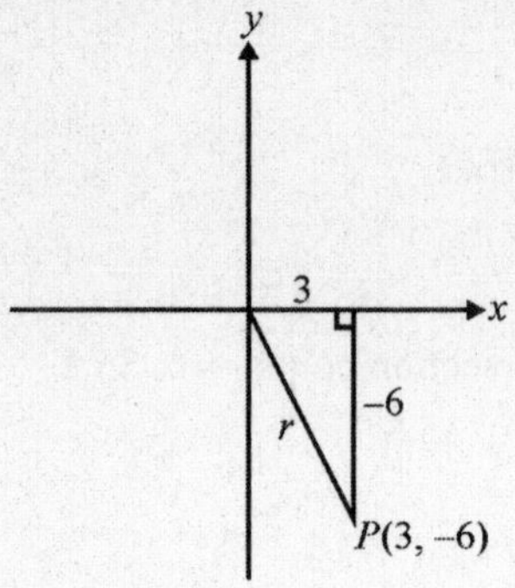


$r^2 = 3^2 + 6^2$

$r^2 = 45$

$r = \sqrt{45} = 3\sqrt{5}$

$\sin\theta = \frac{-6}{3\sqrt{5}} = \frac{-2}{\sqrt{5}} = \frac{-2\sqrt{5}}{5}$

$\cos\theta = \frac{3}{3\sqrt{5}} = \frac{1}{\sqrt{5}} = \frac{\sqrt{5}}{5}$

$\tan\theta = \frac{-6}{3} = -2$

**4.** **a)** $450° = 450 \times \frac{\pi}{180} = \frac{5\pi}{2}$ rad

**b)** $\frac{5\pi}{8}$ rad $= \frac{5\pi}{8} \times \frac{180}{\pi} = 112.5°$

**c)** $-\frac{19\pi}{4}$ rad $= -\frac{19\pi}{4} \times \frac{180}{\pi} = -855°$

**5.** The second hand rotates through an angle of $\frac{5}{12} \times 2\pi = \frac{5\pi}{6}$ rad.

The arc length is $a = r\theta$, thus,

$a = 15 \times \frac{5\pi}{6} = \frac{15\pi}{2}$ cm.

**6.** An angle of $\frac{-18\pi}{5}$ has a principle angle of $\frac{-18\pi}{5} + \frac{20\pi}{5} = \frac{2\pi}{5}$.

Since this angle is between 0 and $\frac{\pi}{2}$, it is in quadrant I and it is also the reference angle.

**7.** **a)** $\cos\left(\frac{11\pi}{6}\right) = \frac{\sqrt{3}}{2}$

**b)** $\csc 240° = \frac{1}{\sin 240°}$

$= \frac{1}{-\frac{\sqrt{3}}{2}} = -\frac{2}{\sqrt{3}} = -\frac{2\sqrt{3}}{3}$

**c)** $\tan\left(-\frac{16\pi}{3}\right) = \tan\left(-\frac{16\pi}{3} + \frac{18\pi}{3}\right)$

$= \tan\left(\frac{2\pi}{3}\right) = -\sqrt{3}$

**8.** $\sin\left(\frac{4\pi}{3}\right) = -\frac{\sqrt{3}}{2}$

The second angle, $\theta$, $0 \le \theta < 2\pi$, with a sine ratio of $-\frac{\sqrt{3}}{2}$ is $\theta = \frac{5\pi}{3}$.

**9.** The angle $\theta$ is in quadrant IV. Since $\cot\theta = \frac{-5}{3} = \frac{x}{y} = \frac{5}{-3}$, a representative triangle can be drawn with $x = 5$ and $y = -3$.

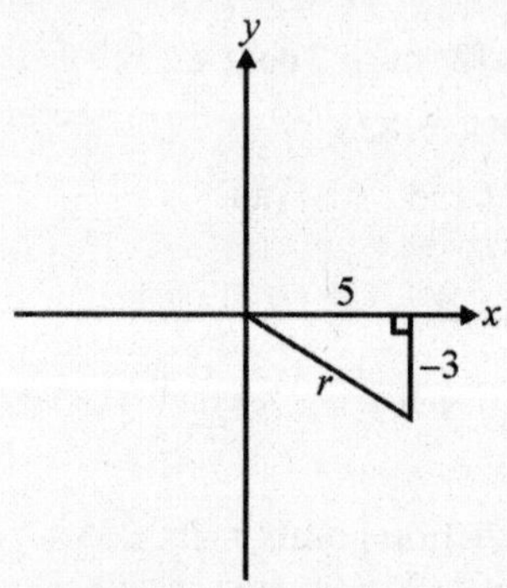


$r^2 = 5^2 + (-3)^2$

$r^2 = 34$

$r = \sqrt{34}$

$\sin\theta = \frac{y}{r} = \frac{-3}{\sqrt{34}} = \frac{-3\sqrt{34}}{34}$

$\cos\theta = \frac{x}{r} = \frac{5}{\sqrt{34}} = \frac{5\sqrt{34}}{34}$

 

**10. a)** $2\sin\theta - 1 = 0$

$2\sin\theta = 1$

$\sin\theta = \frac{1}{2}$

$\theta = 30°, 150°$

**b)** $2\cos^2\theta + \cos\theta = 1$

$2\cos^2\theta + \cos\theta - 1 = 0$

$(2\cos\theta - 1)(\cos\theta + 1) = 0$

$\cos\theta = \frac{1}{2}$ or $\cos\theta = -1$

$\theta = 60°, 180°, 300°$

**11. a)** $2\sin^2\theta + 3\sin\theta + 1 = 0$

$(2\sin\theta + 1)(\sin\theta + 1) = 0$

$\sin\theta = -\frac{1}{2}$ or $\sin\theta = -1$

$\theta = \frac{7\pi}{6}, \frac{3\pi}{2}, \frac{11\pi}{6}$

**b)** If $0 \le \theta < 2\pi$, then $0 \le 2\theta < 4\pi$.

$2\sin(2\theta)\cos(2\theta) = \cos(2\theta)$

$2\sin(2\theta)\cos(2\theta) - \cos(2\theta) = 0$

$\cos(2\theta)(2\sin(2\theta) - 1) = 0$

$\cos(2\theta) = 0$ or $\sin(2\theta) = \frac{1}{2}$

$2\theta = \frac{\pi}{2}, \frac{3\pi}{2}, \frac{5\pi}{2}, \frac{7\pi}{2}, \frac{\pi}{6}, \frac{5\pi}{6}, \frac{13\pi}{6}, \frac{17\pi}{6}$

$\theta = \frac{\pi}{4}, \frac{3\pi}{4}, \frac{5\pi}{4}, \frac{7\pi}{4}, \frac{\pi}{12}, \frac{5\pi}{12}, \frac{13\pi}{12}, \frac{17\pi}{12}$

**12.** $2\sin^2\theta - 2\sin\theta + \sqrt{3}\sin\theta - \sqrt{3} = 0$

$2\sin\theta(\sin\theta - 1) + \sqrt{3}(\sin\theta - 1) = 0$

$(2\sin\theta + \sqrt{3})(\sin\theta - 1) = 0$

$\sin\theta = -\frac{\sqrt{3}}{2}$ or $\sin\theta = 1$

$\theta = \frac{\pi}{2}, \frac{4\pi}{3}, \frac{5\pi}{3}$

The general solution is

$\frac{\pi}{2} + 2n\pi, \frac{4\pi}{3} + 2n\pi, \frac{5\pi}{3} + 2n\pi, n \in I.$

**13.** $9\cos^2\theta - 1 = 0$

$\cos^2\theta = \frac{1}{9}$

$\cos\theta = \pm\frac{1}{3}$

$\theta \doteq 70.5°, 109.5°, 250.5°, 289.5°$

**14.** $10\cos^2\theta + 3\cos\theta - 1 = 0$

$(5\cos\theta - 1)(2\cos\theta + 1) = 0$

$\cos\theta = \frac{1}{5}$ or $\cos\theta = -\frac{1}{2}$

$\theta \doteq 1.4, 2.1, 4.2, 4.9$

**15.** Graph $Y_1 = 3\cos(2x) - 1$, $Y_2 = \sin^2 x$.

Solutions are $x$-values at the intersection points. They are $x \doteq 32.3°, 147.7°, 212.3°, 327.7°$.

## *Lesson 12 Reciprocal, Quotient, and Pythagorean Identities*

### CLASS EXERCISES ANSWERS AND SOLUTIONS

**1.**

| | LHS | RHS |
|---|---|---|
| | $\dfrac{\sec\theta\sin\theta\cot\theta}{\tan\theta\csc\theta}$ | $\cos\theta$ |
| • Reciprocal and Quotient identities | $\dfrac{\left(\frac{1}{\cos\theta}\right)\sin\theta\left(\frac{\cos\theta}{\sin\theta}\right)}{\left(\frac{\sin\theta}{\cos\theta}\right)\left(\frac{1}{\sin\theta}\right)}$ | |
| • Multiply and cancel | $\dfrac{1}{\left(\frac{1}{\cos\theta}\right)}$ | |
| • Divide | $\cos\theta$ <br> = RHS | |

**2.**

| | LHS | RHS |
|---|---|---|
| | $\dfrac{\sin^2 x}{1+\cos x}$ | $1-\cos x$ |
| • Pythagorean identity | $\dfrac{1-\cos^2 x}{1+\cos x}$ | |
| • Factor | $\dfrac{(1-\cos x)(1+\cos x)}{1+\cos x}$ | |
| • Cancel | $1-\cos x$ <br> = RHS | |

**3.**

| | LHS | RHS |
|---|---|---|
| | $\dfrac{\sec x}{\sin x}-\dfrac{\sin x}{\cos x}$ | $\cot x$ |
| • Find a common denominator | $\dfrac{\sec x\cos x}{\sin x\cos x}-\dfrac{\sin^2 x}{\sin x\cos x}$ | |
| • Combine numerators | $\dfrac{\sec x\cos x-\sin^2 x}{\sin x\cos x}$ | |
| • Reciprocal identity | $\dfrac{\left(\frac{1}{\cos x}\right)\cos x-\sin^2 x}{\sin x\cos x}$ | |
| • Cancel (cos $x$) | $\dfrac{1-\sin^2 x}{\sin x\cos x}$ | |
| • Pythagorean identity | $\dfrac{\cos^2 x}{\sin x\cos x}$ | |
| • Cancel (cos $x$) | $\dfrac{\cos x}{\sin x}$ | |
| • Quotient identity | $\cot x$ <br> = RHS | |

### PRACTICE EXERCISES ANSWERS AND SOLUTIONS

**1.** $\cos x\tan x=\sin x$

| | LHS | RHS |
|---|---|---|
| • To verify for $x=60°$: | | |
| • Substitute 60° for $x$ | $\cos 60°\tan 60°$ | $\sin 60°$ |
| • Evaluate each side separately | $\frac{1}{2}\times\sqrt{3}$ | |
| • Simplify | $\dfrac{\sqrt{3}}{2}$ | $\dfrac{\sqrt{3}}{2}$ |
| | LHS = | RHS |

To verify graphically:

• Graph: $Y_1=\cos(x)\times\tan(x)$

$Y_2=\sin(x)$

To prove algebraically:

| | LHS | RHS |
|---|---|---|
| | $\cos x\tan x$ | $\sin x$ |
| • Quotient identity | $(\cos x)\left(\dfrac{\sin x}{\cos x}\right)$ | |
| • Simplify | $\sin x$ <br> = RHS | |

**2.**

$$\frac{\cos x}{1-\sin x}=\frac{1+\sin x}{\cos x}$$

To verify for $x=60°$:

| | LHS | RHS |
|---|---|---|
| • Substitute 60° for $x$ | $\dfrac{\cos 60°}{1-\sin 60°}$ | $\dfrac{1+\sin 60°}{\cos 60°}$ |
| • Evaluate each side separately | $\dfrac{\frac{1}{2}}{1-\frac{\sqrt{3}}{2}}$ | $\dfrac{1+\frac{\sqrt{3}}{2}}{\frac{1}{2}}$ |
| • Find a common denominator | $\dfrac{\frac{1}{2}}{\frac{2-\sqrt{3}}{2}}$ | $\dfrac{\frac{2+\sqrt{3}}{2}}{\frac{1}{2}}$ |
| • Multiply by the reciprocal | $\frac{1}{2}\times\dfrac{2}{\left(2-\sqrt{3}\right)}$ | $\dfrac{\left(2+\sqrt{3}\right)}{2}\times\dfrac{2}{1}$ |

| | LHS | RHS |
|---|---|---|
| • Simplify | $\frac{1}{2-\sqrt{3}}$ | $2+\sqrt{3}$ |
| • Rationalize the denominator on LHS | $\frac{1}{(2-\sqrt{3})}\times\frac{(2+\sqrt{3})}{(2+\sqrt{3})}$ | $2+\sqrt{3}$ |
| | $\frac{(2+\sqrt{3})}{4-3}$ | $2+\sqrt{3}$ |
| | $2+\sqrt{3}$ | $2+\sqrt{3}$ |
| | LHS = | RHS |

To verify graphically:

Graph: $Y_1=\cos(x)\div(1-\sin(x))$

$Y_2=(1+\sin(x))\div\cos(x)$

To prove algebraically:

| | LHS | RHS |
|---|---|---|
| | $\frac{\cos x}{1-\sin x}$ | $\frac{1+\sin x}{\cos x}$ |
| • Multiply numerator and denominator by $(1+\sin x)$ | $\frac{\cos x}{(1-\sin x)}\times\frac{(1+\sin x)}{(1+\sin x)}$ | |
| • Multiply binomials | $\frac{\cos x(1+\sin x)}{1-\sin^2 x}$ | |
| • Pythagorean identity | $\frac{\cos x(1+\sin x)}{\cos^2 x}$ | |
| • Simplify | $\frac{1+\sin x}{\cos x}$ = RHS | |

**3.**

| | LHS | RHS |
|---|---|---|
| | $\csc A-\sin A$ | $\cot A\cos A$ |
| • Reciprocal identity | $\frac{1}{\sin A}-\sin A$ | $\frac{\cos A}{\sin A}\times\cos A$ • Quotient identity |
| • Find a common denominator | $\frac{1}{\sin A}-\frac{\sin^2 A}{\sin A}$ | $\frac{\cos^2 A}{\sin A}$ • Multiply |
| • Simplify | $\frac{1-\sin^2 A}{\sin A}$ | |
| • Pythagorean identity | $\frac{\cos^2 A}{\sin A}$ = RHS | |

**4.**

| | LHS | RHS |
|---|---|---|
| | $\frac{\cos^2 x}{1+2\sin x-3\sin^2 x}$ | $\frac{1+\sin x}{1+3\sin x}$ |
| • Factor the denominator | $\frac{\cos^2 x}{(1-\sin x)(1+3\sin x)}$ | |
| • Pythagorean identity | $\frac{1-\sin^2 x}{(1-\sin x)(1+3\sin x)}$ | |
| • Factor the numerator | $\frac{(1-\sin x)(1+\sin x)}{(1-\sin x)(1+3\sin x)}$ | |
| • Simplify | $\frac{1+\sin x}{1+3\sin x}$ = RHS | |

**5.**

| | LHS | RHS |
|---|---|---|
| | $\frac{\csc x}{\tan x+\cot x}$ | $\cos x$ |
| • Reciprocal and quotient identities | $\frac{\frac{1}{\sin x}}{\frac{\sin x}{\cos x}+\frac{\cos x}{\sin x}}$ | |
| • Find a common denominator | $\frac{\frac{1}{\sin x}}{\frac{\sin^2 x}{\sin x\cos x}+\frac{\cos^2 x}{\sin x\cos x}}$ | |
| • Simplify | $\frac{\frac{1}{\sin x}}{\frac{\sin^2 x+\cos^2 x}{\sin x\cos x}}$ | |
| • Pythagorean identity | $\frac{\frac{1}{\sin x}}{\frac{1}{\sin x\cos x}}$ | |
| • Multiply by the reciprocal | $\frac{1}{\sin x}\times\frac{\sin x\cos x}{1}$ | |
| • Simplify | $\cos x$ = RHS | |

**6.** $\csc x \tan x = \sec x$

- To verify for $x = \frac{\pi}{6}$:

| | LHS | RHS |
|---|---|---|
| • Substitute $\frac{\pi}{6}$ for $x$ | $\csc\frac{\pi}{6}\tan\frac{\pi}{6}$ | $\sec\frac{\pi}{6}$ |
| • Evaluate each side separately: | $(2)\left(\frac{1}{\sqrt{3}}\right)$ | $\frac{2}{\sqrt{3}}$ |
| • Simplify | $\frac{2}{\sqrt{3}}$ | $\frac{2}{\sqrt{3}}$ |

LHS = RHS

To verify graphically:

Graph: $Y_1 = \csc x \tan x \left(\text{enter} \frac{1}{(\sin x)}(\tan(x))\right)$

$Y_2 = \sec x \left(\text{enter} \frac{1}{\cos(x)}\right)$

To prove algebraically:

| | LHS | RHS | |
|---|---|---|---|
| | $\csc x \tan x$ | $\sec x$ | |
| • Reciprocal and quotient identities | $\frac{1}{\sin x} \cdot \frac{\sin x}{\cos x}$ | $\frac{1}{\cos x}$ | • Reciprocal identity |
| • Simplify | $\frac{1}{\cos x}$ <br> = RHS | | |

**7.** $\sec x = \frac{\csc x}{\cot x}$

- To verify for $x = \frac{\pi}{6}$

| | LHS | RHS |
|---|---|---|
| • Substitute $\frac{\pi}{6}$ for $x$ | $\sec\frac{\pi}{6}$ | $\dfrac{\csc\frac{\pi}{6}}{\cot\frac{\pi}{6}}$ |
| • Evaluate each side separately | $\frac{2}{\sqrt{3}}$ | $\frac{2}{\sqrt{3}}$ |

LHS = RHS

To verify graphically:

Graph: $Y_1 = \sec x \left(\text{enter} \frac{1}{\cos(x)}\right)$

$Y_2 = \frac{\csc x}{\cot x} \left(\text{enter} \left(\frac{1}{\sin(x)}\right) \div \left(\frac{1}{\tan(x)}\right)\right)$

To prove algebraically:

| | LHS | RHS | |
|---|---|---|---|
| | $\sec x$ | $\frac{\csc x}{\cot x}$ | |
| • Reciprocal identity | $\frac{1}{\cos x}$ | $\dfrac{\frac{1}{\sin x}}{\frac{\cos x}{\sin x}}$ | • Reciprocal and quotient identities |
| | | $\frac{1}{\sin x} \times \frac{\sin x}{\cos x}$ | • Multiply by the reciprocal |
| | | $\frac{1}{\cos x}$ <br> = LHS | • Simplify |

**8.**

**LHS:**

$\sin^4\theta - \cos^4\theta$

- Factor: $(\sin^2\theta - \cos^2\theta)(\sin^2\theta + \cos^2\theta)$
- Pythagorean identity: $(\sin^2\theta - \cos^2\theta)(1)$
- Pythagorean identity: $1 - \cos^2\theta - \cos^2\theta$

$1 - 2\cos^2\theta$
= RHS

**RHS:**
$1 - 2\cos^2\theta$

**9.**

**LHS:**

$\dfrac{\tan x + \sin x}{\cot x + \csc x}$

- Reciprocal and quotient identities: $\dfrac{\frac{\sin x}{\cos x} + \sin x}{\frac{\cos x}{\sin x} + \frac{1}{\sin x}}$
- Find a common denominator: $\dfrac{\frac{\sin x}{\cos x} + \frac{\sin x \cos x}{\cos x}}{\frac{\cos x}{\sin x} + \frac{1}{\sin x}}$
- Simplify: $\dfrac{\frac{\sin x + \sin x \cos x}{\cos x}}{\frac{\cos x + 1}{\sin x}}$
- Multiply by the reciprocal: $\dfrac{(\sin x + \sin x \cos x)(\sin x)}{\cos x(\cos x + 1)}$
- Factor: $\dfrac{\sin x(1 + \cos x)(\sin x)}{\cos x(\cos x + 1)}$
- Simplify: $\dfrac{\sin^2 x}{\cos x}$

 

**RHS:**

| | |
|---|---|
| | $\sin^2 x \sec x$ |
| • Reciprocal identity | $\sin^2 x\left(\dfrac{1}{\cos x}\right)$ |
| • Simplify | $\dfrac{\sin^2 x}{\cos x}$ |
| | $=$ RHS |

**10.**

**LHS:**

| | |
|---|---|
| | $\tan A+\cot A$ |
| • Quotient identities | $\dfrac{\sin A}{\cos A}+\dfrac{\cos A}{\sin A}$ |
| • Find a common denominator | $\dfrac{\sin^2 A}{\sin A\cos A}+\dfrac{\cos^2 A}{\sin A\cos A}$ |
| • Simplify | $\dfrac{\sin^2 A+\cos^2 A}{\sin A\cos A}$ |
| • Pythagorean identity | $\dfrac{1}{\sin A\cos A}$ |
| | $=$ RHS |

**RHS:**

| | |
|---|---|
| | $\csc A\sec A$ |
| • Quotient identities | $\dfrac{1}{\sin A}\cdot\dfrac{1}{\cos A}$ |
| • Simplify | $\dfrac{1}{\sin A\cos A}$ |

**11.** $\sec x-\cos x=\sin x\tan x$

• To verify for $x=\dfrac{\pi}{6}$:

| | LHS | RHS |
|---|---|---|
| • Substitute $\dfrac{\pi}{6}$ for $x$ | $\sec\dfrac{\pi}{6}-\cos\dfrac{\pi}{6}$ | $\sin\dfrac{\pi}{6}\tan\dfrac{\pi}{6}$ |
| • Calculate both sides separately | $\dfrac{2}{\sqrt{3}}-\dfrac{\sqrt{3}}{2}$ | $\dfrac{1}{2}\times\dfrac{1}{\sqrt{3}}$ |
| • Find a common denominator | $\dfrac{4}{2\sqrt{3}}-\dfrac{3}{2\sqrt{3}}$ | $\dfrac{1}{2\sqrt{3}}$ |
| • Simplify | $\dfrac{1}{2\sqrt{3}}$ | $\dfrac{1}{2\sqrt{3}}$ |
| | LHS $=$ | RHS |

To verify graphically:

Graph: $Y_1=\sec x-\cos x$ (Enter $1\div\cos x-\cos x$)

$Y_2=\sin x\times\tan x$

To prove algebraically:

**LHS:**

| | |
|---|---|
| | $\sec x-\cos x$ |
| • Reciprocal identities | $\dfrac{1}{\cos x}-\cos x$ |
| • Find a common denominator | $\dfrac{1}{\cos x}-\dfrac{\cos^2 x}{\cos x}$ |
| • Simplify | $\dfrac{1-\cos^2 x}{\cos x}$ |
| • Pythagorean identity | $\dfrac{\sin^2 x}{\cos x}$ |

**RHS:**

| | |
|---|---|
| • Reciprocal identities | $\sin x\tan x$ |
| | $\sin x\times\dfrac{\sin x}{\cos x}$ |
| • Multiply | $\dfrac{\sin^2 x}{\cos x}$ |
| | $=$ LHS |

**12.** $\sin^2 x+\cos^2 x\tan^2 x=2\sin^2 x$

• To verify for $x=\dfrac{\pi}{6}$

| | LHS | RHS |
|---|---|---|
| • Substitute $\dfrac{\pi}{6}$ for $x$ | $\sin^2\left(\dfrac{\pi}{6}\right)+\cos^2\left(\dfrac{\pi}{6}\right)\tan^2\left(\dfrac{\pi}{6}\right)$ | $2\sin^2\left(\dfrac{\pi}{6}\right)$ |
| • Calculate both sides separately | $\left(\dfrac{1}{2}\right)^2+\left(\dfrac{\sqrt{3}}{2}\right)^2\left(\dfrac{1}{\sqrt{3}}\right)^2$ | $2\left(\dfrac{1}{2}\right)^2$ |
| | $\dfrac{1}{4}+\left(\dfrac{3}{4}\right)\left(\dfrac{1}{3}\right)$ | $2\left(\dfrac{1}{4}\right)$ |
| • Simplify | $\dfrac{1}{4}+\dfrac{1}{4}$ | $\dfrac{2}{4}$ |
| | $\dfrac{2}{4}$ | |
| | $\dfrac{1}{2}$ | $\dfrac{1}{2}$ |
| | LHS $=$ | RHS |

To verify graphically:

Graph: $Y_1=(\sin(x))^2+\cos(x)(\tan(x))^2$

$Y_2=(2\sin(x))^2$

To prove algebraically:

| | LHS | RHS |
|---|---|---|
| | $\sin^2 x + \cos^2 x \tan^2 x$ | $2\sin^2 x$ |
| • Quotient identities | $\sin^2 x + \cos^2 x\left(\dfrac{\sin^2 x}{\cos^2 x}\right)$ | |
| • Simplify | $\sin^2 x + \sin^2 x$ | |
| | $2\sin^2 x$ | |
| | $= \text{RHS}$ | |

**13.**

**LHS:**

$$\frac{1+\sin x}{1-\sin x}$$

**RHS:**

$$(\sec x + \tan x)^2$$

| | |
|---|---|
| $\sec^2 x + 2\sec x \tan x + \tan^2 x$ | • Expand |
| $\dfrac{1}{\cos^2 x} + 2\times\dfrac{1}{\cos x}\times\dfrac{\sin x}{\cos x} + \dfrac{\sin^2 x}{\cos^2 x}$ | • Reciprocal and quotient identities |
| $\dfrac{1}{\cos^2 x} + \dfrac{2\sin x}{\cos^2 x} + \dfrac{\sin^2 x}{\cos^2 x}$ | • Simplify |
| $\dfrac{1+2\sin x+\sin^2 x}{\cos^2 x}$ | • Simplify |
| $\dfrac{1+2\sin x+\sin^2 x}{1-\sin^2 x}$ | • Pythagorean identity |
| $\dfrac{(1+\sin x)(1+\sin x)}{(1+\sin x)(1-\sin x)}$ | • Factor |
| $\dfrac{1+\sin x}{1-\sin x}$ | • Simplify |
| $= \text{LHS}$ | |

**14.**

| | LHS | RHS |
|---|---|---|
| | $\dfrac{\tan^2 x+1}{\cot^2 x+1}$ | $\tan^2 x$ |
| • Pythagorean identity | $\dfrac{\sec^2 x}{\csc^2 x}$ | $\dfrac{\sin^2 x}{\cos^2 x}$ |
| • Reciprocal identity | $\dfrac{\dfrac{1}{\cos^2 x}}{\dfrac{1}{\sin^2 x}}$ | |
| | $\dfrac{1}{\cos^2 x}\times\dfrac{\sin^2 x}{1}$ | |
| • Simplify | $\dfrac{\sin^2 x}{\cos^2 x}$ | |
| | $= \text{RHS}$ | |

**15.**

| | LHS | RHS | |
|---|---|---|---|
| | $\dfrac{\sin^3\theta}{\cos\theta-\cos^3\theta}$ | $\tan\theta$ | |
| • Factor | $\dfrac{(\sin\theta)(\sin^2\theta)}{\cos\theta(1-\cos^2\theta)}$ | $\dfrac{\sin\theta}{\cos\theta}$ | • Quotient identity |
| • Pythagorean identity | $\dfrac{(\sin\theta)(\sin^2\theta)}{\cos\theta(\sin^2\theta)}$ | | |
| • Simplify | $\dfrac{\sin\theta}{\cos\theta}$ | | |
| | $= \text{RHS}$ | | |

**16. D**

| | |
|---|---|
| | $\dfrac{\cos^2\theta\sin^2\theta+\cos^4\theta}{\cot^2\theta}$ |
| • Factor and Quotient identities | $\dfrac{\cos^2\theta(\sin^2\theta+\cos^2\theta)}{\dfrac{\cos^2\theta}{\sin^2\theta}}$ |
| • Pythagorean identity | $\dfrac{\cos^2\theta(1)}{\dfrac{\cos^2\theta}{\sin^2\theta}}$ |
| • Multiply by reciprocal | $\cos^2\theta\times\dfrac{\sin^2\theta}{\cos^2\theta}$ |
| • Simplify | $\sin^2\theta$ |

## *Lesson 13 Sum, Difference, and Double-Angle Identities*

### CLASS EXERCISES ANSWERS AND SOLUTIONS

1. Since we want to find the value of $\sin\left(\dfrac{\pi}{2}+\theta\right)$, we will use the identity that involves the sine of the sum of angles:
$\sin(A+B) = \sin A\cos B + \cos A\sin B$

In our example, $A = \dfrac{\pi}{2}$ and $B = \theta$, so we can rewrite the expression as:

$$\sin\left(\frac{\pi}{2}+\theta\right) = \sin\left(\frac{\pi}{2}\right)\cos\theta + \cos\left(\frac{\pi}{2}\right)\sin\theta$$

 

We are given

$\sin\theta = \frac{4}{5}$ and $\cos\theta = -\frac{3}{5}$, and we know that

$\sin\left(\frac{\pi}{2}\right) = 1.$

We get:

$$\sin\left(\frac{\pi}{2}+\theta\right) = \sin\left(\frac{\pi}{2}\right)\cos\theta + \cos\left(\frac{\pi}{2}\right)\sin\theta$$

$$= (1)\left(-\frac{3}{5}\right) + (0)\left(\frac{4}{5}\right)$$

$$= -\frac{3}{5}$$

2. Look at the four identities. Which one has the configuration "cos cos + sin sin"?

Only one identity has that arrangement. It is $\cos(A-B) = \cos A \cos B + \sin A \sin B$.

If we compare this configuration to the question, we see that $A = 2x$ and $B = 7x$.

$$\cos 2x \cos 7x + \sin 2x \sin 7x = \cos(2x - 7x)$$

$$= \cos(-5x)$$

So, the answer is $\cos(-5x)$.

3.

| | LHS | RHS |
|---|---|---|
| | $\cos(2A)$ | $2\cos^2 A - 1$ |
| • Double-angle identity | $= \cos^2 A - \sin^2 A$ | |
| • Pythagorean identity | $= \cos^2 A - (1 - \cos^2 A)$ | |
| • Simplify | $= 2\cos^2 A - 1$<br>= RHS | |

4. To verify the identity graphically, we will graph the functions $Y_1 = \frac{(\sin x + \cos x)^2}{\sin 2x}$ and $y = 1 + \csc(2x)$ to see if their graphs overlap.

Note that to graph the function $y = 1 + \csc(2x)$, we will rewrite is as $Y_2 = 1 + \frac{1}{\sin(2x)}$.

Press [Y=] to bring up the Y= editor.

Press [(] [SIN] [X,T,θ,n] [)] [+] [COS] [X,T,θ,n] [)] [)] [^] [2] [÷] [SIN] [2] [X,T,θ,n] [)] [ENTER] to enter the first function.

To enter the second function press [1] [÷] [TAN] [X,T,θ,n] [)] [ENTER]

Press [GRAPH] to display both graphs.

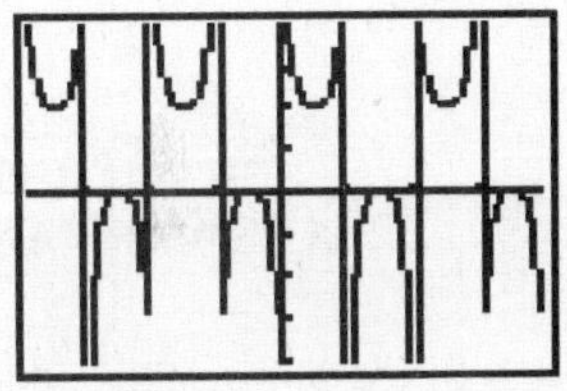

The calculator has been set to radian mode and the viewing window is set to ZTrig.

Since the graphs of $Y_1 = \frac{(\sin x + \cos x)^2}{\sin(2x)}$ and $Y_2 = 1 + \csc(2x)$ overlap, we will conclude that the identity appears to holds true. Now, we will prove the identity algebraically.

**LHS:**

| | |
|---|---|
| | $\frac{(\sin x + \cos x)^2}{\sin(2x)}$ |
| • Expand | $= \frac{\sin^2 x + 2\sin x \cos x + \cos^2 x}{\sin(2x)}$ |
| • Rearrange | $= \frac{(\sin^2 x + \cos^2 x) + 2\sin x \cos x}{\sin(2x)}$ |
| • Pythagorean and double-angle identities | $= \frac{1 + \sin 2x}{\sin(2x)}$ |
| • Split into two fractions | $= \frac{1}{\sin(2x)} + \frac{\sin 2x}{\sin 2x}$ |
| • Reciprocal identity and cancel | $= \csc(2x) + 1$<br>= RHS |

**RHS:**

$1 + \csc 2x$

**5.**

| | LHS | RHS |
|---|---|---|
| | $\frac{2\tan x}{1+\tan^2 x}$ | $\sin 2x$ |
| • Pythagorean identity | $=\frac{2\tan x}{\sec^2 x}$ | |
| • Quotient and reciprocal identities | $=\frac{\frac{2\sin x}{\cos}}{\frac{1}{\cos^2 x}}$ | |
| • Multiply by the reciprocal | $=\frac{2\sin x}{\cos}\times\frac{\cos^2 x}{1}$ | |
| • Cancel | $=2\sin x\cos x$ | |
| • Double-angle identity | $=\sin(2x)$ | |
| | $=\text{RHS}$ | |

## PRACTICE EXERCISES ANSWERS AND SOLUTIONS

**1.** Since $\sin A\cos B-\cos A\sin B=\sin(A-B)$, then $\sin(5k)-\cos(5k)\sin k=\sin(5k-k)$, where $A=5k$, $B=k$.
$\sin(5k-k)=\sin(4k)$

**2.**

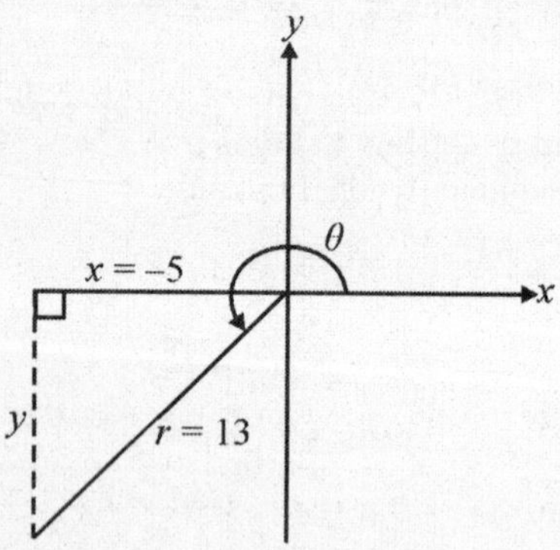


$\cos\theta=\frac{x}{r}$

Since $\cos\theta=\frac{-5}{13}$, then a representative right triangle can be drawn with $x=-5$ and $y=13$.

Apply the Pythagorean theorem:

$$x^2+y^2=r^2$$
$$(-5)^2+y^2=13^2$$
$$25+y^2=169$$
$$y^2=144$$
$$y=12$$

In quadrant III, the value of $y$ is negative, thus $y=\frac{-12}{13}$.

$$\sin\theta=\frac{y}{r}$$
$$\sin\theta=\frac{-12}{13}$$

Since $\cos(A-B)=\cos A\cos B+\sin A\sin B$, then

$$\cos\left(\frac{\pi}{3}-\theta\right)=\cos\left(\frac{\pi}{3}\right)\cos\theta+\sin\left(\frac{\pi}{3}\right)\sin\theta,$$

where $A=\frac{\pi}{3}$ and $B=\theta$.

Substitute:

$$\cos\left(\frac{\pi}{3}-\theta\right)=\left(\frac{1}{2}\right)\left(\frac{-5}{13}\right)+\left(\frac{\sqrt{3}}{2}\right)\left(\frac{-12}{13}\right)$$
$$\cos\left(\frac{\pi}{3}-\theta\right)=\frac{-5}{26}+\frac{-12\sqrt{3}}{26}$$
$$\cos\left(\frac{\pi}{3}-\theta\right)=\frac{-5-12\sqrt{3}}{26}$$

**3.** $\frac{\sin x}{\sin(2x)}=\frac{1}{2}\sec x$

**a)** To verify for $x=60°$

| | LHS | | RHS |
|---|---|---|---|
| • Substitute $x=60°$ | $\frac{\sin 60°}{\sin 120°}$ | | $\frac{1}{2}\sec 60°$ |
| • Evaluate each side separately | $\frac{\frac{\sqrt{3}}{2}}{\frac{\sqrt{3}}{2}}$ | | $\frac{1}{2}(2)$ |
| • Simplify | 1 | | 1 |
| | LHS | = | RHS |

**b)** To verify graphically:

Graph: $Y_1=\frac{\sin x}{\sin(2x)}$

$Y_2=\frac{1}{2}\sec x$

**c)** To prove algebraically:

| | LHS | RHS | |
|---|---|---|---|
| | $\frac{\sin x}{\sin(2x)}$ | $\frac{1}{2}\sec x$ | |
| • Double-angle identity | $\frac{\sin x}{2\sin x\cos x}$ | $\frac{1}{2}\left(\frac{1}{\cos x}\right)$ | • Reciprocal identity |
| • Simplify | $\frac{1}{2\cos x}$ | $\frac{1}{2\cos x}$ | |
| | LHS = | RHS | |

**4.** 105° can be written as $(60°+45°)$.

Since $\sin(A+B)=\sin A\cos B+\cos A\sin B$,

then

$\sin(60°+45°)=\sin 60°\cos 45°+\cos 60°\sin 45°$,

where $A=60°$ and $B=45°$.

$$\sin(60°+45°)=\left(\frac{\sqrt{3}}{2}\right)\left(\frac{\sqrt{2}}{2}\right)+\left(\frac{1}{2}\right)\left(\frac{\sqrt{2}}{2}\right)$$

$$\sin(60°+45°)=\frac{\sqrt{6}}{4}+\frac{\sqrt{2}}{4}$$

$$\sin(105°)=\frac{\sqrt{6}+\sqrt{2}}{4}$$

**5.**

**LHS:**

$$\frac{1}{\sin^4 x-\cos^4 x}$$

• Factor $\dfrac{1}{(\sin^2 x+\cos^2 x)(\sin^2 x-\cos^2 x)}$

• Pythagorean Identity $\dfrac{1}{1(\sin^2 x-\cos^2 x)}$

**RHS:**

$$-\sec(2x)$$

• Reciprocal identity $\dfrac{-1}{\cos(2x)}$

• Double-Angle identity $\dfrac{-1}{\cos^2 x-\sin^2 x}$

• Multiply numerator and denominator by –1

$\dfrac{1}{\sin^2 x-\cos^2 x}$

= LHS

**6.**

**LHS:**

$$\frac{1-\tan^2 x}{1+\tan^2 x}$$

• Quotient identity $\dfrac{1-\dfrac{\sin^2 x}{\cos^2 x}}{1+\dfrac{\sin^2 x}{\cos^2 x}}$

• Common denominator $\dfrac{\dfrac{\cos^2 x}{\cos^2 x}-\dfrac{\sin^2 x}{\cos^2 x}}{\dfrac{\cos^2 x}{\cos^2 x}+\dfrac{\sin^2 x}{\cos^2 x}}$

• Simplify $\dfrac{\dfrac{\cos^2 x-\sin^2 x}{\cos^2 x}}{\dfrac{\cos^2 x+\sin^2 x}{\cos^2 x}}$

• Multiply by the reciprocal $\dfrac{(\cos^2 x-\sin^2 x)}{\cos^2 x}\times\dfrac{\cos^2 x}{(\cos^2 x+\sin^2 x)}$

• Pythagorean identity and simplify $\cos^2 x-\sin^2 x$

= RHS

**RHS:**

$$\cos(2x)$$

• Double-angle identity $\cos^2 x-\sin^2 x$

**7.** Since $\sin(A-B)=\sin A\cos B-\cos A\sin B$,

then $\sin\left(\theta-\frac{\pi}{3}\right)=\sin\theta\cos\frac{\pi}{3}-\cos\theta\sin\frac{\pi}{3}$,

where $A=\theta$ and $B=\frac{\pi}{3}$.

$$\sin\left(\theta-\frac{\pi}{3}\right)=\sin\theta\left(\frac{1}{2}\right)-\cos\theta\left(\frac{\sqrt{3}}{2}\right)$$

$$\sin\left(\theta-\frac{\pi}{3}\right)=\frac{1}{2}\sin\theta-\frac{\sqrt{3}}{2}\cos\theta$$

**8.**

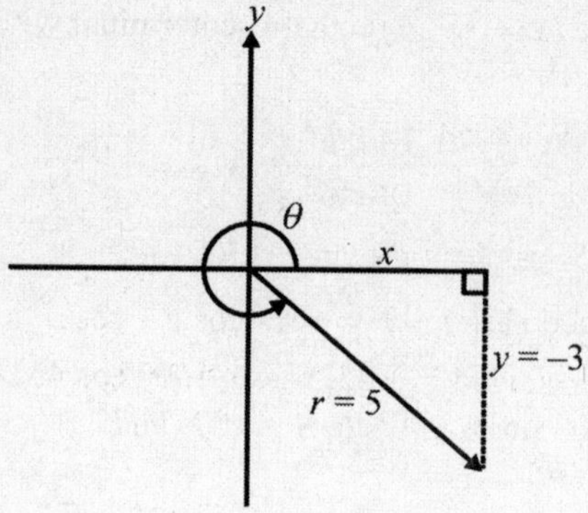


$\sin\theta = \frac{y}{r}$

Since $\sin\theta = \frac{-3}{5}$, the possible values are $y = -3$ and $r = 5$.

Apply the Pythagorean theorem:

$$x^2 + y^2 = r^2$$
$$x^2 + (-3)^2 = 5^2$$
$$x^2 + 9 = 25$$
$$x^2 = 16$$
$$x = \pm 4$$

In quadrant IV, the value of $x$ is positive, thus, $x = 4$.

$\cos\theta = \frac{x}{r}$

$\cos\theta = \frac{4}{5}$

Since $\cos(A+B) = \cos A\cos B - \sin A\sin B$, then $\cos(60° + \theta) = \cos 60° \cos\theta - \sin 60° \sin\theta$, where $A = 60°$ and $B = \theta$.

Substitution:

$$\cos(60° + \theta) = \left(\frac{1}{2}\right)\left(\frac{4}{5}\right) - \left(\frac{\sqrt{3}}{2}\right)\left(\frac{-3}{5}\right)$$

$$\cos(60° + \theta) = \frac{4}{10} + \frac{3\sqrt{3}}{10}$$

$$\cos(60° + \theta) = \frac{4 + 3\sqrt{3}}{10}$$

**9.** $\frac{2\sin x}{\sin 2x} = \sec x$

**a)** To verify for $x = \frac{\pi}{4}$:

| | **LHS** | **RHS** | |
|---|---|---|---|
| • Substitute $\frac{\pi}{4}$ for $x$ | $\frac{2\sin\frac{\pi}{4}}{\sin\frac{\pi}{2}}$ | $\sec\frac{\pi}{4}$ | |
| • Calculate each side separately | $\frac{2\left(\frac{\sqrt{2}}{2}\right)}{1}$ | $\frac{2}{\sqrt{2}}$ | |
| • Simplify | $\sqrt{2}$ | $\frac{2}{\sqrt{2}} \cdot \frac{\sqrt{2}}{\sqrt{2}}$ | • Rationalize denominator |
| | | $\frac{2\sqrt{2}}{2}$ | |
| | | $\sqrt{2}$ | |

LHS = RHS

**b)** To verify graphically:

Graph: $Y_1 = \frac{2\sin x}{\sin 2x}$

$Y_2 = \sec x$

**c)** To prove algebraically:

| | **LHS** | **RHS** | |
|---|---|---|---|
| | $\frac{2\sin x}{\sin(2x)}$ | $\sec x$ | |
| • Double-angle identity | $\frac{2\sin x}{2\sin x\cos x}$ | $\frac{1}{\cos x}$ | • Reciprocal identity |
| • Simplify | $\frac{1}{\cos x}$ | | |
| | = RHS | | |

 

**10.** $\dfrac{11\pi}{12}$ can be written as $\left(\dfrac{8\pi}{12}+\dfrac{3\pi}{12}\right)$ or

$\left(\dfrac{2\pi}{3}+\dfrac{\pi}{4}\right)$

Since $\cos(A+B)=\cos A\cos B-\sin A\sin B$,

then $\cos\left(\dfrac{2\pi}{3}+\dfrac{\pi}{4}\right)=\cos\dfrac{2\pi}{3}\cos\dfrac{\pi}{4}-\sin\dfrac{2\pi}{3}\sin\dfrac{\pi}{4}$,

where $A=\dfrac{2\pi}{3}$ and $B=\dfrac{\pi}{4}$.

$$\cos\left(\frac{2\pi}{3}+\frac{\pi}{4}\right)=\left(\frac{-1}{2}\right)\left(\frac{\sqrt{2}}{2}\right)-\left(\frac{\sqrt{3}}{2}\right)\left(\frac{\sqrt{2}}{2}\right)$$

$$\cos\left(\frac{2\pi}{3}+\frac{\pi}{4}\right)=\frac{-\sqrt{2}}{4}-\frac{\sqrt{6}}{4}$$

$$\cos\left(\frac{11\pi}{12}\right)=\frac{-\sqrt{2}-\sqrt{6}}{4}$$

**11.**

**LHS:**

| | |
|---|---|
| | $1+\sin 2x$ |
| • Double-angle identity | $1+2\sin x\cos x$ |

**RHS**

| | |
|---|---|
| | $(\sin x+\cos x)^2$ |
| • Expand | $\sin^2 x+2\sin x\cos x+\cos^2 x$ |
| • Rearrange | $\sin^2 x+\cos^2 x+2\sin x\cos x$ |
| • Pythagorean identity | $1+2\sin x\cos x$ |
| | = LHS |

**12.**

| | LHS | RHS |
|---|---|---|
| | $\cos B\cos(A-B)-$ $\sin B(\sin A-B)$ | $\cos A$ |
| • Difference identity | $\cos B(\cos A\cos B+\sin A\sin B)$ $-\sin B(\sin A\cos B-\cos A\sin B)$ | |
| • Expand | $\cos A\cos^2 B+\cos B\sin A\sin B$ $-\cos B\sin A\sin B+\cos A\sin^2 B$ | |
| • Simplify | $\cos A\cos^2 B+\cos A\sin^2 B$ | |
| • Factor | $\cos A\left(\cos^2 B+\sin^2 B\right)$ | |
| • Pythagorean Identity | $\cos A$ | |
| | = RHS | |

**13.** The positive angle that is coterminal with (–165°) is

$(-165°+360°)=195°$.

$\sin(-165°)=\sin 195°$

195° can be written as $(150°+45°)$

Since $\sin(A+B)=\sin A\cos B+\cos A\sin B$,

then sin (150° + 45°) = sin 150° cos 45° + cos 150° sin 45°, where $A$ = 150° and $B$ = 45°.

$$\sin(150°+45°)=\left(\frac{1}{2}\right)\left(\frac{\sqrt{2}}{2}\right)+\left(\frac{-\sqrt{3}}{2}\right)\left(\frac{\sqrt{2}}{2}\right)$$

$$\sin(195°)=\frac{\sqrt{2}-\sqrt{6}}{4}$$

$$\sin(-165°)=\frac{\sqrt{2}-\sqrt{6}}{4}$$

**14.** For angle $\alpha$:

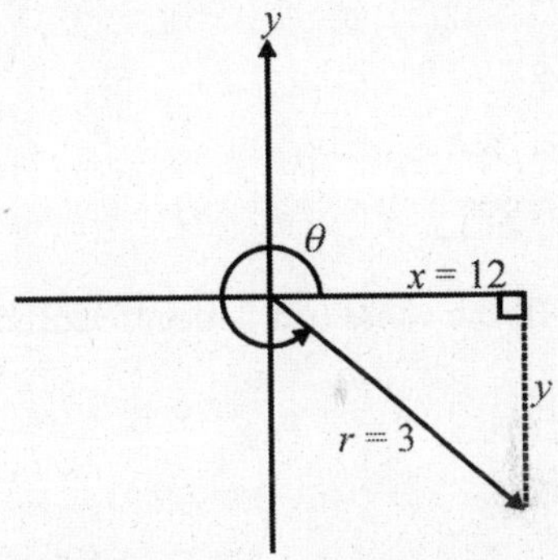


$\cos\alpha=\dfrac{x}{r}$

Since $\cos\alpha=\dfrac{12}{13}$, possible values are $x$ = 12 and $r$ =13

Apply the Pythagorean theorem:

$$x^2+y^2=r^2$$
$$12^2+y^2=13^2$$
$$144+y^2=169$$
$$y^2=25$$
$$y=\pm 5$$

In quadrant IV, the value of $y$ is negative, thus, $y=-5$.

$\sin\alpha=\dfrac{y}{r}$

$\sin\alpha=\dfrac{-5}{13}$

For angle $\beta$:

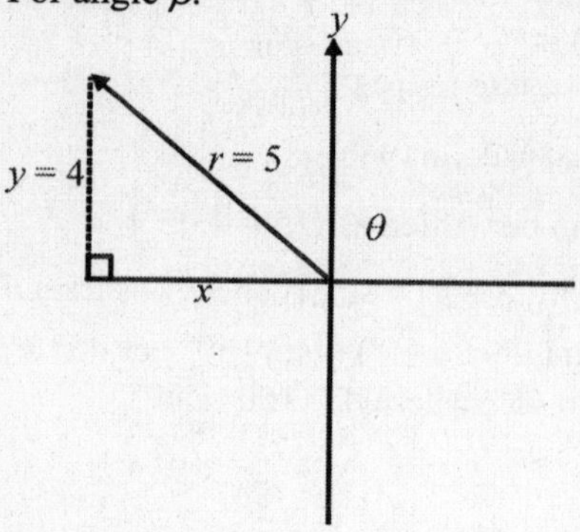


$\sin\beta = \frac{y}{r}$

Since $\sin\beta = \frac{y}{5}$, possible values are $y = 4$ and $r = 5$.

Apply the Pythagorean theorem:

$x^2 + y^2 = r^2$

$x^2 + 4^2 = 5^2$

$x^2 + 16 = 25$

$x^2 = 9$

$x = 3$

In quadrant II, the value of $x$ is negative, thus $x = -3$.

$\cos\beta = \frac{x}{r}$

$\cos\beta = \frac{-3}{5}$

Since $\cos(A+B) = \cos A\cos B - \sin A\sin B$, then $\cos(\alpha+\beta) = \cos\alpha\cos\beta - \sin\alpha\sin\beta$, where $A = \alpha$ and $B = \beta$.

$\cos(\alpha+\beta) = \left(\frac{12}{13}\right)\left(\frac{-3}{5}\right) - \left(\frac{-5}{13}\right)\left(\frac{4}{5}\right)$

$\cos(\alpha+\beta) = \frac{-36}{65} + \frac{20}{65}$

$\cos(\alpha+\beta) = \frac{-16}{65}$

**15.** $\cos^2 x = \frac{1+\cos(2x)}{2}$

**a)** To verify for $x = 30°$:

| | **LHS** | **RHS** |
|---|---|---|
| • Substitute 30° for $x$ | $\cos^2(30°)$ | $\frac{1+\cos 60°}{2}$ |
| • Calculate each side separately | $\left(\frac{\sqrt{3}}{2}\right)^2$ | $\frac{1+\frac{1}{2}}{2}$ |
| • Simplify | $\frac{3}{4}$ | $\frac{\frac{3}{2}}{2}$ |
| | $\frac{3}{4}$ | $\frac{3}{2}\times\frac{1}{2}$ |
| | $\frac{3}{4}$ | $\frac{3}{4}$ |
| | LHS = | RHS |

**b)** To verify graphically:

Graph: $Y_1 = \cos^2 x$

$Y_2 = \frac{1+\cos(2x)}{2}$

**c)** To prove algebraically

| **LHS** | **RHS** | |
|---|---|---|
| $\cos^2 x$ | $\frac{1+\cos(2x)}{2}$ | |
| | $\frac{1+\cos^2 x - \sin^2 x}{2}$ | • Double-angle identity |
| | $\frac{1-\sin^2 x + \cos^2 x}{2}$ | • Rearrange terms |
| | $\frac{\cos^2 x + \cos^2 x}{2}$ | • Pythagorean identity |
| | $\frac{2\cos^2 x}{2}$ | • Simplify |
| | $\cos^2 x$ = RHS | • Simplify |

**16.**

| | **LHS** | **RHS** |
|---|---|---|
| | $\frac{\cos(2x)+1}{\sin(2x)}$ | $\frac{\cos x}{\sin x}$ |
| • Double-angle identity | $\frac{\cos^2 x - \sin^2 x + 1}{2\sin x\cos x}$ | |
| • Rearrange terms | $\frac{\cos^2 x + 1 - \sin^2 x}{2\sin x\cos x}$ | |
| • Pythagorean identity | $\frac{\cos^2 x + \cos^2 x}{2\sin x\cos x}$ | |

 

- Simplify $\frac{2\cos^2 x}{2\sin x\cos x}$

$\frac{\cos x}{\sin x}$

$= \text{RHS}$

**17.**

**LHS:**

$$\sin(x+y)\sin(x-y)$$

Sum and Difference Identities

$(\sin x\cos y+\cos x\sin y)(\sin x\cos y-\cos x\sin y)$

Multiply

$\sin^2 x\cos^2 y-\sin x\cos y\cos x\sin y$
$\quad +\sin x\cos y\cos x\sin y-\cos^2 x\sin^2 y$

Simplify

$\sin^2 x\cos^2 y-\cos^2 x\sin^2 y$

Pythagorean identity

$(1-\cos^2 x)\cos^2 y-\cos^2 x(1-\cos^2 y)$

Multiply

$\cos^2 y-\cos^2 x\cos^2 y-\cos^2 x+\cos^2 x\cos^2 y$

Simplify

$\cos^2 y-\cos^2 x$

$= \text{RHS}$

**RHS:**

$\cos^2 y-\cos^2 x$

**18.**

**LHS:**

$1-\tan^2 x$

- Quotient identity $1-\frac{\sin^2 x}{\cos^2 x}$
- Find a common denominator $\frac{\cos^2 x}{\cos^2 x}-\frac{\sin^2 x}{\cos^2 x}$
- Simplify $\frac{\cos^2 x-\sin^2 x}{\cos^2 x}$

**RHS:**

$\frac{1+\tan^2 x}{\sec 2x}$

- Pythagorean and reciprocal identities $\frac{\sec^2 x}{\frac{1}{\cos 2x}}$
- Reciprocal and double-angle identity $\frac{\frac{1}{\cos^2 x}}{\frac{1}{\cos^2 x-\sin^2 x}}$
- Multiply by reciprocal $\frac{1}{\cos^2 x}\times\frac{\cos^2 x-\sin^2 x}{1}$

$\frac{\cos^2 x-\sin^2 x}{\cos^2 x}$

$= \text{LHS}$

**19.** $\cos A\cos B+\sin A\sin B=\cos(A-B)$
If $A=3.5$, then,
$\cos 3.5\cos B+\sin 3.5\sin B=\cos(3.5-B)$
If $\cos(3.5-B)=\cos(-1.7)$, then $3.5-B=-1.7$.
$-B=-1.7-3.5$
$B=1.7+3.5$
$B=5.2$

**20.** If $\cos Y°+\sin^2 114°=\cos^2 114°$,
then $\cos Y°=\cos^2 114°-\sin^2 114°$.
Since $\cos 2A=\cos^2 A-\sin^2 A$,
then $\cos(2\times 114°)=\cos^2 114°-\sin^2 114°$,
where $A=114°$.
If $\cos Y°=\cos(2\times 114°)$,
then $Y°=2\times 114°$
Thus, $Y°=228°$, $Y=228$.

## *Lesson 14 Using Identities to Solve Trigonometric Equations*

### CLASS EXERCISES ANSWERS AND SOLUTIONS

**1.** Before we can solve this equation, we must first simplify it. We want to simplify the equation so that it only contains one kind of trigonometric ratio, sine or cosine.

Notice that the left side resembles one of the sum and difference identities. We use this to rewrite the equation.

$\sin 3\theta \cos 2\theta - \cos 3\theta \sin 2\theta = -0.406$

$\sin(3\theta - 2\theta) = -0.406$ • Difference identity

$\sin\theta = -0.406$ • Simplify

This is a much easier equation to solve.

To find the first solution, $\theta_1$, enter $\sin^{-1}(-0.406)$ into the calculator.

If you entered this properly and if your calculator is set to radians, you will get the first solution, $\theta_1 \doteq -0.42$ rad.

Since –0.42 falls outside of the domain $0 \le \theta < 2\pi$, we need to find a coterminal angle by adding $2\pi$ to –0.42.

We get $2\pi + (-0.42) = 5.86$.

To find the second solution, $\theta_2$, we can use the formula for sines, $\theta_2 = \pi - \theta_1$.

We get:

$\theta_2 = \pi - (-0.42)$

$\theta_2 = 3.56$

So, the solutions to the equation $1+\cot^2\theta + 1 - \cos^2\theta = 2$ are: $x = 3.56$, $5.86$ rad.

**2.** Before we can solve this equation, we must simplify it. The equation contains both cotangent and cosine functions. Our goal is to simplify it so that it only contains only one kind of trigonometric function.

$1+\cot^2\theta + 1 - \cos^2\theta = 2$

$\csc^2\theta + \sin^2\theta = 2$ • Pythagorean identities

$\dfrac{1}{\sin^2\theta} + \sin^2\theta = 2$ • Reciprocal identity

$1 + \sin^4\theta = 2\sin^2\theta$ • Multiply both sides by $\sin^2\theta$

$\sin^4\theta - 2\sin^2\theta + 1 = 0$ • Set left side equal to zero

The original equation has been rewritten in terms of sine. It is now ready to solve.

To solve this equation, we need to factor the left-hand side. When we do this, we get $(\sin^2\theta - 1)(\sin^2\theta - 1) = 0$.

We can conclude that $\sin^2\theta - 1 = 0$.

$\sin^2\theta - 1 = 0$

$\sin^2\theta = 1$ • Add 1 to both sides

$\sin\theta = \pm 1$ • Find the square root of both sides

So, we know that either $\sin\theta = 1$ or $\sin\theta = -1$.

On the unit circle this is equivalent to $y = \pm 1$.

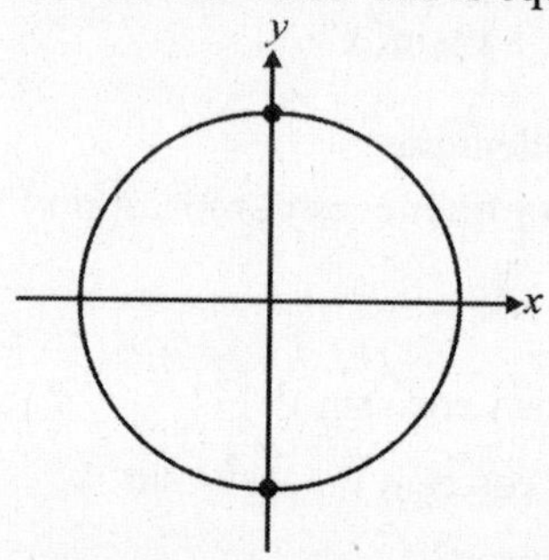


From our work with the unit circle, we know that the solutions are $\theta = \dfrac{\pi}{2}, \dfrac{3\pi}{2}$.

So, the solutions to the equation $1+\cot^2\theta + 1 - \cos^2\theta = 2$ are $\dfrac{\pi}{2}$ and $\dfrac{3\pi}{2}$ rad.

**3.** Our goal here is to find an equivalent equation, using the trigonometric identities we have learned. This equivalent equation should only have one trigonometric function. Recall that if the equation has a second-degree term, look for a replacement involving one of the Pythagorean identities. There is one involving $\csc^2\theta$. It is $1+\cot^2\theta = \csc^2\theta$.

To solve, proceed as follows:

$1+\cot^2 x + 3 - 5\cot x = 0$ • Replace $\csc^2 x$ with $1+\cot^2 x$

$\cot^2 x - 5\cot x + 4 = 0$ • Simplify and rearrange terms

$(\cot x - 4)(\cot x - 1) = 0$ • Factor

$\cot x - 4 = 0$ or $\cot x - 1 = 0$ • Solve for $\cot x$

If $\cot x - 4 = 0$, then $\cot x = 4$.

$\tan x = \dfrac{1}{4}$

$\tan^{-1}\left(\dfrac{1}{4}\right) \doteq 14.0°$ • From a calculator

$x_1 \doteq 14.0°$ • For tangent

 

$x_2 \doteq 180° + 14.0° \doteq 194.0°$ — ratios, $x_2 = 180° + x_1$

If $\cot x - 1 = 0$, then
$\cot x = 1.$
$\tan x = 1$
$x_3 = 45°$

- From a calculator or unit circle
- $x_4 = 180° + x_3$

$x_3 = 45°$
$x_4 = 180° + 45.0° = 225.0°$

The solutions to the equation $\csc^2 x + 3 - 5\cot x = 0,\ 0 \le x < 360°$, are:
$x = 14.0°, 45°, 194.0°, 225°$

## PRACTICE EXERCISES ANSWERS AND SOLUTIONS

**1.**

$$\frac{\sin x}{\tan x} = 0.650$$

- Quotient identity

$$\frac{\sin x}{\frac{\sin x}{\cos x}} = 0.650$$

- Multiply by the reciprocal

$$\sin x \cdot \frac{\cos x}{\sin x} = 0.650$$

$$\cos x = 0.650$$
$$x_1 = \cos^{-1}(0.650) \doteq 49.5°$$
$$x_2 = 360° - x_1$$
$$x_2 \doteq 360° - 49.5°$$
$$x_2 \doteq 310.5°$$

Thus, $x \doteq 49.5°, 310.5°$.

**2.**

$$4\tan x + \cot x = 5$$

- Reciprocal identity

$$4\tan x + \frac{1}{\tan x} = 5$$

- Multiply both sides by $\tan x$

$$4\tan^2 x + 1 = 5\tan x$$
$$4\tan^2 x - 5\tan x + 1 = 0$$

- Factor

$$(4\tan x - 1)(\tan x - 1) = 0$$

$\tan x = \frac{1}{4}$ or $\tan x = 1$

$x_1 = \tan^{-1}\left(\frac{1}{4}\right) \doteq 14.0°$ — $x_1 = \tan^{-1}(1) = 45.0°$

$x_2 = 180° - x_1$ — $x_2 = 180° - x_1$
$x_2 \doteq 180° + 14.0°$ — $x_2 = 180° - 45.0°$
$x_2 \doteq 194.0°$ — $x_2 = 225.0°$

Thus, $x \doteq 14.0°, 45.0°, 194.0°, 225.0°$.

**3.** $\csc x = 2\sin x + 1$

- Reciprocal Identity

$$\frac{1}{\sin x} = 2\sin x + 1$$

- Multiply both sides by $\sin x$

$$1 = 2\sin^2 x + \sin x$$
$$2\sin^2 x + \sin x - 1 = 0$$

- Factor

$$(2\sin x - 1)(\sin x + 1) = 0$$

$\sin x = \frac{1}{2}$ or $\sin x = -1$

$x_1 = \sin^{-1}\left(\frac{1}{2}\right) = 30°$ — $x = \sin^{-1}(-1) = 270°$

$x_2 = 180° - x_1$
$x_2 = 180° - 30°$
$x_2 = 150°$

Thus, $x = 30°, 150°, 270°$.

**4.**

$$\cos(2x)\cos x + \sin(2x)\sin x = 0.330$$

- Difference identity

$$\cos(2x - x) = 0.330$$
$$\cos x = 0.330$$
$$x_1 = \cos^{-1}(0.330) \doteq 70.7°$$
$$x_2 = 360° - x_1$$
$$x_2 \doteq 360° - 70.7°$$
$$x_2 \doteq 289.3°$$

Thus $x \doteq 70.7°, 289.3°$.

**5.**

$$\sec\theta \csc\theta \cot\theta = 4$$

- Quotient and Reciprocal identities

$$\frac{1}{\cos\theta} \times \frac{1}{\sin\theta} \times \frac{\cos\theta}{\sin\theta} = 4$$

$$\frac{1}{\sin^2\theta} = 4$$
$$\sin^2\theta = \frac{1}{4}$$
$$\sin\theta = \pm\frac{1}{2}$$

$\sin\theta = \frac{-1}{2}$ $\quad$ $\sin\theta = \frac{1}{2}$

$\theta_1 = \sin^{-1}\left(\frac{-1}{2}\right) \doteq -0.524$ $\quad$ $\theta_1 = \sin^{-1}\left(\frac{1}{2}\right) \doteq 0.524$

Find the positive coterminal angle with –0.524:
$-0.524 + 2\pi \doteq 5.760$

$\theta_2 = \pi - \theta$

$\theta_2 \doteq \pi - 0.524$

$\theta_2 \doteq 2.617$

$\theta_2 = \pi - \theta_1$

$\theta_2 \doteq \pi - (-0.524)$

$\theta_2 \doteq 3.666$

Thus, $\theta \doteq 0.52$ rad, 2.62 rad, 3.67 rad, 5.76 rad.
(Notice that the exact value solutions are
$\theta = \frac{\pi}{6}, \frac{5\pi}{6}, \frac{7\pi}{6}, \frac{11\pi}{6}$)

**6.**

$\cos(3\theta)\cos(2\theta) + \sin(3\theta)\sin(2\theta) = -0.560$

- Difference identity $\quad \cos(3\theta - 2\theta) = -0.560$

$\cos\theta = -0.560$

$\theta_1 = \cos^{-1}(-0.560) \doteq 2.165$

$\theta_2 = 2\pi - \theta_1$

$\theta_2 \doteq 2\pi - 2.165$

$\theta_2 \doteq 4.118$

Thus, $\theta \doteq 2.17$ rad, 4.12 rad.

**7.**

$2 - 3\cos^2 x - 2\sin x = 0$

- Pythagorean identity $\quad 2 - 3(1 - \sin^2 x) - 2\sin x = 0$
- Simplify $\quad 2 - 3 + 3\sin^2 x - 2\sin x = 0$

$3\sin^2 x - 2\sin x - 1 = 0$

$(3\sin x + 1)(\sin x - 1) = 0$

$\sin x = \frac{-1}{3}$ $\quad$ or $\quad$ $\sin x = 1$

$x_1 = \sin^{-1}\left(\frac{-1}{3}\right)$ $\quad$ $x = \sin^{-1}(1) = 90°$

$\doteq -19.5°$

Find the positive angle coterminal with –19.5°:
–19.5° + 360 = 340.5°.

$x_2 = 180° - x_1$

$x_2 \doteq 180° - (-19.5°)$

$x_2 \doteq 199.5°$

Thus $x \doteq 90°$, 199.5°, 340.5°

**8.**

$\sec^2 x + 3\tan x - 5 = 0$

- Pythagorean identity $\quad 1 + \tan^2 x + 3\tan x - 5 = 0$

$\tan^2 x + 3\tan x - 4 = 0$

$(\tan x - 1)(\tan x + 4) = 0$

$\tan x = 1$ $\quad$ or $\quad$ $\tan x = -4$

$x_1 = \tan^{-1}(1) = 45°$ $\quad$ $x_1 = \tan^{-1}(-4) = -76.0°$

$x_2 = 180° + x$

$x_2 = 180° + 45°$

$x_2 = 225°$

Find the positive angle coterminal with –76.0°:
–76.0° + 360° = 284.0°

$x_2 = 180° + x$

$x_2 \doteq 180° + (-76.0°)$

$x_2 \doteq 104.0°$

Thus, $x = 45.0°$, 104.0°, 225.0°, 284.0°.

**9.**

$2 - \sqrt{3}\sec\theta + 2\sec\theta - \sqrt{3}\sec^2\theta = 0$

- Factor by grouping

$(2 - \sqrt{3}\sec\theta) + \sec\theta(2 - \sqrt{3}\sec^2\theta) = 0$

$(2 - \sqrt{3}\sec\theta)(1 + \sec\theta) = 0$

$\sec\theta = \frac{2}{\sqrt{3}}$ $\quad$ or $\quad$ $\sec\theta = -1$

$\cos\theta = \frac{\sqrt{3}}{2}$ $\quad$ or $\quad$ $\cos\theta = -1$

$\theta = \frac{\pi}{6}, \frac{11\pi}{6}$ $\quad$ or $\quad$ $\theta = \pi$

Thus, $\theta = \frac{\pi}{6}, \pi, \frac{11\pi}{6}$.

**10.**

$4 - \csc^2 x - \cot x = 3, \ 0 \le x < 360°$

- Pythagorean identity $\quad 4 - (1 + \cot^2 x) - \cot x = 3$

$4 - 1 - \cot^2 x - \cot x - 3 = 0$

- Multiply both sides by –1 $\quad -\cot^2 x - \cot x = 0$
- Factor $\quad \cot^2 x + \cot x = 0$

$\cot x(\cot x + 1) = 0$

$\cot x = 0$ $\quad$ or $\quad$ $\cot x = -1$

- Take the reciprocals to get the primary ratios $\quad$ $\tan x$ is undefined $\quad$ or $\quad$ $\tan x = -1$

$x = 90°, 270°$ $\quad$ $x = 135°, 315°$

Thus, x = 90°, 135° 270°, 315°.

**11.**

$$3\tan^2\theta + 5\sec\theta + 1 = 0,\ \ 0 \le \theta < 2\pi$$

- Reciprocal and Quotient identities

$$3\left(\frac{\sin^2\theta}{\cos^2\theta}\right) + 5\left(\frac{1}{\cos\theta}\right) + 1 = 0$$

- Multiply both sides by $\cos^2\theta$

$$3\sin^2\theta + 5\cos\theta + \cos^2\theta = 0$$

- Pythagorean identity

$$3\left(1-\cos^2\theta\right) + 5\cos\theta + \cos^2\theta = 0$$

$$3 - 3\cos^2\theta + 5\cos\theta + \cos^2\theta = 0$$

$$-2\cos^2\theta + 5\cos\theta + 3 = 0$$

- Multiply both sides by –1

$$2\cos^2\theta - 5\cos\theta - 3 = 0$$

$$(2\cos\theta + 1)(\cos\theta - 3) = 0$$

$\cos\theta = \frac{-1}{2}$ or $\cos\theta = 3$

$\theta = \frac{2\pi}{3}, \frac{4\pi}{3}$

Since $\cos\theta$ is never greater than 1, there is no solution.

Thus, $\theta = \frac{2\pi}{3}, \frac{4\pi}{3}$.

**12.**

$$2\sin\theta - \tan\theta + 2\cos\theta - 1 = 0,\ \ 0 \le \theta < 2\pi$$

- Quotient identity

$$2\sin\theta - \frac{\sin\theta}{\cos\theta} + 2\cos\theta - 1 = 0$$

- Multiply both sides by $\cos\theta$

$$2\sin\theta\cos\theta - \sin\theta + 2\cos^2\theta - \cos\theta = 0$$

$$\sin\theta(2\cos\theta - 1) + \cos\theta(2\cos\theta - 1) = 0$$

$$(2\cos\theta - 1)(\sin\theta + \cos\theta) = 0$$

$\cos\theta = \frac{1}{2}$ or $\frac{\sin\theta}{\cos\theta} = -1$

$\tan\theta = -1$

$\theta = \frac{\pi}{3}, \frac{5\pi}{3}$ $\qquad$ $\theta = \frac{3\pi}{4}, \frac{7\pi}{4}$

Thus $\theta = \frac{\pi}{3}, \frac{3\pi}{4}, \frac{5\pi}{3}, \frac{7\pi}{4}$.

## *Lesson 15 Graphing Sine and Cosine Functions*

### PRACTICE EXERCISES ANSWERS AND SOLUTIONS

**1.** The pattern repeats every change of 5 units in the domain. Thus, the period is 5.

**2.** The pattern repeats every change of $\pi$ units in the domain. Thus, the period is $\pi$.

**3.** The pattern repeats every change of $0.5\pi$ units in the domain. Thus, the period is $0.5\pi$.

**4.** **a)**

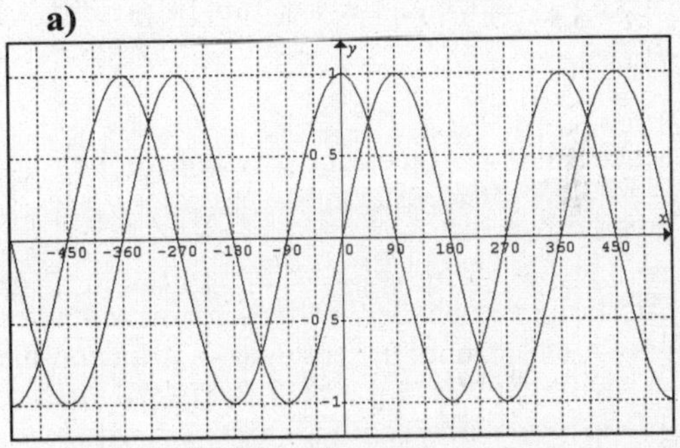


**b)** In the domain $-540° \le x \le 540°$ the functions are equal at the following points

$\left(-495°, \frac{-\sqrt{2}}{2}\right), \left(-315°, \frac{\sqrt{2}}{2}\right), \left(-135°, \frac{-\sqrt{2}}{2}\right),$
$\left(45°, \frac{\sqrt{2}}{2}\right), \left(225°, \frac{-\sqrt{2}}{2}\right), \left(405°, \frac{\sqrt{2}}{2}\right)$

**c)** In the domain $-540° \le x \le 540°$, $\sin x - \cos x = 0$ when $x = -495°, -315°, -135°, 45°, 225°, 405°$.

**d)** $\sin x$ has a maximum value of 1
When $\sin x = 1$, $\cos x = 0$

**e)** $\cos x$ has a minimum value of –1
When $\cos x = -1$, $\sin x = 0$.

**5.** To obtain $y = \cos\ \theta$, $y = \sin\ \theta$ could be translated
$\frac{3\pi}{2}$ units to the right, or $\frac{\pi}{2}$ units to the left.

**6.** $\sin x = \cos(x - 90°),\ \ c = 90°$

$\sin x = \cos(x + 270°),\ \ c = -270°$

**7.**

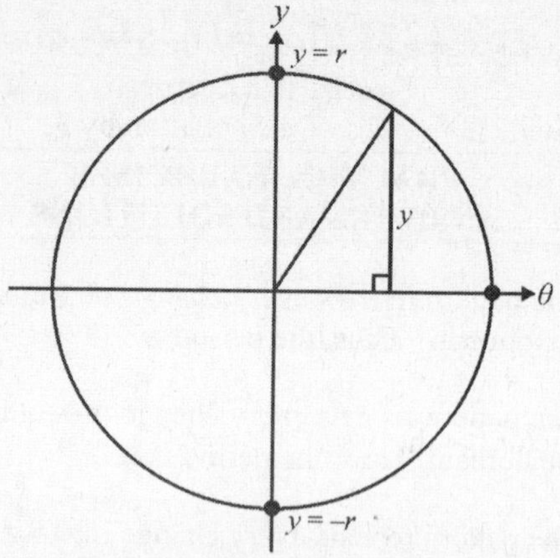


The largest value for $y = -r$ occurs where $y = r$. Since $y$ is on the circle, $y$ cannot be greater than $r$.

Since $\sin\theta = \frac{y}{r}$, then $\sin\theta$ will never be greater than 1.

Similarly, the smallest value for $y$ occurs when $y = -r$, as $y$ cannot be less than $-r$ if it is on the circle.

Since $\sin\theta = \frac{y}{r}$, then $\sin\theta$ will never be less than −1.

## *Lesson 16 Transformed Sine and Cosine Functions—Amplitude*

### CLASS EXERCISES ANSWERS AND SOLUTIONS

**1.** If we rewrite $y = 3\cos(\theta + 25°)$ as $y = 3\cos 1(\theta - (-25°)) + 0$, we can see that $a = 3$, $b = 1$, $c = -25°$, and $d = 0$.

**2.** In this question, $\theta$ does not appear by itself—it has the coefficient 4.

To rewrite the function in the form $y = a\sin b(\theta - c) + d$, we need to factor 4 from the term $(4\theta - \pi)$.

Factoring out the 4 gives: $y = \sin 4\left(\theta - \frac{\pi}{4}\right) - 5$

From this, we can see that $a = 1$, $b = 4$, $c = \frac{\pi}{4}$ and $d = 5$.

**3.** The amplitude is $|-3| = 3$.

The range is $-|3| \le y \le |3| \to -3 \le y \le 3$.

The graph is the graph of $y = 3\cos\theta$ reflected in the $\theta$-axis.

Since $-360° \le \theta \le 360°$, two periods are required.

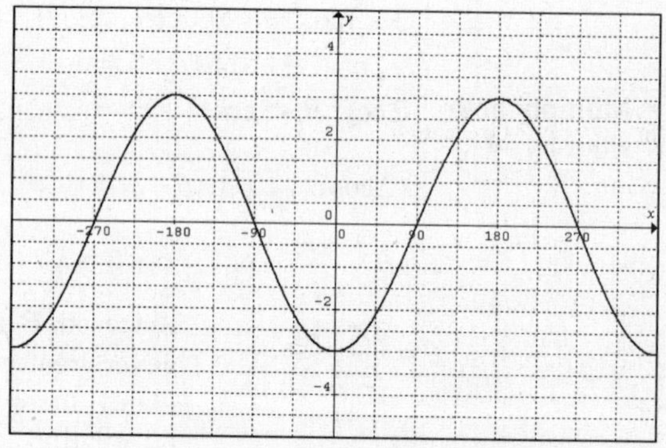


### PRACTICE EXERCISES ANSWERS AND SOLUTIONS

**1.** $y = \sin 2\left(\theta - \frac{7\pi}{4}\right) + 2$

$a = 1, b = 2, c = \frac{7\pi}{4}, d = 2$

**2.** $y = 4 + \cos 2(3\theta + \pi)$

Rewrite the equation and factor out the 3:

$y = \cos 2(3)\left(\theta + \frac{\pi}{3}\right) + 4$

$y = \cos 6\left(\theta + \frac{\pi}{3}\right) + 4$

$a = 1, b = 6, c = \frac{-\pi}{3}, d = 4$

**3.** $y = -\sin\left(2\theta - \frac{3\pi}{5}\right) + 7$

Factor out the 2:

$y = -1\sin 2\left(\theta - \frac{3\pi}{10}\right) + 7$

$a = -1, b = 2, c = \frac{3\pi}{10}, d = 7$

 

**4.** $y = -2\cos 5\left(\theta + \dfrac{\pi}{2}\right) + 1$

$a = -2, b = 5, c = \dfrac{-\pi}{2}, d = 1$

**5.** $y = \cos(3\theta - 60°) - \pi$

Factor out the 3:
$y = \cos 3(\theta - 20°) - \pi$
$a = 1, b = 3, c = 20°, d = -\pi$

**6.** $y = 3\sin\left(2\theta + \dfrac{\pi}{2}\right) - 4$

Factor out the 2:
$y = 3\sin 2\left(\theta + \dfrac{\pi}{4}\right) - 4$

$a = 3, b = 2, c = \dfrac{-\pi}{4}, d = -4$

**7.** $y = 1 - 3\cos(4\theta)$

Rewrite the equation:
$y = -3\cos(4\theta) + 1$
$a = -3, b = 4, c = 0, d = 1$

**8.** $y = \cos 2(4\theta + 100°) + 3$

Factor out the 4:
$y = \cos 2(4)(\theta + 25°) + 3$
$y = \cos 8(\theta + 25°) + 3$
$a = 1, b = 8, c = -25°, d = 3$

**9.** $y - 2 = 3\sin\left(\dfrac{1}{5}x + 60°\right)$

Rewrite the equation:
$y = 3\sin\left(\dfrac{1}{5}x + 60°\right) + 2$

Factor out the $\dfrac{1}{5}$:

$y = 3\sin\dfrac{1}{5}(x + 300°) + 2$

$a = 3, b = \dfrac{1}{5}, c = -300°, d = 2$

**10.** $3y = 5\cos\left(\dfrac{2}{3}\theta + \pi\right) - 4$

Divide both sides of the equation by 3:

$$\frac{3y}{3} = \frac{5\cos\left(\frac{2}{3}\theta + \pi\right)}{3} - \frac{4}{3}$$

$$y = \frac{5}{3}\cos\left(\frac{2}{3}\theta + \pi\right) - \frac{4}{3}$$

Factor out the $\dfrac{2}{3}$:

$$y = \frac{5}{3}\cos\frac{2}{3}\left(\theta + \frac{3\pi}{2}\right) - \frac{4}{3}$$

$a = \dfrac{5}{3}$, $b = \dfrac{2}{3}$, $c = \dfrac{-3\pi}{2}$, $d = \dfrac{-4}{3}$

**11.** $y = 4\sin x$, $0° \le x \le 360°$
amplitude = $|a| = |4| = 4$
Since this is a sine function that is stretched about the x-axis by a factor of 4, the significant points within the given domain are
(0, 0) (90°, 4) (180°, 0), (270°, –4) (360°, 0).

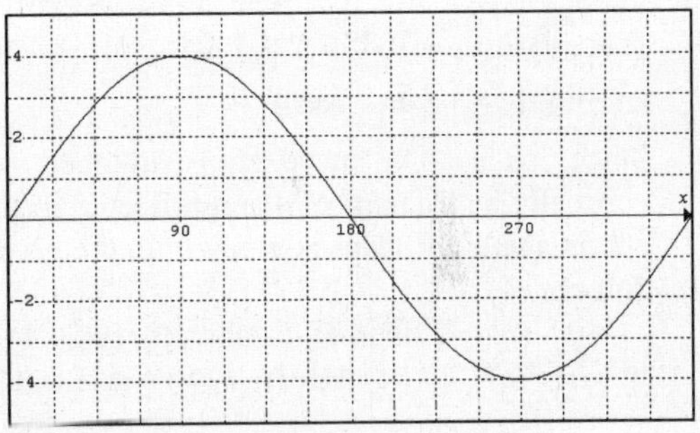


**12.** $y = 3\cos\theta$, $-\pi \le \theta \le \pi$
amplitude = $|a| = |3| = 3$
Since this is a cosine function that is stretched vertically about the $\theta$-axis by a factor of 3, the significant points within the given domain are
$(-\pi, -3)$, $\left(\dfrac{-\pi}{2}, 0\right)$, $(0, 3)$, $\left(\dfrac{\pi}{2}, 0\right)$, $(\pi, -3)$.

The graph is shown below.

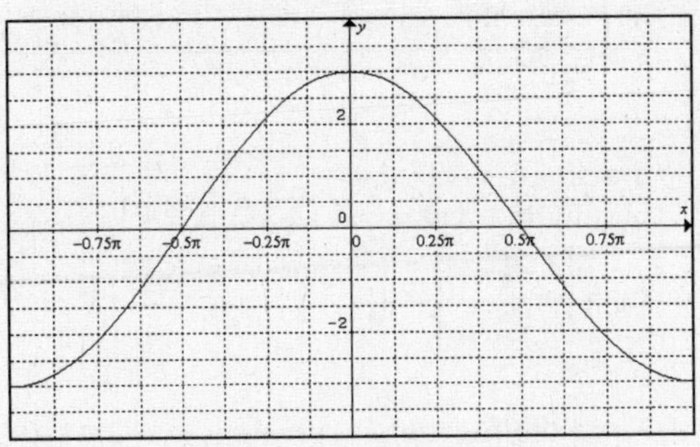

**13.** $y = -2\cos x$, $-2\pi \le x \le 2\pi$

amplitude = $|a| = |-2| = 2$

Since this is a cosine function that is stretched vertically about the $x$-axis by a factor of 2 and reflected in the $x$-axis, the significant point within the given domain are

$(-2\pi,-2)$, $\left(\frac{-3\pi}{2},0\right)$, $(-\pi,2)$, $\left(\frac{-\pi}{2},0\right)$,

$(0,-2)$, $\left(\frac{\pi}{2},0\right)$, $(\pi,2)$, $\left(\frac{3\pi}{2},0\right)$, $(2\pi,-2)$.

The graph is shown below.

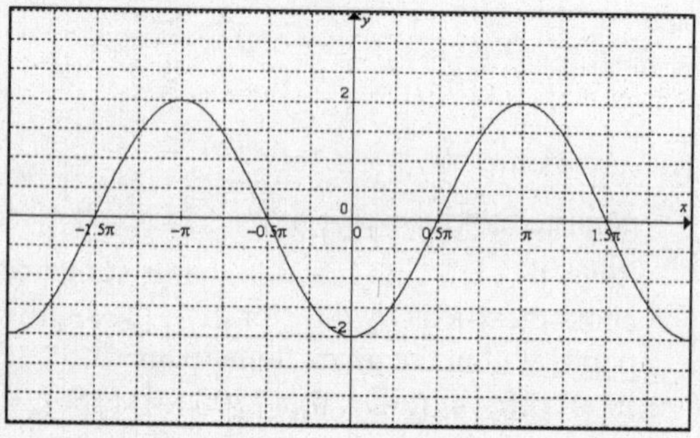


**14.** $y = -4\sin x$, $-360° \le x \le 360°$

amplitude = $|a| = |-4| = 4$

Since this is a sine curve that is stretched vertically by a factor of 4 and reflected in the $x$-axis, the significant points within the given domain are

(–360°, 0), (–270°, –4), (–180°, 0), (90°, 4),
(0°, 0), (90°, –4), (180°, 0), (270°, 4), (360°, 0).

The graph is shown below.

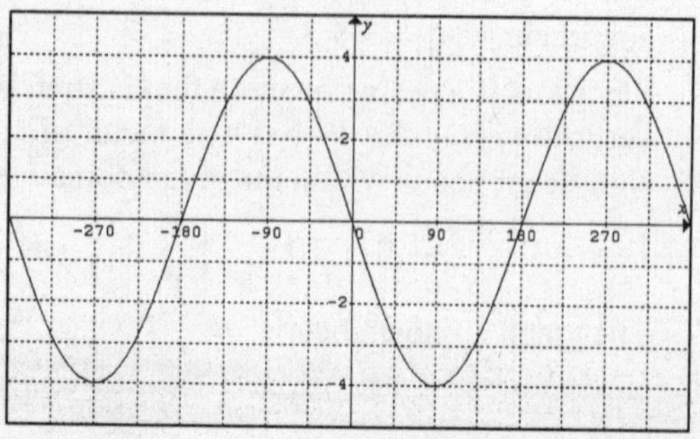


**15.** $y = a\sin\theta$

maximum $y$-value: 4
minimum $y$-value: –4

$$|a| = \left|\frac{4-(-4)}{2}\right| = 4$$

Thus, the function is $y = 4\sin\theta$, $-2\pi \le \theta \le 2\pi$.

**16.** $y = a\sin\theta$

maximum $y$-value: 3
minimum $y$-value: –3

$$|a| = \left|\frac{3-(-3)}{2}\right| = 3$$

Since the function is a reflection of $y = \sin\theta$ in the $\theta$-axis, $a = -3$.

Thus, the function is $y = -3\sin\theta$, $0° \le \theta \le 720°$.

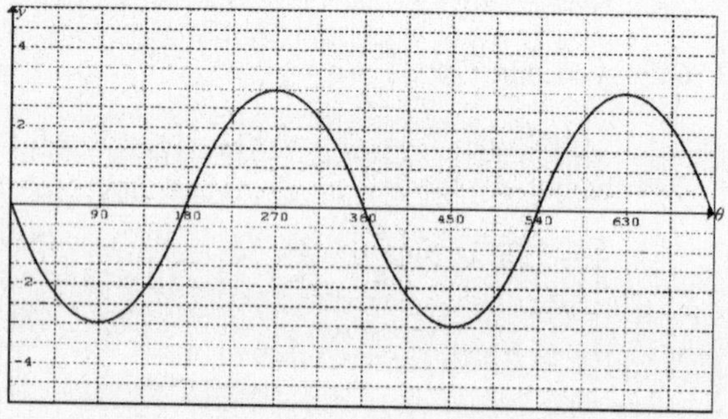


**17.** $y = a\cos\theta$

maximum $y$-value: 3
minimum $y$-value: –3

$$|a| = \left|\frac{3-(-3)}{2}\right| = 3$$

Thus, the function is $y = 3\cos\theta$, $0 \le \theta \le 2\pi$.

**18.** $y = a\cos\theta$

maximum $y$-value: 0.5
minimum $y$-value: –0.5

$$|a| = \left|\frac{0.5-(0.5)}{2}\right| = 0.5$$

Since the graph is a reflection of $y = \cos\theta$ in the $\theta$-axis, $a = -0.5$.

Thus, the function is $y = -0.5\cos\theta$, $-180° \le \theta \le 360°$.

**19.** $y = -13.2\cos 5(\theta - \pi) + 2$

$|a| = |13.2| = 13.2$ the amplitude is 13.2.

| | | | |
|---|---|---|---|
| 1 | 3 | . | 2 |

 

## *Lesson 17 Transformed Sine and Cosine Functions—Vertical Translation*

### CLASS EXERCISES ANSWERS AND SOLUTIONS

1. The amplitude is $|a| = |-3| = 3$
The vertical displacement is $d = -2$.

Draw a new axis at $y = -2$.

Draw guidelines at the maximum $y$-value, $y = -2 + 3 = 1$, and at the minimum $y$-value, $y = -2 - 3 = -5$.

Since this is a cosine function that is stretched by a factor of 3, reflected in the $x$-axis, and translated down 2 units, the significant points are at (0, –5), (90, –2), (180, 1), (270, –2), and (360, –5).

Once these are plotted, a smooth curve can be drawn through the points.

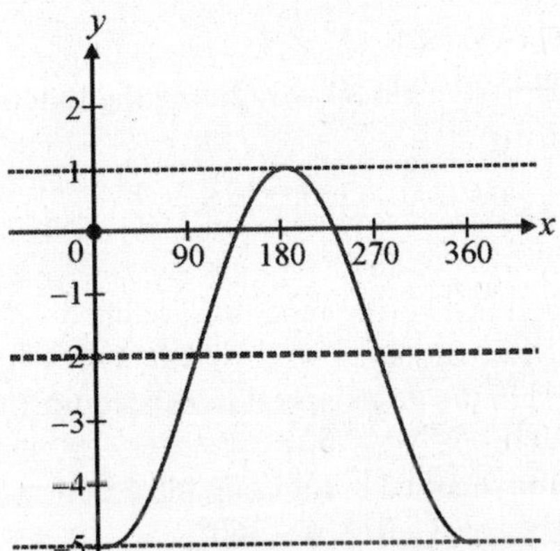


### PRACTICE EXERCISES ANSWERS AND SOLUTIONS

1. $y = \cos x - 3$, $-360° \leq x \leq 360°$
amplitude : $|a| = |1| = 1$

The vertical displacement is $d = -3$. Thus, the new horizontal axis is at $y = -3$.
maximum $y$-value: $-3 + 1 = -2$
minimum $y$-value: $-3 - 1 = -4$

Since this is a cosine function translated down 3 units, the significant points within the given domain are:
(–360°, –2), (–270°, –3), (–180°, –4), (–90°, –3), (0°, –2), (90°,–3) (180°,–4), (270°,– 3), (360°,– 2)

The graph is shown below.

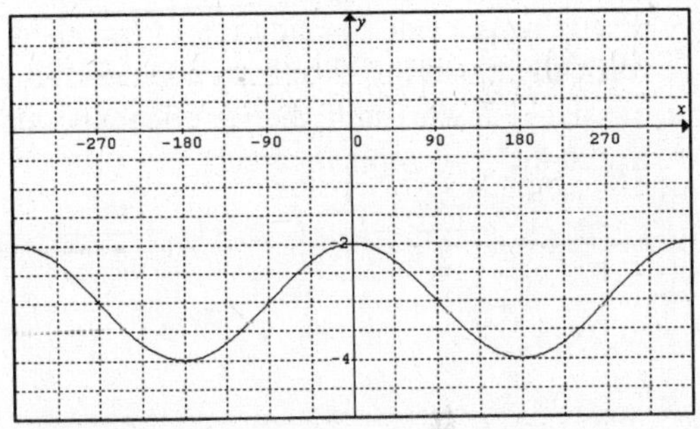


2. $y = \sin\theta + 5$, $0 \leq \theta \leq 2\pi$
amplitude: $|a| = |1| = 1$

The vertical displacement is $d = 5$. Thus, the new horizontal axis is at $y = 5$.
maximum $y$-value: $5 + 1 = 6$
minimum $y$-value: $5 - 1 = 4$

Since this is a sine function translated up 5 units, the significant points within the given domain are:

$(0,5)$, $\left(\frac{\pi}{2},6\right)$, $(\pi,5)$, $\left(\frac{3\pi}{2},4\right)$, $(2\pi,5)$

The graph is shown below.

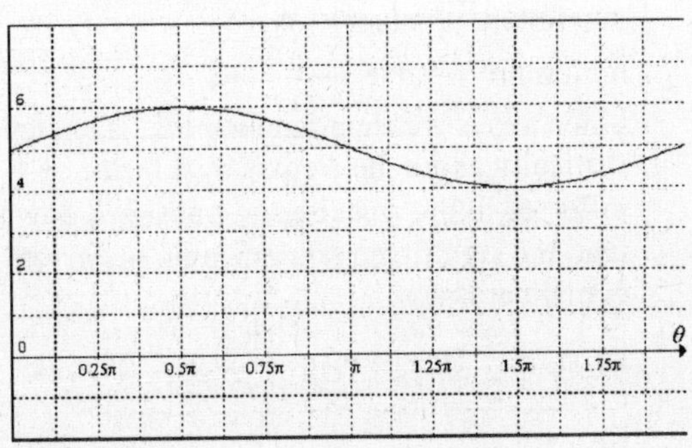

**3.** $y = 3\cos\theta - 1$, $-2\pi \le \theta \le 2\pi$
amplitude: $|a| = |3| = 3$

The vertical displacement is $d = -1$. Thus, the new horizontal axis is at $y = -1$.
maximum $y$-value: $-1 + 3 = 2$
minimum $y$-value: $-1 - 3 = -4$

Since this is a sine function that is stretched vertically about the $\theta$-axis by a factor of 3 and translated down 1 unit, the significant points within the given domain are:
$(-2\pi, 2)$, $\left(\frac{-3\pi}{2}, -1\right)$, $(-\pi, -4)$, $\left(\frac{\pi}{2}, -1\right)$,
$(0, 2)$, $\left(\frac{\pi}{2} - 1\right)$, $(\pi, -4)$, $\left(\frac{3\pi}{2}, -1\right)$, $(2\pi, 2)$

The graph is shown below.

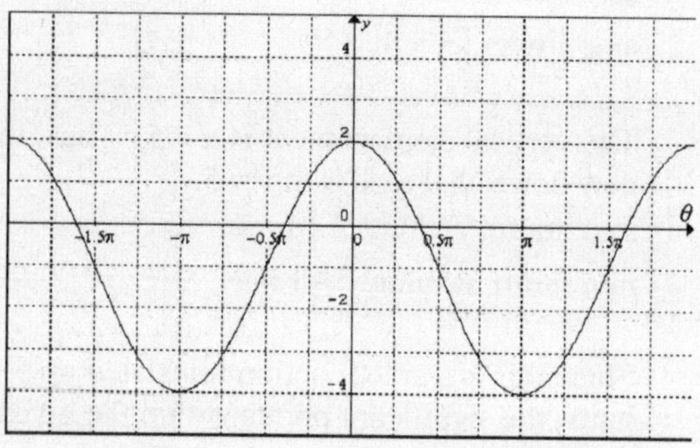


**4.** $y = -4\sin\theta + 1$, $0 \le \theta \le 3\pi$
amplitude: $|a| = |-4| = 4$
The vertical displacement is $d = 1$. Thus, the new horizontal axis is at $y = 1$.
maximum $y$-value: $1 + 4 = 5$
minimum $y$-value: $1 - 4 = -3$
Since this is a cosine function that is stretched vertically about the $\theta$-axis by a factor of 4, reflected in the $\theta$-axis, and translated down 1 unit, the significant points within the given domain are:
$(0, 1)$, $\left(\frac{\pi}{2} - 3\right)$, $(\pi, 1)$, $\left(\frac{3\pi}{2}, 5\right)$, $(2\pi, 1)$,
$\left(\frac{5\pi}{2}, -3\right)$, $(3\pi, 1)$

The graph is shown below.

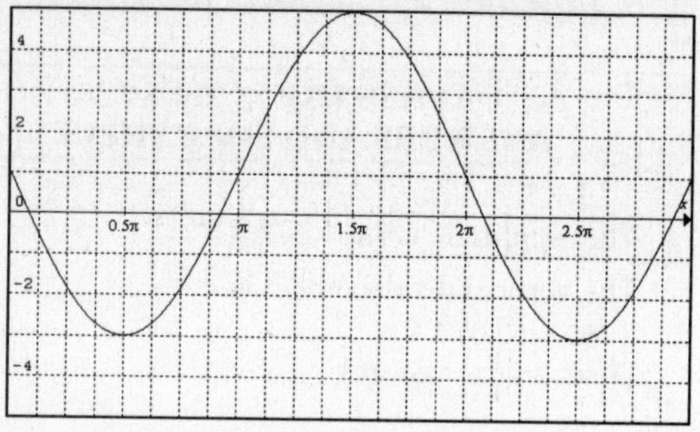


**5.** The graph is a sine function stretched vertically about the $\theta$-axis by a factor of 2 and translated up 3 units.

Thus, $a = 2$ and $d = 3$, giving the function
$y = 2\sin\theta + 3$, $-2\pi \le \theta \le 2\pi$

**6.** The graph is a sine function stretched vertically about the $\theta$-axis a factor of $\frac{1}{2}$, reflected in the $\theta$-axis and translated down 3 units.

Thus, $a = -\frac{1}{2}$ and $d = -3$, giving the function

$y = -\frac{1}{2}\sin\theta - 3$, $-360° \le \theta \le 360°$.

**7.** The graph is a cosine function stretched vertically about the $\theta$-axis by a factor of 4, reflected in the $\theta$-axis, and translated up 1 unit.

Thus, $a = -4$ and $d = 1$, giving the function
$y = -4\cos\theta + 1$, $0° \le \theta \le 360°$.

**8.** The graph is a cosine function stretched vertically about the $\theta$-axis by a factor of 3, reflected in the $\theta$-axis, and translated down 2 units.

Thus, $a = -3$ and $d = -2$, giving the function
$y = -3\cos\theta - 2$, $-2\pi \le \theta \le 2\pi$.

**9.** $y = -5\sin\theta + 0.3$

The graph of the function $y = \sin\theta$ has been stretched vertically about the $\theta$-axis by a factor of 5, reflected in the $\theta$-axis, and translated up 0.3 units.

 

**10.** $y=\frac{5}{2}\cos\theta-4$

The graph of the function $y=\cos\theta$ has been stretched vertically about the $\theta$-axis by a factor of $\frac{5}{2}$ and translated down 4 units.

**11.** $y=3\sin\theta+2.5$

$|a|=3$ and $d=2.5$

range: $d-|a|\le y\le d+|a|$

$2.5-3\le y\le 2.5+3$

$-0.5\le y\le 5.5$

**12.** $y=-5\cos\theta+3.5$

$|a|=|-5|=5$ and $d=3.5$

range: $d-|a|\le y\le d+|a|$

$3.5-5\le y\le 3.5+5$

$-1.5\le y\le 8.5$

**13.** $y=0.6\sin\theta-2.4$

$|a|=0.6$ and $d=-2.4$

range: $d-|a|\le y\le d+|a|$

$-2.4-0.6\le y\le -2.4+0.6$

$-3\le y\le -1.8$

**14.** $y=3.2\sin\theta+2.1$

$|a|=3.2$ and $d=2.1$

range: $d-|a|\le y\le d+|a|$

$2.1-3.2\le y\le 2.1+3.2$

$-1.1\le y\le 5.3$

**15.** $y=2.8\cos\theta+d$, $1.4\le y\le 7.0$

Since $d-|a|\le y\le d+|a|$,

then $d-|a|=1.4$ and $d+|a|=7.0$.

If $|a|=2.8$, then $d-2.8=1.4$ and

$d+2.8=7.0$.

Thus, $d=2.8+1.4$ or $d=-2.8+7.0$.

$d=4.2$

| 4 | . | 2 | |
|---|---|---|---|

**16.** $y=a\sin\theta-6.1$

The maximum value of $y$ is $d+|a|=-3.2$

since $d=6.1$, $-6.1+|a|=-3.2$.

$|a|=-3.2+6.1$

Thus, $|a|=2.9$.

| 2 | . | 9 | |
|---|---|---|---|

## *Lesson 18 Transformed Sine and Cosine Functions—Horizontal Translation (Phase Shift)*

### CLASS EXERCISES ANSWERS AND SOLUTIONS

**1.** Begin by sketching the function $y=\cos\,\theta$ showing the 5 significant points.

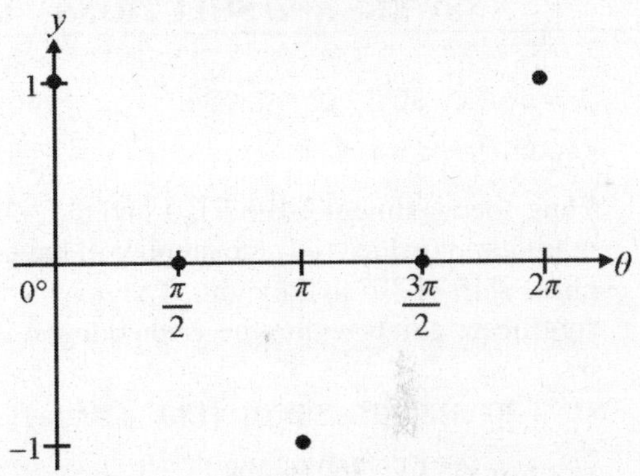


The graph of $y=\cos\left(\theta+\frac{\pi}{4}\right)$ will have a phase shift of $\frac{\pi}{4}$ units left for each significant point.

Thus, simply move the 5 significant points $\frac{\pi}{4}$ units to the left and sketch the graph.

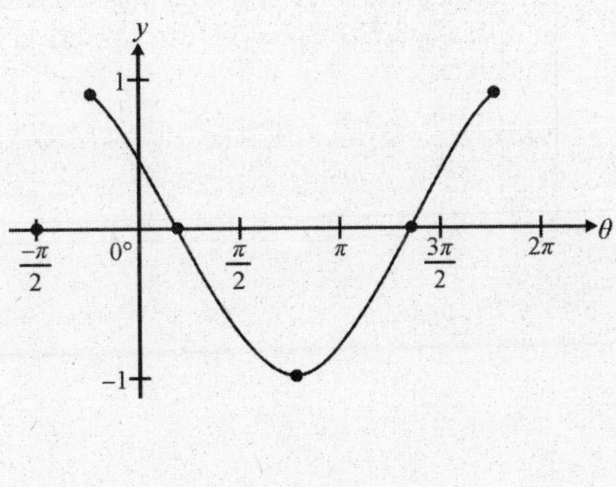

Note: The $x$-intercepts for this one period are $\frac{\pi}{4}$ and $\frac{5\pi}{4}$.

The maximum function value of 1 occurs when $\theta = -\frac{\pi}{4}$ or $\frac{7\pi}{4}$.

The minimum function value of –1 occurs when $\theta = \frac{3\pi}{4}$.

**2.** Since $a = -2$, $c = -\pi$, and $d = 3$, the graph would be stretched vertically about the $\theta$-axis by a factor of 2, reflected in the $\theta$-axis, and translated $\pi$ units left and 3 units up.

## PRACTICE EXERCISES ANSWERS AND SOLUTIONS

**1.** $y = \cos(x - 30°)$, $0° \le x \le 360°$

amplitude: $|a| = 1$

Thus, the maximum $y$-value is 1 and the minimum $y$-value is –1, the graph will have a phase shift of 30° to the right. The new significant points within the given domain will be:

$(0° + 30°, 1)$ $(90° + 30°, 0)$ $(180° + 30°, -1)$ $(270° + 30°, 0)$ $(360° + 30°, 1)$

or $(30°, 1)$, $(120°, 0)$, $(210° - 1)$, $(300°, 0)$, $(390°, 1)$

The graph is shown below.

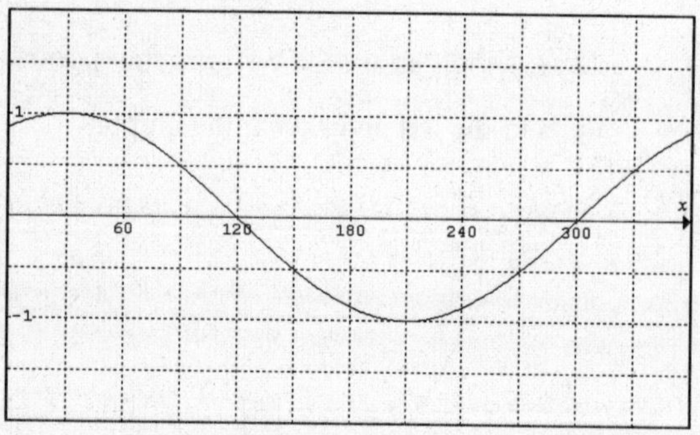


**2.** $y = \sin\left(\theta + \frac{\pi}{2}\right)$, $0° \le \theta \le 2\pi$

amplitude: $|a| = 1$

As $d = 0$, the maximum $y$-value is 1 and the minimum $y$-value is –1. The graph will have a phase shift of $\frac{\pi}{2}$ rad to the left. The new significant points within the given domain will be:

$\left(\frac{\pi}{2} - \frac{\pi}{2}, 1\right)$ $\left(\pi - \frac{\pi}{2}, 0\right)$ $\left(\frac{3\pi}{2} - \frac{\pi}{2}, -1\right)$

$\left(2\pi - \frac{\pi}{2}, 0\right)$ $\left(\frac{5\pi}{2} - \frac{\pi}{2}, -1\right)$

or $(0, 1)$ $\left(\frac{\pi}{2}, 0\right)$ $(\pi, -1)$ $\left(\frac{3\pi}{2}, 0\right)$ $(2\pi, 1)$

The graph is shown below.

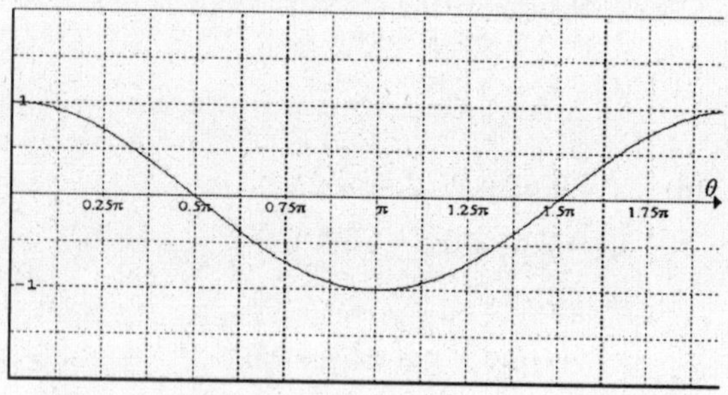


**3.** $y = \cos(\theta - \pi)$, $-2\pi \le \theta \le 2\pi$

amplitude: $|a| = 1$

As $d = 0$, the maximum $y$-value is 1 and the minimum $y$-value is –1.

The graph will have a phase shift of $\pi$ rad to the right. The new significant points within the given domain will be:

$(-3\pi + \pi, -1)$, $\left(\frac{-5\pi}{2} + \pi, 0\right)$, $(-2\pi + \pi, 1)$,

$\left(\frac{-3\pi}{2} + \pi, 0\right)$, $(-\pi + \pi, -1)$, $\left(\frac{-\pi}{2} + \pi, 0\right)$,

$(0 + \pi, 1)$, $\left(\frac{\pi}{2} + \pi, 0\right)$, $(\pi + \pi, -1)$

or

$(-2\pi, -1)$, $\left(\frac{-3\pi}{2}, 0\right)$, $(-\pi, 1)$, $\left(\frac{-\pi}{2}, 0\right)$,

$(0, -1)$, $\left(\frac{\pi}{2}, 0\right)$, $(\pi, 1)$, $\left(\frac{3\pi}{2}, 0\right)$, $(2\pi, -1)$

 

The graph is shown below.

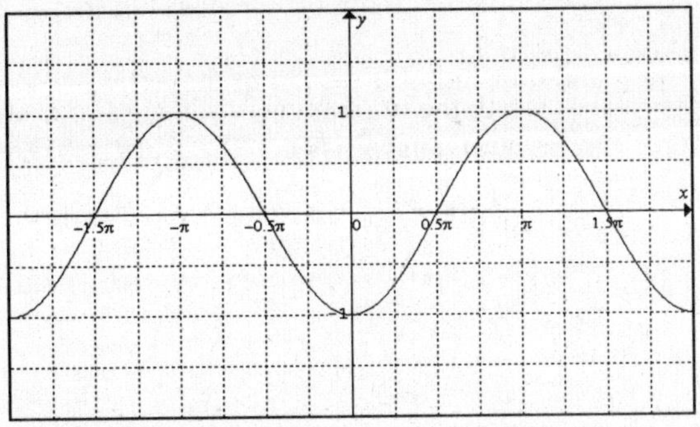


**4.** $y = 2\cos(\theta - \pi),\ 0° \leq \theta \leq 2\pi$

amplitude: $|a| = |2| = 2$

As $d = 0$, the maximum $y$-value is 2 and the minimum $y$-value is –2.

The graph will have a phase shift of $\pi$ rad to the right. The new significant points within the given domain will be:

$(-\pi+\pi, -2), \left(\frac{-\pi}{2}+\pi, 0\right), (0+\pi, 2),$

$\left(\frac{\pi}{2}+\pi, 0\right), (\pi+\pi, -2)$

or $(0, -2), \left(\frac{\pi}{2}, 0\right), (\pi, 2), \left(\frac{3\pi}{2}, 0\right), (2\pi, -2)$

The graph is shown below.

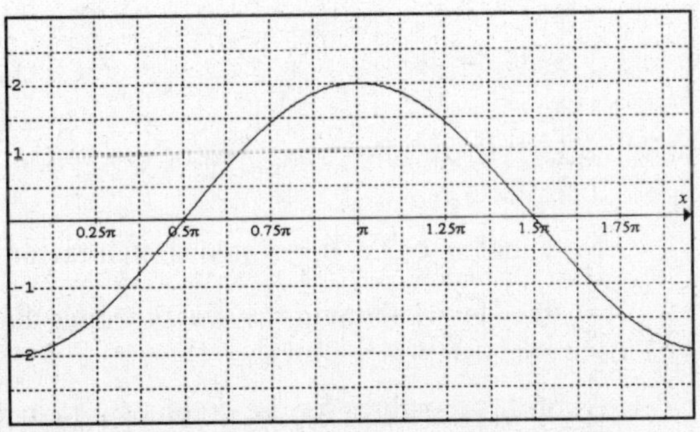


**5.** The graph is a sine function with an amplitude of 1. It has been reflected in the $\theta$-axis and translated $\frac{3\pi}{4}$ rad to the right.

Thus, $a = -1$ and $c = \frac{3\pi}{4}$, giving the function

$y = -\sin\left(\theta - \frac{3\pi}{4}\right),\ 0 \leq \theta \leq 2\pi.$

**6.** The graph is a sine function that has been stretched vertically (about the $\theta$-axis) by a factor of 3 and translated 30° to the right.

Thus, $a = 3$ and $c = 30°$, giving the function $y = 3\sin(\theta - 30°),\ 0° \leq \theta \leq 360°.$

**7.** The graph is a cosine function that has been stretched vertically (about the $\theta$-axis) by a factor of 2, and translated 60° to the left.

Thus, $a = -2$ and $c = -60°$, giving the function $y = -2\cos(\theta + 60°),\ 0 \leq \theta \leq 360°.$

**8.** The graph is a cosine function that has been stretched (vertically about the $\theta$-axis) by a factor of 2 and translated $\frac{\pi}{2}$ rad to the right.

Thus, $a = -2$ and $c = \frac{\pi}{2}$, giving the function

$y = 2\cos\left(\theta - \frac{\pi}{2}\right),\ -\pi \leq \theta \leq \pi.$

or

The graph is a cosine function that has been stretched vertically about the $\theta$-axis by a factor of 2 reflected in the $\theta$-axis and translated $\frac{\pi}{2}$ rad to the left.

Thus, $a = -2$ and $c = -\frac{\pi}{2}$, giving the function

$y = -2\cos\left(\theta + \frac{\pi}{2}\right),\ -\pi \leq \theta \leq \pi.$

**9.** $y = -2\sin\left(\theta - \frac{\pi}{3}\right)$

The graph of the function $y = \sin\theta$ has been stretched vertically about the $\theta$-axis by a factor of 2, reflected in the $\theta$-axis, and translated $\frac{\pi}{3}$ rad to the right.

**10.** $y = \frac{1}{2}\cos(\theta + 20°)$

The graph of the function $y = \cos\theta$ has been stretched vertically about the $\theta$-axis by a factor of $\frac{1}{2}$, and translated 20° to the left.

**11.** $y = -0.4\sin(\theta + 40°) - 3$

The graph of the function $y = \cos\theta$ has been stretched vertically about the $\theta$-axis by a factor of 0.4, reflected in the $\theta$-axis, translated 40° to the left, and then translated 3 units down.

**12.** $y = -\frac{4}{3}\cos\left(\theta - \frac{\pi}{12}\right) + 5$

The graph of the function $y = \cos\theta$ has been stretched vertically about the $\theta$-axis by a factor of $\frac{4}{3}$, reflected in the $\theta$-axis, translated $\frac{\pi}{12}$ units to the right, and translated 5 units up.

**13.** $y = 0.3\sin(\theta + 0.7) - 2.4$
The graph of the function $y = \sin\theta$ has been stretched vertically about the $\theta$-axis by a factor of 0.3, translated 0.7 rad to the left, and translated 2.4 units down.

**14.** $y = 3\cos(\theta - 1) - 2.6$
The graph of the function $y = \cos\theta$ has been stretched vertically about the $\theta$-axis by a factor of 3, translated 1 rad to the right, and translated 2.6 units down.

**15. D**

The graph of $y = 2\sin\theta$ has been translated $\frac{1}{2}$ rad to the left and 3 units up.

**16.** The graph of $y = \cos x$ can be obtained by translating the graph of $y = \sin x$ $\frac{3\pi}{2}$ rad to the right.

If $\cos x = \sin(x - M)$,

then $\cos x = \sin\left(x - \frac{3\pi}{2}\right)$.

Thus, $M = \frac{3\pi}{2} \doteq 4.71$.

**17.** The graph of $y = \sin\theta$ can be obtained by reflecting the graph of $y = \cos\theta$ in the $\theta$-axis and translating it $\frac{\pi}{2}$ rad to the left.

If $\sin\theta = -\cos(\theta + N)$,

then $\sin\theta = -\cos\left(\theta + \frac{\pi}{2}\right)$.

Thus, $N = \frac{\pi}{2} \doteq 1.57$.

## *Lesson 19 Transformed Sine and Cosine Functions—Horizontal Stretch (Change in Period)*

### CLASS EXERCISES ANSWERS AND SOLUTIONS

**1.** First, find the period.

In degrees, the period is $\frac{360°}{3} = 120°$.

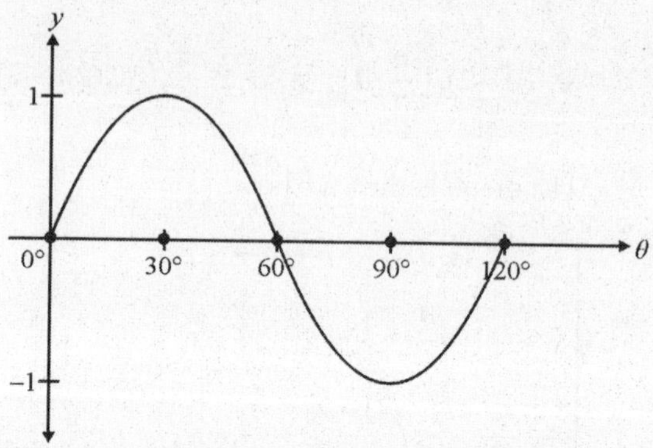


Note that since the horizontal stretch factor is $\frac{1}{3}$, the significant points have $\theta$-values that are $\frac{1}{3}$ of the $\theta$-values for the graph of $y = \sin\ \theta$.

For example, the maximum $y$-value occurs at 30° instead of 90°.

**2.** The vertical stretch and reflection in the $x$-axis gives $a = -3$.

Since the period is $5\pi$, $\frac{2\pi}{b} = 5\pi$, and, thus,

$b = \frac{2}{5}$.

Therefore, the function is $y = -3\cos\left(\frac{2}{5}\theta\right)$.

 

3. For an amplitude of 4, $a = 4$ or $a = -4$.

For period = 270°, $\frac{360}{b} = 270$, thus

$b = \frac{360}{270} = \frac{4}{3}$.

Therefore, a possible function is $y = 4\sin\frac{4}{3}x$.

## PRACTICE EXERCISES ANSWERS AND SOLUTIONS

1. $y = \cos(2\theta)$

$\text{period} = \frac{2\pi}{|b|} = \frac{2\pi}{|2|} = \pi$

2. $y = \sin(-3\theta)$

$\text{period} = \frac{2\pi}{|b|} = \frac{2\pi}{|-3|} = \frac{2\pi}{3}$

3. $y = \cos\left(\frac{2}{3}\theta\right)$

$\text{period} = \frac{2\pi}{|b|} = \frac{2\pi}{\left|\frac{2}{3}\right|} = 2\pi \times \frac{3}{2} = 3\pi$

4. $y = \sin\left(\frac{3}{4}\theta\right)$

$\text{period} = \frac{2\pi}{|b|} = \frac{2\pi}{\left|\frac{3}{4}\right|} = 2\pi \times \frac{4}{3} = \frac{8\pi}{3}$

5. $y = \sin(5x)$

$\text{period} = \frac{360°}{|b|} = \frac{360°}{|5|} = 72°$

6. $y = \cos(-6x)$

$\text{period} = \frac{360°}{|b|} = \frac{360°}{|-6|} = \frac{360°}{6} = 60°$

7. $y = \sin(-0.4x)$

$\text{period} = \frac{360°}{|b|} = \frac{360°}{|-0.4|} = \frac{360°}{0.4} = 900°$

8. $y = \cos(1.2x)$

$\text{period} = \frac{360°}{|b|} = \frac{360°}{|1.2|} = 300°$

9. $y = -2\sin(4\theta)$

$\text{amplitude} = |a| = |-2| = 2$

phase shift = 0

$\text{period} = \frac{2\pi}{|b|} = \frac{2\pi}{|4|} = \frac{\pi}{2}$

10. $y = 3\cos(0.5\theta)$

$\text{amplitude} = |a| = |3| = 3$

phase shift = 0

$\text{period} = \frac{2\pi}{|b|} = \frac{2\pi}{|0.5|} = 4\pi$

11. $y = 2\sin[3(\theta - \pi)] + 4$

$\text{amplitude} = |a| = |2| = 2$

phase shift = $\pi$ rad to the right

$\text{period} = \frac{2\pi}{|b|} = \frac{2\pi}{|3|} = \frac{2\pi}{3}$

12. $y = -2\cos\left[\frac{1}{3}\left(\theta + \frac{\pi}{4}\right)\right] - 3$

$\text{amplitude} = |a| = |-2| = 2$

phase shift = $\frac{\pi}{4}$ rad to the left

$\text{period} = \frac{2\pi}{|b|} = \frac{2\pi}{\left|\frac{1}{3}\right|} = 2\pi \times \frac{3}{1} = 6\pi$

13. For an amplitude of 4, $|a| = 4$.

For a period of 180°, $\frac{360°}{|b|} = 180°$, thus,

$|b| = \frac{360°}{180°} = 2$.

Since $a$ and $b$ are positive, the function is $y = 4\sin(2x)$.

14. For an amplitude of $\frac{1}{2}$, $|a| = \frac{1}{2}$.

For a period of $3\pi$, $\frac{2\pi}{|b|} = 3\pi$, thus, $|b| = \frac{2\pi}{3\pi} = \frac{2}{3}$.

Since $a$ and $b$ are positive, the function is $y = \frac{1}{2}\cos\left(\frac{2}{3}\theta\right)$.

**15.** For an amplitude of $\frac{3}{4}$, $|a| = \frac{3}{4}$.

For a period of $\frac{2\pi}{3}$, $\frac{2\pi}{|b|} = \frac{2\pi}{3}$, thus, $|b| = 3$.

Since $a$ and $b$ are positive, the function is $y = \frac{3}{4}\sin(3\theta)$.

**16.** For an amplitude of 2.4, $|a| = 2.4$.

For a period of 600°, $\frac{360°}{|b|} = 600°$, thus,

$|b| = \frac{360°}{600°} = \frac{3}{5}$.

Since $a$ and $b$ are positive, the function is $y = 2.4\cos\left(\frac{3}{5}x\right)$.

**17.** $y = \cos(2x)$, $0° \le x \le 360°$

The period is $\frac{360°}{|b|} = \frac{360°}{2} = 180°$.

Since the horizontal stretch factor is $\frac{1}{2}$, the significant points have $x$-values that are $\frac{1}{2}$ the $x$-values of the graph of $y = \cos x$.

The significant points within the given domain are:

$(0°, 1), \left(\frac{90°}{2}, 0\right), \left(\frac{180°}{2}, -1\right), \left(\frac{270°}{2}, 0\right),$
$\left(\frac{360°}{2}, 1\right), \left(\frac{450°}{2}, 0\right), \left(\frac{540°}{2}, -1\right),$
$\left(\frac{630°}{2}, 0\right), \left(\frac{720°}{2}, 1\right)$

or $(0°,1)(45°,0)(90°,-1)(135°,0)(180°,1)$
$(225°,0)(270°,-1)(315°,0)(360°,1)$

The graph is shown below.

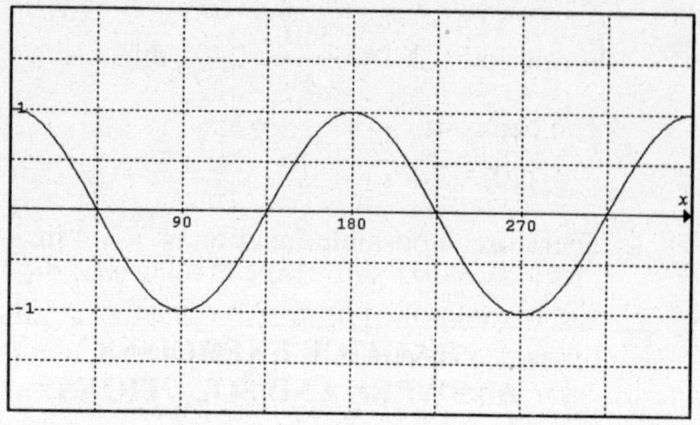


**18.** $y = \sin\left(\frac{1}{2}\theta\right)$, $0 \le \theta \le 4\pi$

The period is $\frac{2\pi}{|b|} = \frac{2\pi}{\frac{1}{2}} = 4\pi$.

Since the horizontal stretch factor is 2, the significant points have $\theta$-values that are 2 times the $\theta$-values of the graph of $y = \sin\theta$.

The significant points within the given domain are:

$(0\times 2, 0), \left(\frac{\pi}{2}\times 2, 1\right), (\pi\times 2, 0), \left(\frac{3\pi}{2}\times 2, -1\right),$
$(2\pi\times 2, 0)$

or $(0, 0)$, $(\pi, 1)$, $(2\pi, 0)$, $(3\pi, -1)$, $(4\pi, 0)$

The graph is shown below.

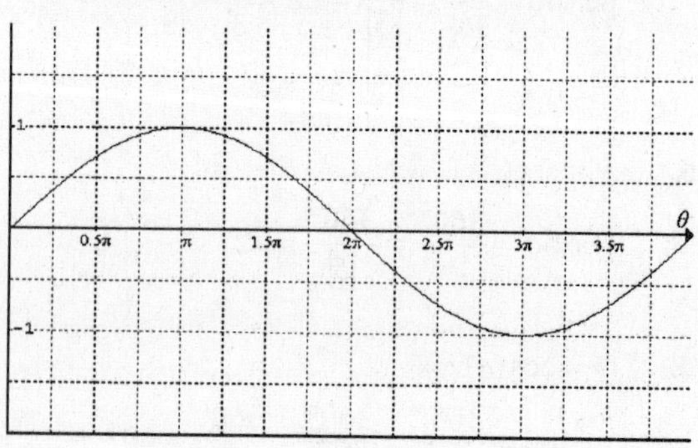

**19.** $y = \cos\left(\frac{1}{3}\theta\right)$, $-6\pi \le \theta \le 6\pi$

The period is $\frac{2\pi}{|b|} = \frac{2\pi}{\frac{1}{3}} = 6\pi.$

Since the horizontal stretch factor is 3, the significant points have $\theta$-values that are 3 times the $\theta$-values of the graph of $y = \cos\theta$.

The significant points within the given domain are:

$(-2\pi \times 3, 1), \left(\frac{-3\pi}{2} \times 3, 0\right), (-\pi \times 3, -1), \left(-\frac{\pi}{2} \times 3, 0\right),$
$(0 \times 3, 1), \left(\frac{\pi}{2} \times 3, 0\right), (\pi \times 3, -1), \left(\frac{3\pi}{2} \times 3, 0\right), (2\pi \times 3, 1)$

or $(-6\pi, 1), \left(\frac{-9\pi}{2}, 0\right), (-3\pi, -1), \left(\frac{-3\pi}{2}, 0\right),$
$(0, 1), \left(\frac{3\pi}{2}, 0\right), (3\pi - 1), \left(\frac{9\pi}{2}, 0\right), (6\pi, 1)$

The graph is shown below.

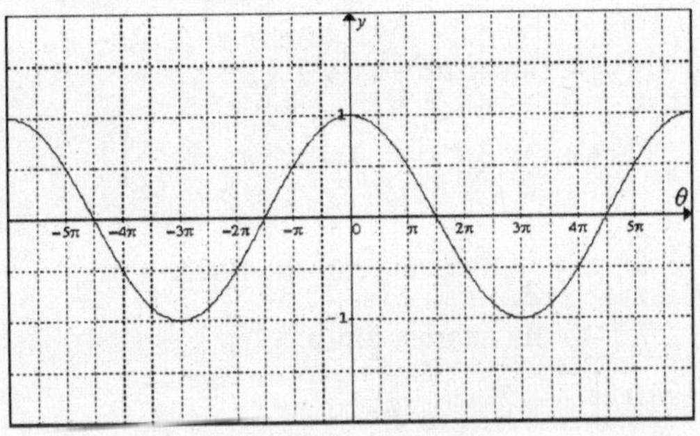


**20.** $y = \frac{1}{2}\sin(2x)$, $0° \le x \le 720°$

The amplitude is $|a| = \frac{1}{2}$.

The period is $\frac{360°}{|b|} = \frac{360°}{2} = 180°.$

Since the horizontal stretch factor is $\frac{1}{2}$, the significant points have $x$-values that are $\frac{1}{2}$ the $x$-values of the graph of $y = \sin x$.
The significant points within the given domain are:

$(0°, 0), \left(\frac{90°}{2}, 1\right), \left(\frac{180°}{2}, 0\right), \left(\frac{270°}{2}, -1\right),$
$\left(\frac{360°}{2}, 0\right), \ldots \left(\frac{1440°}{2}, 0\right)$

or
$(0°, 0), (45°, 1), (90°, 0), (135°, -1),$
$(180°, 0), \ldots (720°, 0)$

The graph is shown below.

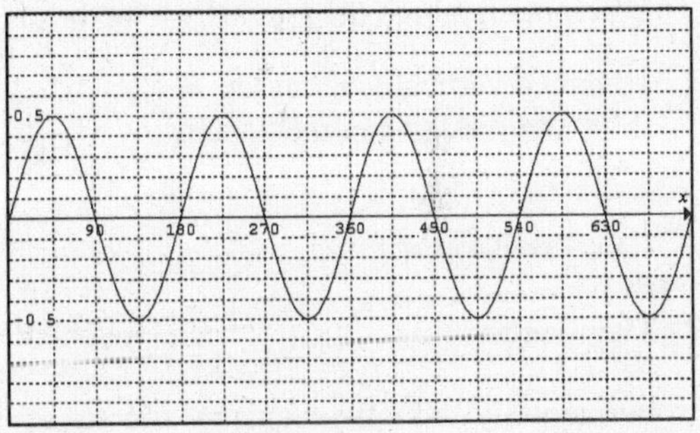


**21.** The graph is a sine function that has been stretched vertically about the $\theta$-axis by a factor of 3 and stretched horizontally about the $y$-axis by a factor of $\frac{1}{2}$.

Thus, $a = 3$ and $b = 2$, giving the function $y = 3\sin(2\theta)$, $-360° \le \theta \le 360°$.

**22.** The graph is a cosine function that has been stretched vertically about the $\theta$-axis by a factor of 2, stretched horizontally about the $y$-axis by a factor of 2, and reflected in the $\theta$-axis.

Thus, $a = -2$ and $b = \frac{1}{2}$, giving the function

$y = -2\cos\left(\frac{1}{2}\theta\right)$, $-360° \le \theta \le 360°$.

**23.** The graph is a sine function that has been stretched vertically about the $\theta$-axis by a factor of 0.5, stretched horizontally about the $y$-axis by a factor of 2, and reflected in the $\theta$-axis.

Thus, $a = 0.5$ and $b = \frac{1}{2}$, giving the function

$y = -0.5\sin\left(\frac{1}{2}\theta\right),\ \theta \in R.$

**24.** The graph is a cosine function that has been stretched vertically about the $\theta$-axis by a factor of 2.5, and stretched horizontally about the $y$-axis by a factor of $\frac{2}{3}$.

Thus, $a = 2.5$ and $b = \frac{3}{2}$, giving the function

$y = -2.5\sin\left(\frac{3}{2}\theta\right),\ \theta \in R.$

**25.** $y = -2\sin(3\theta)$

The function $y = \sin\theta$ has been stretched horizontally about the $y$-axis by a factor of $\frac{1}{3}$, stretched vertically about the $\theta$-axis by a factor of 2, and reflected in the $\theta$-axis.

**26.** $y = \frac{1}{2}\cos(4x)$

The function $y = \cos x$ has been stretched horizontally about the $y$-axis by a factor of $\frac{1}{4}$, and stretched vertically about the $x$-axis by a factor of $\frac{1}{2}$.

**27.** $y = -0.2\sin\left(\frac{1}{4}\theta\right)$

The function $y = \sin\theta$ has been stretched horizontally about the $y$-axis by a factor of 4, stretched vertically about the $\theta$-axis by a factor of 0.2 and reflected in the $\theta$-axis.

**28.** $y = \frac{5}{3}\cos(10x) + 5$

The function $y = \cos x$ has been stretched horizontally about the $y$-axis by a factor of $\frac{1}{10}$, stretched vertically about the $x$-axis by a factor of $\frac{5}{3}$, and translated 5 units up.

**29.** $y = 4\sin(6\theta) - 1.5$

The function $y = \sin\theta$ has been stretched horizontally about the $y$-axis by a factor of $\frac{1}{6}$, stretched vertically about the $\theta$-axis by a factor of 4, and translated 1.5 units down.

**30.** $y = -8\cos(0.4x) - 3.5$

The function $y = \cos\theta$ has been stretched horizontally about the $y$-axis by a factor of $\frac{5}{2}$, stretched vertically about the $x$-axis by a factor of 8, reflected in the $x$-axis, and translated 3.5 units down.

**31.** $y = 4.5\sin(3\pi\theta - \pi) + 2.1$

Factor out $3\pi$: $y = 4\sin\left[3\pi\left(\theta - \frac{1}{3}\right)\right] + 2.1$

As $c = \frac{1}{3}$, the value of the phase shift is $\frac{1}{3}$ or 0.3, to the nearest tenth.

| | | | |
|---|---|---|---|
| 0 | . | 3 | |

**32.** $y = 1.7\cos(5\pi\theta - 2\pi) + 4.1$

Factor out $5\pi$: $y = 1.7\cos\left(5\pi\left(\theta - \frac{2}{5}\right)\right) + 4.1$

As $b = 5\pi$, the period is $\frac{2\pi}{|b|} = \frac{2\pi}{5\pi} = \frac{2}{5} = 0.4.$

| | | | |
|---|---|---|---|
| 0 | . | 4 | |

 

## *Lesson 20 Graphing the Tangent Function, Summary of Transformations for Sine, Cosine and Tangent*

### CLASS EXERCISES ANSWERS AND SOLUTIONS

1.

| $\theta$ | 0 | 30 | 45 | 60 | 90 |
|---|---|---|---|---|---|
| **tan $\theta$** | 0 | $\frac{1}{\sqrt{3}}$ | 1 | $\sqrt{3}$ | undefined |
| **Decimal value** | 0 | 0.58 | 1 | 1.73 | undefined |

| $\theta$ | 120 | 135 | 150 | 180 |
|---|---|---|---|---|
| **tan $\theta$** | $-\sqrt{3}$ | −1 | $\frac{-1}{\sqrt{3}}$ | 0 |
| **Decimal value** | −1.73 | −1 | −0.58 | 0 |

| $\theta$ | 210 | 225 | 240 | 270 |
|---|---|---|---|---|
| **tan $\theta$** | $\frac{1}{\sqrt{3}}$ | 1 | $\sqrt{3}$ | undefined |
| **Decimal value** | 0.58 | 1 | 1.73 | undefined |

| $\theta$ | 300 | 315 | 330 | 360 |
|---|---|---|---|---|
| **tan $\theta$** | $-\sqrt{3}$ | −1 | $\frac{-1}{\sqrt{3}}$ | 0 |
| **Decimal value** | −1.73 | −1 | −0.58 | 0 |

2.

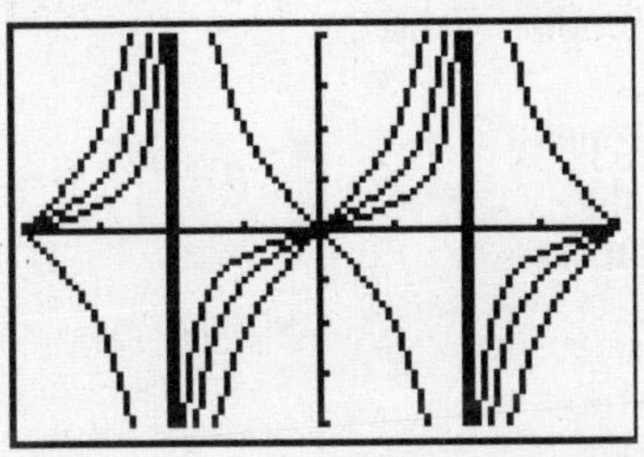

The value of $|a|$ is a vertical stretch factor about the $\theta$-axis. If $a < 0$, the graph is a reflection in the $\theta$-axis, when compared to $a > 0$.

3.

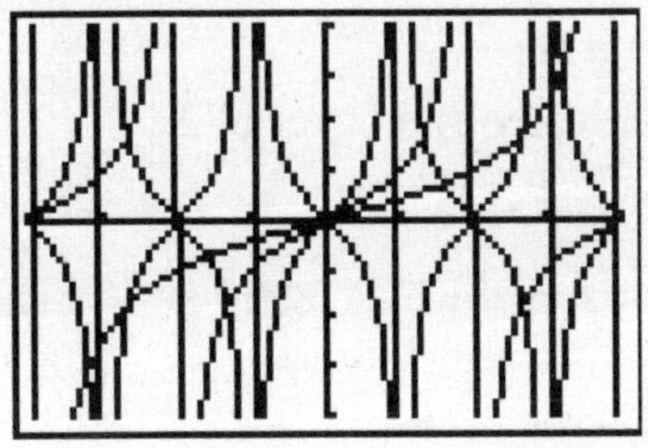

The value of $|b|$ is a horizontal stretch factor. If $b < 0$, the graph is a reflection in the $y$-axis, when compared to $b < 0$. The period is $\frac{\pi}{|b|}$.

4.

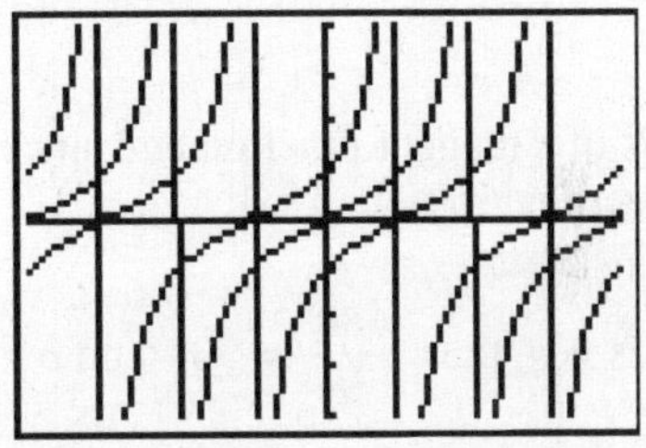

As with previous functions, the value of $c$ determines a horizontal translation of $c$ units (radians or degrees, depending on the units for $\theta$), which for trig functions, is called a phase shift. If $c > 0$, the phase shift is right, if $c < 0$, the phase shift is left.

5.

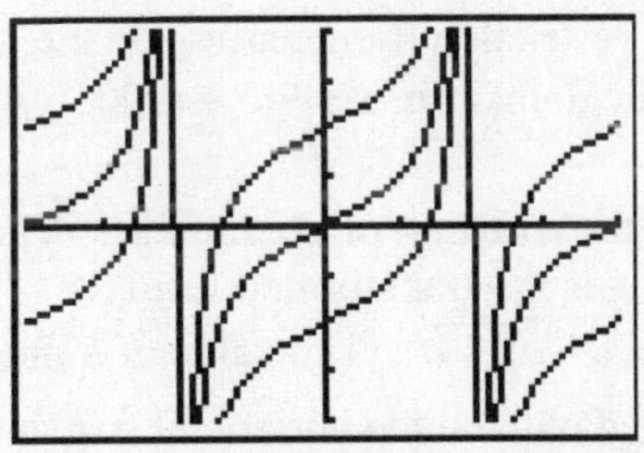

Again, as with previous functions, the value of $d$ determines a vertical translation of $d$ units. If $d > 0$, the translation is up, if $d < 0$, the translation is down.

**6.** The graph would be;

- stretched vertically by a factor of 2
- reflected in the $\theta$-axis
- stretched horizontally by a factor of $\frac{1}{4}$. The period is $\frac{\pi}{\frac{1}{4}} = 4\pi$.
- translated $\frac{\pi}{4}$ radians right. The phase shift is $\frac{\pi}{4}$ radians.
- translated 3 units up. The vertical displacement is 3.

**7.** The graph is of a tangent function and since there is no vertical stretch, $a = 1$.

The period is 90°, thus, $\frac{180°}{|b|} = 90°$ and $b = 2$

($b$ is positive because the graph is not reflected in the $y$-axis).

The graph is translated 45° to the right.

The graph is not translated vertically, so $d = 0$.

Thus, the function is $y = \tan 2(x - 45°)$.

There is no restriction (boundaries) on the domain. The domain is $x \neq 90° + n(90°),\ n \in I$.

**8.** Notice that the equation of the function is not quite in the general transformed form, $y = a\cos[b(\theta - c)] + d$. The value of $b$ needs to be factored out and the horizontal translation needs to be written as ($\theta - c$). This gives

$y = -2\cos 3\left(\theta + \frac{\pi}{3}\right) + 6$ and then,

$y = -2\cos 3\left(\theta - \left(-\frac{\pi}{3}\right)\right) + 6.$

amplitude $= |a| = |-2| = 2$

period $= \frac{2\pi}{|b|} = \frac{2\pi}{|3|} = \frac{2}{3}\pi$

phase shift $= c = -\frac{\pi}{3}$, which is $\frac{\pi}{3}$ left

vertical displacement = 6, which is 6 units up

**9.** Vertical stretch gives $a = 4$.

Period $= \frac{4}{3}\pi \rightarrow \frac{\pi}{|b|} = \frac{4}{3}\pi$. Thus, $|b| = \frac{3}{4}$

Since there is a reflection in the $y$-axis, $b = -\frac{3}{4}$

Translation (phase shift) $\frac{\pi}{6}$ left $\rightarrow c = -\frac{\pi}{6}$

Vertical displacement $\rightarrow d = -2$

Therefore, the function is

$y = 4\tan\left[-\frac{3}{4}\left(\theta - \left(-\frac{\pi}{6}\right)\right)\right] - 2$

## PRACTICE EXERCISES ANSWERS AND SOLUTIONS

**1.**

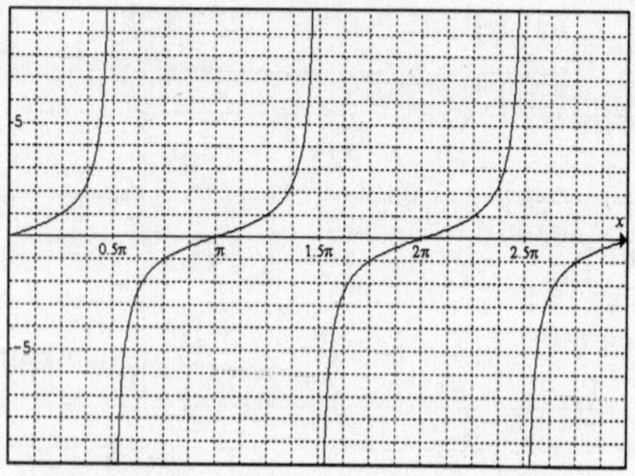


**2.** For $y = \tan\theta,\ 0 \leq \theta \leq 3\pi$:

| | | |
|---|---|---|
| a) | Domain: | $\theta \neq \frac{\pi}{2}, \frac{3\pi}{2}, \frac{5\pi}{2}$, $0 \leq \theta \leq 3\pi$ |
| b) | Range: | $y \in R$ |
| c) | $\theta$-intercepts: | $0, \pi, 2\pi, 3\pi$ |
| d) | $y$-intercepts: | 0 |
| e) | Period: | $\pi$ |
| f) | Equations of vertical asymptotes: | $\theta = \frac{\pi}{2}, \theta = \frac{3\pi}{2}, \theta = \frac{5\pi}{2}$ |

 

**3.** $y = 3\cos 4\left(\theta + \frac{\pi}{4}\right) - 1$

amplitude = $|a| = |3| = 3$

period = $\frac{2\pi}{|b|} = \frac{2\pi}{|4|} = \frac{\pi}{2}$

phase shift $c = \frac{-\pi}{4}$

The phase shift is $\frac{\pi}{4}$ rad to the left.

vertical displacement = $d = -1$, which is 1 unit down

**4.** $y = -2\sin\frac{1}{2}\left(\theta + \frac{\pi}{3}\right) - 5$

amplitude = $|a| = |-2| = 2$

period = $\frac{2\pi}{|b|} = \frac{2\pi}{\frac{1}{2}} = 4\pi$

phase shift = $c = \frac{-\pi}{3}$

The phase shift is $\frac{\pi}{3}$ rad to the left.

vertical displacement = $d = -5$, which is 5 units down

**5.** $y = -\frac{2}{3}\tan\frac{1}{4}(x + 20°) - 1$

amplitude = not applicable

period = $\frac{\pi}{|b|} = \frac{2\pi}{\frac{1}{4}} = 4\pi$

phase shift = $c = -20°$

The phase is 20°to the left.

vertical displacement = $d = -1$, which is one unit down

**6.** $y = -4\sin(3\theta - \pi) - 2.5$

factor out the 3 $y = -4\sin\left[3\left(\theta - \frac{\pi}{3}\right)\right] - 2.5$

amplitude = $|a| = |-4| = 4$

period = $\frac{2\pi}{|b|} = \frac{2\pi}{3}$

phase shift = $c = \frac{\pi}{3}$

The phase shift is $\frac{\pi}{3}$ rad to the right.

vertical displacement = $d = -2.5$, which is 2.5 units down

**7.** $y = \frac{-3}{2}\tan\left(3\theta - \frac{\pi}{3}\right) + 2$

factor out the 3 $y = \frac{-3}{2}\tan\left[3\left(\theta - \frac{\pi}{9}\right)\right] + 2$

amplitude = not applicable

period = $\frac{\pi}{|b|} = \frac{\pi}{3}$

phase shift = $c = \frac{\pi}{9}$

The phase shift is $\frac{\pi}{9}$ rad to the right.

vertical displacement = $d = 2$, which is 2 units up

**8.** $y = -2\cos\left(\frac{3}{4}\theta + \frac{\pi}{2}\right) - 5$

factor out the $\frac{3}{4}$ $y = -2\cos\left[\frac{3}{4}\left(\theta + \frac{2\pi}{3}\right)\right] - 5$

amplitude = $|a| = |-2| = 2$

period = $\frac{2\pi}{|b|} = \frac{2\pi}{\frac{3}{4}} = \frac{8\pi}{3}$

phase shift = $c = \frac{-2\pi}{3}$

The phase shift is $\frac{2\pi}{3}$ rad to the left.

vertical displacement = $d = -5$, which is 5 units down

**9.** $y = \frac{1}{2}\sin\left(\frac{5}{6}x - 15°\right) - 3$

factor out the $\frac{5}{6}$ $\quad y = \frac{1}{2}\sin\left[\frac{5}{6}(x - 18°)\right] - 3$

amplitude = $|a| = \left|\frac{1}{2}\right| = \frac{1}{2}$

period = $\frac{360°}{|b|} = \frac{360°}{\frac{5}{6}} = 432°$

phase shift = $c = 18°$
The phase shift is 18° to the right.

vertical displacement = $d = -3$, which is 3 units down

**10.** $y = -4\cos\left(\frac{2}{5}x + 30°\right) - 1$

factor out the $\frac{2}{5}$ $\quad y = -4\cos\left[\frac{2}{5}(x + 75°)\right] - 1$

amplitude = $|a| = |-4| = 4$

period = $\frac{360°}{|b|} = \frac{360°}{\frac{2}{5}} = 900°$

phase shift = $c = -75°$
The phase shift is 75° to the left.

vertical displacement = $d = -1$, which is 1 unit down

**11.** $y = 3\tan 2\left(\theta + \frac{\pi}{2}\right) - 4$

amplitude = not applicable

period = $\frac{\pi}{|b|} = \frac{\pi}{2}$

phase shift = $c = -\frac{\pi}{2}$

The phase shift is $\frac{\pi}{2}$ to the left.

vertical displacement = $d = -4$, which is 4 units down

**12.** $y - 2 = 3\tan\left(\frac{1}{5}x + 60°\right)$

Rewrite the equation and factor out the $\frac{1}{5}$:

$y = 3\tan\left[\frac{1}{5}(x + 300°)\right] + 2$

amplitude = not applicable

period = $\frac{180°}{|b|} = \frac{180°}{\frac{1}{5}} = 900°$

phase shift = $c = -300°$
The phase shift is 300° to the left.

vertical displacement = $d = 2$, which is 2 units up

**13.** $y = \tan\theta - 2$, $0 \le \theta \le 2\pi$

period = $\pi$

vertical displacement = $d = -2$. The graph is. translated 2 units down

The graph is shown below.

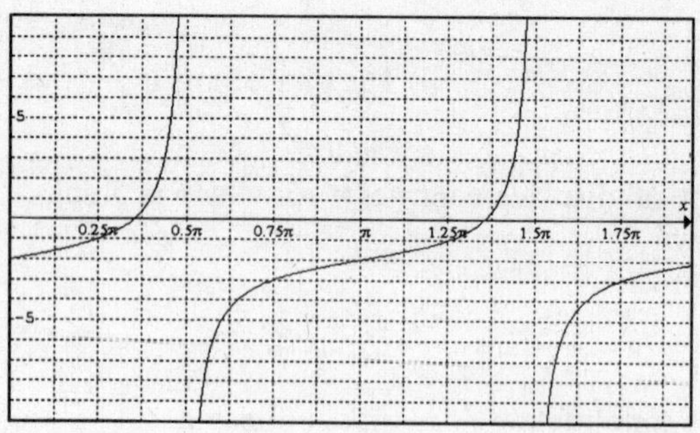


**14.** $y = -3\tan x$, $-180° \le x \le 180°$

period = 180°

Since $|a| = |3| = 3$, the graph is stretched vertically about the $x$-axis by a factor of 3. Since $a$ is negative, the graph is reflected about the $x$-axis.

The graph is shown below.

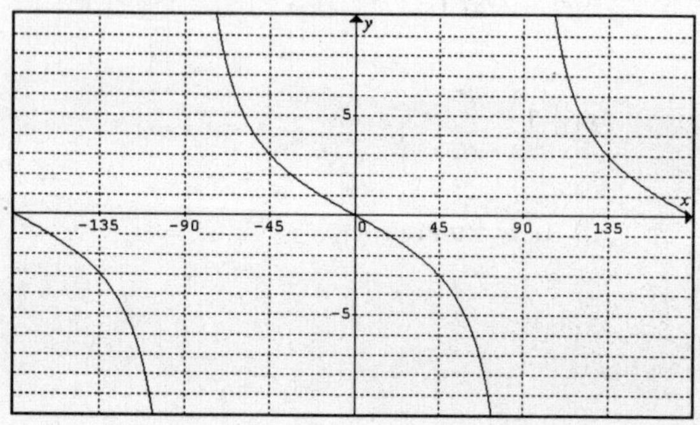

**15.** The graph is a tangent function that has been stretched horizontally by a factor of $\frac{1}{2}$.
Thus, $b = 2$, giving the function
$y = \tan(2\theta),\ 0 \leq \theta \leq 2\pi.$

**16.** The graph is a cosine function that has been stretched vertically about the $x$-axis by factor of 2, then translated 45° to the left and 1 unit down
Thus, $a = 2$, $c = -45°$, and $d = -1$, giving the function
$y = 2\cos(x + 45°) - 1,\ x \in R.$

**17.** The graph is a sine function that has been stretched vertically about the x-axis by a factor of $\frac{1}{2}$, stretched horizontally about the $y$-axis by a factor of 2, then translated 180° to the right and 1 unit down.
Thus, $a = \frac{1}{2}$, $b = \frac{1}{2}$, $c = 180°$, and $d = -1$,
giving the function
$y = \frac{1}{2}\sin\left[\frac{1}{2}(x - 180°)\right] - 1,\ x \in R.$

**18.** **A**

$f(\theta) = 3\sin 2\left(\theta - \frac{\pi}{2}\right) + 1$

$\text{Period} = \frac{2\pi}{|b|} = \frac{2\pi}{2} = \pi$

$g(\theta) = -3\cos\left(2\theta - \frac{\pi}{4}\right) + 1$

Factor out the 2: $g(\theta) = -3\cos\left[2\left(\theta - \frac{\pi}{8}\right)\right] + 1$

$\text{Period} = \frac{2\pi}{|b|} = \frac{2\pi}{2} = \pi$

$h(\theta) = 3\tan(\theta + \pi) - 1$

$\text{Period} = \frac{\pi}{|b|} = \frac{\pi}{1} = \pi$

All three functions have period = $\pi$.

**19.** $y = 3\cos\left(2\theta + \frac{\pi}{4}\right) = 0$

To find the $\theta$-intercept, let $y = 0$ and solve for $\theta$.

$3\cos\left(2\theta + \frac{\pi}{4}\right) = 0$

$\cos\left(2\theta + \frac{\pi}{4}\right) = 0$ • divide both sides by 3

$\left(2\theta + \frac{\pi}{4}\right) = \cos^{-1}(0) = \frac{\pi}{2}$ or $\frac{3\pi}{2}$

$\frac{\pi}{2}$ is the smallest positive value.

Thus, $\left(2\theta + \frac{\pi}{4}\right) = \frac{\pi}{2}.$

$2\theta = \frac{\pi}{4}$

$\theta = \frac{\pi}{8} \doteq 0.4$ rad

| 0 | . | 4 | |
|---|---|---|---|

**20. a)**

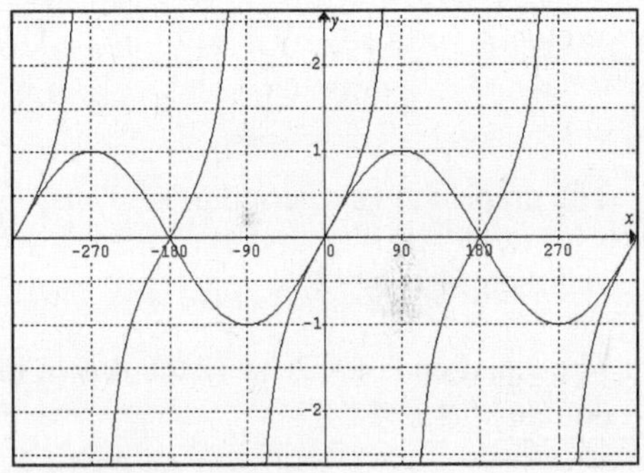


**b)** The solution to the equation $\sin x = \tan x$, $-360° \leq x \leq 360°$, will be the $x$-values of the points of intersection within the given domain.

Thus, $x = -360°, -180°, 0°, 180°, 360°$.

**c)**

$\sin x = \tan x,$
$-360° \leq x \leq 360°$

• Quotient identity $\sin x = \frac{\sin x}{\cos x}$

• Multiply both sides by $\cos x$ $\sin x \cos x = \sin x$

• Factor $\sin x \cos x - \sin x = 0$

$\sin x(\cos x - 1) = 0$

$\sin x = 0$ or $\cos x = 1$

Thus, $x = -360°, -180°, 0°, 180°, 360°$.

## *Lesson 21 Graphing the Reciprocal Trigonometric Functions*

### CLASS EXERCISES ANSWERS AND SOLUTIONS

**1.** The periods of the reciprocal functions are the same as the periods for their corresponding primary functions so, the period of $y = \csc(3\theta)$ is $\frac{2\pi}{|b|} = \frac{2\pi}{3}$.

**2.** The vertical stretch factor $= a = \frac{1}{2}$.

The period $= \frac{360°}{|b|} = \frac{360°}{\frac{2}{3}} = 360° \times \frac{3}{2} = 540°$.

**3.** The graph has the form of $y = a\csc(b\theta)$. Since the sections of the graph are 2 units from the $\theta$-axis, the graph of $y = \csc\theta$ has been stretched vertically by a factor of 2. Thus, $a = 2$.

The graph shows a period of $\pi$. Thus, $\frac{2\pi}{|b|} = \pi$, and therefore, $b = 2$.

The equation is $y = 2\csc(2\theta)$ drawn in the domain $0 \le \theta \le 2\pi$.

### PRACTICE EXERCISES ANSWERS AND SOLUTIONS

**1.** $y = \sec(2\theta),\ 0 \le \theta \le 3\pi$

Period $= \frac{2\pi}{|b|} = \frac{2\pi}{|2|} = \pi$

Since the horizontal stretch factor is $\frac{1}{2}$, the vertical asymptotes will be $\frac{1}{2}$ the value of the vertical asymptotes of $y = \sec\theta$.

Thus, the vertical asymptotes are:

$$\theta = \frac{\frac{\pi}{2}}{2},\ \theta = \frac{\frac{3\pi}{2}}{2},\ \theta = \frac{\frac{5\pi}{2}}{2},\ \theta = \frac{\frac{7\pi}{2}}{2},$$

$$\theta = \frac{\frac{9\pi}{2}}{2},\ \theta = \frac{\frac{11\pi}{2}}{2},\ \theta = \frac{\frac{13\pi}{2}}{2},\ \theta = \frac{\frac{15\pi}{2}}{2}$$

or $\theta = \frac{\pi}{4}, \frac{3\pi}{4}, \frac{5\pi}{4}, \frac{7\pi}{4}, \frac{9\pi}{4}, \frac{11\pi}{4}, \frac{13\pi}{4}, \frac{15\pi}{4}$

If $\cos(2\theta) = 1$ (which occurs at $\theta = 0, \pi, 2\pi, 3\pi$), then $\sec(2\theta)$ will also equal 1.

If $\cos(2\theta) = -1$ (which occurs at $\theta = \frac{\pi}{2}, \frac{3\pi}{2}, \frac{5\pi}{2}$), then $\sec(2\theta)$ will also equal $-1$.

The graph is shown below.

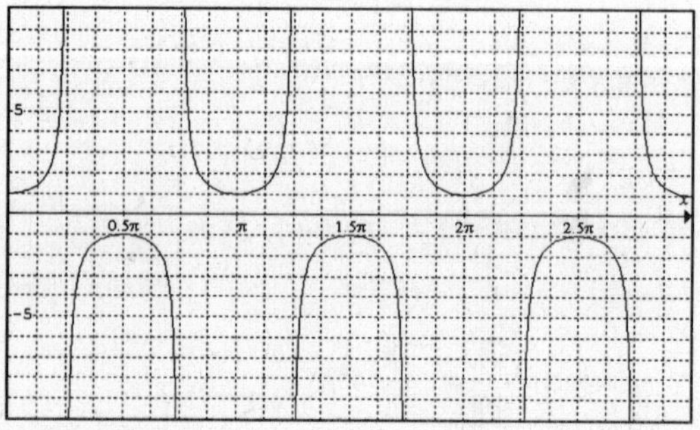


 

**2.** $y = \csc\left(\frac{1}{2}\theta\right)$, $-2\pi \le \theta \le 2\pi$

Period $= \frac{2\pi}{|b|} = \frac{2\pi}{\left|\frac{1}{2}\right|} = 4\pi$

Since the horizontal stretch factor is 2, the vertical asymptotes will be twice the value of the vertical asymptotes of $y = \csc\theta$.
Thus, the vertical asymptotes are
$\theta = 2\times 0$, $\theta = 2\times -\pi$, $\theta = 2\times\pi$
or $\theta = 0$, $\theta = -2\pi$, $\theta = 2\pi$.

If $\sin\left(\frac{1}{2}\theta\right) = 1$ (which occurs at $\theta = \pi$),

then $\csc\left(\frac{1}{2}\theta\right)$ will also equal 1.

If $\sin\left(\frac{1}{2}\theta\right) = -1$ (which occurs at $\theta = -\pi$),

then $\csc\left(\frac{1}{2}\theta\right)$ will also equal –1.

The graph is shown below.

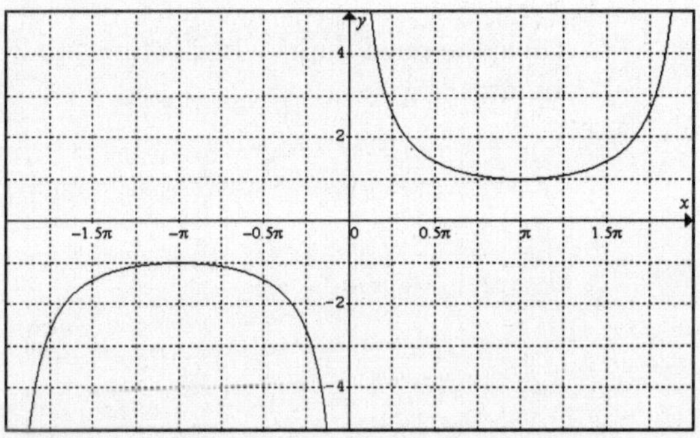


**3.** $y = 3\csc(0.25\theta)$
Domain: $\theta \ne 4n\pi$, $n \in I$

Period: $\frac{2\pi}{|b|} = \frac{2\pi}{|.25|} = 8\pi$

$y$- intercept: none
Range: $y \ge 3$ or $y \le -3$
$x$-intercepts: none

**4.** $y = 3\sec(0.4\theta)$
Domain: $\theta \ne 1.25\pi + 2.5n\pi$, $n \in I$

Period: $\frac{2\pi}{|b|} = \frac{2\pi}{|.4|} = 5\pi$

$y$-intercept: 3
Range: $y \ge 3$ or $y \le -3$
$x$-intercepts: none

**5.** $y = \frac{1}{3}\cot(2.5\theta)$

Domain: $\theta \ne \frac{2n\pi}{5}$, $n \in I$

Period: $\frac{2\pi}{|b|} = \frac{2\pi}{|2.5|} = 0.4\pi$

$y$- intercept: none
Range: $y \in R$
$x$-intercepts: $0.4n\pi$, $y \in I$

**6.** The graph is a csc function that has been stretched horizontally about the $y$-axis by a factor of 2.

Thus, $b = \frac{1}{2}$, giving the function

$y = \csc\left(\frac{1}{2}\theta\right)$, $-4\pi \le \theta \le 4\pi$.

**7.** The graph is a sec function that has been stretched horizontally about the $y$-axis by a factor of $\frac{1}{2}$, and stretched vertically about the $\theta$-axis by a factor of 3.

Thus, $a = 3$ and $b = 2$, giving the function
$y = 3\sec(2\theta)$, $-2\pi \le \theta \le 2\pi$.

**8.** **C**
$y = 0.5\csc(2\theta)$

The csc function has been stretched vertically about the $\theta$-axis by a factor of 0.5.
Thus, the range will be $y \ge 0.5$ or $y \le -0.5$.

**9.**
1. $y = 2\cot\theta$ Period: $\frac{\pi}{|b|} = \frac{\pi}{1} = \pi$
2. $y = 3\csc(0.5\theta)$ Period: $\frac{2\pi}{|b|} = \frac{2\pi}{|0.5|} = 4\pi$
3. $y = 3\sec\theta$ Period: $\frac{2\pi}{|b|} = \frac{2\pi}{|1|} = 2\pi$
4. $y = \sin(4\theta)$ Period: $\frac{2\pi}{|b|} = \frac{2\pi}{|4|} = \frac{\pi}{2}$

The order of the periods from the smallest period to the largest is functions 4, 1, 3, 2.

| | | | |
|---|---|---|---|
| 4 | 1 | 3 | 2 |

## *Lesson 22 Modelling Problems Using Sinusoidal Functions*

### CLASS EXERCISES ANSWERS AND SOLUTIONS

1. **a)** The highest point reached by the stone is 28 cm, therefore the diameter of the tire is 28 cm.

   **b)** The axle is located halfway between the lowest point, 0 cm, and the highest point, 28 cm. Therefore, the axle is located 14 cm above the ground.

   **c)** The graph shows a period of 4, therefore, the time for one complete revolution is 4 s.

   **d)** The amplitude is $\frac{28-0}{2}=14$. Since $a<0$, $a=-14$.

   The period is 4, thus $\frac{2\pi}{b}=4$, and $b=\frac{\pi}{2}$.

   Since the horizontal axis is at $y = 14$, $d = 14$.

   The equation is $h(t)=-14\cos\left(\frac{\pi}{2}t\right)+14$.

   **e)** When $t = 2.5$,

   $$h(2.5)=-14\cos\left(\frac{\pi}{2}(2.5)\right)+14$$
   $$\doteq -14(-.707)+14$$
   $$\doteq 23.9$$

   The height of the stone is approximately 23.9 cm.

   **f)** The graph can be considered to be the graph of a sine function that is translated 1 unit to the right. Thus, the value of $c$ is 1. The value of $a$ is now +14. The values for $b$ and $d$ remain as 14, just as before.

   The equation is

   $h(t)=14\sin\left(\frac{\pi}{2}(t-1)\right)+14$.

### PRACTICE EXERCISES ANSWERS AND SOLUTIONS

1. $d(t)=2\sin\left(\frac{\pi}{10}t\right)+6$

   $a=2$, $b=\frac{\pi}{10}$, $c=0$, $d=6$

   **a)** Maximum value: $d+|a|=6+2=8$

   The pendulum will swing 8 m from the side of the clock tower.

   **b)** Period: $\frac{2\pi}{|b|}=\frac{2\pi}{\left|\frac{\pi}{10}\right|}=20$

   One complete cycle is 20 seconds, thus, the pendulum will take 20 seconds to return to its farthest distance from the edge of the tower.

   **c)** 2 minutes $=2\times 60=120$ seconds

   Since one complete cycle is 20 seconds, 6 complete cycles (6 × 20) will occur in 120 seconds.

   Thus, the pendulum swings 6 times in 2 minutes.

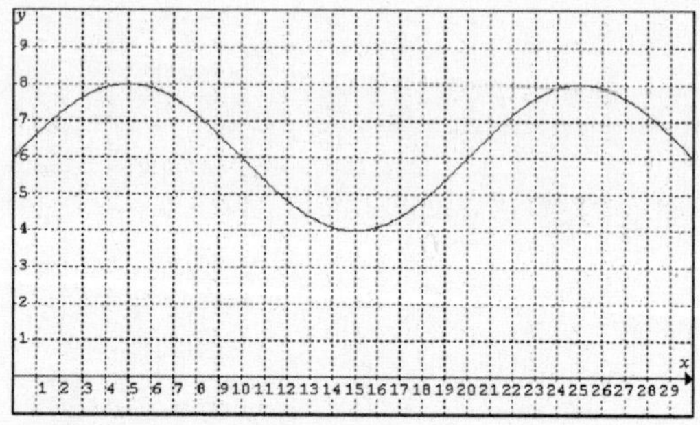

2. **a)** Maximum depth: 18 m
   Minimum depth: 8 m

   Amplitude $=\left|\frac{\max-\min}{2}\right|=\left|\frac{18-8}{2}\right|=5$

   **b)** The range of the function is $8\le h(t)\le 18$.

   **c)** The time taken from the maximum to the minimum depth is $\frac{1}{2}$ of a cycle, or $\frac{1}{2}$ of a period.

   Thus, the period will be (2 × 6) or 12 hours.

 

**d)** $a = 5$

$$d = \frac{\text{max} + \text{min}}{2} = \frac{8+18}{2} = 13$$

The maximum depth is at 7:00 A.M., thus, $c = 7$.

Since the period is given by $\frac{2\pi}{|b|}$, $\frac{2\pi}{|b|} = 12$. Thus, $b = \frac{\pi}{6}$.

Substituting into the equation:
$h(t) = a \cos[b(t-c)] + d$ gives

$$h(t) = 5\cos\left[\frac{\pi}{6}(t-7)\right] + 13$$

**e)** 5:00 P.M. is represented by
$t = 12h + 5h = 17h$.

$$h(t) = 5\cos\left[\frac{\pi}{6}(17-7)\right] + 13$$

$$h(t) = 5\cos\left[\frac{\pi}{6}(10)\right] + 13$$

$$h(t) = 5\cos\left(\frac{10\pi}{6}\right) + 13$$

$$h(t) = 5(0.5) + 13$$

$$h(t) = 15.5$$

The depth of the water at 5.00 P.M. is 15.5 m.

**f)** If the depth of the water is 10 m, then $h(t) = 10$.

To solve algebraically:

$$10 = 5\cos\left[\frac{\pi}{6}(t-7)\right] + 13$$

$$-3 = 5\cos\left(\frac{\pi}{6}(t-7)\right)$$

$$\frac{-3}{5} = \cos\left[\frac{\pi}{6}(t-7)\right]$$

$$\cos^{-1}\left(\frac{-3}{5}\right) = \frac{\pi}{6}(t-7)$$

$\frac{\pi}{6}(t-7) \doteq 2.214$ or $\frac{\pi}{6}(t-7) \doteq \pi - 2.214$

$(t-7) \doteq 2.214 \times \frac{6}{11}$ $\qquad$ $\frac{\pi}{6}(t-7) \doteq 4.069$

$(t-7) \doteq 4.069 \times \frac{6}{\pi}$

$t \doteq 2.214 \times \frac{6}{11} + 7$ $\qquad$ $t \doteq 4.067 \times \frac{6}{\pi} + 7$

$t \doteq 11.23$ $\qquad$ $t \doteq 14.78$

11.23 is the FIRST time after 9:00 A.M. that the depth of water is 10 m.
0.23 hours $= 0.23 \times 60 = 13.7$ min
The time will be 11:14 A.M.

To solve graphically:

Graph: $Y_1 = 5\cos\left[\frac{\pi}{6}(t-7)\right] + 13$

$Y_2 = 10$

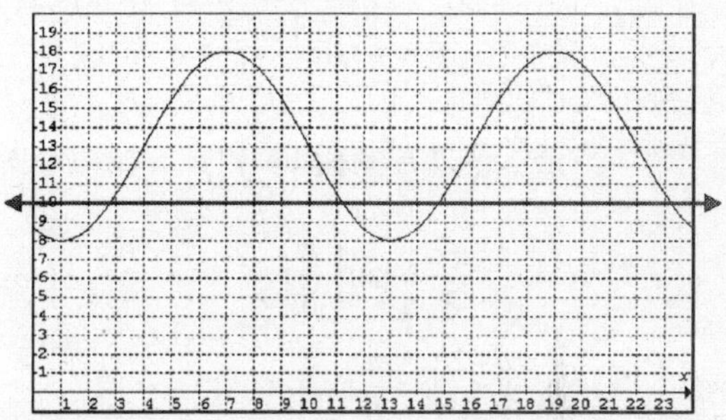

The $x$-value of the first intersection point after 9:00 A.M. is 11.23, which represents 11:14 A.M., as shown in the algebraic solution.

**3.** **a)** maximum height: 7 m
minimum height: 1 m

$$\text{amplitude} = \left|\frac{\text{max} - \text{min}}{2}\right| = \left|\frac{7-1}{2}\right| = 3$$

**b)** The cycle takes 12 s, thus, the period is 12 s.

**c)** Since the period begins at $t = 0$, there is no phase shift and the sine function is reflected in the $t$-axis.

Thus, $a = -3$ and $c = 0$.

As $d = \frac{\text{max} + \text{min}}{2} = \frac{7+1}{2} = 4$,

the period is
$\frac{2\pi}{|b|}$, $\frac{2\pi}{|b|} = 12$. Thus, $b = \frac{\pi}{6}$.

Substituting into the equation:

$h(t) = a\sin[b(t-c)] + d$ gives

$$h(t) = -3\sin\left(\frac{\pi}{6}t\right) + 4$$

**d)** To determine the height 5 seconds after the ride commences, determine $h(5)$.

$$h(t) = -3\sin\left(\frac{\pi}{6}(5)\right) + 4$$

$$h(t) = -3\sin\left(\frac{5\pi}{6}\right) + 4$$

$$h(t) = -3(0.5) + 4$$

$$h(t) = 2.5$$

The height 2.5 seconds after the ride commences is 2.5 m.

**e)** If the height is 6 m, then $h(t) = 6$.

To solve algebraically:

$$6 = -3\sin\left[\frac{\pi}{6}t\right] + 4$$

$$2 = -3\sin\left(\frac{\pi}{6}t\right)$$

$$\frac{-2}{3} = \sin\left[\frac{\pi}{6}t\right]$$

$$\sin^{-1}\left(\frac{-2}{3}\right) = \frac{\pi}{6}t$$

$\frac{\pi}{6}t \doteq -0.7298$ or $\frac{\pi}{6}t \doteq \pi - (0.7298)$

$t \doteq -1.39$ or $\frac{\pi}{6}t \doteq 3.87$

$t \doteq 7.39$

When $t \doteq 7.4$ s, they are 6 m above the ground for the first time.

To solve graphically:

Graph: $y_1 = -3\sin\left[\frac{\pi}{6}x\right] + 4$

$y_2 = 6$

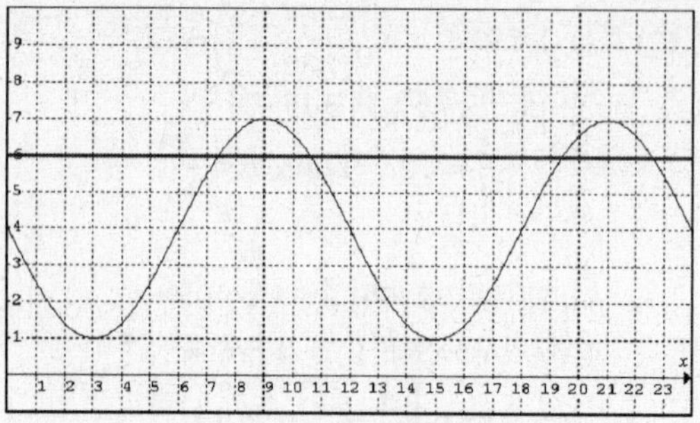

The $x$-value of the first intersection point is approximately 7.4 s.

**4.** **a)** amplitude

$$|a| = \left|\frac{\max - \min}{2}\right| = \left|\frac{17-8}{2}\right| = 4.5$$

Since the $a < 0$, $a = -4.5$, which means the graph has been reflected in the horizontal axis and the beginning of one cycle will occur at a minimum value. One minimum value is at day 355. Since the period is 366 days, another minimum value will occur at day (355 – 366) or day $-11$, Thus, $c = 11$.

Since the period is given by $\frac{2\pi}{|b|}$, $\frac{2\pi}{|b|} = 366$.

Thus, $b = \frac{\pi}{183}$

$$d = \frac{\max + \min}{2} = \frac{17+8}{2} = 12.5$$

Substituting into the equation:

$$h(t) = -4.5\cos\left[\frac{\pi}{183}(t+11)\right] + 12.5$$

**b)** March 7 is day 66.
To determine the number of hours of daylight on day 66:

$$h(t) = -4.5\cos\left(\frac{\pi}{183}(66+11)\right)+12.5$$
$$h(t) = -4.5\cos\left(\frac{\pi}{183}(77)\right)+12.5$$
$$h(t) \doteq -4.5\cos(1.322)+12.5$$
$$h(t) \doteq -4.5(0.2464)+12.5$$
$$h(t) \doteq 11.391$$

The number of hours of daylight on March 7 is approximately 11.4 hours.

**c)** The most hours of daylight will occur when $h(t) = 17$.

To solve algebraically:

$$17 = -4.5\cos\left[\frac{\pi}{183}(t-11)\right]+12.5$$
$$4.5 = -4.5\cos\left(\frac{\pi}{183}(t+11)\right)$$
$$-1 = \cos\left[\frac{\pi}{183}(t+11)\right]$$
$$\frac{\pi}{183}(t+11) = \cos^{-1}(-1)$$
$$\frac{\pi}{183}(t+11) = \pi$$
$$t+11 = \frac{\pi\times 183}{\pi}$$
$$t = \frac{\pi\times 183}{\pi}-11$$
$$t = 172$$

The most hours of daylight will occur on day 172, which represents June 21.

To solve graphically:

Graph: $Y_1 = -4.5\cos\left[\left(\frac{\pi}{183}\right)(x+11)\right]+12.5$

$Y_2 = 17$

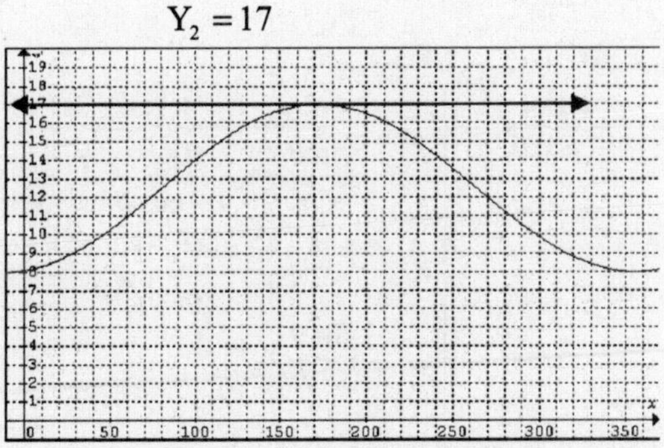


The $x$-value of the point of intersection is 172.

**d)** Graph: $Y_1 = -4.5\cos\left[\left(\frac{\pi}{183}\right)(x+11)\right]+12.5$

$Y_2 = 10$

The $x$-values of the points of intersections are 46 and 298

Thus, there will be 10 hours of daylight on day 46, which represents February 15, and day 298, which represents October 25.

**5. a)**

$$y = a\sin(bx+c)+d$$
$$y = 3.44\sin(1.64x+2.28)+19.32$$

**b)**

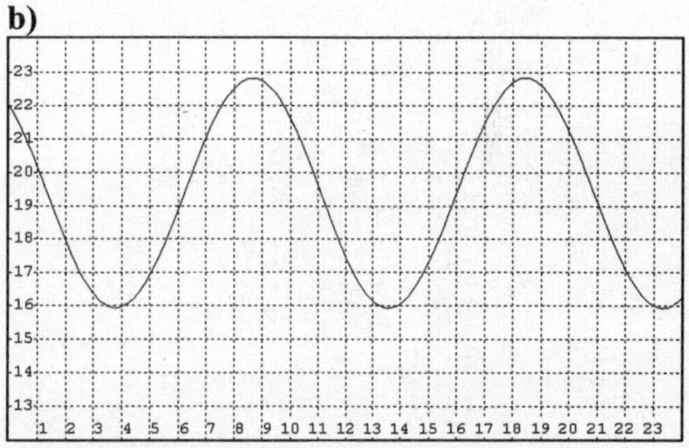


**c)** Graph: $Y_1 = 3.44\sin(0.64x+2.28)+19.32$

Calculate the value of $y$ when $x = 17$.
The value of $y$ will be approximately 21.39.
Thus, the pressure after 17 hours is approximately 21.39 PSI.

**d)** Graph: $Y_1 = 3.44\sin(0.64+2.28)+19.32$

$Y_2 = 18$

The $x$-values of the points of intersection, rounded of the nearest tenth, are 1.95, 5.62, and 11.7.

The $y$-values during the first 12 hours will be less than 18 for the following time periods:

$5.62-1.95 = 3.67$
$12-11.72 = 0.28$

The total amount of time during the first 12 hours that the pressure is less than 18 PSI will be $3.67+0.28 = 3.95$ hours.

## *Practice Test*

### ANSWERS AND SOLUTIONS

1.

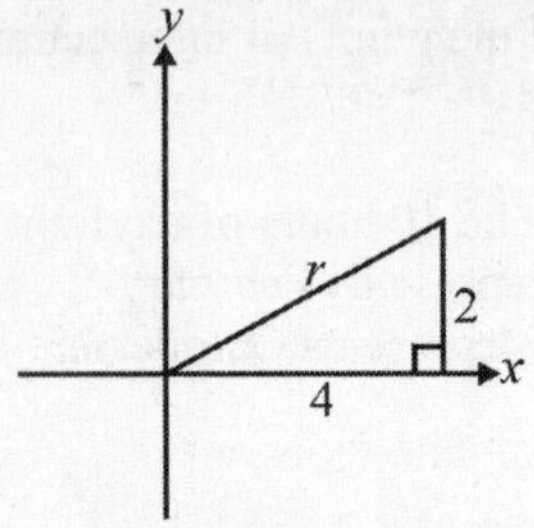


$r^2 = 4^2 + 2^2$

$r^2 = 20$

$r = \sqrt{20} = 2\sqrt{5}$

$\sin\theta = \frac{2}{2\sqrt{5}} = \frac{1}{\sqrt{5}} = \frac{\sqrt{5}}{5}$

$\cos\theta = \frac{4}{2\sqrt{5}} = \frac{2}{\sqrt{5}} = \frac{2\sqrt{5}}{5}$

$\tan\theta = \frac{2}{4} = \frac{1}{2}$

2. **a)** $\cos\frac{14\pi}{3} = \cos\frac{2\pi}{3} = -\frac{1}{2}$

**b)** $\sec 300° = \frac{1}{\cos 300°} = \frac{1}{\frac{1}{2}} = 2$

**c)** $\cot\left(-\frac{11\pi}{3}\right) = \frac{1}{\tan\left(-\frac{11\pi}{3}\right)}$

$= \frac{1}{\tan\left(\frac{\pi}{3}\right)} = \frac{1}{\sqrt{3}} = \frac{\sqrt{3}}{3}$

3. $\cos^2\theta\sin\theta = \cos\theta\sin\theta$

$\cos^2\theta\sin\theta - \cos\theta\sin\theta = 0$

$\cos\theta\sin\theta(\cos\theta - 1) = 0$

$\cos\theta = 0$ or $\sin\theta = 0$ or $\cos\theta = 1$

$\theta = 0°, 90°, 180°, 270°$

4. Note that if $0 \le \theta < 2\pi$, then $0 \le \left(\frac{1}{2}\theta\right) < \pi$.

$2\cos\left(\frac{1}{2}\theta\right)\sin\left(\frac{1}{2}\theta\right) - \cos\left(\frac{1}{2}\theta\right) + 2\sin\left(\frac{1}{2}\theta\right) - 1 = 0$

$\cos\left(\frac{1}{2}\theta\right)\left(2\sin\left(\frac{1}{2}\theta\right) - 1\right) + 1\left(2\sin\left(\frac{1}{2}\theta\right) - 1\right) = 0$

$\left(\cos\left(\frac{1}{2}\theta\right) + 1\right)\left(2\sin\left(\frac{1}{2}\theta\right) - 1\right) = 0$

$\cos\left(\frac{1}{2}\theta\right) = -1$ or $\sin\left(\frac{1}{2}\theta\right) = \frac{1}{2}$

no solution $\qquad \frac{1}{2}\theta = \frac{\pi}{6}$

$\theta = \frac{\pi}{3}$

(There is no solution to $\cos\left(\frac{1}{2}\theta\right) = -1$ because $0 \le \left(\frac{1}{2}\theta\right) < \pi$.)

5. Graph $Y_1 = 2\sin\theta + 1$, $Y_2 = \tan\theta$

The solutions are the $x$-values at intersection points, which are $x = 1.24$ rad, $3.47$ rad.

6. **a)**

| **LHS** | **RHS** |
|---|---|
| $\frac{1-\tan\theta}{1-\cot\theta}$ | $-\tan\theta$ |
| $\frac{1-\tan\theta}{1-\frac{1}{\tan\theta}} \times \frac{\tan\theta}{\tan\theta}$ | |
| $\frac{(1-\tan\theta)\tan\theta}{\tan\theta - 1}$ | |
| $-\tan\theta$ | |

**b)**

| **LHS** | **RHS** |
|---|---|
| $2\csc\theta$ | $\frac{\sin\theta}{1-\cos\theta} + \frac{\sin\theta}{1+\cos\theta}$ |
| | $\frac{\sin\theta(1+\cos\theta) + \sin\theta(1-\cos\theta)}{1-\cos^2\theta}$ |
| | $\frac{2\sin\theta}{\sin^2\theta}$ |
| | $\frac{2}{\sin\theta}$ |
| | $2\csc\theta$ |

**7.**

| LHS | RHS |
|---|---|
| $\dfrac{1+\cos 2x}{\sin 2x}$ | $\cot x$ |
| $\dfrac{1+2\cos^2 x-1}{2\sin x\cos x}$ | |
| $\dfrac{2\cos^2 x}{2\sin x\cos x}$ | |
| $\dfrac{\cos x}{\sin x}$ | |
| $\cot x$ | |

**8.** $\sin\left(-\dfrac{5\pi}{12}\right)=\sin\left(-\dfrac{2\pi}{12}-\dfrac{3\pi}{12}\right)$

$=\sin\left(-\dfrac{\pi}{6}\right)\cos\left(\dfrac{\pi}{4}\right)-\cos\left(-\dfrac{\pi}{6}\right)\sin\left(\dfrac{\pi}{4}\right)$

$=\left(-\dfrac{1}{2}\right)\left(\dfrac{\sqrt{2}}{2}\right)-\left(\dfrac{\sqrt{3}}{2}\right)\left(\dfrac{\sqrt{2}}{2}\right)$

$=\dfrac{-\sqrt{2}-\sqrt{6}}{4}$

**9.** $4\sin\theta\cos\theta-1=0$

$2(2\sin\theta\cos\theta)=1$

$2\sin(2\theta)=1$

$\sin(2\theta)=\dfrac{1}{2}$

$2\theta=\dfrac{\pi}{6},\dfrac{5\pi}{6},\dfrac{13\pi}{6},\dfrac{17\pi}{6}$

$\theta=\dfrac{\pi}{12},\dfrac{5\pi}{12},\dfrac{13\pi}{12},\dfrac{17\pi}{12}$

**10.** $\cos x=\sin(x-c)$

From the graphs, $c=-\dfrac{\pi}{2}$ or $c=\dfrac{3\pi}{2}$.

**11. a)** $y=2\sin(x-30)°+2$

**b)** $y=3\cos\left(2\left(x+\dfrac{\pi}{4}\right)\right)-\dfrac{1}{2}$

**12. a)** amplitude: 4
period: $8\pi$

phase shift: $\dfrac{\pi}{6}$ right
vertical displacement: 2 down

**b)** amplitude: $\dfrac{1}{3}$

period: $\dfrac{\pi}{2}$
phase shift: 40° right
vertical displacement: 3 down

**c)** amplitude: 2
period: $3\pi$

phase shift: 90° left (factor the $\dfrac{2}{3}$ out)
vertical displacement: 2 down

**d)** amplitude: $\dfrac{1}{3}$
period: $8\pi$
phase shift: none
vertical displacement: none

**13. a)** amplitude $=\dfrac{1.8-0.1}{2}=0.85$

**b)** range: $0.6\le y\le 1.8$

**c)** period: 12 h

**d)** $h(t)=0.85\cos\left[\dfrac{\pi}{6}(t-9)\right]+0.95$

**e)** $h(17)=0.85\cos\left[\dfrac{\pi}{6}(17-9)\right]+0.95$

$h(17)\doteq 0.5\,\text{m}$

**14. B**

Since the period is $\dfrac{3\pi}{2}$, then $\dfrac{2\pi}{|b|}=\dfrac{3\pi}{2}$.

Solving for $|b|$ gives $|b|=\dfrac{4}{3}$.

$y=\dfrac{1}{4}\cos\left(\dfrac{4}{3}\theta-\dfrac{\pi}{3}\right)+2\rightarrow$

$y=\dfrac{1}{4}\cos\dfrac{4}{3}\left(\theta-\dfrac{\pi}{4}\right)+2$

**15. A**

Since the sine ratio is negative and the tangent ratio is positive, $A$ and $B$ are in quadrant III.

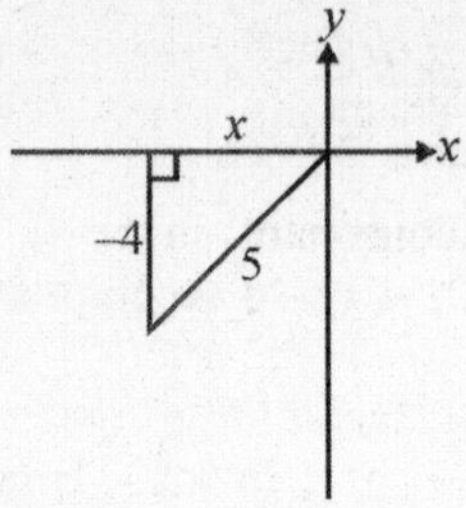


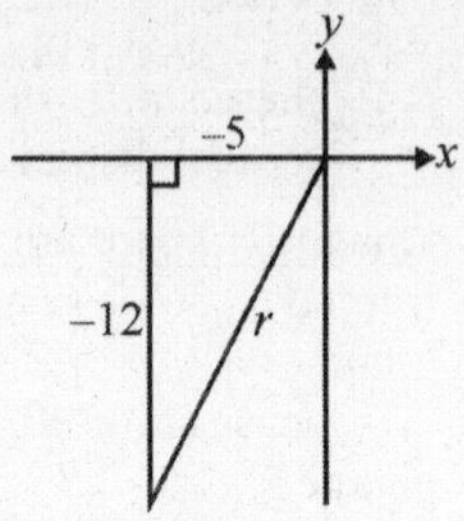


$x=-\sqrt{5^2-(-4)^2}=-3 \quad r=\sqrt{5^2+12^2}=13$

Thus, $\cos A=-\frac{3}{5}$, $\sin B=-\frac{12}{13}$, and $\cos B=-\frac{5}{13}$.

$$\begin{aligned}\cos(A-B)&=\cos A\cos B+\sin A\sin B\\&=\left(-\frac{3}{5}\right)\left(-\frac{5}{13}\right)+\left(-\frac{4}{5}\right)\left(-\frac{12}{13}\right)\\&=\frac{15}{65}+\frac{48}{65}\\&=\frac{63}{65}\end{aligned}$$

**16.** $a=r\theta$

$$14=(5)(\theta)\rightarrow\theta=\frac{14}{5}=2.8\text{ rad}$$

$$2.8\text{ rad}=2.8\times\frac{180}{\pi}\doteq 160°$$

| | | | |
|---|---|---|---|
| 1 | 6 | 0 | |

# PERMUTATIONS AND COMBINATIONS

## *Lesson 1 The Fundamental Counting Principle*

### CLASS EXERCISES ANSWERS AND SOLUTIONS

**1.** Consider how we would list every possible code. We can think of this task as having three stages.

Stage 1: Select the 1st digit
Stage 2: Select the 2nd digit
Stage 3: Select the 3rd digit

Stage 1: How many ways are there to accomplish stage 1? How many ways can we select the first digit? Since there are 10 digits to choose from—the digits 0 to 9—there are 10 ways to accomplish stage 1.

Stage 2: There are 10 ways to accomplish stage 2 since there are 10 digits to choose from.

Stage 3: There are 10 ways to accomplish stage 3.

To completely list every possible code, we would have to accomplish stage 1 *and* stage 2 *and* stage 3.

By the Fundamental Counting Principle, we know that the number of ways to accomplish all three stages is 10 × 10 × 10 =1 000.

So, there are 1 000 possible codes.

 

2. The task of taking this test has six stages.
Stage 1: Answer the 1st question
Stage 2: Answer the 2nd question
Stage 3: Answer the 3rd question
Stage 4: Answer the 4th question
Stage 5: Answer the 5th question
Stage 6: Answer the 6th question

Since there are 5 alternatives for each question, there are 5 ways to accomplish stage 1, there are 5 ways to accomplish stage 2, there are 5 ways to accomplish stage 3, and so on.

So, the total number of different sets of answers is $5 \times 5 \times 5 \times 5 \times 5 \times 5 = 15\,625$.

There are 15 625 ways for a student to answer the test.

3. The task of selecting a bank machine code consists of four stages.
Stage 1: Select the 1st digit
Stage 2: Select the 2nd digit
Stage 3: Select the 3rd digit
Stage 4: Select the 4th digit

Stage 1: There are 10 choices for the first digit. So, there are 10 ways to accomplish stage 1.

Stage 2: Once we select the first digit in stage 1, we need to select a *different* digit in stage 2. Thus, not all 10 digits are available in stage 2. Since there are 9 digits left to select from, we know that there are 9 ways to accomplish stage 2.

Stage 3: We need to select a digit from the remaining 8 digits. So, there are 8 ways to accomplish stage 3.

Stage 4: With 7 digits remaining, we know there are 7 ways to accomplish stage 4.

By the Fundamental Principle Counting, we know that the number of ways to accomplish all 4 stages is $10 \times 9 \times 8 \times 7 = 5\,040$. There are 5 040 ways to select a four-digit bank code so that the digits are different.

4. This task consists of six stages.
Stage 1: Select the 1st character
Stage 2: Select the 2nd character
Stage 3: Select the 3rd character
Stage 4: Select the 4th character
Stage 5: Select the 5th character

Stage 1: There are 21 consonants in the alphabet, so there are 21 ways to accomplish stage 1.

Stage 2: This letter must be a vowel. There are 5 vowels, so there are 5 ways to accomplish stage 2.

Stage 3: There are 21 ways to accomplish stage 3.

Stage 4: We must select an odd digit. Since there are 5 odd digits (1, 3, 5, 7, and 9), there are 5 ways to accomplish stage 5.

Stage 5: Once we have selected an odd digit in stage 4, we cannot select the same digit in stage 5. We must select a digit from the remaining 4 odd digits. So, there are 4 ways to accomplish stage 5.

By the Fundamental Principle Counting, we know that the number of ways to accomplish all four stages is $21 \times 5 \times 21 \times 5 \times 4 = 44\,100$.

There are 44 100 different possible licence plates in this particular country.

5. This task consists of five stages.
Stage 1: Select person to stand on the far left
Stage 2: Select person to stand second to the left
Stage 3: Select person to stand in the middle
Stage 4: Select person to stand second to the right
Stage 5: Select person to stand on the far right

Stage 1: Either Alice or Bob must stand on the far left, so there are 2 ways to accomplish stage 1.

Stage 2: Only Cathy, Don, or Ed can stand here, so there are 3 ways to accomplish stage 2.

Stage 3: Once a person has been selected in stage 2, there are only 2 remaining people who can stand in the middle. This means there are 2 ways to accomplish stage 3.

Stage 4: Once people have been designated in stages 2 and 3, there will only be 1 person remaining to stand here. So, there is 1 way to accomplish stage 4.

Stage 5: Either Alice or Bob must stand on the far right. Once one of them has been selected to stand on the far left, there will only be one person remaining to stand on the far right. There is 1 way to accomplish stage 5.

By the Fundamental Principle Counting, we know that the number of possible arrangements is $2 \times 3 \times 2 \times 1 \times 1 = 12$.

## PRACTICE EXERCISES ANSWERS AND SOLUTIONS

**1.** Using the Fundamental Counting Principle, multiply the number of choices for each stage.

$3 \times 4 \times 2 = 24$ possible meals

**2.** $12 \times 11 \times 10 = 1\,320$ executives can be formed. (Note that each position must be a different person.)

**3.** $14 \times 4 \times 6 = 336$ different outfits

**4. a)** $5 \times 5 \times 5 = 125$ "words"

**b)** $1 \times 5 \times 5 = 25$ "words"

**c)** $5 \times 3 \times 5 = 75$ "words"

**5.** $2 \times 10 \times 10 \times 10 \times 5 = 10\,000$ even 5-digit numbers begin with a 7 or 8

**6.** $52 \times 6 = 312$ possible outcomes

**7.** $1 \times 26 \times 26 = 676$ "words" that begin with L
$26 \times 26 \times 1 = 676$ "words" that end with P

However, there are $1 \times 26 \times 1$ "words" that begin with L and end with P. These are counted in the "words" that begin with L and in those that end in P.

Thus, there are $676 + 676 - 26 = 1\,326$ "words" that begin with L or end with P.

**8.** There are 4 outcomes that accomplish task *A*—draw a queen.

There are 26 outcomes that accomplish task *B*—draw a red card.

However, there are 2 duplicate cards in task *A* and task *B* (Queen of Hearts and the Queen of Diamonds).

Thus, the total number of outcomes of drawing a red card or a queen is: $4 + 26 - 2 = 28$.

**9.** $9 \times 8 \times 7 \times 6 \times 5 = 15\,120$ orders that Ayla can play the 5 CDs

**10.** $2 \times 26 \times 26 \times 26 = 35\,152$ station names

**11.** $6 \times 6 \times 9 = 324$ possibilities

**12. a)** $7 \times 6 \times 5 \times 4 = 840$ committees

**b)** $1 \times 6 \times 5 \times 4 = 120$ committees if Bee is president

**c)** $1 \times 6 \times 5 \times 4 = 120$ committees if Gary is president

$6 \times 1 \times 5 \times 4 = 120$ committees if Gary is vice president

Thus, $120 + 120 = 240$ committees if Gary is president or vice president.

**d)** $6 \times 5 \times 4 \times 3 = 360$ committees with Don not included

**13.** Determine the total the number of outcomes for each task and allow for any duplicates.

There are 4 jacks, 4 eights, and 13 spades.

However, there is a Jack of Spades and an eight of spades.

Thus, there are $4 + 4 + 13 - 2 = 19$ outcomes with a jack or an eight or a spade.

**14.** $3 \times 3 = 9$ packages of a pair of boots and skis

**15.** $5 \times 7 = 35$ different routes from town *A* to town *C* such that he passes through town *B*

**16.** $7 \times 9 = 63$ pairs of a mother character and a father character

 

**17.** **a)** $7\times7\times7\times7\times7=16\ 807$ "words" with no restrictions

**b)** $1\times1\times7\times7\times7=343$ "words" begin with HT

**c)** $7\times1\times1\times1\times7=49$ "words" with TEB in the middle

**d)** $1\times7\times7\times7\times7=2\ 401$ "words" begin with T
$7\times7\times7\times7\times1=2\ 401$ "words" end with T

However, $1\times7\times7\times7\times1=343$ "words" begin with T and end with T. These are counted in both the previous groups.

Thus, there are $2\ 401+2\ 401-343=4\ 459$ "words" begin or end with T.

**18.** **C**
Stage 1: 4 choices $\rightarrow$ 4, 5, 6, or 7
Stage 2: 4 choices
Stage 3: 4 choices
Stage 4: 4 choices

$4\times4\times4\times4=256$

**19.** **C**
Stage 1: 4 choices $\rightarrow$ A, B, C, D
Stage 2: 3 choices
Stage 3: 2 choices
Stage 4: 5 choices $\rightarrow$ 0, 2, 4, 6, 8
Stage 5: 4 choices
Stage 6: 3 choices

$4\times3\times2\times5\times4\times3=1\ 440$

## *Lesson 2 Factorial Notation*

### CLASS EXERCISES ANSWERS AND SOLUTIONS

**1.** When we rewrite the factorials as products and then cancel the factors in the numerator and denominator, we get:

$$\frac{3!}{5!}=\frac{3\times2\times1}{5\times4\times3\times2\times1}=\frac{1}{5\times4}=\frac{1}{20}$$

So, $\frac{3!}{5!}=\frac{1}{20}$.

**2.**

$$\frac{n!}{(n-2)!}=110$$

Rewrite factorials as products

$$\frac{n\times(n-1)\times(n-2)\times(n-3)\times\ldots\times2\times1}{(n-2)\times(n-3)\times\ldots\times2\times1}=110$$

Cancel terms
$n\times(n-1)=110$

Simplify
$n^2-n=110$

Set equation equal to zero
$n^2-n-110=0$

Factor
$(n+10)(n-11)=0$

$$n=-10,\ 11$$
$$n=11$$

There are two answers, however, we cannot use factorial notation with negative numbers, so $-10$ is not included in the solution. Thus, the solution is only $n=11$.

### PRACTICE EXERCISES ANSWERS AND SOLUTIONS

**1.** $\frac{0!}{4!}=\frac{1}{4\times3\times2\times1}=\frac{1}{24}$
(Note that by definition, $0!=1$.)

**2.** $\frac{7!}{6!}=\frac{7\times6!}{6!}=7$

**3.** $\frac{14!}{16!}=\frac{14!}{16\times15\times14!}=\frac{1}{16\times15}=\frac{1}{240}$

**4.** $\frac{6!}{9!}=\frac{6!}{9\times8\times7\times6!}=\frac{1}{9\times8\times7}=\frac{1}{504}$

**5.** $\frac{65!}{66!}=\frac{65!}{66\times65!}=\frac{1}{66}$

**6.** $\frac{26!}{24!}=\frac{26\times25\times24!}{24!}=26\times25=650$

**7.** $\frac{7!}{10!}=\frac{7!}{10\times9\times8\times7!}=\frac{1}{10\times9\times8}=\frac{1}{720}$

**8.** $\frac{5!}{0!} = 5! \; (0! = 1)$
$= 120$

**9.** $\frac{345!}{344!} = \frac{345 \times 344!}{344!} = 345$

**10.** $\frac{(k+1)!}{k!} = \frac{(k+1)(k!)}{k!} = k+1$ (when reduced)

**11.** $\frac{(n+3)!}{n!} = \frac{(n+3)(n+2)(n+1)(n!)}{n!}$
$= (n+3)(n+2)(n+1)$ (reduced)

**12.** $\frac{(m-2)!}{(m-4)!} = \frac{(m-2)(m-3)(m-4)!}{(m-4)!}$
$= (m-2)(m-3)$
$= m^2 - 5m + 6$

**13.** $\frac{(n+3)!}{(n+5)!} = \frac{(n+3)!}{(n+5)(n+4)(n+3)!}$
$= \frac{1}{(n+5)(n+4)}$
$= \frac{1}{n^2 + 9n + 20}$

**14.** $\frac{(n+1)!}{(n-2)!} = \frac{(n+1)(n)(n-1)(n-2)!}{(n-2)!}$
$= (n+1)(n)(n-1)$
$= n^3 - n$

**15.** $\frac{(n!)(n+1)!}{(n-1)!(n-1)!} = \frac{(n)(n-1)!(n+1)(n)(n-1)!}{(n-1)!(n-1)!}$
$= (n)(n+1)(n)$
$= n^2(n+1)$
$= n^3 + n$

**16.** $\frac{(n+5)!}{(n+3)!} = 56$

$\frac{(n+5)(n+4)(n+3)!}{(n+3)!} = 56$ • Expand

$(n+5)(n+4) = 56$ • Simplify

$n^2 + 9n + 20 = 56$ • Multiply the binomials

$n^2 + 9n - 36 = 0$ • Subtract 56 from both sides

$(n+12)(n-3) = 0$ • Factor

$n = -12$ or $n = 3$ • Solve

$n = -12$ is extraneous since $(n + 5)$ and $(n + 3)$ would then be negative.

Thus, $n = 3$ is the only solution.

**17.** $(n+9)! = 342(n+7)!$

$\frac{(n+9)!}{(n+7)!} = 342$

$\frac{(n+9)(n+8)(n+7)!}{(n+7)!} = 342$

$n^2 + 17n + 72 = 342$

$n^2 + 17n - 270 = 0$

$(n+27)(n-10) = 0$

$n = -27$ or $10$

The value of $n$ must be at least $-7$, thus $n = 10$.

**18.** $\frac{(n-33)!}{(n-32)!} = \frac{1}{19}$

$\frac{(n-33)!}{(n-32)(n-33)!} = \frac{1}{19}$ • Expand the factorial notation for the larger number

$\frac{1}{(n-32)} = \frac{1}{19}$ • Simplify

$n - 32 = 19$ • Equate the cross-products

$n = 51$ • Add 32 to both sides

 

**19.** $\dfrac{(n-5)!}{(n-7)!}=6$

$\dfrac{(n-5)(n-6)(n-7)!}{(n-7)!}=6$ • Expand the factorial notation for the larger number

$(n-5)(n-6)=6$ • Simplify

$n^2-11n+30=6$ • Multiply the binomials

$n^2-11n+24=0$ • Subtract 6 from both sides

$(n-8)(n-3)=0$ • Factor

$n=8$ or $n=3$ • Solve

$n = 3$ is an extraneous solution since $(n-5)$ and $(n-7)$ will then be negative. Thus, $n = 8$ is the only solution.

**20.** $\dfrac{(n+2)!}{n!}=42$

$\dfrac{(n+2)(n+1)n!}{n!}=42$ • Expand the factorial notation for the larger number

$(n+2)(n+1)=42$ • Simplify

$n^2+3n+2=42$ • Multiply the binomials

$n^2+3n-40=0$ • Subtract 42 from both sides

$(n+8)(n-5)=0$ • Factor

$n=-8$ or $n=5$

$n=-8$ is an extraneous solution since $(n+2)$ and $n$ will then be negative. Thus, $n = 5$ is the only solution.

**21.** $\dfrac{(n-15)!}{(n-14)!}=\dfrac{1}{12}$

$\dfrac{(n-15)!}{(n-14)(n-15)!}=\dfrac{1}{12}$ • Expand the factorial notation for the larger number

$\dfrac{1}{n-14}=\dfrac{1}{12}$ • Simplify

$n-14=12$ • Equate the cross-products

$n=26$ • Add 14 to both sides

**22.** $\dfrac{(n-1)!}{(n+1)!}=\dfrac{1}{72}$

$\dfrac{(n-1)!}{(n+1)(n)(n-1)!}=\dfrac{1}{72}$ • Expand

$\dfrac{1}{n^2+n}=\dfrac{1}{72}$ • Simplify

$72=n^2+n$ • Equate the cross-products

$n^2+n-72=0$ • Subtract 72 from both sides

$(n+9)(n-8)=0$ • Factor

$n=-9$ or $n=8$

$n=-9$ is extraneous, thus $n = 8$.

| | | | |
|---|---|---|---|
| 8 | | | |

**23.** $\dfrac{n!}{(n+2)!}=\dfrac{1}{6n}$

$\dfrac{n!}{(n+2)(n+1)(n!)}=\dfrac{1}{6n}$ • Expand

$\dfrac{1}{(n+2)(n+1)}=\dfrac{1}{6n}$ • Simplify

$(n+2)(n+1)=6n$ • Equate the cross-products

$n^2+3n+2=6n$ • Multiply the binomials

$n^2-3n+2=0$ • Subtract $6n$ from both sides

$(n-1)(n\ \ 2)=0$ • Factor

$n=1$ or $n=2$

Thus, the largest value of $n$ is 2.

| | | | |
|---|---|---|---|
| 2 | | | |

## *Lesson 3 Classifying Counting Problems*

### CLASS EXERCISES ANSWERS AND SOLUTIONS

**1.** Let's answer the three questions.

1. How many objects do we begin with?
We begin with a standard deck of playing cards, so we begin with 52 cards.

2. How many objects are selected from them?
Five cards are selected from the deck.

3. Does the order of the selected objects matter?
In poker, the order of the cards in your hand does not matter, so the answer to question 3 is NO, the order of the selected objects does not matter.

Since order does not matter, this is classified as a combinations question.

**2.** Answer the three questions.

1. How many objects do we begin with?
We begin with the letters in the word TRIANGLE. So, we begin with 8 letters.

2. How many objects are selected from them?
Three letters are selected to make a 3-letter "word."

3. Does the order of the selected objects matter?

Once 3 letters have been selected, does their order matter? For example, if the 3 selected letters are T, N, and E, does it matter if they are arranged as TEN or ENT? YES, it matters—these are 2 different "words." So, the order of the selected objects matters.

Since the order of the selected objects matters, this is a permutations question.

**3.** Answer the three questions.

1. How many objects do we begin with?
We begin with 25 pizza toppings.

2. How many objects are selected from them?
We select 4 toppings.

3. Does the order of the selected objects matter?
Once 4 toppings have been selected, does the order matter? For example, if the selected toppings are ham, pineapple, mushrooms, and peppers, does it matter which is added first, second, and so on? NO, it does not matter; the pizza will be the same.

Since the order of the selected objects does not matter, this is a combinations question.

**4.** Answer the three questions.

1. How many objects do we begin with?
We begin with 6 people.

2. How many objects are selected from them?
We select all 6 people for the photo.

3. Does the order of the selected objects matter?
YES, the order matters. If the 6 people are Al, Bea, Cal, Dee, Ed, and Faye, the arrangement Al-Bea-Cal-Dee-Ed-Faye is different than the arrangement Dee-Ed-Al-Faye-Bea-Cal.

Since the order of the selected objects matters, this is a permutations question.

### PRACTICE EXERCISES ANSWERS AND SOLUTIONS

**1.** Permutation

**2.** Combination

**3.** Permutation

**4.** Combination

**5.** Combination

**6.** Combination

**7.** Combination

**8.** Combination

**9.** Permutation,
$9 \times 8 \times ... 3 \times 2 \times 1 = 9! = 362\ 880$
(Fundamental Counting Principle)

**10.** Permutation, $3 \times 2 \times 1 = 6$
(Fundamental Counting Principle)

**11.** Permutation, 6 different letters
$6 \times 5 \times 4 \times 3 \times 2 \times 1 = 6! = 720$
(Fundamental Counting Principle)

**12.** Permutation $40 \times 39 = 1\ 560$
(Fundamental Counting Principle)

## *Lesson 4 Permutations of* n *Objects*

### CLASS EXERCISES
### ANSWERS AND SOLUTIONS

**1.** This question is almost identical to Example 1.

There are 4 children to choose from for chair 1.

Once this first child is seated, there are 3 children remaining for chair 2.

There are 2 children to choose from for chair 3.

And, there is 1 child to choose from for chair 4.

Altogether, there are $4 \times 3 \times 2 \times 1$ (or 4!) arrangements.

There are 24 possible arrangements for seating 4 children in a row.

**2.** We have 6 letters and we are selecting all 6 of them to arrange. The order in which the letters are arranged matters. For example, we consider the word CETJOB to be different than JTBOEC, since the ordering of the letters is different.

Since the 6 letters in the word OBJECT are all different (or distinct), we can say that the number of different permutations of 6 distinct letters is 6! or 720.

So, 720 different 6-letter words can be made from the letters in the word OBJECT.

**3.** In this question, we begin with 11 different players, and each of the 11 is to be assigned a different position. So, order does matter.

The number of permutations of 11 players is 11!, or 39 916 800, which is the number of different arrangements.

### PRACTICE EXERCISES
### ANSWERS AND SOLUTIONS

**1.** $5! = 120$

**2.** $7! = 5\ 040$

**3.** $5! = 120$

**4.** $4! = 24$

**5.** $5! = 120$

**6.** $4! = 24$

**7.** $8! = 40\ 320$

**8.** $9! = 362\ 880$

**9.** $7! = 5\ 040$

**10.** $9! = 362\ 880$

## *Lesson 5 Permutations of* n *Objects Taken* r *at a Time*

### CLASS EXERCISES
### ANSWERS AND SOLUTIONS

**1.**

$${}_6P_2 = \frac{6!}{(6-2)!}$$
- Substitute $n = 6$ and $r = 2$

$$= \frac{6!}{4!}$$
- Rewrite as a quotient of products

$$= \frac{6 \times 5 \times 4!}{4!}$$
- Cancel out common factors

$$= 6 \times 5$$
$$= 30$$
- Evaluate

**2.** We have 12 girls and we will select 4 girls and assign each a different role. We are essentially permuting the girls into the 4 roles.

The number of ways to permute 4 girls out of 12 girls is ${}_{12}P_4 = 11\ 880$.

There are 11 880 different assignments possible.

**3.** We are taking 7 books and permuting them 7 at a time.

This can be done in ${}_7P_7$ ways.

When we evaluate ${}_7P_7$, we get:

$${}_7P_7 = \frac{7!}{(7-7)!}$$
• Substitute $n = 7$ and $r = 7$

$$= \frac{7!}{0!}$$
• Simplify

$$= 7!$$

Recall that $0! = 1$.

So, 7 books can be arranged in 7! ways.

As you can see, generally, $n$ objects can be permuted ${}_nP_n = n!$ ways.

**4.** We want to count the number of words that can be made when one or more of the letters are selected. We are interested in counting the number of words with 1 letter or 2 letters or 3 letters or 4 letters.

So, let's determine the number of 1-letter words, 2-letter words, 3-letter words, and 4-letter words.

1-letter words: To make 1-letter words, we select 1 letter from the 4. The number of permutations of 4 letters taken 1 at a time is ${}_4P_1$.

2-letter words: We make 2-letter words when we select 2 letters and permute them. This can be done ${}_4P_2$ ways.

3-letter words: There are ${}_4P_3$ different 3-letter words.

4-letter words: There are ${}_4P_4$ different 4-letter words.

Since we are counting the number of words with 1 letter or 2 letters or 3 letters or 4 letters, by addition, we get;

$${}_4P_1 + {}_4P_2 + {}_4P_3 + {}_4P_4 = 4 + 12 + 24 + 24$$
$$= 64$$

So, there are 64 different words that can be formed using one or more letters from the word WIND.

## PRACTICE EXERCISES ANSWERS AND SOLUTIONS

Using the formula ${}_nP_r = \frac{n!}{(n-r)!}$ for the following questions.

**1.** ${}_5P_3 = \frac{5!}{(5-3)} = \frac{5!}{2!} = 60$

**2.** ${}_{11}P_3 = \frac{11!}{(11-3)!} = \frac{11!}{8!} = 990$

**3.** ${}_{73}P_1 = \frac{73!}{(73-1)!} = \frac{73!}{72!} = 73$

**4.** ${}_{15}P_0 = \frac{15!}{(15-0)!} = \frac{15!}{15!} = 1$

**5.** ${}_{10}P_2 = \frac{10!}{(10-2)!} = \frac{10!}{8!} = 90$

**6.** ${}_6P_4 = \frac{6!}{(6-4)!} = \frac{6!}{2!} = 360$

**7.** ${}_3P_3 = \frac{3!}{(3-3)!} = \frac{3!}{0!} = 6$ (remember that $0! = 1$)

**8.** ${}_{100}P_2 = \frac{100!}{(100-2)!} = \frac{100!}{98!} = 9\ 900$

**9.** From a group of 10 swimmers, 3 swimmers are being arranged.

$${}_{10}P_3 = \frac{10!}{(10-3)!} = \frac{10!}{7!} = 720$$

 

**10.** From the 10 CDs, Pauline is arranging 6 of them.

$${}_{10}P_6 = \frac{10!}{(10-6)!} = \frac{10!}{4!} = 151\ 200$$

**11.** From the group of 8 players, 4 are arranged so that each of these 4 gets one prize.

Thus, ${}_8P_4 = \frac{8!}{(8-4)!} = \frac{8!}{4!} = 1\ 680.$

**12.** There are 5 letters. Since the word can be of any length, the 5 letters are arranged one at a time or 2 at a time or 3 at a time or 4 at a time or 5 at a time.

Thus, the number of "words" is

$${}_5P_1 + {}_5P_2 + {}_5P_3 + {}_5P_4 + {}_5P_5 = 5 + 20 + 60 + 120 + 120$$
$$= 325$$

**13.** The 8 letters are arranged 6 at a time, thus, the number of arrangements is:

$${}_8P_6 = \frac{8!}{(8-6)!} = \frac{8!}{2!} = 20\ 160$$

**14.** The 13 soccer players are being arranged 11 at a time. The number of arrangements is:

$${}_{13}P_{11} = \frac{13!}{(13-11)!} = \frac{13!}{2!} = 3\ 113\ 510\ 400$$

**15. D**

Using the formula

$${}_nP_r = \frac{n!}{(n-r)!}:$$

$${}_8P_3 = \frac{8!}{(8-3)!} = \frac{8!}{5!} = 336$$

**16. A**

$${}_{n+1}P_2 = \frac{(n+1)!}{((n+1)-2)!} = \frac{(n+1)!}{(n-1)!}$$
$$= \frac{(n+1)(n)(n-1)!}{(n-1)!} = n(n+1)$$

## *Lesson 6 Permutations With Restrictions*

### CLASS EXERCISES
### ANSWERS AND SOLUTIONS

**1.** We will sort out the letters to see how much duplication there is.

| LETTER | FREQUENCY |
|---|---|
| M | 1 |
| I | 4 |
| S | 4 |
| P | 2 |
| Total | 11 |

Altogether, there are 11 letters, with 1 M, 4 Is, 4 Ss, and 2 Ps.

If the 11 letters were distinct, there would be 11! different permutations, but we need to account for the 4 repeated Is, 4 repeated Ss, and 2 repeated Ps.

So, we divide: $\frac{11!}{4!4!2!} = 34\ 650$

There are 34 650 different arrangements possible using the letters in the word MISSISSIPPI.

**2.** If we were asked to determine how many arrangements for planting 10 distinct trees, there would be 10! orderings. But, not all of trees are distinct.

| TREE | FREQUENCY |
|---|---|
| Pine | 3 |
| Elm | 4 |
| Birch | 2 |
| Willow | 1 |
| Total | 10 |

If we represent each tree by its first letter (P=Pine, E=Elm, B=Birch, W=Willow), we have a question that is very similar to the last question.

For example, one possible order to plant the trees would be EEBWPEPPBE and another order would be PEPEEEBPBW.

How many arrangements are there of these 10 letters?

The number of possible arrangements for planting the trees is $\frac{10!}{4!3!2!} = 12\ 600.$

**3.** In this question, we are dealing not only with permutations of non-distinct objects, but also with the added restriction that the permutations begin with an R.

We will first sort the letters:

| LETTER | FREQUENCY |
|---|---|
| R | 2 |
| E | 3 |
| D | 2 |
| Total | 7 |

Using an approach similar to what was used in previous questions, consider all of the letters as being different, two of which could be selected for the first position.

The result would be 2 letter choices for the first position, from which one is chosen, leaving 6! arrangements for the remaining positions.

Thus, if the letters were all different, the answer would be 2 × 6!. Since there are 2 Rs, 3 Es, and 2 Ds, we need to eliminate the effect of ordering these letters. The result is $\frac{2\times 6!}{2!3!2!}$, which is equal to 60.

There are 60 permutations of the letters in RED DEER in which each new word begins with the letter R.

**4. a)** The number of arrangements of the 7 books with volumes 1, 3, 5, and 7 in order is $\frac{7!}{4!}=7\times6\times5=210$

**b)** The number of arrangements of the 7 books with volumes 1 and 2 out of order is equal to: Total number of arrangements – Number of arrangements in order

$=7!-\frac{7!}{2!}=5\ 040-2\ 520=2\ 520$

**5.** Again, we are faced with a permutation question where some objects are grouped together. Let's create the different groups, with Ann, Bob, and Emma all in one group.

| ABE | C | D | F |
|---|---|---|---|

Altogether, there are 4 distinct groups. Here is one possible arrangement of these groups:

| D | F | C | ABE |
|---|---|---|---|

There are 4! permutations of these 4 distinct groups and there are 3! permutations of Ann, Bob, and Emma within their group.

By the Fundamental Counting Principle, the number of permutations is $4!\times3!=24\times6=144$.

**6.** The restriction in this question is that the 2 Cs are not together.

To solve this, we will first determine the number of possible arrangements without the restriction.

Since the letters of CALCULUS are not all distinct, we will sort out the letters to see how much duplication exists.

| LETTER | FREQUENCY |
|---|---|
| C | 2 |
| A | 1 |
| L | 2 |
| U | 2 |
| S | 1 |
| Total | 8 |

Altogether, there are 8 letters with 2 Cs, 1 A, 2 Ls, 2 Us, and 1 S.

If the 8 letters were distinct, there would be 8! different permutations, but we need to account for the repeated letters.

So, we divide by the number of arrangements of the repeated letters: $\frac{8!}{2!2!2!}=5\ 040$

There are 5 040 different arrangements of the letters in the word CALCULUS.

Now, we must determine how many arrangements have the 2 Cs together.

We create a box with 2 Cs, and 6 more blocks for each of the remaining letters.

| CC | A | L | L | U | U | S |
|---|---|---|---|---|---|---|

There are 7 boxes altogether, but not all of the blocks are distinct. There are 2 boxes with Ls and 2 boxes with Us.

So, the number of permutations of the 7 boxes is $\frac{7!}{2!2!}=1\ 260$.

 

There are 1 260 permutations of the letters such that the 2 Cs are together.

There are 5 040 different arrangements of the letters in the word CALCULUS and there are 1 260 arrangements of the letters such that the 2 Cs are together.

So, the number of arrangements where the 2 Cs are not together is:
5 040 – 1 260 = 3 780

## PRACTICE EXERCISES ANSWERS AND SOLUTIONS

1. There are 8 letters in total.

   There are 3 As and 2 Ns.

   Thus, there are $\frac{8!}{3!2!}=3\ 360$ arrangements.

2. There are 9 gifts to be arranged.

   There are 3 identical music books and 4 identical tuning forks.

   Thus, there are $\frac{9!}{4!3!}=2\ 520$ different distributions of the gifts.

3. Since the last two balls must be green, there is only 1 choice for each of the last two draws.

   For the remaining 8 balls, there are 3 red, 3 green, and 2 yellow balls.

   These remaining 8 balls can be arranged in:
   $\frac{8!}{3!3!2!}=560$ orders

4. Arrange the girls in the 1st, 3rd, 5th, 7th, and 9th positions. There are 5! = 120 arrangements.

   Arrange the boys in the 2nd, 4th, 6th, and 8th positions. There are 4! = 24 arrangements.

   Since the girls and boys are arranged together there are $120\times 24=2\ 880$ arrangements.

5. Here you can consider arranging 11 Ws, 4 Ls, and 5 Ts.

   Thus, there are $\frac{20!}{11!4!5!}=21\ 162\ 960$ arrangements.

6. One approach is to use spaces to represent each location in the arrangement and use the Fundamental Counting Principle to determine the number of arrangements.

   There are 7 letters, thus 7 spaces.
   $\_\times\_\times\_\times\_\times\_\times\_\times\_$

   The 1st space must have a T, thus there is only 1 choice for that space.
   $\underline{1}\times\_\times\_\times\_\times\_\times\_\times\_$

   The last space must be a vowel; thus, there are 3 choices (2 Us and 1 A) for that space.
   $\underline{1}\times\_\times\_\times\_\times\_\times\_\times\underline{3}$

   The remaining 5 letters can be arranged in the 5 remaining 5 spaces. The result is,
   $\underline{1}\times\underline{5}\times\underline{4}\times\underline{3}\times\underline{2}\times\underline{1}\times\underline{3}$

   or $1\times 5!\times 3$

   Allowing for like objects gives
   $\frac{1\times 5!\times 3}{2!2!}=90$ arrangements.

7. Consider arranging 6 Hs and 9 Ts.

   This gives $\frac{15!}{6!9!}=5\ 005$ arrangements.

8. There are 7 letters, and 2 are vowels (O and E). The vowels must be kept in the order EO.

   Number of arrangements =
   $\frac{\text{Number of arrangements of all 7 letters}}{\text{Number of arrangements of the vowels}}$
   $=\frac{7!}{2!}=2\ 520$

9. There are 7 letters in total.

   There are 2 Ns and 2 As.

   Thus, there are $\frac{7!}{2!2!}=1\ 260$ arrangements.

10. First, consider the letters to be all different. The first letter must be an N, thus there are two choices for this space. For the other spaces, there are 6 letters to arrange. Allowing for the 2 Ns and 2 As there are $\frac{2\times 6!}{2!2!}=360$ arrangements.

**11.** The answer is the number of arrangements of 4 dots, 4 dashes, and 3 spaces in a line.

This can be done in $\frac{11!}{4!4!3!} = 11\ 550$ ways.

**12.** The number of arrangements without the Ferrari, Audi, and BMW in that order will equal the total number of permutations minus the total number of permutations with the 3 cars in that order.

Total number of permutations = 7!

Total number of permutations with the 3 cars in order =

$$\frac{\text{Total number of permutations}}{\text{Number permutations for the 3 cars}}$$
$$= \frac{7!}{3!}$$

Total number of arrangements with the 3 cars not in order $= 7! - \frac{7!}{3!} = 4\ 200.$

**13.** Consider the husbands as one object. Thus, there are 4 objects (3 wives plus this group).

These can be arranged in $4! = 24$ ways.

The group of husbands can be arranged in $3! = 6$ ways.

Thus, the total number of arrangements is $24 \times 6 = 144$ (using the Fundamental Counting Principle).

**14.** With no restrictions, there are $8! = 40\ 320$ arrangements of the 8-person committee.

Calculate the number of arrangements with the president and vice-president together.

Consider them as 1 object. Thus, there are 7 objects to arrange (6 individuals plus this 1 group of two).

There are $7! = 5\ 040$ arrangements of these 7 objects and $2! = 2$ arrangements for the group of 7.

Thus, $5\ 040 \times 2 = 10\ 080$ arrangements have the president and vice-president together.

Therefore, $40\ 320 - 10\ 080 = 30\ 240$ arrangements are possible if the president and vice-president are not together.

**15.** Since the first letter is a P, there is only one choice for the first space. Thus, only the remaining 11 letters can be arranged. These 11 letters consist of 4 Ss and 2 Es.

The total number of arrangements for these 11 letter is $\frac{11!}{4!2!} = 831\ 600.$

Consider the number of arrangements that have the 2 Es together.

Consider the 2 Es as a single object.

There are 10 objects to arrange (9 other letters plus the group of 2 Es).

These 10 objects include 4 Ss.

There are $\frac{10!}{4!} = 151\ 200$ arrangements with the 2 Es together.

Note that you do not need to multiply this answer by 2! because the objects in the group are like objects.

Thus, there are $831\ 600 - 151\ 200 = 680\ 400$ arrangements with P first and the 2 Es not together.

**16.** There are 4 objects to arrange (each couple can be considered as one object).

For these 4 objects there are $4! = 24$ arrangements.

There are 2! arrangements of each of the 4 groups.

Thus, the total number of arrangements is $24 \times 2! \times 2! \times 2! \times 2! = 384.$

 

**17.** Let a win be represented by a W.
Let a loss be represented by a L.
Let a tie be represented by a T.

First, calculate how many arrangements of the 8 Ls and 4 Ts are possible.

There are $\frac{12!}{8!4!} = 495$ arrangements.

One of these arrangements would be:
L L L L L L L L T T T T

When the 6 Ws are included, they must be separated by at least one other letter.

There are 13 positions for the 1st W (this W could be 1st, last, or in any of the 11 spaces between the letters shown).

The 2nd W has only 12 possible positions.
The 3rd W has only 11 possible positions.
The 4th W has only 10 possible positions.
The 5th W has only 9 possible positions.
The 6th W has only 8 possible positions.

Thus, using the arrangement of 8 Ls and 4 Ts shown above, there are
$\frac{13\times12\times11\times10\times9\times8}{6!} = 1\,716$ possible insertions for the 6 Ws.

Note: divide by 6! because the wins are considered identical.

For every one of 495 arrangements of the 8 Ls and 4 Ts, there are 1 716 possible insertions of the 6 Ws.

Thus, the number of possible arrangements of the game results is $495\times1\,716 = 849\,420$.

**18.** Arrange the girls in the 1st, 3rd, 5th, and 7th spaces. This gives 4! = 29 arrangements.
Arrange the boys in the 2nd, 4th, and 6th spaces. This gives 3! = 6 arrangements.

Since the boys and girls are arranged together, there are $24\times6 = 144$ arrangements.

**19.** Consider the 4 Os as a single object.

Thus, there are 7 objects to arrange (the 6 other letters plus the group of 4 Os).

This gives 7! arrangements.

Now, allow for the 2 Ns, which gives
$\frac{7!}{2} = 2\,520$ arrangements.

Note that in the group held together, the objects are alike and, thus, you do not multiply by 4!.

**20.** First consider arranging the 6 letters that are not Os.

This gives $\frac{6!}{2!} = 360$ arrangements of letters, 2 of which are alike.

One of these arrangements would be:
M N T N U S

The group of 3 Os and the single O must be separated by at least one letter.

The group of 3 Os have 7 possible positions.

They could be first, last, or in any of the 5 spaces between the letters shown.

Once this group is positioned, the single O has 6 possible locations. Using the arrangement shown, the groups of 3 Os and the single O gives $7\times6 = 42$ arrangements.

This is the same for each of the 360 arrangements of the other 6 letters.

Thus, there are $360\times42 = 15\,120$ arrangements.

**21.** There are 11 letters, with 4 vowels to consider as a single group, plus
2 Ms
2 Ts
1 H
1 C
1 S

There are 8 objects to arrange (4 vowels plus 7 other letters).

This gives 8! = 40 320 arrangements.

Allow for the like objects of 2 Ms and 2 Ts.

This gives $\frac{8!}{2!2!} = 10\ 080$ arrangements.

Now, arrange the group being held together (the 4 vowels).

In this group of 4, there are 2 like objects, thus $\frac{4!}{2!} = 12$ arrangements.

Therefore, the total number of arrangements is $10\ 080 \times 12 = 120\ 960$.

**22.** Let W represent a Win, L a loss, and T a tie.

Arrange the 6 Ws and 3 Ts.

There are $\frac{9!}{6!3!} = 84$ arrangements.

One of these would be:
W W W W W W T T T

Consider inserting the Ls so that they are separated by at least one other letter.

The 1st L has 10 possible locations: 1st, last or any of the 8 spaces between the letters shown.
The 2nd L has 9 possible positions.
The 3rd L has 8 possible positions.
The 4th L has 7 possible positions.

Using the arrangement shown, there are $10 \times 9 \times 8 \times 7 = 5040$ arrangements possible for the 4 Ls.

This would be the same for each of the 84 possible arrangements of the 6 Ws and 3 Ts.

Thus, the total number of arrangements is $84 \times 5\ 040 = 423\ 360$.

**23. D**
There are 8 letters including 4 vowels, which must be kept together.

The 4 vowels act as a single group.

Thus there are 5 objects to arrange, which gives 5! possible arrangements.

Now, arrange the group of 4 vowels. There are 2 like objects.

Thus, there are $\frac{4!}{2!}$ possible arrangements for the vowels.

Thus, the total number of arrangements is $5! \times \frac{4!}{2!} = \frac{5! \times 4!}{2!}$.

**24. C**
First, you need to find the number of arrangements with the vowels in alphabetical order, and then subtract that number from the total number of arrangements.

The number of arrangement in order = $\frac{\text{Total arrangements}}{\text{Number of ways to arrange vowels}} = \frac{6!}{3!}$

Thus, the number of arrangements not in order $= 6! - \frac{6!}{3!} = 600$.

**25.** The total number of arrangements without restrictions is $\frac{8!}{3!}$ or 6 720. Now, you need to determine the number of arrangements that have P and R together. One arrangement is
PR A A B O L A

The number of arrangements with P and R together is $\frac{7!}{3!} \times 2! = 1\ 680$

Thus, the number of arrangements where P and R are not together is $6\ 720 - 1\ 680 = 5\ 040$.

| 5 | 0 | 4 | 0 |
|---|---|---|---|

 

**26.** The number of arrangements that begin with an E is $\frac{2\times5!}{2!}=120.$

The number of arrangements that begin with a T is $\frac{1\times5!}{2!}=60.$

Total arrangements that begin with an E or a T is
$=120+60=180$

| 1 | 8 | 0 | |
|---|---|---|---|

## *Lesson 7 Combinations of* n *Objects Taken* r *at a Time*

### CLASS EXERCISES ANSWERS AND SOLUTIONS

**1.** i) How many objects do we begin with? We begin with 25 sculptures.

ii) How many objects are selected from them? We will select 2 sculptures to donate.

iii) Does the order of the selected objects matter? No, the order does not matter.

Since order does not matter, this is a combination question.

We are selecting 2 sculptures from 25 and order does not matter. There are ${}_{25}C_2$ combinations.

When we evaluate ${}_{25}C_2$, we get ${}_{25}C_2=300$.

There are 300 different combinations for selecting 2 sculptures from 25 sculptures, thus 300 different donations.

**2.** This problem consists of two stages.

Stage 1: Give 4 paintings to her son
Stage 2: Give 3 paintings to her granddaughter

Let's first consider stage 1.

Stage 1: Mrs. Klemke begins with 10 paintings and selects 4 of them to give to her son. Since the order in which she gives the paintings to her son does not matter, this is a combination question.

There are ${}_{10}C_4=210$ combinations for selecting 4 out of 10 paintings.

Now that she has selected 4 paintings for her son, we can move to stage 2 and determine how she can select 3 paintings for her granddaughter.

Stage 2: Since she has already selected 4 paintings for her son, there are only 6 paintings remaining to select from. She will select 3 paintings from these 6. There are ${}_6C_3=20$ combinations for selecting 3 out of 6 paintings.

In order to solve this problem, we must consider accomplishing stage 1 and stage 2. There are 210 combinations in stage 1 and 20 combinations in stage 2.

By the Fundamental Counting Principle, there are 210 × 20 or 4 200 combinations for both stages. Mrs. Klemke has 4 200 possible distributions for giving 4 paintings to her son and 3 paintings to her granddaughter.

**3.** This problem consists of two stages.
Stage 1: Select the 3 girls
Stage 2: Select the 3 boys

Let's first accomplish stage 1.

Stage 1: We begin with 7 girls and must select 3 of them. Since the order in which the girls are selected does not matter, this is a combination question.

There are ${}_7C_3=35$ combinations for selecting 3 out of 7 girls.

Stage 2: We begin with 8 boys and must select 3 of them. Since the order in which we select the boys does not matter, this is a combination question. There are ${}_8C_3=56$ combinations for selecting 3 out of 8 boys.

In order to solve this problem, we must accomplish stage 1 and stage 2. There are 35 combinations for accomplishing stage 1 and 56 combinations for accomplishing stage 2.

By the Fundamental Counting Principle, there are 35 × 56 = 1 960 combinations for accomplishing both stages.

Thus, there are 1 960 possible teams.

## PRACTICE EXERCISES ANSWERS AND SOLUTIONS

**1.** ${}_7C_5 = \frac{7!}{(7-5)!5!} = 21$

**2.** ${}_{11}C_3 = \frac{11!}{(11-3)!3!} = 165$

**3.** ${}_{55}C_1 = \frac{55!}{(55-1)!1!} = 55$

**4.** ${}_{33}C_{33} = \frac{33!}{(33-33)!33!} = 1$
(Remember that 0! = 1.)

**5.** ${}_7C_2 = \frac{7!}{(7-2)!2!} = 21$

**6.** ${}_7C_5 = \frac{7!}{(7-5)!5!} = 21$

**7.** ${}_{15}C_4 = \frac{15!}{(15-4)!4!} = 1\,365$

**8.** ${}_9C_9 = \frac{9!}{(9-9)!9!} = 1$

**9.** ${}_{12}C_5 = \frac{12!}{(12-5)!5!} = 792$ selections

**10.** ${}_{52}C_5 = \frac{52!}{(52-5)!5!} = 2\,598\,960$
hands are possible

**11.** Stage 1, select two entrées:
${}_5C_2 = \frac{5!}{(5-2)!2!} = 10$

There are 10 combinations for selecting 2 entrées.

Stage 2, select 3 side dishes:
${}_6C_3 = \frac{6!}{(6-3)!3!} = 20$

There are 20 combinations for selecting 3 side dishes.

Stage 3, select one dessert:
${}_4C_1 = \frac{4!}{(4-1)!1!} = 4$

There are 4 combinations for selecting one dessert.

Since the meal consists of stage 1, stage 2, and stage 3, using the Fundamental Counting Principle, the number of different meals is $10 \times 20 \times 4 = 800$.

**12.** Stage 1 is selecting 3 hearts. There are ${}_{13}C_3 = \frac{13!}{(13-3)!3!} = 286$ combinations for doing this.

Stage 2 is selecting 2 cards from the remaining 39 cards (clubs, diamonds, and spades), which gives ${}_{39}C_2 = \frac{39!}{(39-2)!2!} = 741$ combinations.

The poker hand consists of stage 1 and stage 2.

Using the Fundamental Counting Principle the number of hands is $286 \times 741 = 211\,926$.

**13.** ${}_{49}C_6 = \frac{49!}{(49-6)!6!} = 13\,983\,816$
(order does not matter)

**14.** ${}_{12}C_8 = \frac{12!}{(12-8)!8!} = 495$

**15.** Stage 1, select 1 vice-principal:
${}_3C_1 = \frac{3!}{(3-1)!1!} = 3$

There are 3 combinations for selecting 1 vice-principal.

Stage 2, select 3 teachers:
${}_5C_3 = \frac{5!}{(5-3)!3!} = 10.$

There are 10 combinations for selecting 3 teachers.

 

Stage 3, select 4 students:

$_8C_4 = \frac{8!}{(8-4)!4!} = 70$

There are 70 combinations for selecting 4 students.

Since the committee consists of stage 1, stage 2, and stage 3, using the Fundamental Counting Principle, the number of committees is $3 \times 10 \times 70 = 2\ 100$.

**16.** Stage 1, select 2 diamonds:

This gives $_{13}C_2 = \frac{13!}{(13-2)!2!} = 78$ combinations.

Stage 2, select 3 cards from the remaining 26 cards (13 clubs and 13 hearts):

This gives $_{26}C_3 = \frac{26!}{(26-3)!3!} = 2\ 600$ combinations.

The poker hand consists of stage 1 and stage 2. Using the Fundamental Courting Principle, the number of poker hands is $78 \times 2\ 600 = 202\ 800$.

**17.** There are $_{31}C_2 = \frac{31!}{(31-2)!2!} = 465$ different 2-scoop combinations.

**18.** One classmate does not get a donut.

Thus, he must select 1 person from the group of 24.

This gives $_{24}C_1 = \frac{24!}{(24-1)!1!} = 24$ combinations.

**19.** There are 4 suits (clubs, spades, hearts, diamonds) each with 13 cards.

There are $_{13}C_5$ combinations for selecting 5 cards from any one of the 4 suits.

The total number of flushes

$= 4 \times {_{13}C_5}$

$= 4 \times \frac{13!}{(13-5)!5!}$

$= 5\ 148$

**20.** In stage 1, the 1st child receives 2 cars.

The number of different selections of these cars is $_8C_2 = \frac{8!}{(8-2)!2!} = 28.$

In stage 2, the 2nd child receives 2 cars from the remaining 6 cars.

The number of different selections of these cars is $_6C_2 = \frac{6!}{(6-2)!2!} = 15.$

In stage 3, the 3rd child receives 2 cars from the remaining 4 cars.

The number of different selections of these cars is $_4C_2 = \frac{4!}{(4-2)!2!} = 6.$

In stage 4, the 4th child receives 2 cars from the remaining 2 cars.

The number of different selections of these cars is $_2C_2 = \frac{2!}{(2-2)!2!} = 1.$

Since all 4 stages must be accomplished, using the Fundamental Counting Principle, there are $28 \times 15 \times 6 \times 1 = 2\ 520$ possible distributions.

**21.** Stage 1, select the 2 cards:
There are 4 suits; choose 1, which is $_4C_1$ selections.
Each suit has 13 cards; choose 2, which is $_{13}C_2$ selections.
Total number of selections: $_4C_1 \times {_{13}C_2}$

Stage 2, select the 3 cards:
There are 3 suits remaining; choose 1, which is $_3C_1$ selections.
Each suit has 13 cards; choose 3, which is $_{13}C_3$ selections.
Total number of selections: $_3C_1 \times {_{13}C_3}$

Total number of 5-card hands:
(stages 1 and 2)
$= {_4C_1} \times {_{13}C_2} \times {_3C_1} \times {_{13}C_3}$
$= 4 \times 78 \times 3 \times 286$
$= 267\ 696$

**22.** Stage 1, choose two of one kind:
There are 13 kinds; choose 1, which is ${}_{13}C_1$ selections.
Each kind has 4 cards; choose 2, which is ${}_4C_2$ selections.
Total number of selections: ${}_{13}C_1 \times {}_4C_2$

Stage 2, choose three of the other kind:
There are 12 other kinds; choose 1, which is ${}_{12}C_1$ selections.

Each kind has 4 cards; choose 3, which is ${}_4C_3$ selections.
Total number of 5-card hands:

(stages 1 and 2)
$= {}_{13}C_1 \times {}_4C_2 \times {}_{12}C_1 \times {}_4C_3$
$= 13 \times 6 \times 12 \times 4$
$= 3\ 744$

**23.** **D**
Stage 1, select 2 kings: ${}_4C_2 = 6$ selections
Stage 2, select 2 aces: ${}_4C_2 = 6$ selections
Stage 3, select 1 card from the remaining cards that do not include kings or aces: ${}_{44}C_1 = 44$ selections

Total number of hands $= 6 \times 6 \times 44 = 1\ 584$

**24.** **C**
Stage 1, select first group: ${}_{15}C_3 = 455$ selections
Stage 2, select second group: ${}_{12}C_5 = 792$ selections
Stage 3, select third group: ${}_7C_7 = 1$ selection

Total selections for all 3 groups
$= 455 \times 792 \times 1 = 360\ 360$

**25.** Stage 1, select 2 women: ${}_4C_2 = 6$ selections
Stage 2, select 4 men: ${}_5C_4 = 5$ selections

Total number of committees of the required number of men and women $= 6 \times 5 = 30$

| 3 | 0 | | |
|---|---|---|---|

**26.** Stage 1, select 3 aces: ${}_4C_3 = 4$ selections
Stage 2, select remaining 2 cards from 48 other cards: ${}_{48}C_2 = 1\ 128$

Total number of hands $= 4 \times 1\ 128 = 4\ 512$

| 4 | 5 | 1 | 2 |
|---|---|---|---|

## *Practice Quiz*

### ANSWERS AND SOLUTIONS

**1.** Number of different outfits $= 5 \times 4 \times 3 = 60$

**2.** Number of numbers
$= 2 \times 10 \times 10 \times 10 \times 5 = 10\ 000$

**3. a)** Number of words $= {}_6P_5 = 720$

**b)** Number of words $= (1 \times 5 \times 4 \times 3 \times 2) \times 2 = 240$

**4.**
$$\begin{aligned}(n+6)! &= 72(n+4)!\\ (n+6)(n+5)(n+4)! &= 72(n+4)!\\ (n+6)(n+5) &= 72\\ n^2+11n+30 &= 72\\ n^2+11n-42 &= 0\\ (n-6)(n+7) &= 0\end{aligned}$$

$n = 6$ or $n = -7$ but $n \geq -4$, thus $n = 6$.

**5.** Number of words
$= {}_4P_1 + {}_4P_2 + {}_4P_3 + {}_4P_4$
$= 4 + 12 + 24 + 24$
$= 64$

**6.** Number of arrangements $= \dfrac{7!}{2!2!} = 1\ 260$

**7. a)** Number of arrangements $= 5! = 120$

**b)** Number of arrangements $= \dfrac{3 \times 2 \times 5!}{3!} = 120$

**8.** There are $\dfrac{12!}{10!2!} = 66$ arrangements of the 10 wins and 2 ties.

There are 13 locations (between, before, after) for the losses to occur.

 

As the losses are considered identical, there are $\frac{13 \times 12 \times 11}{3!} = 429$ arrangements.

In total, the number of arrangements for the results is $66 \times 429 = 28\ 314$.

**9.** Number of hands
$= ({}_{13}C_3)({}_{13}C_2)$
$= (286)(78)$
$= 22\ 308$

**10.** Number of numbers $= \frac{2 \times 4 \times 3 \times 2 \times 1}{2!} = 24$

**11.** Team $A$ may have 0, 1, 2, 3, 4, 5, or 6 goals.
Team $B$ may have 0, 1, 2, or 3 goals.

The number of possible half-time scores
$= 7 \times 4 = 28$

**12.** Number of arrangements $= {}_4P_4 \times {}_5P_4 = 2\ 880$

**13.** Note that for each coin there are 2 choices: either select the coin or not. You need to subtract because selecting no coins is not considered a sum of money.

Number of different sums
$= 2 \times 2 \times 2 \times 2 \times 2 - 1 = 31$

**14.** Number of 3-digit even numbers = those that end in 0 + those that end in 2 or 4.
$= (4 \times 3 \times 1) + (3 \times 3 \times 2) = 30$

## *Lesson 8 Applications of Combinations and Permutations*

### CLASS EXERCISES ANSWERS AND SOLUTIONS

**1.** This question is very similar to the handshake question, but instead of selecting 2 people to shake hands, we are selecting 2 teams to play a game of soccer. Also, instead of each pair of teams playing just 1 game, they are playing 3 games.

So, we are interested in counting combinations of 10 teams taken 2 at a time. There are ${}_{10}C_2 = 45$ such combinations.

If each team were to play every other team once, there would be 45 games. But, the question tells us that each team plays every other team 3 times. So, we need to multiply this answer by 3.

We get $3 \times 45 = 135$. There are 135 games in all.

**2.** Although this question seems quite similar to the last 2 questions, it is a bit different.

In the first 2 questions, the order did not matter.

However, order matters in the case of gift-giving, because one of the people is the giver and the other is the receiver.

Since order matters, we are dealing with a permutation question.

So, to count the number of gifts given, we need to count the number of permutations of 7 participants taken 2 at a time. There are ${}_7P_2 = 42$ such permutations.

So, there are 42 gifts given in the exchange.

**3.** To solve this problem, we need to recognize that a triangle can be formed by selecting any 3 points from the 9 and connecting them. Does the order matter here?

No, it does not matter what order we select the points—we will get the same triangle in each case.

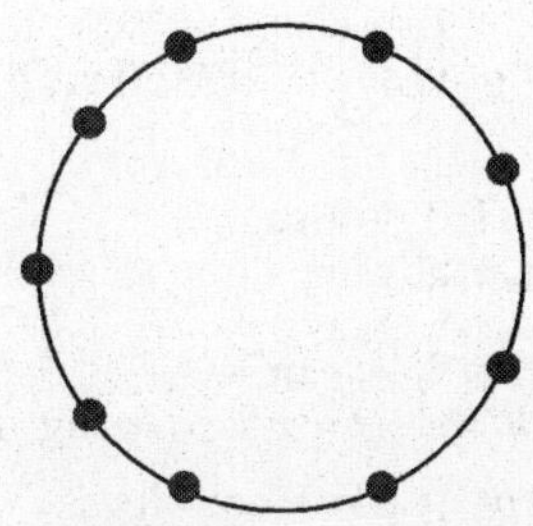

Since order does not matter, we are interested in counting combinations of 3 points selected from 9. There are ${}_9C_3 = 84$ such combinations.

So, it is possible to form 84 different triangles using 9 points on the circumference of a circle.

**4.** In this question, to get from point $A$ to point $B$, you must travel 2 blocks down, 3 blocks right and 1 block back.

If we let D denote 1 block down, R denote 1 block right, and B denote 1 block back, we can think of a route as consisting of 2 Ds, 3 Rs and 1 B.

For example, one possible route could be DRRBRD (as shown here).

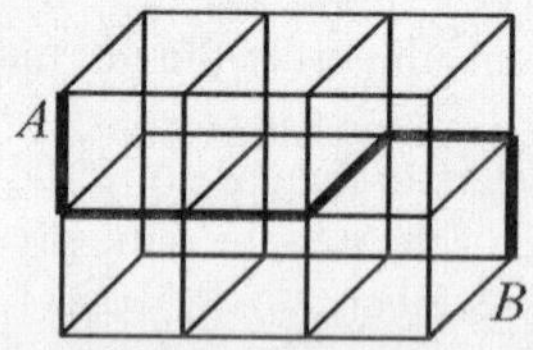


How many different arrangements are possible for the movements of 2 Ds, 3 Rs and 1 B?

There are $\dfrac{6!}{3!2!1!} = 60$ arrangements and thus, 60 different routes.

## PRACTICE EXERCISES ANSWERS AND SOLUTIONS

**1.** ${}_{20}C_2 = \dfrac{20!}{(20-2)!2!} = 190$

There are 190 different groups possible.

**2.** Each group of 5 points will produce a pentagon.

Thus, there are ${}_{12}C_5 = \dfrac{12!}{(12-5)!5!} = 792$ pentagons possible.

**3.** Using the 7 vertices of the heptagon as points, ${}_7C_2 = \dfrac{7!}{(7-2)!2!} = 21$ lines can be created.

Since 7 of these lines are sides, there are $21 - 7 = 14$ diagonals (the lines that are not sides are diagonals).

**4.** First, calculate the number of routes from $A$ to $C$.

$\dfrac{6!}{4!2!} = 15$ (stage 1)

Then, calculate the number of routes from $C$ to $B$.

$\dfrac{6!}{3!3!} = 20$ (stage 2)

The complete routes must have both stage 1 and stage 2. Thus, using the Fundamental Counting Principle, there are $15 \times 20 = 300$ different routes.

**5.** Each group of 2 represents a different game.

Thus, there are ${}_{13}C_2 = \dfrac{13!}{(13-2)!2!} = 78$ different games possible.

If each team played every other team once, there would be 78 games. Since every team plays every other team twice, there will be $78 \times 2 = 156$ games.

**6.** There are ${}_9C_2 = \dfrac{9!}{(9-2)!7!} = 36$ groups of 2.

Each group of 2 represents 2 gift exchanges. For example, Mary gives to Jane and Jane gives to Mary.

Thus, there are $36 \times 2 = 72$ gifts given. Notice that this answer can also be obtained from ${}_9P_2$ because order is counted.

 

7. Every group of 4 points will produce a quadrilateral. Thus, there are ${}_nC_4$ quadrilaterals.

Here, ${}_nC_4 = 495$. Solve for $n$:

$$\frac{n!}{(n-4)!4!} = 495$$
$$\frac{n(n-1)(n-2)(n-3)(n-4)!}{(n-4)!24} = 495$$
$$\frac{n(n-1)(n-2)(n-3)}{24} = 495$$
$$n(n-1)(n-2)(n-3) = 11\,880$$

This is a 4th degree equation.

One method to solve this equation is by graphing.

Graph $Y_1 = n(n-1)(n-2)(n-3)$ and $Y_2 = 11\,880$.

Determine the point of intersection. This occurs when $n = 12$. Thus, 12 points will produce 495 quadrilaterals. Alternatively, you could try values of $n$ to solve ${}_nC_4 = 495$.

8. To move from point $A$ to point $B$, each move consists of 6 "blocks" left and 4 "blocks" down. The moves can be represented by the number of arrangements of 6 Ls and 4 Ds.

This is $\frac{10!}{6!4!} = 210$.

9. Using the 8 vertices of the octagon as points there are ${}_8C_2 = \frac{8!}{(8-2)!2!} = 28$ lines.

Since 8 of these lines are sides, there are $28 - 8 = 20$ diagonals. (The lines that are not sides are diagonals.)

10. Every group of 6 points will produce a hexagon. Thus, ${}_{15}C_6 = \frac{15!}{(15-6)!6!} = 5\,005$ hexagons can be formed.

11. Every group of 2 people represents a handshake.

Thus, there are ${}_{18}C_2 = \frac{18!}{(18-2)!2!} = 153$ different handshakes.

12. Let D denote a move of 1 block down, R denote a move of 1 block right, and F denote a move of 1 block forward.

The number of arrangements of 2 Ds, 6 Rs and 1 F in a line is the answer.

This is $\frac{9!}{2!6!1!} = 252$.

13. Let $n$ represent the number of teams. Each group of 2 represents a game situation. Thus, there are ${}_nC_2$ different game possibilities. Since each team played every other team twice, there were $2 \times {}_nC_2$ games.

$$2 \times {}_nC_2 = 110$$
$${}_nC_2 = 55$$
$$\frac{n!}{(n-2)!2!} = 55$$
$$\frac{n(n-1)(n-2)!}{(n-2)!2} = 55$$
$$\frac{n(n-1)}{2} = 55$$
$$n(n-1) = 110$$
$$n^2 - n = 110$$
$$n^2 - n - 110 = 0$$
$$(n-11)(n+10) = 0$$
$$n-11 = 0 \text{ or } n+10 = 0$$
$$n = 11 \text{ or } n = -10$$

$n = 11$ since $n = -10$ is extraneous ($n$ must be positive)

Thus, there were 11 teams.

14. Each group of 2 represents a possible match.

Thus, there will be ${}_{40}C_2 = \frac{40!}{(40-2)!2!} = 780$ matches.

**15.** Using the 10 vertices of the decagon as points, the number of lines that can be produced is

$${}_{10}C_2 = \frac{10!}{(10-2)!2!} = 45.$$

Since 10 of these lines are sides, there are $45 - 10 = 35$ diagonals. (The lines that are not sides are diagonals.)

**16.** Let E denote travelling 1 block east, and let N denote travelling 1 block north. The number of different routes is the number of arrangements of 3 Es and 4 Ns in a line. This is $\frac{9!}{3!4!} = 126.$

**17.** Since 4 points are required to form a quadrilateral, the number of quadrilaterals is

$${}_7C_4 = \frac{7!}{(7-4)!4!} = 35.$$

**18. A**

A polygon with n sides has ${}_nC_2 - n$ diagonals.

$$\begin{aligned}
{}_nC_2 - n &= 35\\
\frac{n!}{(n-2)!2!} - n &= 35\\
\frac{(n)(n-1)}{2} - n &= 35\\
n^2 - n - 2n &= 70\\
n^2 - 3n - 70 &= 0\\
(n-10)(n+7) &= 0\\
n = 10 \text{ or } n &= -7
\end{aligned}$$

(which is extraneous because $n$ must be positive)

**19. C**

Each of the boys gives 5 gifts.

The 5 boys give a total of $5 \times 5 = 25$ gifts.

Similarly, the girls give 25 gifts.

The total number of gifts given is $25 + 25 = 50$.

## *Lesson 9 Tricky Pathway Problems*

### CLASS EXERCISES ANSWERS AND SOLUTIONS

**1.** Because this question involves an irregular-shaped grid, it cannot be easily solved using permutations of the letters representing directions. Instead, we will solve it by determining the number of paths to each intersection.

Once again, we will label each intersection, starting from point $A$.

There is only 1 way to move from point $A$ to each of the intersections that are labelled with a 1.

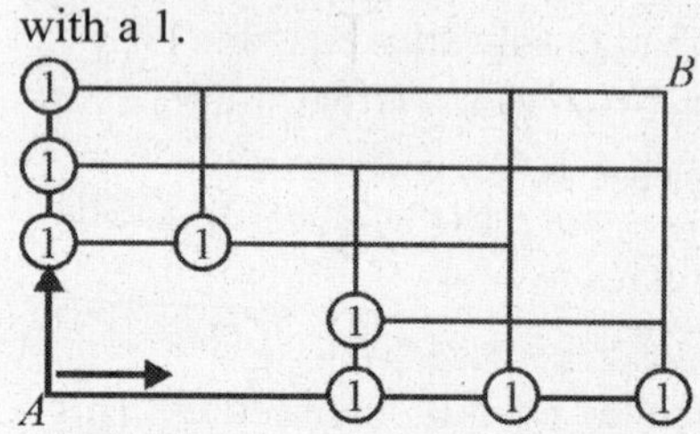

To label the remaining intersections, we will simply add the values at the two intersections that are directly connected to it.

We get the following:

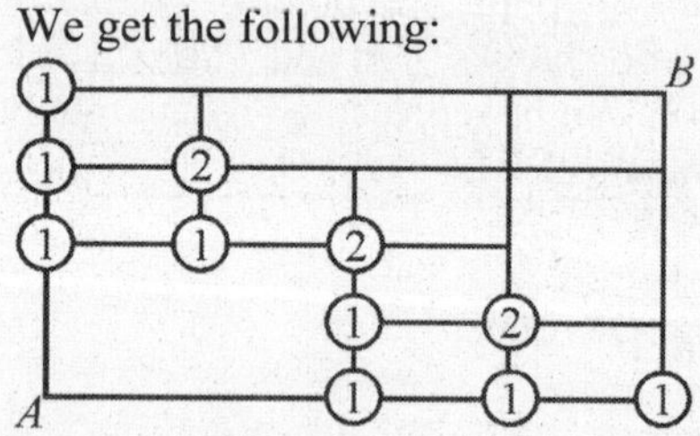

Continuing in the same manner, we get:

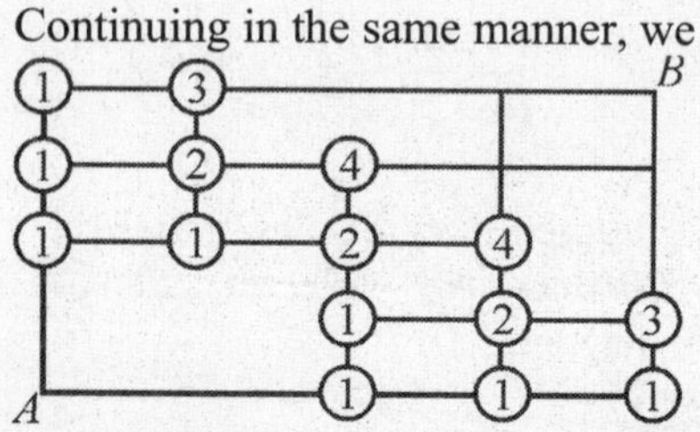

And then this:

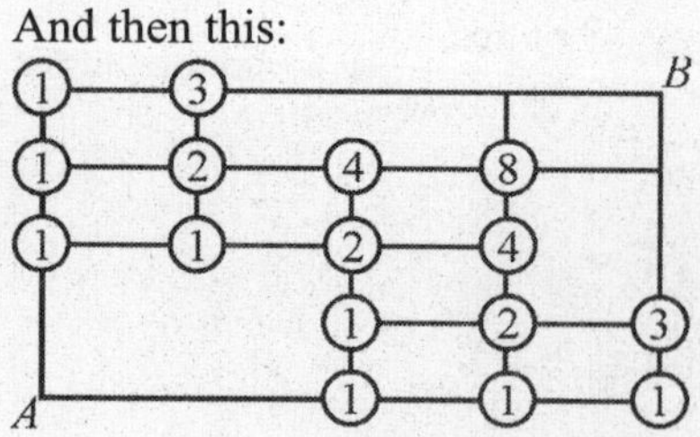

And finally this:

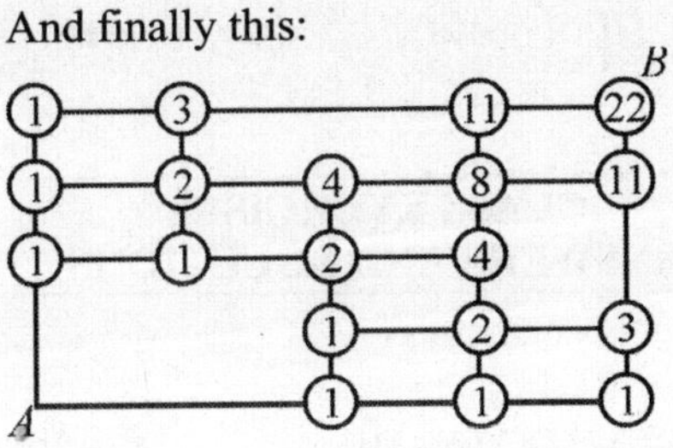


There are 22 routes from point *A* to point *B*.

## PRACTICE EXERCISES ANSWERS AND SOLUTIONS

**1.**

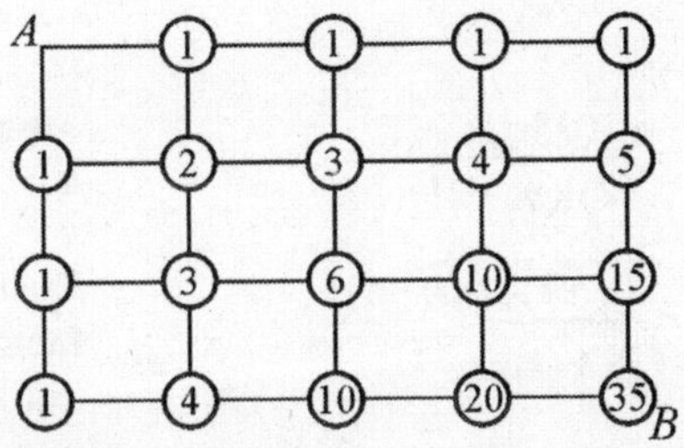


There are 35 routes.

**2.**

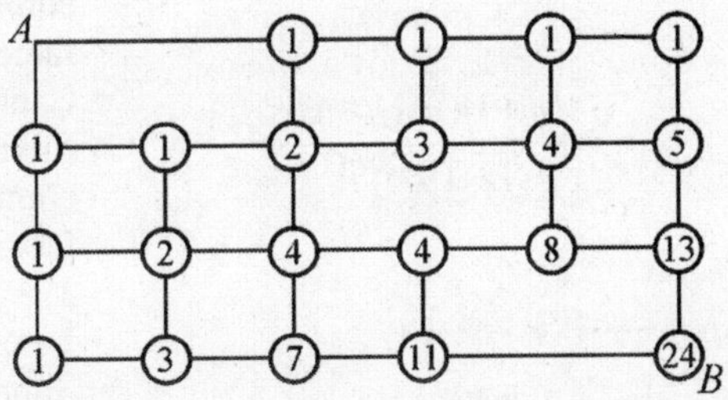


There are 24 routes.

**3.**

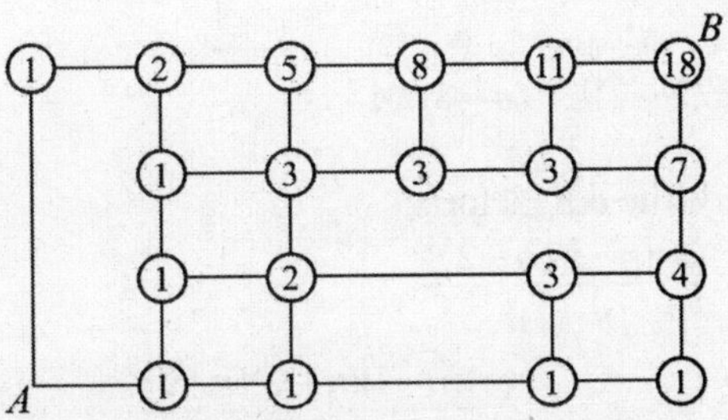


There are 18 routes.

**4.**

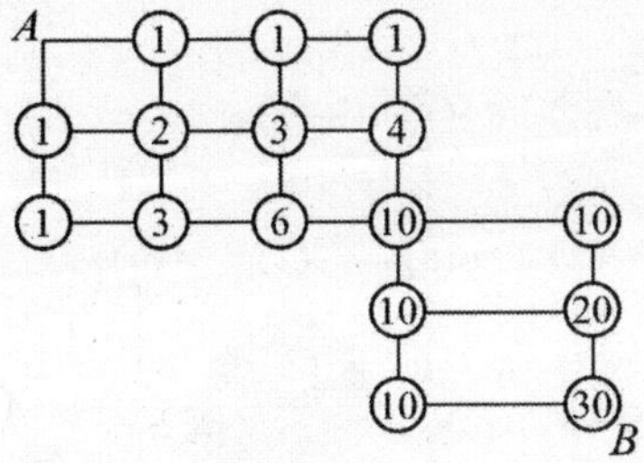


There are 30 routes.

**5.**

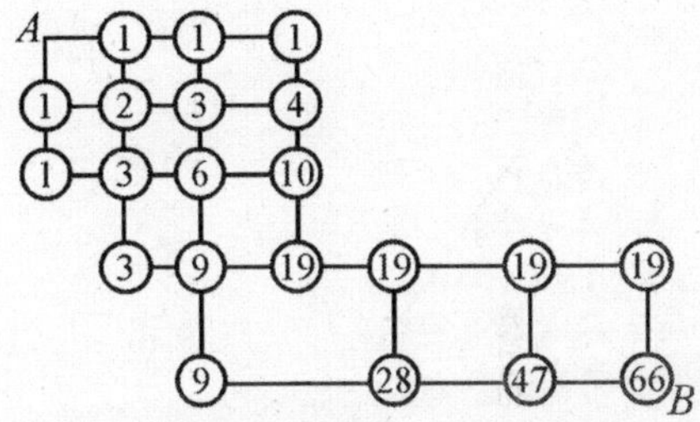


There are 66 routes.

**6.**

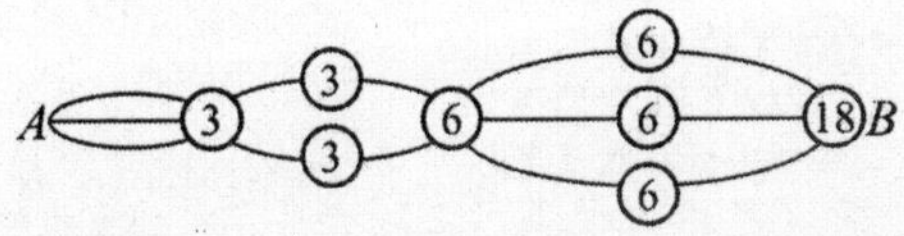


Note that in this case, the number of routes is $3 \times 2 \times 3 = 18$.

**7.**

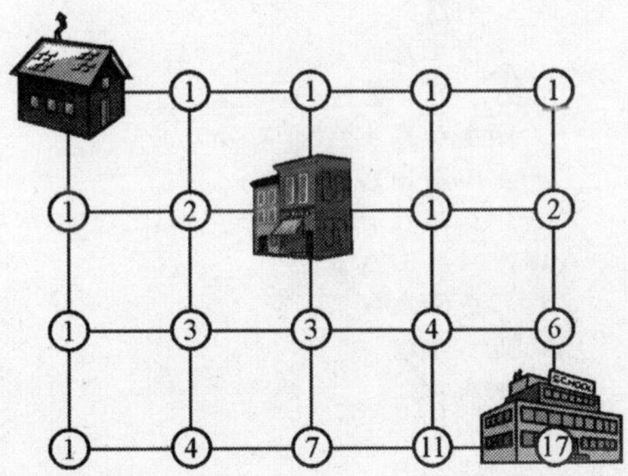


There are 17 paths.

**8.**

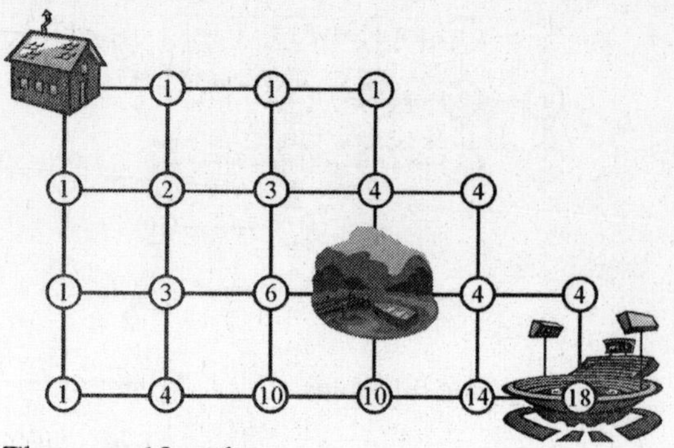

There are 18 paths.

**9.**

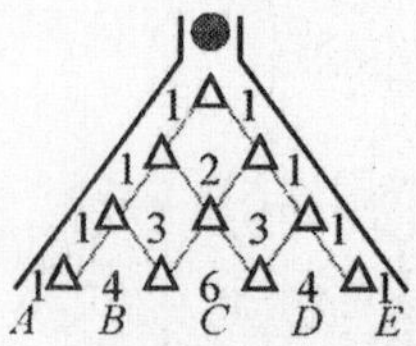


**10.**

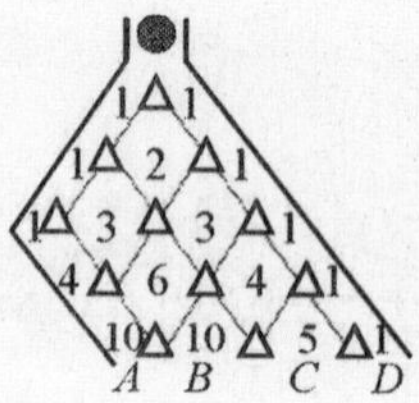


**11.**

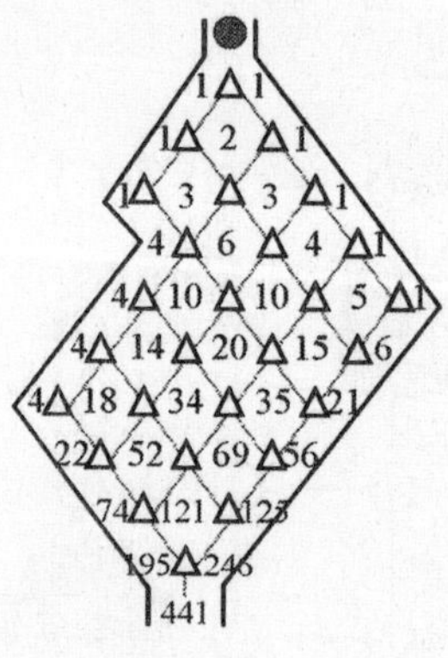


## *Lesson 10 Algebraic Calculations With $_nP_r$ and $_nC_r$*

### CLASS EXERCISES
### ANSWERS AND SOLUTIONS

**1.**

$$_{n+1}C_{n-1} = \frac{(n+1)!}{((n+1)-(n-1))!(n-1)!}$$

- Apply the formula for $_nC_r$

$$= \frac{(n+1)!}{(n+1-n+1)!(n-1)!}$$

- Remove inner brackets in $((n+1)-(n-1))$

$$= \frac{(n+1)!}{(2)!(n-1)!}$$

- Simplify $n+1-n+1=2$

$$= \frac{(n+1)(n)(n-1)!}{(2)!(n-1)!}$$

- Write out factorials until numerator and denominator have an equal factorial

$$= \frac{(n+1)(n)\cancel{(n-1)!}}{(2)\cancel{(n-1)!}}$$

- Cancel (divide out common factorials) $2! = 2 \times 1 = 2$

$$= \frac{n^2+n}{2}$$

- Simplify

**2.** Applying the formulas for $_nP_r$ and $_nC_r$:

$$\frac{n!}{(n-2)!} = \frac{n!}{(n-3)!3!}$$

Write out factorials

$$\frac{(n)(n-1)(n-2)!}{(n-2)!} = \frac{(n)(n-1)(n-2)(n-3)!}{(n-3)!\,3!}$$

Cancel out the common factors on each side:

$$\frac{(n)(n-1)\cancel{(n-2)}!}{\cancel{(n-2)}!} = \frac{(n)(n-1)(n-2)\cancel{(n-3)}!}{\cancel{(n-3)}!\,(3)(2)(1)}$$

 

Simplify:

$$\frac{(n)(n-1)}{1}=\frac{(n)(n-1)(n-2)}{(3)(2)(1)}$$

Divide both sides by $(n)(n-1)$:

$$\frac{\cancel{(n)}\,\cancel{(n-1)}}{1}=\frac{\cancel{(n)}\,\cancel{(n-1)}\,(n-2)}{(3)(2)(1)}$$

Simplify:

$$\frac{1}{1}=\frac{(n-2)}{6}$$

Equate the cross-products:

$n-2=6$

Solve:

$n=8$

3. Apply the formulas for ${}_nP_r$ and ${}_nC_r$

$$\frac{n!}{(n-3)!}=2\left(\frac{n!}{(n-4)!4!}\right)$$

Divide both sides by $n!$

$$\frac{1}{(n-3)!}=2\left(\frac{1}{(n-4)!4!}\right)$$

Write out factorials until denominators have an equal factorial

$$\frac{1}{(n-3)(n-4)!}=\frac{2}{(n-4)!(4\times3\times2\times1)}$$

Multiply both sides by $(n-4)!$

$$\frac{1}{(n-3)}=\frac{2}{(4\times3\times2\times1)}$$

Simplify

$$\frac{1}{(n-3)}=\frac{\cancel{2}^{1}}{(12\times\cancel{2}\times1)}$$

Equate the cross-products

$n-3=12$

Solve

$n=15$

## PRACTICE EXERCISES ANSWERS AND SOLUTIONS

For all problems, use the formulas

$${}_nP_r=\frac{n!}{(n-r)!}\text{ and }{}_nC_r=\frac{n!}{(n-r)!r!}$$

1.

$$({}_7P_4)-({}_7C_3)=\frac{7!}{(7-4)!}-\frac{7!}{(7-3)!3!}=\frac{7!}{3!}-\frac{7!}{4!3!}=805$$

2.

$${}_{n-1}P_2=\frac{(n-1)!}{((n-1)-2)!}=\frac{(n-1)!}{(n-3)!}=\frac{(n-1)(n-2)(n-3)!}{(n-3)!}$$
$$=n^2-3n+2$$

3.

$${}_nP_3=\frac{n!}{(n-3)!}=\frac{n\times(n-1)\times(n-2)\times(n-3)!}{(n-3)!}$$
$$=n^3-3n^2+2n$$

4.

$${}_{n+3}C_{n+1}=\frac{(n+3)!}{((n+3)-(n+1))!(n+1)!}=\frac{(n+3)!}{(2)!(n+1)!}$$
$$=\frac{(n+3)(n+2)(n+1)!}{2(n+1)!}=\frac{n^2+5n+6}{2}$$

5.

$$\frac{{}_nP_{n-2}}{{}_{n+2}P_n}=\frac{\dfrac{n!}{(n-(n-2))!}}{\dfrac{(n+2)!}{((n+2)-n)!}}=\frac{\dfrac{n!}{2!}}{\dfrac{(n+2)!}{2!}}$$
$$=\frac{n!}{2}\cdot\frac{2}{(n+2)!}=\frac{n!}{(n+2)!}$$
$$=\frac{n!}{(n+2)(n+1)(n)!}=\frac{1}{n^2+3n+2}$$

6.

$$({}_{n+1}C_n)({}_nP_{n-1})=\frac{(n+1)!}{((n+1)-n)!n!}\times\frac{n!}{(n-(n-1))!}$$
$$=\frac{(n+1)!}{1!n!}\times\frac{n!}{1!}=(n+1)!$$

**7.**

$${}_nC_0 + {}_nC_1 + {}_nC_{n-1} + {}_nC_n$$

$$= \frac{n!}{(n-0)!0!} + \frac{n!}{(n-1)!1!} + \frac{n!}{(n-(n-1))!(n-1)!} + \frac{n!}{(n-n)!(n)!}$$

$$= \frac{n!}{n! \times 1} + \frac{(n)(n-1)!}{(n-1)! \times 1} + \frac{(n)(n-1)!}{(1)(n-1)!} + \frac{n!}{(0!)(n)!}$$

$$= 1 + n + n + 1 = 2n + 2$$

**8.**

$${}_4C_0 + {}_4C_1 + {}_4C_2 + {}_4C_3 + {}_4C_4$$

$$= \frac{4!}{(4-0)!0!} + \frac{4!}{(4-1)!1!} + \frac{4!}{(4-2)!2!} + \frac{4!}{(4-3)!3!} + \frac{4!}{(4-4)!4!}$$

$$= \frac{4!}{4! \times 1} + \frac{4!}{3! \times 1!} + \frac{4!}{2! \times 2!} + \frac{4!}{1! \times 3!} + \frac{4!}{1 \times 4!}$$

$$= 1 + 4 + 6 + 4 + 1 = 16$$

**9.**

$${}_5C_0 + {}_5C_1 + {}_5C_2 + {}_5C_3 + {}_5C_4 + {}_5C_5$$

$$= \frac{5!}{(5-0)!0!} + \frac{5!}{(5-1)!1!} + \frac{5!}{(5-2)!2!} + \frac{5!}{(5-3)!3!} + \frac{5!}{(5-4)!4!} + \frac{5!}{(5-5)!5!}$$

$$= \frac{5!}{5! \times 1} + \frac{5!}{4! \times 1!} + \frac{5!}{3!2!} + \frac{5!}{2!3!} + \frac{5!}{1!4!} + \frac{5!}{1 \cdot 5!}$$

$$= 1 + 5 + 10 + 10 + 5 + 1 = 32$$

**10.**

$$\sum_{r=0}^{6} {}_6C_r = {}_6C_0 + {}_6C_1 + {}_6C_2 + {}_6C_3 + {}_6C_4 + {}_6C_5 + {}_6C_6$$

$$= \frac{6!}{(6-0)!0!} + \frac{6!}{(6-1)!1!} + \frac{6!}{(6-2)!2!} + \frac{6!}{(6-3)!3!} + \frac{6!}{(6-4)!4!} + \frac{6!}{(6-5)!5!} + \frac{6!}{(6-6)!6!}$$

$$= \frac{6!}{6! \times 1} + \frac{6!}{5! \times 1!} + \frac{6!}{4!2!} + \frac{6!}{3!3!} + \frac{6!}{2!4!} + \frac{6!}{1! \times 5!} + \frac{6!}{1 \times 6!}$$

$$= 1 + 6 + 15 + 20 + 15 + 6 + 1 = 64$$

**11.**

$${}_nP_4 = 12({}_nC_3)$$

$$\frac{n!}{(n-4)!} = 12\left(\frac{n!}{(n-3)!3!}\right)$$

$$\frac{(n)(n-1)(n-2)(n-3)(n-4)!}{(n-4)!} = 12\left(\frac{(n)(n-1)(n-2)(n-3)!}{(n-3)!3!}\right)$$

$$(n)(n-1)(n-2)(n-3) = 12\left(\frac{(n)(n-1)(n-2)}{6}\right)$$

$$n - 3 = 2$$

$$n = 5$$

- Divide both sides by $(n)(n-1)(n-2)$

**12.**

$${}_nP_4 = 4({}_{n-1}P_3)$$

$$\frac{n!}{(n-4)!} = 4\left(\frac{(n-1)!}{((n-1)-3)!}\right)$$

$$\frac{(n)(n-1)(n-2)(n-3)(n-4)!}{(n-4)!} = 4\left(\frac{(n-1)(n-2)(n-3)(n-4)!}{(n-4)!}\right)$$

$$(n)(n-1)(n-2)(n-3) = 4(n-1)(n-2)(n-3)$$

$$n = 4$$

- Divide both sides by $(n-1)(n-2)(n-3)$

**13.**

$${}_nC_6 = {}_nC_5$$

$$\frac{n!}{(n-6)!6!} = \frac{n!}{(n-5)!5!}$$

$$\frac{n!}{(n-6)!(6)(5!)} = \frac{n!}{(n-5)(n-6)!5!}$$

$$\frac{1}{6} = \frac{1}{n-5}$$

$$n - 5 = 6$$

$$n = 11$$

- Divide and multiply both sides by common factors
- Equate the cross-products

**14.**

$${}_3P_r = 3!$$

$$\frac{3!}{(3-r)!} = 3!$$

$$(3-r)! = 1$$

$3 - r = 0$ or

$3 - r = 1$

$r = 3$ or $r = 2$

- Because 0! = 1 and 1! = 1

**15.**
$$_6C_r = 20$$
$$\frac{6!}{(6-r)!r!} = 20$$
$$\frac{6\cdot5\cdot4\cdot3\cdot2\cdot1}{(6-r)!r!} = 20$$
$$\frac{6\cdot3\cdot2\cdot1}{(6-r)!r!} = 1$$
$$(6)(6) = (6-r)!r!$$

By inspection, $r = 3$.

**16.**
$$_{n+1}C_8 = {}_nC_7$$
$$\frac{(n+1)!}{((n+1)-8)!8!} = \frac{n!}{(n-7)!7!}$$
$$\frac{(n+1)(n)!}{(n-7)!(8)(7)!} = \frac{n!}{(n-7)!7!}$$

$$\frac{n+1}{8} = \frac{1}{1}$$
$$n+1 = 8$$
$$n = 7$$

- Multiply and divide both sides by common factors

**17.**
$$_nP_2 = 10\,100$$
$$\frac{n!}{(n-2)!} = 10\,100$$
$$(n)(n-1) = 10\,100$$
$$n^2 - n - 10\,100 = 0$$
$$(n-101)(n+100) = 0$$

$n = 101$ or $n = -100$

$n$ must be $\geq 2$, thus $n = 101$

**18.**
$$_{n+2}P_4 = 72\left(_nP_2\right)$$
$$\frac{(n+2)!}{((n+2)-4)!} = 72\left(\frac{n!}{(n-2)!}\right)$$
$$\frac{(n+2)(n+1)(n!)}{(n-2)!} = 72\left(\frac{n!}{(n-2)!}\right)$$

$$(n+2)(n+1) = 72$$

- Multiply and divide both sides by common factors

$$n^2 + 3n + 2 = 72$$
$$n^2 + 3n - 70 = 0$$
$$(n+10)(n-7) = 0$$

$n = -10$ or $n = 7$

$n$ must be $\geq 2$, thus $n = 7$

**19.** Prove $_nC_r = {}_nC_{n-r}$

| LHS | RHS |
|---|---|
| $\frac{n!}{(n-r)!r!}$ | $\frac{n!}{(n-(n-r))!(n-r)!}$ |
| | $\frac{n!}{r!(n-r)!}$ |

**20.** $_{23}C_7 = {}_{23}C_b$

From the identity in number 19, $_nC_r = {}_nC_{n-r}$.

Here, $n = 23$, $r = 7$, $b = n - r = 23 - 7 = 16$

**21.** $_nC_{15} = {}_nC_{42}$

Using $_nC_r = {}_nC_{n-r}$,

$r = 15$, $n - r = 42$

$n - 15 = 42$

Thus, $n = 57$

## *Lesson 11 Questions With the Phrases "At Least" or "At Most"*

### CLASS EXERCISES
### ANSWERS AND SOLUTIONS

**1.** The direct approach for this question would be to consider teams with the number of girls being 2 or 3 or 4 or 5 or 6 or 7. This would involve 6 different cases to calculate the sum.

There is a better way.

The team has at least 2 girls or, the only other two possibilities, it has 0 girls, or 1 girl.

If we calculate the number of possible teams in total (no restrictions on the membership) and then subtract the number of teams with 0 girls or 1 girl, the result will be what is required.

Number of teams with at least 2 girls = (Total Number) – (Number with 0 girls or with 1 girl)

Total number possible = $_{16}C_7 = 11\ 440$

- Number of teams if the 7 members were chosen from any of the 16 (boys or girls)

Number with 0 girls = $_8C_7 = 8$

- Number of teams if the 7 members were chosen from the 8 boys

Number with 1 girl = $(_8C_1)(_8C_6)$
$= (8)(28) = 224$

- From 8 girls, choose 1 and from 8 boys, choose 6

Number with at least 2 girls = (Total Number) – (Number with 0 girls or with 1 girl) =
$(_{16}C_7) - (_8C_7 + (_8C_1)(_8C_6))$
$= 11\ 440 - (8 + 224)$
$= 11\ 440 - 232 = 11\ 208$

2. There are 52 cards in a standard deck, 13 of them are clubs and, 52 – 13 = 39 of them are not clubs.

"At most 2 clubs" means either 0 clubs or 1 club or 2 clubs. We will consider each of these three cases separately.

Number of hands with 0 clubs =
$_{39}C_5 = 575\ 757$

- If there are no clubs, then the 5 cards are chosen from the 39 other cards (non-clubs).

Number of hands with 1 club =
$(_{13}C_1)(_{39}C_4)$
$= (13)(82\ 251)$
$= 1\ 069\ 263$

- If there is 1 club, then it is chosen from the 13 clubs and the 4 other cards are chosen from the 39 non-clubs.

Number of hands with 2 clubs =
$(_{13}C_2)(_{39}C_3)$
$= (78)(9\ 139)$
$= 712\ 842$

- If there are 2 clubs, then they are chosen from the 13 clubs and the 3 other cards are chosen from the 39 non-clubs.

Total number of hands =

- The hands have 0 clubs or 1 club or 2 clubs, so the total is the SUM of the different hand types.

$$\text{Total number of hands} = {}_{39}C_5 + (_{13}C_1)(_{39}C_4) + (_{13}C_2)(_{39}C_3)$$
$$= 575\ 757 + 1\ 069\ 263 + 712\ 842$$
$$= 2\ 357\ 862$$

3. There are 3 ways to choose turnips (choose 0, 1, or 2), 4 ways to choose carrots, 3 ways to choose potatoes, and 2 ways to choose parsnips. Choosing none of them is not a selection so, the number of selections of the vegetables is 3 × 4 × 3 × 2 – 1 = 71. There is a subtraction of 1 to indicate the invalid selection.

## PRACTICE EXERCISES ANSWERS AND SOLUTIONS

1. **a)** Number of different teams with at least one boy = (Total number of teams) – (Number of teams with no boys) = $_{13}C_6 - {}_7C_6 = 1\ 716 - 7 = 1\ 709$

**b)** Number of teams with at most one boy = (Number of teams with 0 boys) + (Number of teams with 1 boy)
$= {}_7C_6 + (_6C_1)(_7C_5)$
$= 7 + (6)(21) = 133$

**c)** Number of teams with at least 2 girls = (Total number of teams) – (Number of teams with 0 or 1 girl(s))
$= {}_{13}C_6 - (_6C_6 + (_6C_5)(_7C_1))$
$= 1\ 716 - (1 + 42)$
$= 1\ 716 - (43) = 1\ 673$

**d)** Number of teams with at least 5 girls = Number of teams with 5 or 6 girls
$= {}_7C_5 + {}_7C_6$
$= 21 + 7 = 28$

**e)** Number of teams with at least 6 boys $= {}_6C_6 = 1$

2. **a)** Number of hands with no restrictions
$= {}_{52}C_5 = 2\ 598\ 960$

**b)** Number of hands with at least one ace = (Total number of hands) – (Number of hands with no aces)
$= {}_{52}C_5 - {}_{48}C_5$
$= 2\ 598\ 960 - 1\ 712\ 304 = 886\ 656$

 

**c)** Number of hands with at least 2 hearts
= Number of hands with 2, 3, 4, or 5 hearts
$=({}_{13}C_2)({}_{39}C_3)+({}_{13}C_3)({}_{39}C_2)+({}_{13}C_4)({}_{39}C_1)+{}_{13}C_5$
$=(78)(9\,139)+(286)(741)+(715)(39)+1\,287$
$=953\,940$

**d)** Number of hands with at most 2 hearts
= Number of hands with 0, 1, or 2 hearts
$={}_{39}C_5+({}_{13}C_1)({}_{39}C_4)+({}_{13}C_2)({}_{39}C_3)$
$=575\,757+(13)(82\,251)+(78)(9\,139)$
$=2\,357\,862$

**e)** Number of hands with at least 4 hearts
= Number of hands with 4 or 5 hearts
$=({}_{13}C_4)({}_{39}C_1)+{}_{13}C_5$
$=(715)(39)+1\,287$
$=29\,172$

**f)** Number of hands with at most 2 black cards
= Number of hands with 0, 1, or 2 black cards
$={}_{26}C_5+({}_{26}C_1)({}_{26}C_4)+({}_{26}C_2)({}_{26}C_3)$
$=65\,780+(26)(14\,950)+(325)(2\,600)$
$=1\,299\,480$

**g)** The number of hands with at least one ace and one king. The possibilities are:
1 ace, 1 king, 3 others:
$({}_4C_1)({}_4C_1)({}_{44}C_3)=211\,904$

1 ace, 2 kings, 2 others:
$({}_4C_1)({}_4C_2)({}_{44}C_2)=22\,704$

1 ace, 3 kings, 1 other:
$({}_4C_1)({}_4C_3)({}_{44}C_1)=704$

1 ace, 4 kings:
$({}_4C_1)({}_4C_4)=4$

2 aces, 1 kings, 2 others:
$({}_4C_2)({}_4C_1)({}_{44}C_2)=22\,704$

2 aces, 2 kings, 1 other:
$({}_4C_2)({}_4C_2)({}_{44}C_1)=1\,584$

2 aces 3 kings:
$({}_4C_2)({}_4C_3)=24$

3 aces 1 king, 1 other:
$({}_4C_3)({}_4C_1)({}_{44}C_1)=704$

3 aces 2 kings:
$({}_4C_3)({}_4C_2)=24$

4 aces 1 king:
$({}_4C_4)({}_4C_1)=4$

Total number of hands = sum = 260 360

**h)** The number of hands with at least one ace or one king = (Total number of hands) – (Number of hands with no ace or king)
$={}_{52}C_5-{}_{44}C_5=2\,598\,960-1\,086\,008$
$=1\,512\,952$

**i)** Total number of hands with at least one ace or one black card = (Total number of hands) – (Number of hands with no aces or black cards)
${}_{52}C_5-{}_{24}C_5=2\,598\,960-42\,504=2\,556\,456$

Note: there are 24 cards that are not black or aces

**3.** There are 4 ways to choose 3 apples (0, 1, 2, 3), 5 ways to choose 4 pears, and 2 ways to choose 1 banana. Choosing none is not a valid selection. Thus, the number of selections is $4\times5\times2-1=39$.

**4.** The word will contain 2 A's or 3 A's. If it's 2 A's, then 2 other letters are chosen.

Possibilities: AANN, which gives $\frac{4!}{2!2!}=6$ words

AABN, which gives $\frac{4!}{2!}=12$ words

If it has 3A's, then 1 other letter is chosen.

Possibilities: AAAN, which gives $\frac{4!}{3!}=4$ words

AAAB, which gives $\frac{4!}{3!}=4$ words

The total number of words is
$6+12+4+4=26$.

**5.** He travels 0, 1, or 2 blocks west.

Number of routes when travelling 0 blocks west is ${}_7C_0 = 1$
Number of routes when travelling 1 block west is ${}_7C_1 = 7$
Number of routes when travelling 2 blocks west is ${}_7C_2 = 21$

Total number of routes $= 1 + 7 + 21 = 29$

**6.** **B**
Number of starting lineups with at least one girl = (Total number of lineups) – (number of lineups with no girls)
$= {}_{11}C_6 - {}_7C_6$
$= 462 - 7 = 455$

**7.** **B**
Number of 4-digit numbers with at least one 3 = (Total number of 4-digit numbers) – (Number of 4-digit numbers with no 3s)
$= 9\times10\times10\times10 - 8\times9\times9\times9$
$= 9\,000 - 5\,832 = 3\,168$

**8.** There are 4 ways to select green marbles
3 ways to select red marbles, and
2 ways to select black marbles.

Choosing none is not a valid selection.

The number of selections is $4\times3\times2-1=23$.

| | | | |
|---|---|---|---|
| 2 | 3 | | |

**9.** Number of committees with at most one man = Number with 0 men or 1 man
$= {}_5C_4 + ({}_3C_1)({}_5C_3)$
$= 5 + (3)(10) = 35$

| | | | |
|---|---|---|---|
| 3 | 5 | | |

## *Lesson 12 Binomial Expansion and Pascal's Triangle*

### CLASS EXERCISES ANSWERS AND SOLUTIONS

**1.**

$(1-x)^{13}$
$t_8 = t_{7+1}$
$t_{7+1} = {}_{13}C_7(1)^{13-7}(-x)^7$
$= 1\,716(1)(-x^7)$
$= -1\,716x^7$

- $8^{\text{th}}$ term is $t_{7+1}$
- Obtain ${}_{13}C_7$ from a calculator
- Simplify

The coefficient is –1 716.

**2.**

$\left(2+\frac{x}{2}\right)^{18}$

- Recall that the number of terms is one more than the power so the expanded form here has 19 terms. The middle term is the $\frac{19+1}{2} = 10^{\text{th}}$ term.

$t_{10} = t_{9+1}$
$= {}_{18}C_9(2)^{18-9}\left(\frac{x}{2}\right)^9$
$= 48\,620(2^9)\left(\frac{x^9}{2^9}\right)$
$= 48\,620x^9$

- The $10^{\text{th}}$ term is $t_{9+1}$.
- Apply the formula $t_{k+1} = {}_nC_k x^{n-k}y^k$
- Obtain ${}_{18}C_9$ from a calculator
- Simplify ($2^9$ divides out)

**3.**

For this expansion, the general term is
$t_{k+1} = (y^2)^{9-k}(-2)^k$.
We require $9-k=2$ to get $y^4$.
Therefore, $k=7$.
$t_{7+1} = {}_9C_7(y^2)^2(-2)^7$
$= 36y^4(-128)$
$= -4\,608y^4$

- Apply the formula $t_{k+1} = {}_nC_k x^{n-k}y^k$
- Notice that this is the $8^{\text{th}}$ term

 

**4.**

| | |
|---|---|
| For this expansion, the general term is $t_{k+1}=(2x)^{15-k}\left(\frac{1}{x^2}\right)^k$. | • Apply the formula $t_{k+1}={}_nC_k x^{n-k}y^k$ |
| $t_{k+1}=\left(2^{15-k}x^{15-k}\right)\left(\frac{1}{x^{2k}}\right)$ <br> $t_{k+1}=2^{15-k}x^{15-3k}$ | • Simplify <br> • Subtract exponents when dividing powers (to get $x^{15-3k}$) |
| For the constant term: <br> $15-3k=0$ <br> $k=5$ <br> Therefore, the constant term is: | • We require $x^{15-3k}=x^0$ to eliminate the $x$ variable |
| $t_{5+1}=2^{15-5}$ <br> $=2^{10}=1024$ | • Notice that the constant term is the 6th term. |

**5.**

| | |
|---|---|
| ${}_9C_0+{}_9C_1+{}_9C_2+{}_9C_3+{}_9C_4+....+{}_9C_9$ <br> $=2^9$ | • This is the result if we consider that we can choose 0 books, 1 book, 2 books,…,9 books |
| Number of selections for choosing at least one book <br> $=2^9-1$ <br> $=512-1$ <br> $=511$ | • We must subtract 1 because choosing 0 books is not considered a valid selection |

## PRACTICE EXERCISES ANSWERS AND SOLUTIONS

**1.** Using Pascal's Triangle to determine the coefficients:

$$(x+3)^4=1x^4+4x^3(3)+6x^2(3)^2+4x(3)^3+1(3)$$
$$=x^4+12x^3+54x^2+108x+81$$

**2.** Using Pascal's Triangle to determine the coefficients:

$$\left(\frac{2}{x}-2\right)^5=1\left(\frac{2}{x}\right)^5+5\left(\frac{2}{x}\right)^4(-2)+10\left(\frac{2}{x}\right)^3(-2)^2+10\left(\frac{2}{x}\right)^2(-2)^3+5\left(\frac{2}{x}\right)(-2)^4+1(-2)^5$$
$$=\frac{32}{x^5}+\left(\frac{-160}{x^4}\right)+\frac{320}{x^3}+\left(\frac{-320}{x^2}\right)+\frac{160}{x}+(-32)$$
$$=\frac{32}{x^5}-\frac{160}{x^4}+\frac{320}{x^3}-\frac{320}{x^2}+\frac{160}{x}-32$$

**3.** Using Pascal's Triangle to determine the coefficients:

$$\left(\frac{x}{3}-3^2\right)^4=1\left(\frac{x}{3}\right)^4+4\left(\frac{x}{3}\right)^4\left(-3^2\right)+6\left(\frac{x}{3}\right)^2\left(-3^2\right)^2+4\left(\frac{x}{3}\right)\left(-3^2\right)^3+1\left(-3^2\right)^4$$
$$=\frac{x^4}{81}-\frac{36x^4}{81}+\frac{486x^2}{9}-\frac{2\,916x}{3}+6\,561$$
$$=\frac{x^4}{81}-\frac{4x^4}{9}+54x^2-972x+6\,561$$

**4.** Using the Binomial Theorem to determine the coefficients:

$$\left(\frac{1}{x}+\frac{1}{y}\right)^6$$
$$={}_6C_0\left(\frac{1}{x}\right)^6+{}_6C_1\left(\frac{1}{x}\right)^5\left(\frac{1}{y}\right)+{}_6C_2\left(\frac{1}{x}\right)^4\left(\frac{1}{y}\right)^2+{}_6C_3\left(\frac{1}{x}\right)^3\left(\frac{1}{y}\right)^3+{}_6C_4\left(\frac{1}{x}\right)^2\left(\frac{1}{y}\right)^4+{}_6C_5\left(\frac{1}{x}\right)\left(\frac{1}{y}\right)^5+{}_6C_6\left(\frac{1}{y}\right)^6$$
$$=\frac{1}{x^6}+\frac{6}{x^5y}+\frac{15}{x^4y^2}+\frac{20}{x^3y^3}+\frac{15}{x^2y^4}+\frac{6}{xy^5}+\frac{1}{y^6}$$

5. Using the Binomial Theorem to determine the coefficients:

$$\left(\frac{1}{2x^2}-4x^5\right)^5$$

$$={}_5C_0\left(\frac{1}{2x^2}\right)^5+{}_5C_1\left(\frac{1}{2x^2}\right)^4\left(4x^5\right)$$

$$+{}_5C_2\left(\frac{1}{2x^2}\right)^3\left(4x^5\right)^2+{}_5C_3\left(\frac{1}{2x^2}\right)^2\left(4x^5\right)^3$$

$$+{}_5C_4\left(\frac{1}{2x^2}\right)\left(4x^5\right)^4+{}_5C_5\left(4x^5\right)^5$$

$$=\frac{1}{32x^{10}}+\frac{20x^5}{16x^2}+\frac{160x^{10}}{8x^6}+\frac{640x^{15}}{4x^4}$$

$$+\frac{1\,280x^{20}}{2x^2}+1\,024x^{25}$$

$$=\frac{1}{32x^{10}}+\frac{5}{4x^3}+20x^4+160x^{11}+640x^{18}+1\,024x^{25}$$

6. Using the Binomial Theorem to determine the coefficients:

$$\left(x^2-2\sqrt{x}\right)^8=\left(x^2-2(x)^{\frac{1}{2}}\right)^8$$

$$={}_8C_0(x^2)^8+{}_8C_1(x^2)^7\left(-2(x)^{\frac{1}{2}}\right)+{}_8C_2(x^2)^6$$

$$\left(-2(x)^{\frac{1}{2}}\right)^2+{}_8C_3(x^2)^5\left(-2(x)^{\frac{1}{2}}\right)^3$$

$$+{}_8C_4(x^2)^4\left(-2(x)^{\frac{1}{2}}\right)^4+{}_8C_5(x^2)^3\left(-2(x)^{\frac{1}{2}}\right)^5$$

$$+{}_8C_6(x^2)^2\left(-2(x)^{\frac{1}{2}}\right)^6$$

$$+{}_8C_7(x^2)\left(-2(x)^{\frac{1}{2}}\right)^7+{}_8C_8\left(-2(x)^{\frac{1}{2}}\right)^8$$

$$=x^{16}+\left(-16x^{\frac{29}{2}}\right)+112x^{13}+\left(-448x^{\frac{23}{2}}\right)$$

$$+1\,120x^{10}+\left(-1\,792x^{\frac{17}{2}}\right)+1\,792x^7$$

$$+\left(-1\,024x^{\frac{11}{2}}\right)+256x^4$$

$$=x^{16}-16x^{\frac{29}{2}}+112x^{13}-448x^{\frac{23}{2}}+1\,120x^{10}$$

$$-1\,792x^{\frac{17}{2}}+1\,792x^7-1\,024x^{\frac{11}{2}}+256x^4$$

7. In the $8^{\text{th}}$ row, the numbers are ${}_7C_0$, ${}_7C_1$, ${}_7C_3$, ${}_7C_4$, ${}_7C_5$, ${}_7C_0$, ${}_7C_6$, ${}_7C_7$.

Thus, the numbers are 1, 7, 21, 35, 35, 21, 7, 1.

8. Using $t_{k+1}={}_nC_kx^{n-k}y^k$, the $8^{\text{th}}$ term occurs when $k=7$.

$$t_{7+1}={}_{10}C_7(x)^{10-7}\left(\frac{2}{y}\right)^7$$

$$=120x^3\left(\frac{2}{y}\right)^7$$

$$=\frac{15\,360x^3}{y^7}$$

9. $\left(x+\sqrt{y}\right)^{10}$

The middle term is the $6^{\text{th}}$ term $\left(\frac{11+1}{2}\right)$.

Thus, $k=5$.

Use the formula:

$$t_{k+1}={}_nC_kx^{n-k}y^k$$

$$t_{5+1}={}_{10}C_5(x)^{10-5}\left(\sqrt{y}\right)^5$$

$$=252x^5y^{\frac{5}{2}}$$

10. $\left(\frac{1}{x^3}-2x\right)^8$

Use the formula $t_{k+1}={}_nC_kx^{n-k}y^k$:

$$t_{k+1}={}_8C_k\left(\frac{1}{x^3}\right)^{8-k}(-2x)^k$$

$$={}_8C_k\frac{1}{\left(x^3\right)^{8-k}}(-2)^k\left(x^k\right)$$

$$={}_8C_k\frac{(-2)^kx^k}{x^{24-3k}}$$

$$={}_8C_k(-2)^kx^{4k-24}$$

Make $4k-24=0$. Thus, $k=6$.

$$t_{6+1}={}_8C_6\left(\frac{1}{x^3}\right)^{8-6}(-2x)^6$$

$$t_{6+1}=28\left(\frac{1}{x^3}\right)^2(-2x)^6=28\left(\frac{1}{x^6}\right)(-2)^6\left(x^6\right)=1\,792$$

11. $(2x-y)^{2p-5}$

If there are 18 terms, then $2p-5=17$, $p=11$.

**12.** Terms are of the form $t_{k+1} = {}_8C_k\left(x\right)^{8-k}\left(b\right)^k$.

For this term, $8-k=3$. Thus $k=5$.

$$ {}_8C_5\left(x\right)^3\left(b\right)^5 = 13\,608x^3 $$
$$ b^5 = \frac{13\,608}{{}_8C_5} = \frac{13\,608}{56} = 243 $$
$$ b = \sqrt[5]{243} = 3 $$

**13.** $\left(x^3-2\right)^{10}$

The general term is
$$ t_{k+1} = {}_{10}C_k\left(x^3\right)^{10-k}\left(-2\right)^k = {}_{10}C_k x^{30-3k}\left(-2\right)^k $$

Make $30-3k=18$.

Thus, $3k=12$, $k=4$.
$$ t_{4+1} = {}_{10}C_4\left(x\right)^{30-3(4)}\left(-2\right)^4 $$
$$ t_{4+1} = 210\left(x\right)^{18}\left(16\right) = 3\,360x^{18} $$

**14.** $\left(2x^2-\frac{1}{x^3}\right)^{10}$

The general term is:
$$ t_{k+1} = {}_{10}C_k\left(2x^2\right)^{10-k}\left(-\frac{1}{x^3}\right)^k $$
$$ t_{k+1} = {}_{10}C_k\left(2\right)^{10-k}\left(-1\right)^k\frac{x^{20-2k}}{x^{3k}} $$
$$ t_{k+1} = {}_{10}C_k\left(2\right)^{10-k}\left(-1\right)^k x^{20-5k} $$

Make $20-5k=5$.
Thus, $5k=15$, $k=3$.
$$ t_{3+1} = {}_{10}C_3\left(2\right)^{10-3}\left(-1\right)^3 x^{20-5(3)} $$
$$ t_4 = 120(2)^7(-1)x^5 $$
$$ t_4 = -15\,360x^5 $$

**15.** $\left(x+y\right)^7$ The sum of the coefficients is ${}_7C_0 + {}_7C_1 + {}_7C_2 ... + {}_7C_7$.

The sum of these numbers is $2^7 = 128$.

**16.** The number of choices for toppings is: ${}_9C_0 + {}_9C_1 + {}_9C_2 ... + {}_9C_9$

This includes 1 choice of no toppings, which is not valid. Thus, the total number of choices is $2^9 - 1 = 511$.

**17.** **C**
$\left(x-2\right)^8$

You want the numerical coefficient of the 4th term, which is the one of degree 5.

Thus, $8-k=5$, $k=3$.
$$ t_{3+1} = {}_8C_3\left(x\right)^{8-3}\left(-2\right)^3 $$
$$ = -448x^5 $$

The coefficient is –448.

**18.** **B**
$\left(x^4-\frac{1}{x}\right)^5$

The general term is:
$$ t_{k+1} = {}_5C_k\left(x^4\right)^{5-k}\left(-\frac{1}{x}\right)^k $$
$$ = {}_5C_k\left(-1\right)^k\frac{x^{20-4k}}{x^k} = {}_5C_k\left(-1\right)^k x^{20-5k} $$

Make $20-5k=0$.

Thus, $k=4$.
$$ t_{4+1} = {}_5C_4\left(-1\right)^4 x^{20-5(4)} = 5 $$

**19.** ${}_7C_0 + {}_7C_1 + {}_7C_2 \ldots + {}_7C_7 = 2^7 = 128$

The number of selections for at least 1 filling is $2^7 - 1 = 127$.

| | | | |
|---|---|---|---|
| 1 | 2 | 7 | |

**20.** $\left(x^2-\frac{1}{x}\right)^{10}$

The 5th term occurs when $k=4$.
$$ t_{4+1} = {}_{10}C_4\left(x^2\right)^{10-4}\left(-\frac{1}{x}\right)^4 $$
$$ t_{4+1} = 210\frac{x^{12}}{x^4} = 210x^8 $$

The coefficient is 210.

| | | | |
|---|---|---|---|
| 2 | 1 | 0 | |

## Lesson 13 Probability Problems Involving Permutations and Combinations

### CLASS EXERCISES ANSWERS AND SOLUTIONS

1. The number of arrangements of the players in a line is ${}_{10}P_{10} = 3\ 628\ 800$. The number of arrangements with 3 particular players together is ${}_8P_8 \times {}_3P_3 = 241\ 920$.

   The probability of the 3 players being together is $\dfrac{241\ 920}{3\ 628\ 800} \doteq 0.667$

2. $P$(3 out of the 5 women winning)

   $= \dfrac{\text{Number of outcomes when choosing 3 from 5}}{\text{Number of outcomes when choosing 3 from 25}}$

   $= \dfrac{{}_5C_3}{{}_{25}C_3} = \dfrac{10}{2\ 300} = \dfrac{1}{230}$

   An alternative solution is to determine the number of possible selections of choosing 3 women from 5, and multiply this by the probability of each winning:

   $P$(3 out of the 5 women winning)

   $= {}_5C_3 \times \dfrac{3}{25} \times \dfrac{2}{24} \times \dfrac{1}{23} = \dfrac{60}{13\ 800} = \dfrac{1}{230}$

3. Probability of 2 calculators with good batteries

   $= \dfrac{\text{Number of outcomes of choosing 2 from 5}}{\text{Number of outcomes of choosing 2 from 8}}$

   $= \dfrac{{}_5C_2}{{}_8C_2} = \dfrac{10}{28} = \dfrac{5}{14}$

   Alternatively, considering the calculators are drawn successively, then the probability of 2 calculators with good batteries

   $= P(\text{1st good}) \times$

   $P(\text{2nd good given that the first is good})$

   $= \dfrac{5}{8} \times \dfrac{4}{7} = \dfrac{5}{14}$

### PRACTICE EXERCISES ANSWERS AND SOLUTIONS

1. **a)** $P(\text{R and R}) = \dfrac{{}_3C_2}{{}_{12}C_2} = \dfrac{3}{66} = \dfrac{1}{22}$

   **b)** $P(\text{Y, W}) = \dfrac{4}{12} \times \dfrac{5}{11} = \dfrac{5}{33}$

   **c)** $P(\text{R and Y}) = \dfrac{{}_3C_1 \times {}_4C_1}{{}_{12}C_2} = \dfrac{12}{66} = \dfrac{2}{11}$

   **d)** $P(\text{at least one R}) = 1 - P(\text{no R})$

   $= 1 - \dfrac{{}_9C_2}{{}_{12}C_2} = 1 - \dfrac{36}{66} = \dfrac{5}{11}$

2. **a)** $P(\text{all women}) = \dfrac{{}_6C_5}{{}_{10}C_5} = \dfrac{6}{252} = \dfrac{1}{42}$

   **b)** $P(\text{2 men, 3 women})$

   $= \dfrac{{}_4C_2 \times {}_6C_3}{{}_{10}C_5} = \dfrac{6 \times 20}{252} = \dfrac{10}{21}$

   **c)** $P(\text{at least one man})$

   $= 1 - P(\text{no men})$

   $= 1 - \dfrac{{}_6C_5}{{}_{10}C_5} = 1 - \dfrac{1}{42} = \dfrac{41}{42}$

   **d)** $P(\text{at most one man})$

   $= P(\text{0 men}) + P(\text{1 man})$

   $= \dfrac{{}_6C_5}{{}_{10}C_5} + \dfrac{{}_4C_1 \times {}_6C_4}{{}_{10}C_5}$

   $= \dfrac{6}{252} + \dfrac{4 \times 15}{252} = \dfrac{11}{42}$

3. **a)** $P(\text{3 aces}) = \dfrac{{}_4C_3 \times {}_{48}C_2}{{}_{52}C_5}$

   **b)** $P(\text{5 of same suit}) = \dfrac{{}_4C_1 \times {}_{13}C_5}{{}_{52}C_5}$

   **c)** $P(\text{at least one queen})$

   $= 1 - P(\text{no queen})$

   $= 1 - \dfrac{{}_{48}C_5}{{}_{52}C_5}$

   **d)** $P(\text{at most one red card})$

   $= P(\text{0 red}) + P(\text{1 red}) = \dfrac{{}_{26}C_5}{{}_{52}C_5} + \dfrac{{}_{26}C_1 \times {}_{26}C_4}{{}_{52}C_5}$

 

**4.** **a)** $P(\text{orange and orange})$
$= \frac{{}_8C_2}{{}_{24}C_2} \doteq 0.101$

**b)** $P(\text{white and green}) = \frac{{}_6C_1 \times {}_{10}C_1}{{}_{24}C_2} \doteq 0.217$

**c)** $P(\text{white, orange}) = \frac{6}{24} \times \frac{8}{23} \doteq 0.087$

**d)** $P(\text{both white or both green})$
$= \frac{{}_6C_2}{{}_{24}C_2} + \frac{{}_8C_2}{{}_{24}C_2}$
$\doteq 0.156$

**e)** $P(\text{at least 1 green})$
$= 1 - P(\text{no green})$
$= 1 - \frac{{}_{14}C_2}{{}_{24}C_2} = 0.670$

**f)** $P(\text{at most one white})$
$= P(0 \text{ white}) + P(1 \text{ white})$
$= \frac{{}_{18}C_2}{{}_{24}C_2} + \frac{{}_6C_1 \times {}_{18}C_1}{{}_{24}C_2}$
$= 0.946$

**5.** **a)** $P(\text{win with one ticket})$
$= \frac{1}{{}_{49}C_6} = \frac{1}{13\,983\,816}$

**b)** $P(\text{win with three tickets})$
$= \frac{3}{{}_{49}C_6} = \frac{3}{13\,983\,816} = \frac{1}{4\,661\,272}$

**6.** $P(\text{H H H T in any order}) = \left(\frac{1}{2}\right)^3 \left(\frac{1}{2}\right)^1 \times \frac{4!}{3!} = \frac{1}{4}$

**7.** **a)** $P(5 \text{ correct})$
$= {}_{10}C_5 \times \left(\frac{1}{2}\right)^5 \times \left(\frac{1}{2}\right)^5 = \frac{252}{1\,024} = \frac{63}{256}$

**b)** $P(\text{at least one correct})$
$= 1 - P(0 \text{ correct})$
$= 1 - \left(\frac{1}{2}\right)^{10}$
$= \frac{1\,023}{1\,024}$

**8.** $P(\text{DOG}) = \frac{1}{{}_4P_3} = \frac{1}{24}$

**9.** $P(\text{two the same colour})$
$= \frac{{}_4C_2 + {}_6C_2 + {}_4C_2}{{}_{14}C_2} = \frac{27}{91}$
$\left(\text{or } \frac{4}{14} \times \frac{3}{13} + \frac{6}{14} \times \frac{5}{13} + \frac{4}{14} \times \frac{3}{13}\right)$

**10.** **a)** $P(\text{both rotten}) = \frac{{}_3C_2}{{}_{12}C_2} = \frac{3}{66} = \frac{1}{22}$

**b)** $P(\text{at least one rotten})$
$= 1 - P(0 \text{ rotten})$
$= 1 - \frac{{}_9C_2}{{}_{12}C_2} = 1 - \frac{36}{66} = \frac{30}{66} = \frac{5}{11}$

**c)** $P(\text{one is good and one is rotten})$
$= \frac{{}_9C_1 \times {}_3C_1}{{}_{12}C_2} = \frac{27}{66} = \frac{9}{22}$

**11.** **a)** $P(\text{Bob, Jane, Lin, Helmut win})$
$= \frac{{}_4C_4}{{}_{12}C_4} = \frac{1}{495}$

**b)** $P(\text{Jane 1}^{\text{st}}\text{, Lin 2}^{\text{nd}}\text{, Bob 3}^{\text{rd}}\text{, Helmut 4}^{\text{th}})$
$= \frac{1}{12} \times \frac{1}{11} \times \frac{1}{10} \times \frac{1}{9} = \frac{1}{11\,880}$

**12.** **a)** $P(\text{two boys and two girls})$
$= \frac{{}_6C_2 \times {}_4C_2}{{}_{10}C_4} = \frac{15 \times 6}{210} = \frac{3}{7} \doteq 0.429$

**b)** $P(\text{at least one girl})$
$= 1 - P(0 \text{ girls})$
$= 1 - \frac{{}_6C_4}{{}_{10}C_4} = 1 - \frac{15}{210} = \frac{13}{14} \doteq 0.929$

**c)** $P(\text{Roxanne and two other girls})$
$= \frac{{}_1C_1 \times {}_3C_2 \times {}_6C_1}{{}_{10}C_4}$
$= \frac{18}{210} = \frac{3}{35} \doteq 0.086$

**13.** $P(\text{all six numbers})$
$= \frac{{}_6P_6}{6^6} = \frac{6!}{6^6} = \frac{5}{324}$
$\left(\text{or } \frac{6}{6} \times \frac{5}{6} \times \frac{4}{6} \times \frac{3}{6} \times \frac{2}{6} \times \frac{1}{6}\right)$

**14.** **D**

$P(\text{vowels together}) = \frac{3! \times 3!}{5!} = \frac{3}{10}$

**15. A**
Stephan has 7 possible partners, all equally likely. $P(\text{partner is Alex}) = \frac{1}{7}$.

**16.** $P(\text{TOOTH}) = \dfrac{1}{\frac{5!}{2!2!}} = \dfrac{1}{30} \doteq 0.03$

| | | | |
|---|---|---|---|
| 0 | . | 0 | 3 |

## *Practice Test*

### ANSWERS AND SOLUTIONS

**1.** Number of arrangements $= 5! \times 4! = 2\ 880$

**2.** Number of arrangements $= \dfrac{9!}{2!4!} = 7\ 560$

**3.** Number of committees $= {}_7C_4 = 35$

**4.**

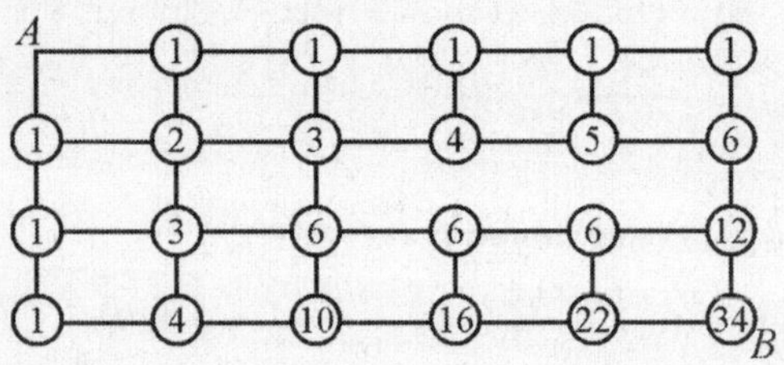


There are 34 paths.

**5.** Number of the triangles $= {}_{10}C_3 = 120$

**6.** Number of 5-card hands

$$= \left({}_4C_3\right)\left({}_{48}C_2\right) + \left({}_4C_4\right)\left({}_{48}C_1\right)$$
$$= (4)(1\ 128) + (1)(48)$$
$$= 4\ 512 + 48 = 4\ 560$$

**7.**

$${}_nC_2 - n = 44$$
$$\frac{(n)(n-1)}{2} - n = 44$$
$$n^2 - n - 2n = 88$$
$$n^2 - 3n - 88 = 0$$
$$(n-11)(n+8) = 0$$
$$n = 11 \text{ or } n = -8$$

Thus, the number of sides is 11.

**8.** Number of teams with at least one girl = (Total number of teams) – (Numbers of teams with no girls)

$$= {}_7C_3 - {}_4C_3$$
$$= 35 - 4 = 31$$

**9.** The paths require 2 moves up (U), 1 back (B), and 4 right (R). The number of paths is $\dfrac{7!}{2!4!} = 105$.

**10.** Number of hands

$$= {}_{39}C_5 + \left({}_{13}C_1\right)\left({}_{39}C_4\right) + \left({}_{13}C_2\right)\left({}_{39}C_3\right)$$
$$= 575\ 757 + (13)(82\ 251) + (66)(9\ 139)$$
$$= 2\ 248\ 194$$

**11.**

$${}_{15}P_2 = {}_7P_k$$
$$\frac{15!}{13!} = \frac{7!}{(7-k)!}$$
$$(\cancel{15})(\cancel{14}) = \frac{(\cancel{7})(6)(\cancel{5})(4)(\cancel{3})(\cancel{2})}{(7-k)!}$$
$$(7-k)! = (6)(4)$$
$$(7-k)! = 2 \cdot 3 \cdot 4 = 4!$$
$$7 - k = 4$$
$$k = 3$$

**12.**

$${}_nP_4 = 60\left({}_{n-1}C_3\right)$$
$$\frac{n!}{(n-4)!} = 60\left(\frac{(n-1)!}{(n-4)!3!}\right)$$
$$n = \frac{60}{6} = 10$$

**13.** Number of selections $= 5 \times 3 \times 2 - 1 = 29$ (as choosing none is not valid, you must subtract 1).

**14.**

$$t_3 = t_{2+1} = {}_7C_2\left(2x^2\right)^5(-y)^2$$
$$= (21)\left(32x^{10}\right)\left(y^2\right) = 672x^{10}y^2$$

**15.**

$$t_{k+1} = {}_{10}C_k\left(x^3\right)^{10-k}\left(-\frac{1}{x^2}\right)^k$$
$$= {}_{10}C_k(-1)^k\frac{x^{30-3k}}{x^{2k}} = {}_{10}C_k(-1)^k x^{30-5k}$$

**16.** From question 15:

$30-5k=15$

$k=3$

The term is the $4^{\text{th}}$ term $(k+1)$.

The coefficient is ${}_{10}C_3(-1)^3=-120$.

**17.** Number of games $= {}_8C_2\times 3=84$

**18.** **a)** $P(2\ As \text{ and } 1\ B) = \dfrac{{}_4C_2\times{}_3C_1}{{}_9C_3}=\dfrac{18}{84}=\dfrac{3}{14}$

**b)** $P(A, B, C \text{ in order}) = \dfrac{4}{9}\times\dfrac{3}{8}\times\dfrac{2}{7}=\dfrac{1}{21}$

**19.** **a)** $P$(David and Jamal win)

$=\dfrac{{}_2C_2\times{}_{23}C_1}{{}_{25}C_3}=\dfrac{23}{2\ 300}=\dfrac{1}{100}$

**b)** $P$(at least one wins)

$=1-P(\text{neither wins})$

$=1-\dfrac{{}_{23}C_3}{{}_{25}C_3}$

$=1-\dfrac{1\ 771}{2\ 300}=\dfrac{529}{2\ 300}=\dfrac{23}{100}$

**20.** **C**

**21.** **A**

$P(2 \text{ with } 6) = \dfrac{1}{6}\times\dfrac{1}{6}\times\dfrac{5}{6}\times\dfrac{5}{6}\times\dfrac{5}{6}\times\dfrac{5!}{3!2!}\doteq 0.160\ 8$

**22.** $n=19$

| 1 | 9 | | |
|---|---|---|---|

**23.** The number of arrangements is

$\dfrac{4\times 6!\times 3}{4!}=4\times 6\times 5\times 3=360.$

| 3 | 6 | 0 | |
|---|---|---|---|

# STATISTICS AND PROBABILITY

## *Lesson 1 Mean, Median, Mode, Range, and Standard Deviation*

### CLASS EXERCISES ANSWERS AND SOLUTIONS

**1.** Enter the "Number of shots" in $L_1$ and the "Frequency" in $L_2$.

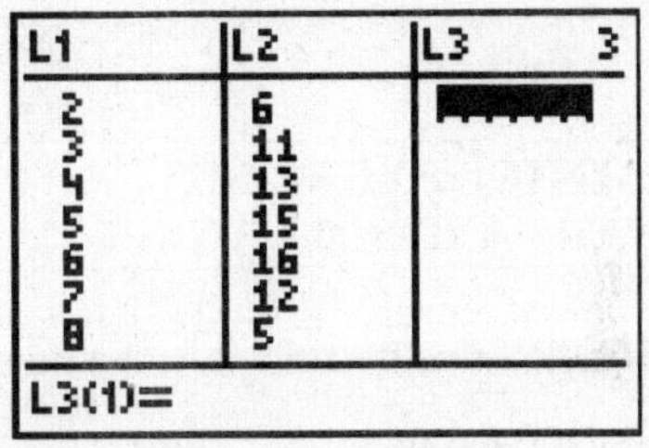


Press STAT ▷ CALC ∇ 1:1-Var Stats ENTER $L_1$, $L_2$.

The main calculator screen should now display 1-Var Stats ENTER $L_1$, $L_2$.

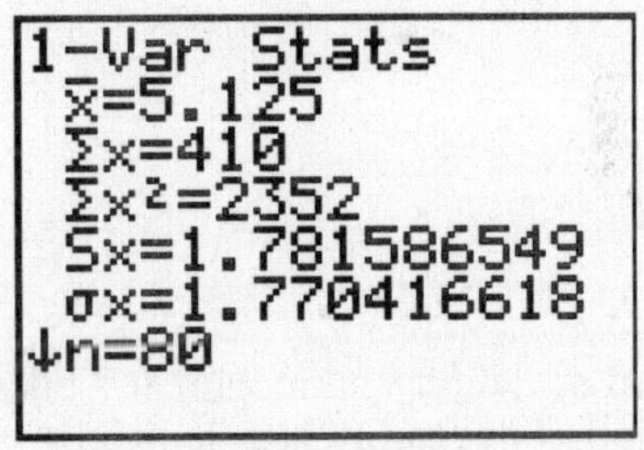


Press ENTER to obtain the calculations.

You should determine that

mean = 5.125

median = 5 ($40.5^{\text{th}}$ score from the distribution)

mode = 6 (Most frequent score)

standard deviation = $\sigma\doteq 1.77$

**2.** Notice that if 10 numbers have a mean of 6, then the sum of the numbers is (10)(6) = 60. This is confirmed by rearranging the formula

$\mu = \dfrac{x_1 + x_2 + \ldots + x_n}{n}$ to

$(n)(\mu) = x_1 + x_2 + \ldots + x_n$, and, for this case,

$x_1 + x_2 + \ldots + x_n = (n)(\mu) = (10)(6) = 60.$

Since 5 is added to each of the 10 numbers, a total of 50 is added to the sum of all the data values. So, the new mean is

$$\mu = \frac{(x_1 + x_2 + \ldots + x_n) + 50}{n} = \frac{60 + 50}{10} = \frac{110}{10} = 11.$$

## PRACTICE EXERCISES ANSWERS AND SOLUTIONS

**1.** Using the calculator as described in the notes:
mean = 5
median = 4
mode = 4 (not shown on the calculator)
range = $\max x - \min x = 11 - 2 = 9$
standard deviation $\doteq 2.49$

**2.** $\dfrac{12 + 8 + x + 7 + 11}{5} = 8$

$\dfrac{38 + x}{5} = 8$

$38 + x = 40$

$x = 2$

**3.** The standard deviation will remain at 7 because the dispersion of marks has not changed.

**4.** **a)** Using the calculator as described in the notes, the mean $\doteq 68.85$ and the standard deviation $\doteq 15.53$.

**b)** One standard deviation below the mean is approximately $68.85 - 15.53 \doteq 53.32$.

One standard deviation above the mean is approximately $68.85 + 15.53 \doteq 84.38$.

There are 13 marks between 53.32 and 84.38.

The percentage is $\dfrac{13}{20} = 0.65 = 65\%$.

**5.** **a)** The weights of the Hawks' team members have the greater mean.

**b)** The weights of the Eagles' team members have the greater standard deviation.

**c)** The Eagles team would likely be better because football requires a variety of sizes of players to fill the positions on the team.

**6.** Using the calculator as described in the notes:
mean $\doteq 14.25$
median = 15
mode = 8 (not shown on the calculator)
range $\max x - \min x = 21 - 8 = 13$
standard deviation $\doteq 4.58$

**7.** **a)** New mean: $11 + 10 = 21$

**b)** The standard deviation will remain at 5 because the dispersion of the numbers has not changed.

**c)** New modes: $7 + 10 = 17$ and $9 + 10 = 19$

**8.** Using the calculator as described in the notes:
**a)** Mean weight $\doteq 2.46$

**b)** Standard deviation $\doteq 0.30$

**c)** One standard deviation below the mean is approximately $2.46 - 0.30 = 2.16$.
One standard deviation above the mean is approximately $2.46 + 0.30 = 2.76$
There are 11 weights between 2.16 and 2.76
The percentage is $\dfrac{11}{16} \doteq 0.6875 \doteq 68.75\%$.

**9.** For a standard deviation of 0, all numbers must be the same. Thus, the numbers are 8, 8, 8, 8, 8. Verify this answer by using your calculator.

**10.** There are an infinite number of possible answers.
For example: 3, 5.5, 5.5, 5.5, 5.5
Verify your answer by using your calculator.

**11.** Using the calculator as described in the notes:
mean: $\doteq 12.4$
median = 13
mode = 14 (not shown on the calculator)
range = $\max x - \min x = 20 - 6 = 14$
standard deviation $\doteq 3.74$

**12.** Using the calculator as described in the notes:
**a)** Mean = 6.7

**b)** Standard deviation $\doteq 2.36$

 

**c)** One standard deviation below the mean is approximately $6.7 - 2.36 = 4.34$.

One standard deviation above the mean is approximately $6.7 + 2.36 = 9.06$.

The number of ratings between 4.34 and 9.06 is $16 + 22 + 13 + 48 + 25 = 124$.

$$\frac{124}{160} = 0.775 = 77.5\%$$

**13.** $$\frac{15+23+6+30+19+y}{6} = 25$$

$$\frac{93+y}{6} = 25$$

$$93 + y = 150$$

$$y = 57$$

**14.** There are an infinite number of possible answers.
For example: 6, 6, 8, 10, 15
Verify your answer by using your calculator.

**15.** Using the calculator as described in the notes:
**a)** Mean = 21.50

**b)** Standard deviation $\doteq 13.08$

**c)** Two standard deviations below the mean is approximately $21.50 - 2(13.08) = -4.66$ (or zero).

Two standard deviations above the mean is approximately $21.50 + 2(13.08) = 47.66$.
There are 9 prices between 0 and 47.66.

$$\frac{9}{10} = 0.9 = 90\%$$

**16.** There are an infinite number of possible answers.
For example: 1, 2, 3, 4, 5, 6, 7, 8, 9, 100

**17. C**
The measure of central tendency that is least affected by one extreme value is the median, because the median is the middle value.

**18.** $$\frac{\text{sum of values}}{14} = 65$$

Thus, the sum of values $= 14 \times 65 = 910$.

$$\frac{\text{sum of values} + 15^{\text{th}} \text{ value}}{15} = 66$$

$$910 + 15^{\text{th}} \text{ value} = 990$$

$$15^{\text{th}} \text{ value} = 80$$

| | | | |
|---|---|---|---|
| 8 | 0 | | |

**19.** $$\frac{\text{sum of values}}{10} = 50$$

sum of values = 500

$$\text{New mean} = \frac{500 + 70}{11} \doteq 51.8$$

| | | | |
|---|---|---|---|
| 5 | 1 | . | 8 |

## *Lesson 2 Probability Distributions and the Normal Curve*

### CLASS EXERCISES ANSWERS AND SOLUTIONS

**1.** If every data value is reduced by $x$, the new mean will also be reduced by $x$, so the new mean will become $\mu - x$.

However, by reducing *every* data value by $x$, the standard deviation is unaffected, since the values retain their dispersion. For example, if two data values differ by 5 units, they will still differ by 5 units after the same amount is subtracted from both values. So, the new standard deviation will be the same as the original standard deviation, $\sigma$.

Thus, the new mean will be $\mu - x$ and the new standard deviation will be $\sigma$.

**2. a)** A score of 78 is 8 marks greater than the mean of 70. However, we want to know how this translates into units of standard deviation. If the standard deviation is 8, then 1 unit of standard deviation is 8. Thus, 78 is 1 standard deviation above the mean.

**b)** A score of 56 is 14 marks below the mean of 70. How many units of standard deviation is 14 marks?

14 marks is $\frac{14}{8}=1.75$ units of standard deviation. Thus, 56 is 1.75 units of standard deviation below the mean.

## PRACTICE EXERCISES ANSWERS AND SOLUTIONS

**1.**

| Number of baskets | Frequency | Probability |
|---|---|---|
| 0 | 0 | 0 |
| 1 | 1 | 0.006 25 |
| 2 | 2 | 0.012 5 |
| 3 | 7 | 0.043 75 |
| 4 | 17 | 0.106 25 |
| 5 | 34 | 0.212 5 |
| 6 | 40 | 0.25 |
| 7 | 32 | 0.2 |
| 8 | 20 | 0.125 |
| 9 | 6 | 0.037 5 |
| 10 | 1 | 0.006 25 |
| | Sum: 160 | |

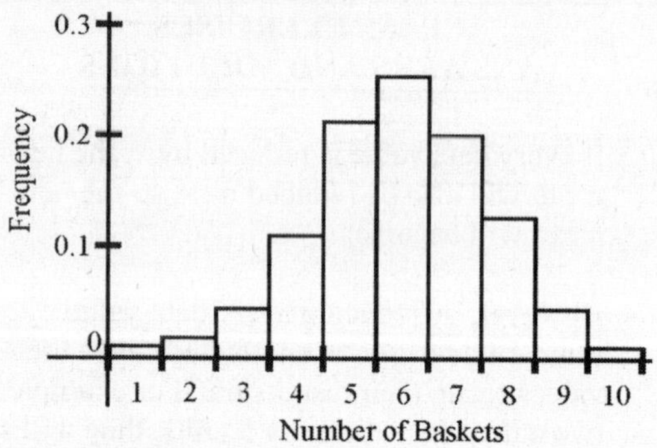


**2.** The height that is 1.8 units of standard deviation above the mean is $140+5(1.8)=140+9=149$ cm.

**3.** The lifetime of a bulb that is 0.25 units of standard deviation below the mean is $1\,100-80(0.25)=1\,100-20=1\,080$ hours.

**4.** A score of 83 is 15 marks above the mean, which is $\frac{15}{6}=2.5$ units of standard deviation above the mean.

**5.** A score of 67 is 17 above the mean, which is $\frac{17}{10}=1.7$ units of standard deviation above the mean.

**6.** 2.25 units of standard deviation is $(2.25)(0.04)=0.09$.

Thus, the price of \$0.93 is \$0.09 below the mean.

The mean is $0.93+0.09=1.02$. The mean price is \$1.02/L.

**7.** A score of 81 is 9 marks above the mean. Since 9 marks is equal to 1.5 units of standard deviation, then the standard deviation is $\frac{9}{1.5}=6$.

**8.**

| Rating | Frequency | Probability |
|---|---|---|
| 1 | 5 | 0.05 |
| 2 | 12 | 0.12 |
| 3 | 19 | 0.19 |
| 4 | 27 | 0.27 |
| 5 | 18 | 0.18 |
| 6 | 11 | 0.11 |
| 7 | 4 | 0.04 |
| 8 | 2 | 0.02 |
| 9 | 1 | 0.01 |
| 10 | 1 | 0.01 |
| | Total: 100 | |

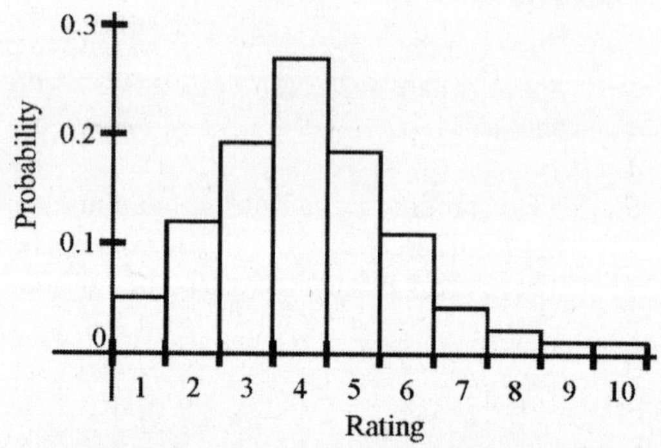


**9.** The value that is 2.1 units of standard deviation below the mean is $47-2.1(3.2)=40.28$.

**10.** A temperature of 30.2°C is 4.2°C above the mean, which is $\frac{4.2}{4.8}=0.875$ units of standard deviation above the mean.

**11.** A score of 14 is 4 below the mean. Since the standard deviation is 2.4, then 14 is $\frac{4}{2.4}=1\frac{2}{3}$ units of standard deviation below the mean.

 

**12.** 1.8 units of standard deviation is $(1.8)(0.05) = \$0.09$.

Thus, the gas sells for $0.75 - 0.09 = \$0.66/\text{L}$.

**13.** 0.8 units of standard deviation is $(0.8)(55) = 44$.

Thus, the data value of 167 is 44 above the mean. The mean is $167 - 44 = 123$.

**14.** A data value of 66 is 10 above the mean. Since 10 is 1.25 units of standard deviation, the standard deviation is $\frac{10}{1.25} = 8$.

**15.** 0.8 units of standard deviation is $(0.8)(12) = 9.6$ cm.

Since Carmen's height is 9.6 cm below the mean, her height is $125 - 9.6 = 115.4$ cm.

**16.** A score of 80 is 15 above the mean. Since the standard deviation is 8, a score of 80 is $\frac{15}{8} = 1.875$ units of standard deviation above the mean.

**17.**

| Age | Frequency | Probability |
|---|---|---|
| 4 | 0 | 0 |
| 5 | 8 | 0.016 |
| 6 | 23 | 0.046 |
| 7 | 55 | 0.11 |
| 8 | 96 | 0.192 |
| 9 | 121 | 0.242 |
| 10 | 102 | 0.204 |
| 11 | 69 | 0.138 |
| 12 | 22 | 0.044 |
| 13 | 4 | 0.008 |
| | Total: 500 | |

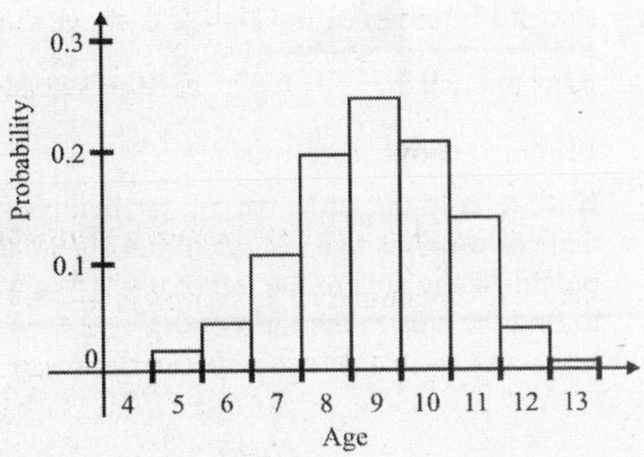


**18.** 1.8 units of standard deviation is $(1.8)(25) = 45$ hours.

Thus, 640 is 45 hours below the mean
The mean is $640 + 45 = 685$ hours.

**19.** 1.3 units of standard deviation is $(1.3)(2) = 2.6$. Since point $x$ is 2.6 units below the mean, the value of $x$ is $15 - 2.6 = 12.4$.

**20.** A score of 62 is 12% above the mean on the original test.

12% is $\frac{12}{8} = 1.5$ units of standard deviation.

On the scaled test, 1.5 units of standard deviation is $(1.5)(12) = 18\%$.

Since the new score is 18% above the mean, the new score will be $60\% + 18\% = 78\%$.

## *Lesson 3 Z-Scores and the Standard Normal Curve*

### CLASS EXERCISES
### ANSWERS AND SOLUTIONS

**1.** A negative $z$-score means that the temperature is below the mean. It is 1.2 units of standard deviation below the mean.

If 5°C is one unit of standard deviation, then 1.2 units of standard deviation will be 6°C, since $1.2 \times 5 = 6$.

So, the temperature is 6°C below the mean. The temperature is 23° – 6° = 17°C.

**2.** To find the $z$-score, we use the formula
$$z = \frac{x - \mu}{\sigma}.$$

When we substitute $\mu = 58$, $\sigma = 4$, and $x = 53$, we get:
$$z = \frac{53 - 58}{4} = \frac{-5}{4} = -1.25$$

The $z$-score is –1.25.

**3.** When we substitute $\mu = 185$, $z = 1.8$, and $x = 194$, we get:

$$1.8 = \frac{194-185}{\sigma}$$

$$1.8 = \frac{9}{\sigma}$$ • Simplify

$$1.8\sigma = 9$$ • Multiply both sides by $\sigma$

$$\sigma = 5$$ • Divide both sides by 1.8

The standard deviation is 5 cm.

**4.** To determine Maria's standing in each subject compared to the rest of the class, we need to calculate her *z*-scores.

To find the *z*-score for each subject, we use the formula $z = \frac{x-\mu}{\sigma}$.

| Math | History |
|---|---|
| Substitute $\mu = 69$, $\sigma = 9$, and $x = 80$ | Substitute $\mu = 66$, $\sigma = 5$, and $x = 76$ |
| $z = \frac{80-69}{9}$ | $z = \frac{76-66}{5}$ |
| $z \doteq 1.22$ | $z = 2$ |

Maria's *z*-score in math is approximately 1.22 and in history it is 2. This means that Maria's math score is approximately 1.22 units of standard deviation above the mean, and her history mark is 2 units of standard deviation above the mean.

So, compared to the rest of her class, Maria's best standing is in history.

**5.** Our goal in solving this problem will be to calculate Natalie's *z*-score before the adjustment, and then find a new mark so that her *z*-score will remain the same after the adjustment.

To find the original *z*-score, we use the formula $z = \frac{x-\mu}{\sigma}$.

When we substitute $\mu = 51$, $\sigma = 8$, and $x = 63$, we get:

$$z = \frac{63-51}{8}$$

$$= 1.5$$

Natalie's *z*-score is 1.5.

When the teacher adjusts the mean and standard deviation, Natalie's standing does not change compared to her class—her *z*-score remains 1.5.

Using the adjusted mean (65) and the adjusted standard deviation (10), we need to find Natalie's adjusted mark so that her corresponding *z*-score is still 1.5.

We substitute our known values into the formula $z = \frac{x-\mu}{\sigma}$.

$$1.5 = \frac{x-65}{10}$$

$$15 = x - 65$$ • Multiply both sides by 10

$$80 = x$$ • Add 65 to both sides

In order for Natalie to retain her *z*-score of 1.5, her adjusted test mark will be 80.

**6.**

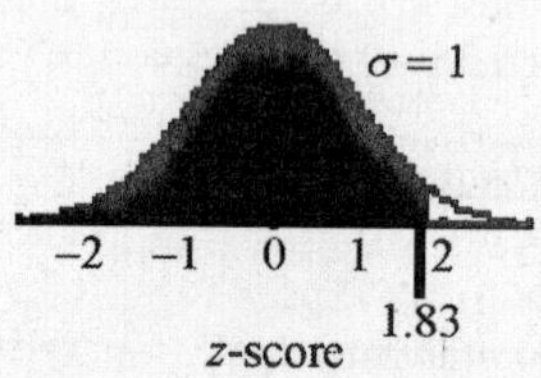


First, we move vertically down the column labeled *z* until we find "1.8."

Then, we move horizontally in that row until we find the column titled "0.03."

The corresponding area is 0.966 4, which means that the area to the left of the *z*-score 1.83 is 0.966 4 (correct to 4 decimal places).

In other words, 96.64% (correct to the nearest 0.01%) of normally distributed data values have *z*-scores less than 1.83.

**7.**

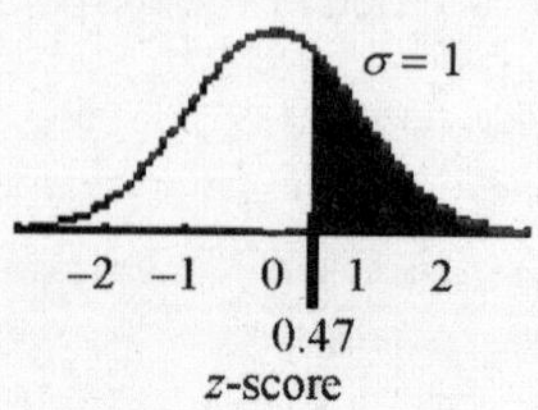


If we refer to the table, we can see that the area that corresponds to a *z*-score of 0.47 is 0.680 8, but this is the area to the left of 0.47. We wish to find the area to the right of 0.47.

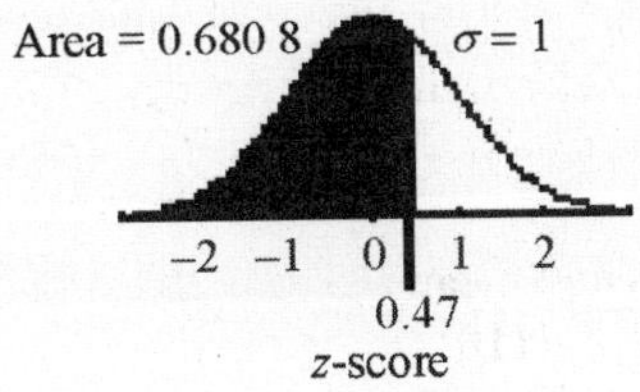


However, we know that the total area under the curve is 1. So, to find the area to the right of the $z$-score 0.47, we subtract 0.680 8 from 1.

We get $1 - 0.680\,8 = 0.319\,2$.

So, the area to the right of the $z$-score 0.47 is 0.319 2, which means that 31.92% of the data values have $z$-scores greater than 0.47.

**8.**

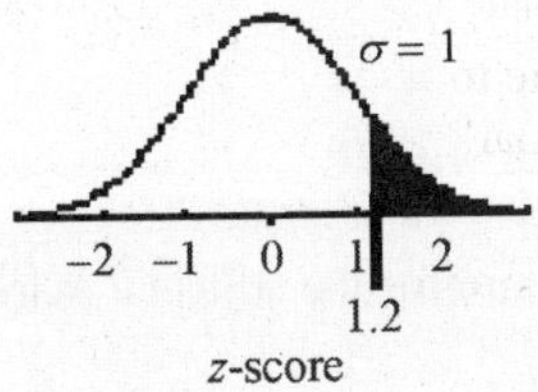


The table gives the area to the left of the $z$-score 1.2 as 0.884 9. So, to find the area to the right of the $z$-score 1.2, we subtract 0.884 9 from 1.

We get $1 - 0.884\,9 = 0.115\,1$.

If the area to the right of the $z$-score 1.2 is 0.115 1, then 11.51% of the test marks have a $z$-score greater than 1.2.

**9.** The expression $\mu + 0.33\sigma$ refers to a data value that is equal to the mean plus 0.33 units of standard deviation. This is equivalent to a $z$-score of 0.33.

Similarly, $\mu + 1.81\sigma$ is a data value that is 1.81 units of standard deviation above the mean. The corresponding $z$-score is 1.81.

So, we are looking for the area under the standard normal curve that lies between $z = 0.33$ and $z = 1.81$.

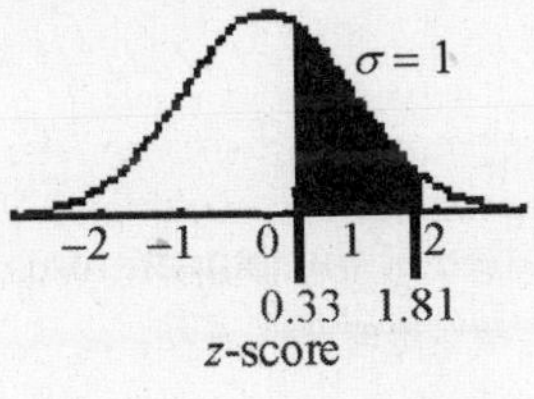


Using a graphing calculator, we enter normalcdf(0.33, 1.81, 0, 1) to give 0.335 6 (rounded to 4 decimal places).

```
normalcdf(0.33,1
.81,0,1)
         .3355522068
```

The area between $z = 0.33$ and $z = 1.81$ is approximately 0.335 6, which means that approximately 33.56% of the test marks are between $\mu + 0.33\sigma$ and $\mu + 1.81\sigma$.

**10.** Using a graphing calculator, we enter normalcdf($-1 \times 10^{99}$,144,148,4.2) to give 0.170 5 (to 4 decimal places).

```
normalcdf(-1*10^
99,144,148,4.2)
         .1704518946
```

Thus, approximately 17% of the students are less than 144 cm tall.

**11.**

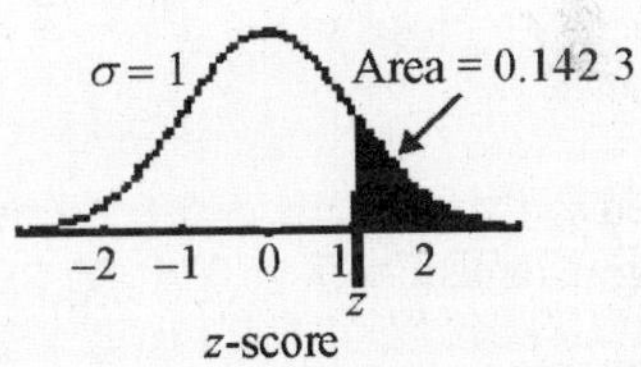


Here, we are told the area to the right of the $z$-score. However, the table only provides areas to the left of the $z$-score.

So, we need to find the area to the left of the $z$-score before continuing.

Since the total area under the standard normal curve is equal to 1, the area to the left of the $z$-score is $1 - 0.142\,3 = 0.857\,7$.

When we refer to the table, the $z$-score that corresponds to an area of 0.857 7 is $z = 1.07$.

## PRACTICE EXERCISES ANSWERS AND SOLUTIONS

1. $z=\frac{x-\mu}{\sigma}$

$z=\frac{0.731-0.78}{0.03}$

$z \doteq -1.67$

The corresponding $z$-score is $-1.67$ to the nearest hundredth.

2. Find Cooper's original $z$-score

$z=\frac{61-56}{10}$

$z=0.5$

Keep the $z$-score of 0.5, but now find $x$ using the new mean and standard deviation.

Again, use $z=\frac{x-\mu}{\sigma}$.

$0.5=\frac{x-67}{12}$

$6=x-67$

$73=x$

Cooper's new mark is 73.

3. $z=\frac{x-\mu}{\sigma}$

$\frac{147-\mu}{4}=3.75$

$147-\mu=15$

$\mu=132$

The mean of the data is 132.

4. Substitute the given values into $z=\frac{x-\mu}{\sigma}$.

$1.5=\frac{13k-25-8k}{k+2}$

$1.5=\frac{5k-25}{k+2}$

$1.5k+3=5k-25$

$3=3.5k-25$

$28=3.5k$

$8=k$

Thus, Paul's score of $13k-25$ is $(13)(8)-25=79$.

5. $z=\frac{x-\mu}{\sigma}$

$-0.8=\frac{x-150}{5}$

$-4.0=x-150$

$146=x$

Thus, Juliet's height is 146 cm.

6. For math:

$z=\frac{82-67}{6}$

$z=2.5$

Thus, the $z$-score for science must also be 2.5.

$2.5=\frac{x-71}{8}$

$20=x-71$

$91=x$

Lyle's science mark was 91.

7. Find Andrea's $z$-score for the adjusted mark.

$z=\frac{x-\mu}{\sigma}$

$z=60-60$

$z=0$

A $z$-score of 0 corresponds to the mean. Since the mean on the original test was 48, Andrea's original score was 48.

8. Substitute the given values into $z=\frac{x-\mu}{\sigma}$.

$2.4=\frac{(8k-1)-(6k+5)}{k+1}$

$2.4k+2.4=2k-6$

$0.4k=-8.4$

$k=-21$

Since the mean is $(6k+5)$, the value of the mean is $6(-21)+5=-121$.

9. $-1.4=\frac{23.3-26.8}{\sigma}$

$-1.4\,\sigma=-3.5$

$\sigma=2.5$

The standard deviation of the temperatures is 2.5°C.

 

**10.** $z = \frac{x-\mu}{\sigma}$

$1.8 = \frac{x-650}{50}$
$90 = x - 650$
$740 = x$

The bulb's lifetime is 740 hours.

**11.** Substitute the given values into $z = \frac{x-\mu}{\sigma}$.

$2.4 = \frac{(4m-9)-(m+21)}{\sigma}$
$2.4\sigma = 3m - 30$
$\sigma = \frac{3m-30}{2.4} = 1.25m - 12.5$

**12.** For the original mark:

$z = \frac{x-\mu}{\sigma}$
$z = \frac{81-88}{5}$
$z = -1.4$

Keep the z-score of $-1.4$, but now find $x$ using the new values for the mean and standard deviation.

$-1.4 = \frac{x-73}{8}$
$-11.2 = x - 73$
$61.8 = x$

Vern's new mark was 62, to the nearest whole number.

**13.**

For History: $z = \frac{76-71}{12} \doteq 0.42$

For Chemistry: $z = \frac{78-75}{8} = 0.375$

For Math: $z = \frac{73-69}{6} \doteq 0.67$

For English: $z = \frac{80-73}{10} = 0.70$

For French: $z = \frac{75-68}{11} \doteq 0.64$

**a)** Trevor's highest *z*-score is in English. Thus, his best subject is English.

**b)** Trevor's lowest *z*-score is in chemistry. Thus, his worst subject is chemistry.

(Note: For questions 14 to 49, the methods shown are calculator solutions. The *z*-score tables could also be used.)

**14.** $\text{normalcdf}(-1\times10^{99}, -1.35, 0, 1) \doteq 0.0885$

**15.** $\text{normalcdf}(-1\times10^{99}, 1.1, 0, 1) \doteq 0.8643$

**16.** $\text{normalcdf}(-2.09, 1\times10^{99}, 0, 1) \doteq 0.9817$

**17.** $\text{normalcdf}(-0.16, 1\times10^{99}, 0, 1) \doteq 0.5636$

**18.** $\text{normalcdf}(-0.83, 1.3, 0, 1) \doteq 0.6999$

**19.** $\text{normalcdf}(0.56, 2.09, 0, 1) \doteq 0.2694$

**20.** $\text{normalcdf}(-1\times10^{99}, -0.39, 0, 1) \doteq 0.3483$

**21.** $\text{normalcdf}(-1.2, -0.46, 0, 1) \doteq 0.2077$

**22.** $\text{normalcdf}(-0.4, 1.65, 0, 1) \doteq 0.6060$

**23.** $\text{normalcdf}(0.6, 2.05, 0, 1) \doteq 0.2541$

**24.** $z = \text{invNorm}(0.9633) \doteq 1.79$

**25.** $z = \text{invNorm}(1-0.1469) \doteq 1.05$

**26.** $z_1 = \text{invNorm}(0.5+0.2224) \doteq 0.59$
$z_2 = \text{invNorm}(0.5+0.2224+0.2321) \doteq 1.69$

**27.** $z = \text{invNorm}(1-0.6628) \doteq -0.42$

**28.** $\text{normalcdf}(-1\times10^{99}, -1.91, 0, 1) \doteq 0.0281$
$z = \text{invNorm}(0.0281+0.3426) \doteq -0.33$

**29.** $z = \text{invNorm}(1-0.8665) \doteq -1.11$

**30.** $z = \text{invNorm}(0.9573) \doteq 1.72$

**31.** $z = \text{invNorm}(1-0.9015) \doteq -1.29$

**32.** $z_1 = \text{invNorm}(1-0.1020-0.1098) \doteq 0.80$
$z_2 = \text{invNorm}(1-0.1020) \doteq 1.27$

**33.** $z = \text{invNorm}(0.5 - 0.3485) \doteq -1.03$

**34.** $z_1 = \text{invNorm}(1 - 0.2810 - 0.4070) \doteq -0.49$

$z_2 = \text{invNorm}(1 - 0.2810) \doteq 0.58$

**35.** $z = \text{invNorm}(0.5 + 0.2734) \doteq 0.75$

**36. D**

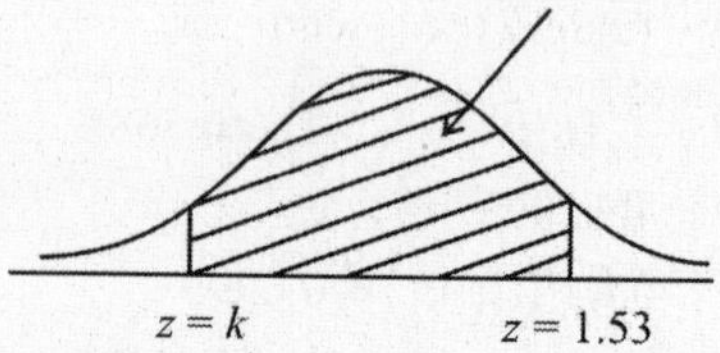


$\text{normalcdf}(-1\times10^{99}, 1.53, 0, 1) \doteq 0.9370$

$z = k = (0.9370 - 0.82) \doteq -1.19$

## *Lesson 4 Applications*

### CLASS EXERCISES ANSWERS AND SOLUTIONS

**1.** We are looking for the proportional area under the normal curve that is to the right of 780.

To solve this problem, we first convert 780 into its corresponding $z$-score.

In this question, $\mu = 730$ and $\sigma = 40$, so we get:

$$z = \frac{780 - 730}{40} = 1.25$$

The $z$-score is 1.25.

Since we want the percentage of light bulbs with a lifetime greater than 780 hours, we need to find the area under the standard normal curve that is to the right of $z = 1.25$.

The table, or a calculator, gives the area corresponding to $z = 1.25$ as 0.894 4, but this is the area to the left of $z = 1.25$. We want the area that lies to the right of $z = 1.25$, so we subtract the given area from 1.

$1 - 0.894\,4 = 0.105\,6$

So, the area to the right of $z = 1.25$ is approximately 0.105 6.

This means that approximately 10.56% of light bulbs have lifetimes greater than 780 hours.

This question can also be done directly using the normalcdf function on the calculator. We enter normalcdf(780, $1 \times 10^{99}$, 730, 40), press ENTER, and get 0.105 6 (to 4 decimal places).

**2.** To express this problem visually, we are looking for the proportional area under the normal curve that is between masses of 2.7 kg and 3.5 kg.

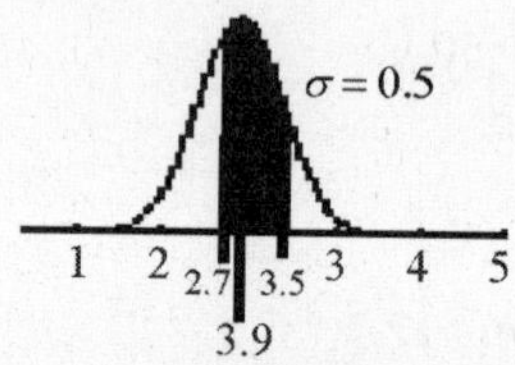


To solve this problem, we first convert 2.7 kg and 3.5 kg into their corresponding $z$-scores.

In this question, $\mu = 2.9$ and $\sigma = 0.5$, so for the value 2.7 kg, we get:

$$z = \frac{2.7 - 2.9}{0.5} = -0.4$$

The first $z$-score is –0.4.

For the value 3.5 kg, the corresponding $z$-score is

$$z = \frac{3.5 - 2.9}{0.5} = 1.2$$

The second $z$-score is 1.2.

So, we are interested in finding the area under the standard normal curve that is between $z = -0.4$ and $z = 1.2$.

According to the table, or from a calculator, the area to the left of $z = 1.2$ is 0.884 9, and the area to the left of $z = -0.4$ is 0.344 6.

So, the area between $z = -0.4$ and $z = 1.2$ is $0.884\,9 - 0.344\,6 = 0.540\,3$.

This means that the probability that a chicken has a mass between 2.7 kg and 3.5 kg is approximately 0.540 3.

 

This question can also be done directly using the normalcdf function on the calculator. We enter normalcdf(2.7,3.5 ,2.9,0.5), press ENTER, and get 0.540 4 (to 4 decimal places), which is 0.000 1 more than the answer obtained from the table. This is not a significant difference.

3. To solve this problem, we will first find the percentage of students who scored lower than 50, and then calculate the number of students who scored lower than 50.

We will first convert 50 into its corresponding *z*-score.

In this question, $\mu = 68$ and $\sigma = 11$, so we get:

$$z = \frac{50-68}{11}$$
$$\doteq -1.64$$

The *z*-score is –1.64 (to 2 decimal places).

Since we want the percentage of marks that are lower than 50, we need to find the area under the standard normal curve that is to the left of $z = -1.64$.

The table, or calculator, tells us that the area corresponding to a *z*-score of 1.64 is 0.050 5, so we know that approximately 5.05% of students got a mark lower than 50 on the exam.

$$5.05\% \text{ of } 20\,000 = 0.050\,5 \times 20\,000 = 1\,010$$

Approximately 1 010 students scored less than 50% on the exam.

This question can also be done using the normalcdf function on the calculator. We enter normalcdf($-1 \times 10^{99}$,50,68,11), press ENTER, and get 0.050 9 (to 4 decimal places). $0.050\,9 \times 20\,000 = 1\,018$. Again, the difference between the calculator answer and the table answer is relatively small.

4. In this question, we are told that 211 students out of 5 000 scored over 90 on the test.

So, the percentage of students who scored over 90 is: $\frac{411}{5\,000} = 0.082\,2 = 8.22\%$.

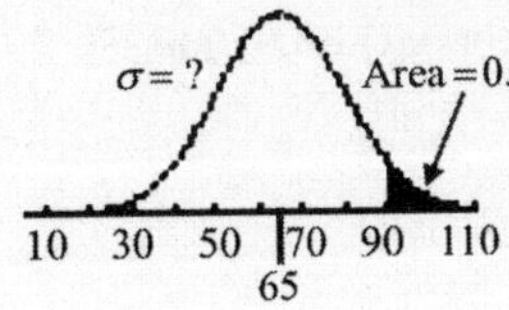

Exam mark

Since the proportional area to the right of 90 is 0.082 2, we also know that the area to the left of 90 will be $1 - 0.082\,2 = 0.917\,8$.

We begin by finding the *z*-score that corresponds to the area of 0.917 8.

According to the table, the closest corresponding *z*-score is 1.39.

Or, by using a calculator, invNorm(0.917 8) $\doteq$ 1.39.

We now know that $z = 1.39$, the actual score, $x = 90$, and the mean, $\mu = 6$, so we can use the formula $z = \frac{x-\mu}{\sigma}$ to determine the standard deviation.

By substituting $z = 1.39$, $x = 90$, $\mu = 65$, we get:

$$1.39 = \frac{90-65}{\sigma}$$

Now, solving for $\sigma$:

$$1.39 = \frac{90-65}{\sigma}$$
$$1.39\sigma = 25$$
$$\sigma = 17.99$$

The standard deviation of the exam marks was 17.99 (to the nearest hundredth).

5. The distribution of speeds is as follows, with the shaded portion representing cars travelling faster than 90 km/h.

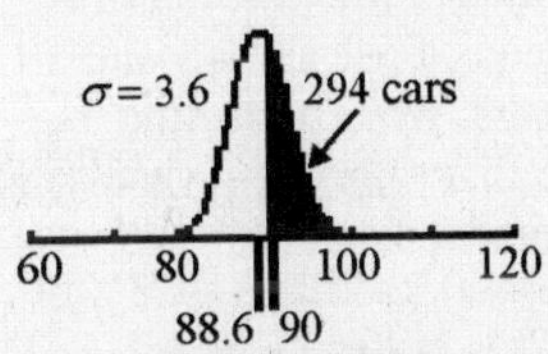

Speed (km/h)

First, we will find the proportional area of the shaded region. To determine the area, we convert the speed, 90 km/h, into the equivalent *z*-score using the formula $z = \frac{x-\mu}{\sigma}$.

By substituting $x = 90$, $\mu = 88.6$, and $\sigma = 3.6$, we get:

$$z = \frac{90 - 88.6}{3.6}$$
$$z \doteq 0.39$$

So, a speed of 90 km/h has a corresponding $z$-score of 0.39.

When we consult the table of areas under the standard normal curve, we find that the area to the left of $z = 0.39$ is 0.651 7.

Therefore, the shaded area to the right of $z = 0.39$ is $1 - 0.651\,7 = 0.348\,3$.

(Or, by using a calculator, the area to the right of $x = 90$ can be found in one step: normalcdf(90, $1 \times 10^{99}$,88.6,3.6) $\doteq$ 0.348 7. (The difference in accuracy is not significant.)

This is the fraction of cars that traveled faster than 90 km/h. We also know that 294 cars travelled faster that 90 km/h.

The question asks for the total number of cars that travelled on the road.

If we let $T$ = the total number of cars on the road, then we can say that $\frac{294}{T} = 0.348\,3$.

When we solve this equation for $T$, we get $T \doteq 844.1$.

So, approximately 844 cars travelled on the road that day.

Using the value from the calculator, 0.348 7 gives $T = \frac{294}{0.348\,7} \doteq 843$.

**6.** Given the normal distribution of monitor lifetimes with a mean of 5.1 and a standard deviation of 0.6 years, we need to find a warranty period so that only 5% (at most) of the monitors will be returned.

In other words, we want to find the lifetime, $x$, such that only 5% of the data is less than this value. On the normal curve, we require an $x$-value such that the proportional area to the left of this value is 5%.

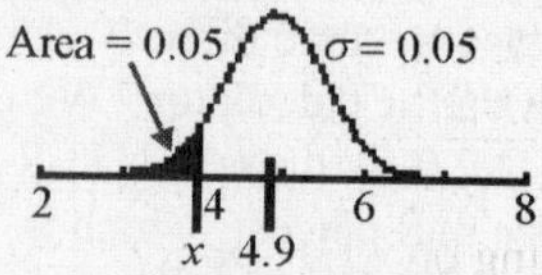


When we consult the table of areas under the standard normal curve to find the $z$-score that corresponds to an area of 0.05, we find that this value does not have a closest $z$-score. Instead, we have the areas 0.049 5 and 0.050 5. Which of these should we use?

Remember that we want, at most, 5% of the monitors to be returned. The area 0.050 5 is equal to 5.05%, which is more than 5%.

So, we select the area 0.049 5, which has a corresponding $z$-score of $-1.65$. This means that 4.95% of the monitors are to the left of a $z$-score of $-1.65$.

Next, we convert the $z$-score into the corresponding data value by using the formula $z = \frac{x - \mu}{\sigma}$.

When we substitute $z = -1.65$, $\mu = 5.1$, and $\sigma = 0.6$, we get:

$$-1.65 = \frac{x - 5.1}{0.6}$$

Solving for $x$:

$$-1.65 = \frac{x - 5.1}{0.6}$$
$$-0.99 = x - 5.1$$
$$x = 4.11$$

So, if the warranty period is 4.11 years, no more than 5% of monitors will be returned under the warranty. Of course, manufacturers seldom give warranties using decimal fractions of years. If the manufacturer wanted to guarantee the monitors using a whole number of years, the warranty would be for 4 years. This should ensure that close to, but less than, 5% of the monitors are returned.

Or, by using a calculator, this question can be done very efficiently.

The required $x$-value can be determined from invNorm(0.05,5.1,0.6). The result is 4.11 (to 2 decimal places), the required warranty period.

 

## PRACTICE EXERCISES ANSWERS AND SOLUTIONS

Note: Answers for the following questions may vary slightly, depending on which solution method is used.

**1.**

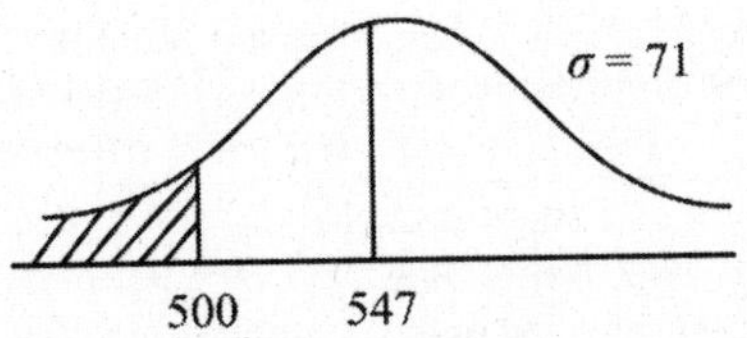


$$z = \frac{500 - 547}{71} \doteq -0.66$$

Using the table, the area to the left of $z = -0.66$ is 0.254 6. Thus, the probability that the city will receive less than 500 mm of rainfall is approximately 0.25.

OR

Using calculator:
$\text{normalcdf}\left(-1\times 10^{99}, 500, 547, 71\right) \doteq 0.25$.

**2.**

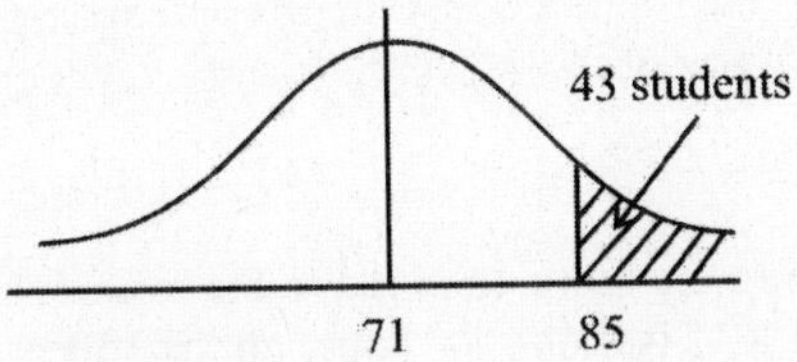


Using $z = \dfrac{x-\mu}{\sigma}$:

$$z = \frac{85-71}{8}$$
$$z = 1.75$$

A mark of 85 corresponds to a $z$-score of 1.75. Using the table or a calculator, the area to the right of $z = 1.75$ can be found as 0.040 1. This is the fraction of students that scored over 85.

If $T$ represents the total number of students that wrote this test, then

$$\frac{43}{T} = 0.040\,1.$$

Solving for $T$, gives $T = 1\,072$ students.

**3. a)**

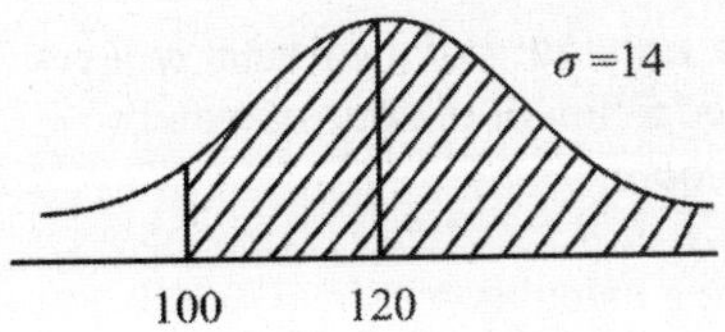


$$z = \frac{100-120}{14}$$
$$z \doteq -1.43$$

Using the table, the area to the right of $z = -1.43$ is $(1 - 0.076\,5) = 0.923\,6$.

Thus, approximately 92.36% of the plants are 100 cm or taller.

OR

Using a calculator:
$\text{normalcdf}\left(100, 1\times 10^{99}, 120, 14\right) \doteq 92.34$

**b)**

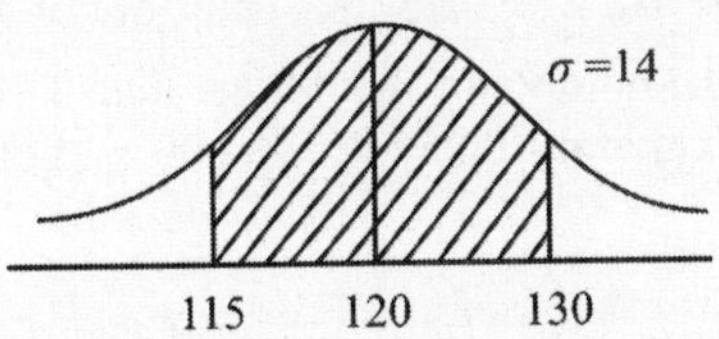


$$z = \frac{115-120}{14} \qquad z = \frac{130-120}{14}$$
$$z \doteq -0.36 \qquad z \doteq 0.71$$

Using the table, the area between $z = -0.36$ and $z = 0.71$ is $(0.761\,1 - 0.359\,4) = 0.401\,7$.

Thus, approximately 40.17% of the plants are between 115 cm and 130 cm.

OR

Using a calculator:
normalcdf(115,130,120,14) $\doteq$ 40.20

**c)**

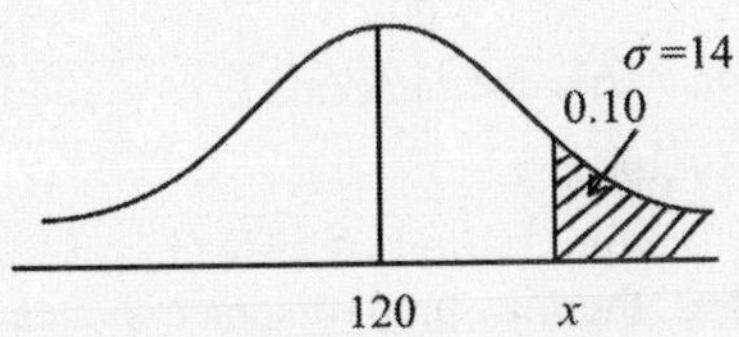


Using the table, the area under the curve is (1.0 – 0.10) = 0.90 for $z$ = 1.29.

$$1.29 = \frac{x-120}{14}$$
$$18.06 = x - 120$$
$$138.06 = x$$

Thus, only 10% of the plants will reach or exceed a height of approximately 138.06cm.

Using a calculator:
$\text{invNorm}(0.9,120,14) \doteq 137.94$

**4.**

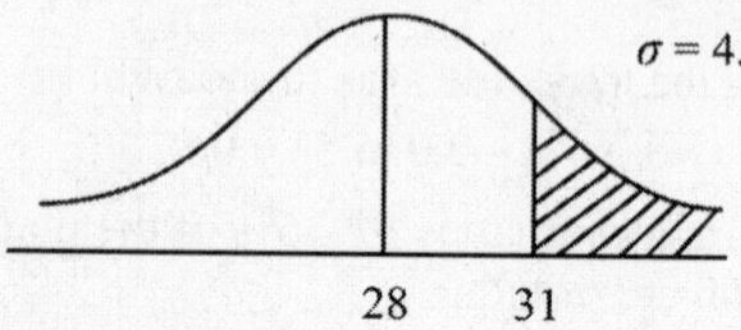


$z = \dfrac{31-28}{4.6}$
$z \doteq 0.65$

Using the table, the area to the right of $z = 0.65$ is $(1-.742\ 2) = 0.257\ 8$. Thus, out of 45 000 fish, the number of fish longer than 31 cm is approximately $0.257\ 8 \times 45\ 000 = 11\ 601$

OR

Using a calculator:
$\text{normalcdf}\left(31, 1\times10^{99}, 28, 4.6\right)\times 45\ 000 \doteq 11\ 571$

**5.**

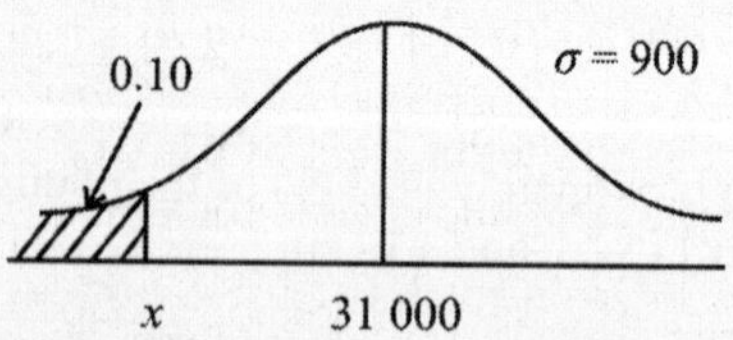


Using the table, the area under the curve is 0.10 for $z = -1.29$.

$-1.29 = \dfrac{x - 31\ 000}{9\ 000}$

$-11\ 610 = x - 31\ 000$

$19\ 390 = x$

Thus, the maximum distance that can be driven to still qualify for the reward is approximately 19 390 km.

OR

Using a calculator:
$(0.10, 31\ 000, 9\ 000) \doteq 19\ 466$

**6.**

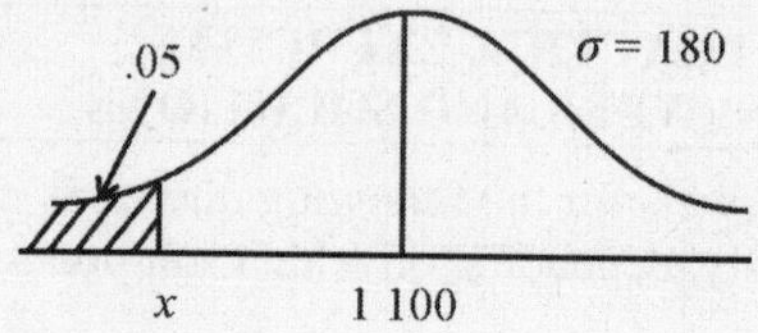


Using the table, the area under the curve is 0.05 for $z = -1.65$.

$-1.65 = \dfrac{x - 1\ 100}{180}$

$-297 = x - 1\ 100$

$803 = x$

Thus, the warranty should be 803 days.

OR

Using a calculator:
$\text{invNorm}(0.05, 1\ 100, 180) \doteq 803$

**7.**

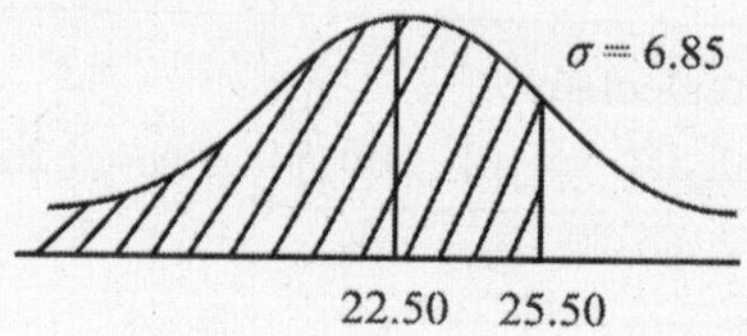


$z = \dfrac{25.00 - 22.50}{6.85} \doteq 0.37$

Using the table, the area to the left of $z = 0.37$ is 0.644 3. Thus, out of 350 customers selected at random, the number expected to spend less than $25.00 is $0.644\ 3 \times 350 \doteq 225.51$, or approximately 226 customers.

OR

Using a calculator:
normalcdf
$\left(-1\times10^{99}, 25, 22.5, 6.85\right)\times 350 \doteq 225$

**8. a)**

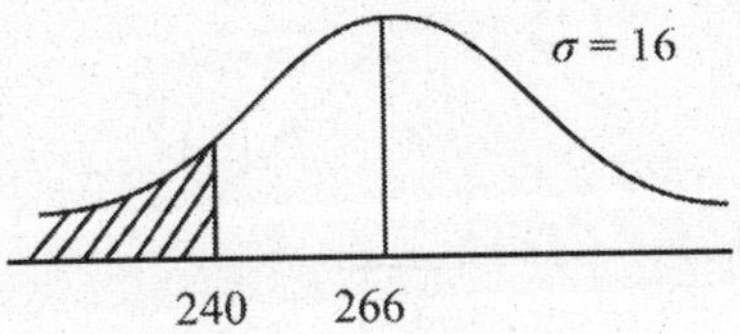


$$z = \frac{240-266}{16}$$
$$z \doteq -1.63$$

Using the table, the area to the left of $z = -1.63$ is 0.051 6. Thus, approximately 5% of the pregnancies last less than 240 days.

OR

Using a calculator:
$\text{normalcdf}\left(-1\times10^{99}, 240, 266, 16\right) \doteq .0521$

**b)**

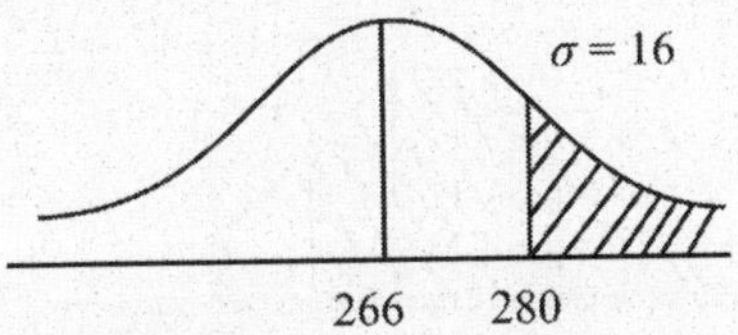


$$z = \frac{280-266}{16}$$
$$z \doteq 0.88$$

Using the table, the area to the right of $z = 0.88$ is $(1-0.8106) = 0.1894$. Thus, approximately 19% of the pregnancies last more than 280 days.

OR

Using a calculator:
$\text{normalcdf}\left(280, 1\times10^{99}, 266, 16\right) \doteq 0.1908$

**9.**

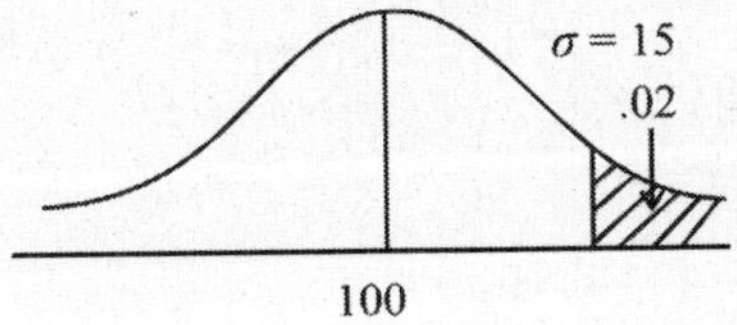


Using the table, the area under the curve is $(1-0.02) = 0.98$ for $z = 2.06$.

$$2.06 = \frac{x-100}{15}$$
$$x = 130.9$$

Thus, 290 of the scores will have a value greater than approximately 131.

OR

Using a calculator:
$\text{invNorm}(.98, 100, 15) \doteq 131$

**10.**

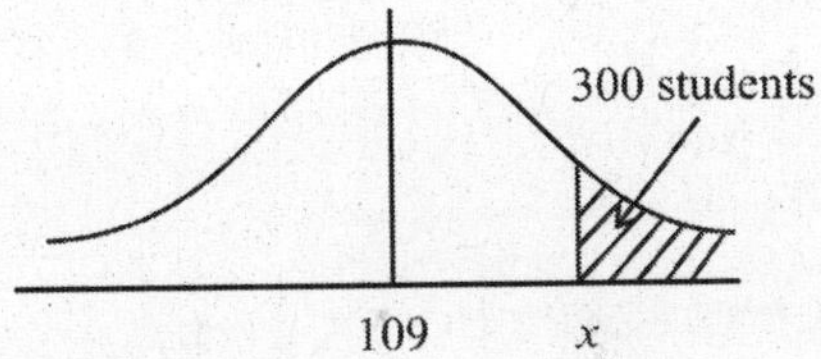


The top 300 represents $\frac{300}{8\ 500} = 0.035\ 3$ or 3.53% of the total.

Thus, the area to the right of $x$ is 0.035 3. Using the table, the $z$-score for $x$ must be 1.81. Using this, calculate $x$.

$$z = \frac{x-\mu}{\sigma}$$
$$1.81 = \frac{x-109}{19}$$
$$34.39 = x - 109$$
$$143.39 = x$$

Thus, to be in the top 300, the score must be at least 144.

**11.**

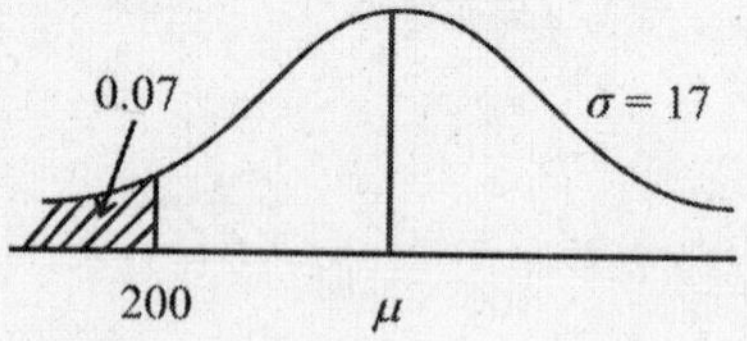


Using the table, the area under the curve is 0.07 for $z = -1.48$.

$$-1.48 = \frac{200-\mu}{17}$$
$$-25.16 = 200-\mu$$
$$225.16 = \mu$$

The mean weight of the potatoes, to the nearest gram, is approximately 225 grams.

OR

Using a calculator:
$z = \text{invNorm}(0.07,0,1) \doteq -1.48$

$$-1.48 = \frac{200-\mu}{17}$$
$$\mu = 225.16$$

**12.**

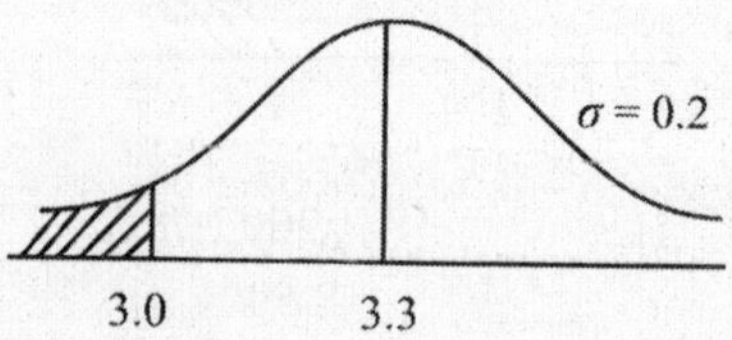


$$z = \frac{3.0-3.3}{0.2}$$
$$z = -1.5$$

Using the table, the area under the curve to the left of $z = -1.5$ is 0.066 8. Thus, out of 2 200 applicants, the number of students that will be rejected is $0.066\,8 \times 2\,200 = 146.96$, or 147.

OR

Using a calculator:
$\text{normalcdf}(-1\times10^{99},3,3.3,0.2)\times 2\,200 \doteq 147$

**13.**

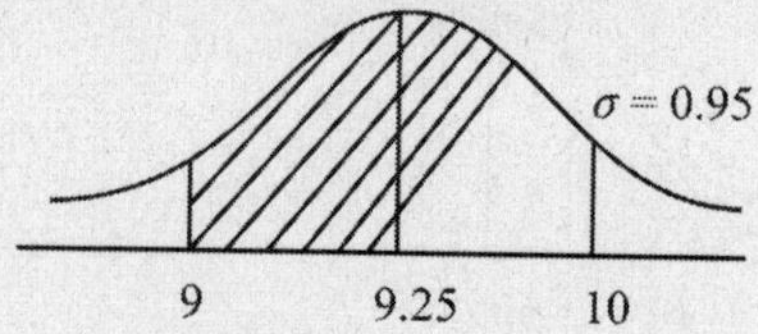


$$z = \frac{9-9.25}{0.95} \qquad z = \frac{10-9.25}{0.95}$$
$$z \doteq -0.26 \qquad z \doteq 0.79$$

Using the table, the area between $z = -0.26$ and $z = 0.79$ is $(0.7852-0.3914) = 0.3938$.

Thus, approximately 39% of the workers receive wages between \$9.00 and \$10.00/h.

OR

Using a calculator:
$\text{normalcdf}(9,10,9.25,0.95) \doteq 0.388\,9$

**14. a)**

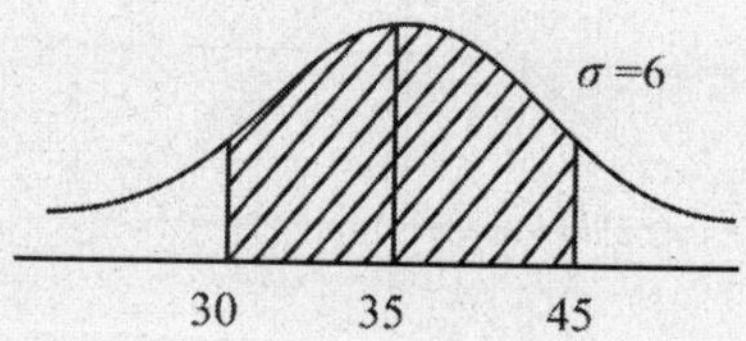


$$z = \frac{30-35}{6} \qquad z = \frac{45-35}{6}$$
$$z \doteq -0.83 \qquad z \doteq 1.67$$

Using the table, the area between $z = -0.83$ and $z = 1.67$ is $(0.952\,5-0.203\,3) = 0.749\,2$.
Thus, approximately 74% of the time milk production is between 30 L and 45 L.

OR

Using a calculator:
$\text{normalcdf}\,(30,45,35,6) \doteq 0.749\,9$

**b)**

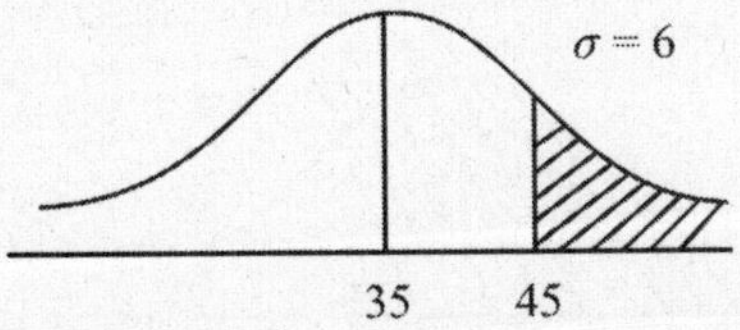


From part (a), 45 has a corresponding $z$-score of 1.67. Using the table, the area to the right of $z = 1.67$ is $(1-.9525) = 0.0475$. Thus, approximately 5% of the time, milk production exceeds 45 L.

OR

Using a calculator:
normalcdf $(45, 1\times10^{99}, 35, 6) \doteq 0.047\,80$

**15.**

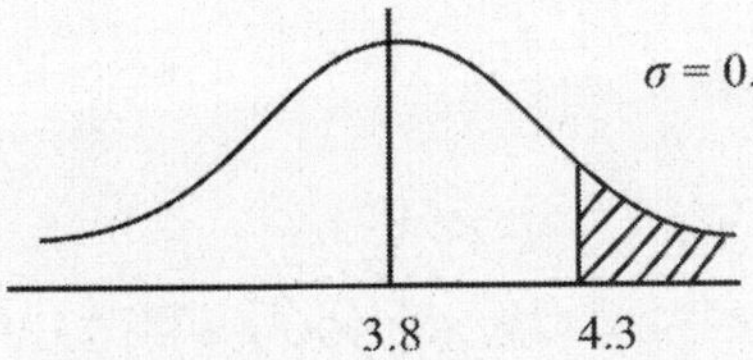


$$z = \frac{4.3-3.8}{0.8}$$
$$z \doteq 0.63$$

Using the table, the area to the right of $z = 0.63$ is $(1-0.735\,7) = 0.264\,3$.

Since 71 fish weigh over 4.3 kg, 71 fish represent 0.264 3 of the population.

Thus
$$\frac{71}{\text{total}} = 0.2643$$
$$\text{total} \doteq 268.63$$

There are approximately 269 salmon in the school.

OR

Using a calculator:
$71 \div \text{normalcdf}(4.3, 1\times10^{99}, 3.8, 0.8) \doteq 267$

**16.**

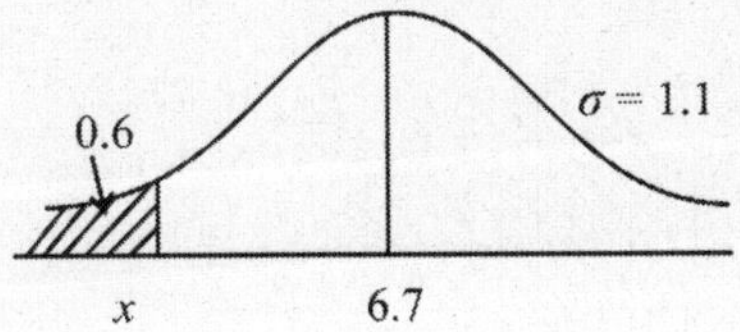


Using the table, the area under the curve is 0.06 for $z = -1.56$
$$-1.56 = \frac{x-6.7}{1.1}$$
$$-1.716 = x - 6.7$$
$$4.984 = x$$

The warranty period is 4 years. (Because the value is so close to 5, the warranty would probably be raised to 5 years).

OR

Using a calculator:
$\text{invNorm}(0.06, 6.7, 1.1) \doteq 4.99$

**17.**

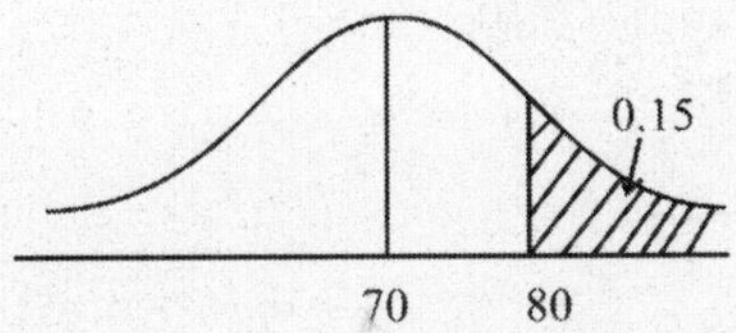


Using the table, the area under the curve is $(1-0.15) = 0.85$ for $z = 1.04$.
$$1.04 = \frac{80-70}{\sigma}$$
$$1.04 \times \sigma = 10$$
$$\sigma \doteq 9.62$$

The standard deviation is approximately 10 minutes.

OR

Using a calculator:
$z = \text{invNorm}(0.85, 0, 1) \doteq 1.04$
$$1.04 = \frac{80-70}{\sigma}$$
$$\sigma \doteq 9.62$$

**18. C**

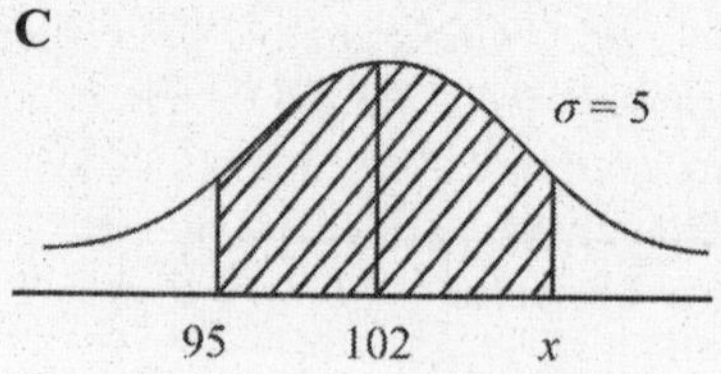


$$z = \frac{95-102}{5}$$
$$z = -1.4$$

Using the table, the area to the left of $z = -1.4$ is 0.080 8.

Since 100 scores represents half the scores, the corresponding area is $\frac{100}{200} = 0.5$.

Thus, the area to the left of $x$ is $0.080\ 8 + 0.5 = 0.580\ 8$.

Using the table, the area under the curve is 0.580 8 for $z = 0.20$.

$$0.20 = \frac{x-102}{5}$$
$$1 = x - 102$$
$$103 = x$$

OR

Using a calculator:
$\text{normalcdf}\left(-1\times10^{99}, 95, 102, 5\right) \doteq 0.080\ 8$
$\text{InvNorm}(.5 + .080\ 8, 102, 5) \doteq 103$

**19. C**

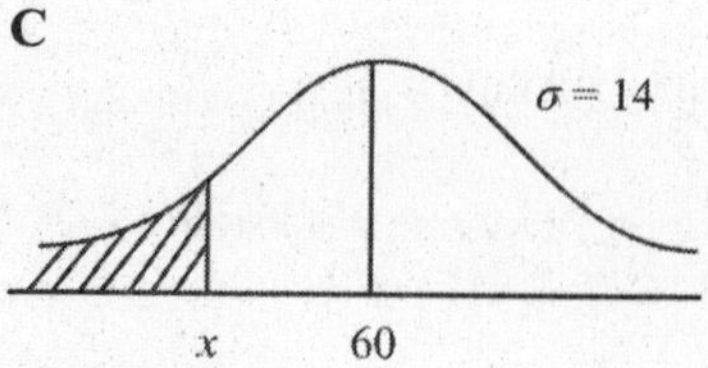


Using the table, the area under the curve is 0.10 for $z = -1.29$.

$$-1.29 = \frac{x-60}{14}$$
$$-18.06 = x - 60$$
$$41.94 = x$$

OR

Using acalculator:
$\text{invNorm}(0{,}10, 60, 14) \doteq 42.06$

The guarantee should be for 42 months.

**20. D**

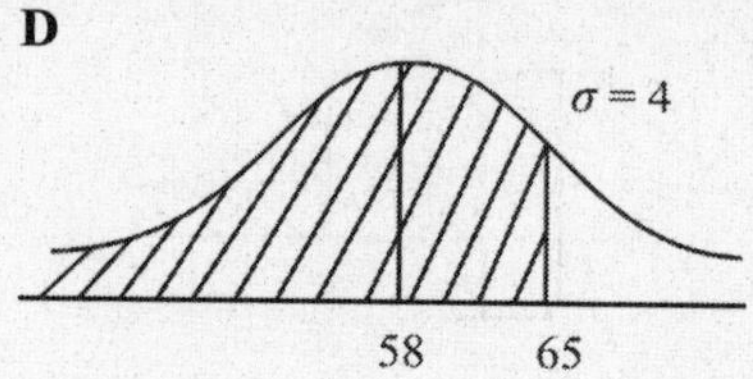


$$z = \frac{65-58}{4}$$
$$z = 1.75$$

Since cars that are not in danger would be doing laps in less than 65 s, using the tables, the area to the left of $z = 1.75$ is 0.959 9.

Thus, out of 200 stock cars, the number of cars that would be out of danger is $0.959\ 9 \times 200 \doteq 191.98$, or approximately 192 cars.

OR

Using a calculator:
$\text{normalcdf}\left(-1\times10^{99}, 65, 58, 4\right)\times 200 \doteq 191.99$

**21. C**

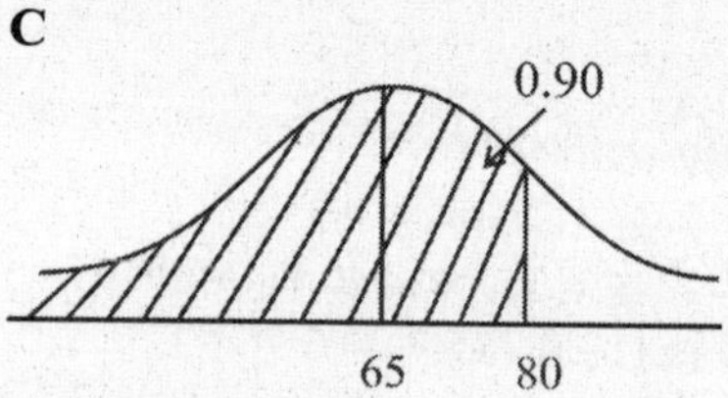


Using the table, the area under the curve is 0.90 for $z = 1.28$.

$$1.28 = \frac{80-65}{\sigma}$$
$$1.28\sigma = 15$$
$$\sigma \doteq 11.72$$

The approximate standard deviation is 11.7.

OR

Using a calculator:
$z = \text{invNorm}(0.90, 0, 1) \doteq 1.28$

$$1.28 = \frac{80-65}{\sigma}$$
$$\sigma \doteq 11.72$$

**22. A**

$\text{normalcdf}\left(70, 1\times10^{99}, 56, 10\right) \doteq 0.08$. Thus, 8% of the students received an A.

 

**23.**

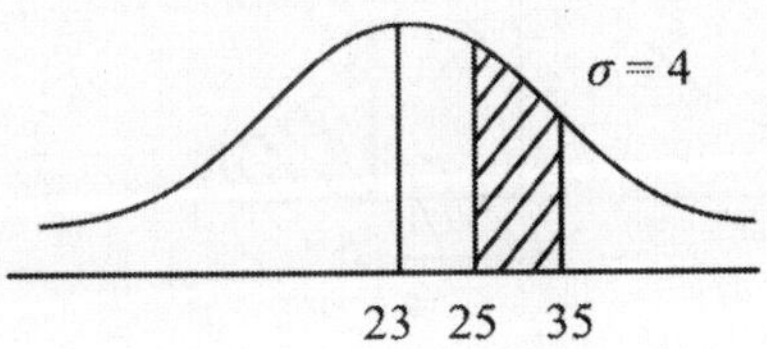


$z = \dfrac{25-23}{4}$ $\quad z = \dfrac{35-25}{4}$

$z = 0.5$ $\quad z = 2.5$

Using the table, the area between $z = 0.5$ and $z = 2.5$ is $(0.993\ 8 - 0.691\ 5) = 0.302\ 3$.

Thus, the probability of a teacher teaching between 25 and 35 years, to the nearest hundredth, is approximately 0.30.

| | | | |
|---|---|---|---|
| 0 | . | 3 | 0 |

OR

Using a calculator:
normalcdf $(25, 35, 23, 4) \doteq 0.31$

| | | | |
|---|---|---|---|
| 0 | . | 3 | 1 |

## *Practice Quiz*

### ANSWERS AND SOLUTIONS

**1.** Use a calculator as described in the notes:
mean = 5.3
median = 4.5
mode(s) = 2.0 (not shown on calculator)
range = max $x$ – min $x$ $= 12 - 2 = 10.0$
standard deviation = 3.1

**2.** $\dfrac{\text{Sum of values}}{9} = 62$

$\text{Sum of values} = 558$

$\dfrac{\text{Sum of values} + 10^{\text{th}} \text{ value}}{10} = 65$

$\dfrac{558 + 10^{\text{th}} \text{ value}}{10} = 65$

$558 + 10^{\text{th}} \text{ value} = 650$

$10^{\text{th}} \text{ value} = 92$

**3.** Use a calculator as described in the notes:

**a)** Mean = 8.2

**b)** Standard deviation $\doteq 3.3$

**c)** One standard deviation below the mean is approximately $8.2 - 3.3 = 4.9$. One standard deviation above the mean is approximately $8.2 + 3.3 = 11.5$.
There are 68 scores between 4.9 and 11.5.

The percentage is $\dfrac{68}{90} \doteq 0.7555 \doteq 75.6\%$

**4.** The number of pages a print cartridge should be good for is $1\ 400 + 90(0.25) = 1\ 422.5$, which would be 1 422.

**5.** 1.2 units of standard deviation is $1.2(25) = 30$. Thus, the data value of 150 is 30 above the mean. The mean is $150 - 30 = 120$.

**6.** On the original test, $z = \dfrac{48-55}{9}$.

$z \doteq -0.78$
The $z$-score in the new test will remain the same.

On the new test, $-0.78 = \dfrac{x-65}{12}$.

$-9.33 = x - 65$
$55.67 \doteq x$

The new score will be approximately 56%.

**7.** Using the table, the area under the curve between $z = 0.5$ and $z = 1.4$ is $(0.919\ 2 - 0.691\ 5) = 0.227\ 7 \doteq 0.23$.

OR

Using a calculator:
$\text{normalcdf}(0.5, 1.4, 0, 1) \doteq 0.23$

**8.** Using the table, the area to the right of $z = 0.230$ is $(1 - 0.591\ 0) = 0.409 \doteq 0.4$.

**9.** $z = \text{invNorm}(0.5 - 0.360) \doteq -1.1$

**10.**

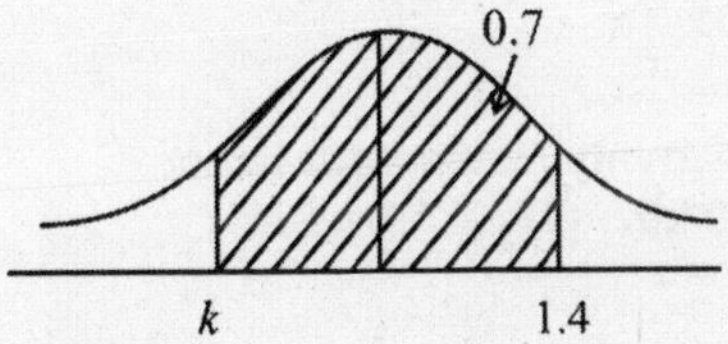


$\text{normalcdf}(-1 \times 10^{99}, 1.4, 0, 1) \doteq 0.919\ 2$

$k = \text{invNorm}(0.919\ 2 - 0.7) \doteq -0.8$

**11.**

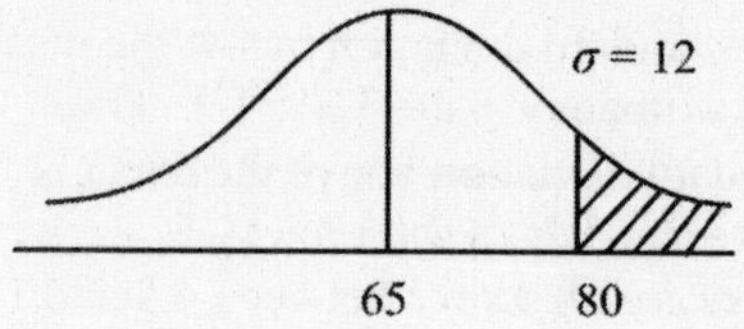


$$z=\frac{80-65}{12}$$
$$z=1.25$$

Using the table, the area to the right of $z=1.25$ is $(1-0.894\ 4)=0.105\ 6$.

Since 450 students received a mark of more than 80, 450 is 0.105 6 of the population.

Thus,

$$\frac{450}{\text{total}}=0.105\ 6$$
$$\text{total}\doteq 4\ 261.36$$

There were approximately 4 261 students who wrote the exam.

OR

Using a calculator:
$450\div\text{normalcdf}\,(80,1\times10^{99},65,12)\doteq 4\ 259$

**12.**

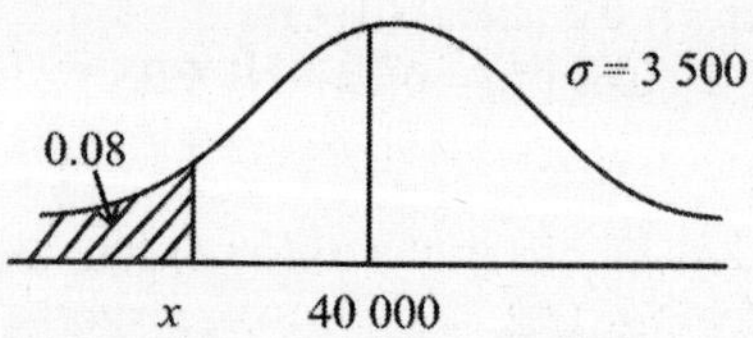


Using the table, when the area under the curve is 0.08, $z=-1.41$.

$$-1.41=\frac{x-40\ 000}{3\ 500}$$
$$-4\ 935=x-40\ 000$$
$$35\ 065=x$$

The warranty period must be set at 35 000 km

OR

Using a calculator:
$\text{invNorm}\,(0.08,40000,3500)\doteq 35000$

**13.**

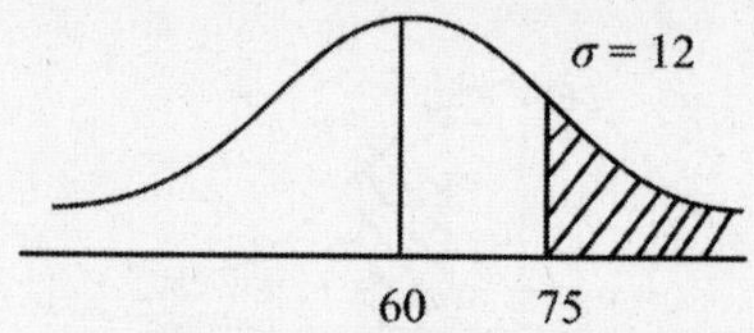


$$z=\frac{60-75}{12}$$
$$z=-1.25$$

Using the table, the area to the left of $z=-1.250$ is 0.105 6. Thus, out of 200 customers, the number expected to spend less than 1 hour in the fitness centre is $0.105\ 6\times200\doteq 21.12$, or 21.

OR

Using a calculator:
$\text{normalcdf}\left(-1\times10^{99},60,75,12\right)\times200\doteq 21$

**14.**

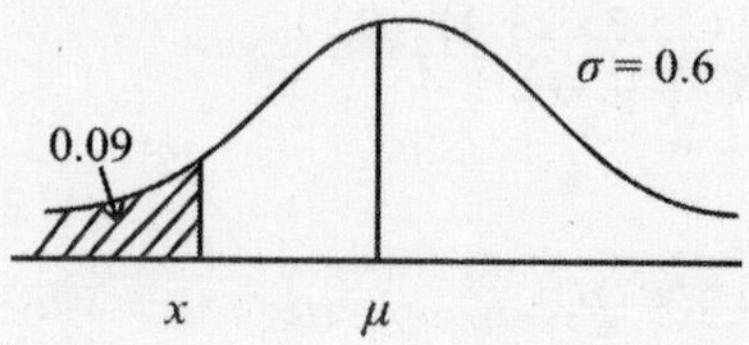


Using the table, when the area under the curve is 0.09, $z=-1.35$.

$$-1.35=\frac{8-\mu}{0.6}$$
$$-0.81=8-\mu$$
$$8.81=\mu$$

The mean diameter of the tomatoes is approximately 8.8 cm.

OR

Using a calculator:
if $z=\text{invNorm}(0.09,0,1)\doteq-1.34$

$$-1.34=\frac{8-\mu}{0.6}$$
$$\mu\doteq 8.81$$

**15.**

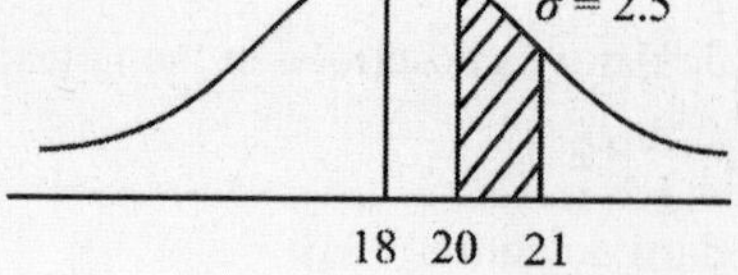


 

$z = \frac{20-18}{2.5}$ $\qquad$ $z = \frac{21-18}{2.5}$

$z = 0.8$ $\qquad$ $z = 1.2$

Using the table, the area between $z = 0.8$ and $z = 1.2$ is $0.884\,9 - 0.788\,1 = 0.096\,8$.

Thus, out of 1 000 batteries, the number that should last between 20 and 21 hours before recharging is $0.096\,8 \times 1\,000 = 96.8$, or 96.

OR

Using a calculator:
$\text{normalcdf}(20, 21, 18, 2.5) \times 1\,000 \doteq 96$

## *Lesson 5 The Binomial Experiment and Binomial Distribution*

### CLASS EXERCISES ANSWERS AND SOLUTIONS

**1.** The number of trials, $n = 10$.
The number of successes, $k = 5$.

The probability of success, $p = \frac{1}{4} = 0.25$.

The probability of failure $= 1 - \frac{1}{4} = \frac{3}{4} = 0.75$.

The probability of 5 successes in 10 trials is $P(5) = {}_{10}C_5(0.25)^5(0.75)^5 \doteq 0.058$.

**2.** Recall that the best approach to the "probability of at least one" situation is usually to determine the probability of none, and subtract from 1.

The number of trials, $n = 5$
The number of successes, $k$ = at least one (i.e., 1, 2, 3, 4, or 5)

The probability of success, $p = 0.70$
The probability of failure $= 1 - 0.70 = 0.30$

The probability that the drug fails for all 5 patients is $P(0) = {}_5C_0(0.70)^0(0.30)^5 = (0.30)^5$.

Therefore, the probability that the drug is effective for at least one patient is $1 - (0.30)^5 = 0.997\,57 \doteq 0.998$.

**3.** $n = 6, p = 0.3, k = 1, 2, 3$

The probability of the player scoring 1, 2, or 3 goals individually in 6 shots is:
binompdf(6, 0.3, {1, 2, 3})
= {.302 526,.324 135,.185 22}

[2nd] [STAT] ▷▷ [MATH] [5:sum(2nd] [ANS] [ENTER] gives 0.811 81.

The probability of the player scoring 1, 2, or 3 goals is approximately 0.811.

**4 a)** The answer could be obtained by finding the binomial probabilities of the marksman hitting his target 0, 2, 3, 4, … 24, 25 times and getting the sum of these probabilities, however, that would be very tedious. The same result can be obtained by using binomcdf(40.0.75,25), which gives 0.054.

**b)** The probability of him hitting the target at least 30 times is:

1 – (the probability of him hitting the target 29 or less times).

The answer can be obtained from
1 – binomcdf(40,0.75,29)

The probability is 0.584.

**5.** binompdf(4, 0.4)= {.129 6, .345 6, .345 6, .153 6, .025 6}.

This is a list of the probabilities of him hitting the fairway 0, 1, 2, 3, or 4 times, respectively, out of 4 drives.

### PRACTICE EXERCISES ANSWERS AND SOLUTIONS

**1.** $P(6) = {}_5C_1\left(\frac{1}{6}\right)^1\left(\frac{5}{6}\right)^4 \doteq 0.401\,9 \doteq 0.402$

OR

$P(6) = \text{binompdf}\left(5, \frac{1}{6}, 1\right) \doteq 0.401\,9 \doteq 0.402$

**2.** $P(3\text{ heads}) = {}_6C_3\left(\frac{1}{2}\right)^3\left(\frac{1}{2}\right)^3 \doteq 0.312\ 5 \doteq 0.313$

OR

$P(3\text{ heads}) = \text{binompdf}(6, 0.5, 3) \doteq 0.312\ 5 \doteq 0.313$

**3.** $P(\text{scoring 4 times}) = {}_5C_4(0.8)^4(0.2)^1$
$\doteq 0.409\ 6 \doteq 0.410$

OR

$P(\text{scoring 4 times}) = \text{binompdf}(5, 0.8, 4)$
$\doteq 0.409\ 6 \doteq 0.410$

**4.** $P(3\text{ goals}) = {}_{10}C_3(0.1)^3(0.9)^7 = 0.057$

OR

$P(3\text{ goals}) = \text{binompdf}(10, 0.1, 3) = 0.057$

**5.**

$$\begin{aligned} P(\text{at least 4 heads}) &= {}_5C_4\left(\frac{1}{2}\right)^4\left(\frac{1}{2}\right)^1 \\ &\quad + {}_5C_5\left(\frac{1}{2}\right)^5\left(\frac{1}{2}\right)^0 \\ &= 0.156\ 25 + 0.031\ 25 \\ &= 0.187\ 50 \\ &\doteq 0.188 \end{aligned}$$

OR

$P(\text{at least 4 heads}) -$
$\text{binompdf}(5, 0.5, \{4, 5\}) \rightarrow \text{ENTER}$
$\text{LIST} \rightarrow \text{MATH} \rightarrow \text{SUM} \rightarrow (\text{ANS}) \rightarrow \text{ENTER}$
$= 0.187\ 5$
$\doteq 0.188$

**6.**

$$\begin{aligned} P(\text{at least 2 6s}) &= {}_4C_2\left(\frac{1}{6}\right)^2\left(\frac{5}{6}\right)^2 + \\ &\quad {}_4C_3\left(\frac{1}{6}\right)^3\left(\frac{5}{6}\right)^1 + {}_4C_4\left(\frac{1}{6}\right)^4\left(\frac{5}{6}\right)^6 \\ &\doteq 0.115\ 7 + 0.015\ 4 + 0.000\ 77 \\ &\doteq 0.131\ 9 \\ &\doteq 0.132 \end{aligned}$$

OR

$P(\text{at least 2 6s}) =$

$\text{binompdf}\left(4, \frac{1}{6}, \{2, 3, 4\}\right) \rightarrow \text{ENTER}$

$\text{LIST} \rightarrow \text{MATH} \rightarrow \text{SUM} \rightarrow (\text{ANS}) \rightarrow \text{ENTER}$
$\doteq 0.131\ 9$
$\doteq 0.132$

**7.**

$$\begin{aligned} P(\text{at least 1 score}) &= 1 - {}_4C_0(0.5)^0(0.5)^4 \\ &\doteq 1 - 0.062\ 5 \\ &= 0.937\ 5 \\ &\doteq 0.938 \end{aligned}$$

OR

$$\begin{aligned} P(\text{at least 1 score}) &= 1 - \text{binompdf}(4, 0.5, 0) \\ &= 1 - 0.062\ 5 \\ &= 0.937\ 5 \\ &\doteq 0.938 \end{aligned}$$

**8.**

$$\begin{aligned} P(\text{effective at least once}) &= 1 - {}_5C_0(0.6)^0(0.4)^5 \\ &= 1 - 0.010\ 24 \\ &= 0.989\ 76 \\ &\doteq 0.990 \end{aligned}$$

OR

$$\begin{aligned} P(\text{effective at least once}) &= 1 - \text{binompdf}(5, 0.6, 0) \\ &= 1 - 0.010\ 24 \\ &\quad 0.989\ 76 \\ &\doteq 0.990 \end{aligned}$$

 

**9.** $P(9 \text{ or } 10 \text{ throws}) = {}_{10}C_9(0.9)^9(0.1)^1$
$+{}_{10}C_{10}(0.9)^{10}+(0.1)^0$
$\doteq 0.387\ 4+0.348\ 7$
$\doteq 0.736\ 1$

OR

$P(9 \text{ or } 10 \text{ throws}) = \text{binompdf}(10,\ 0.9,\ \{9,10\})$
$\rightarrow$ ENTER

LIST $\rightarrow$ MATH $\rightarrow$ SUM $\rightarrow$ (ANS) $\rightarrow$ ENTER
$\doteq 0.736\ 1$

The player is incorrect. In order to be "virtually guaranteed" the probability should be very close to 1. The probability in this case is less than 75%.

**10.** Probability distribution:
${}_4C_0(0.8)^4(0.2)^0,\ {}_4C_1(0.8)^1(0.2)^3,\ {}_4C_2(0.8)^2(0.2)^2,$
${}_4C_3(0.8)^3(0.2)^1,\ {}_4C_4(0.8)^0(0.2)^4$
0.001 6, 0.025 6, 0.153 6, 0.409 6, 0.001 6
OR
binompdf
binompdf $(4, 0.8, \{0,1,2,3,4\}) = .001\ 6,\ .025\ 6,$
.153 6, .409 6, .001 6

**11.** Probability distribution:
$${}_5C_0\left(\frac{1}{2}\right)^5\left(\frac{1}{2}\right)^0 + {}_5C_1\left(\frac{1}{2}\right)^4\left(\frac{1}{2}\right)^1 + {}_5C_2\left(\frac{1}{2}\right)^2\left(\frac{1}{2}\right)^3$$
$$+{}_5C_3\left(\frac{1}{2}\right)^3\left(\frac{1}{2}\right)^2 + {}_5C_4\left(\frac{1}{2}\right)^4\left(\frac{1}{2}\right)^1 + {}_5C_5\left(\frac{1}{2}\right)^0\left(\frac{1}{2}\right)^5$$
0.03125, 0.15625, 0.3125, 0.3125, 0.15625, 0.03125

OR

binompdf(5, 0.5, {0,1,2,3,4,5})
.031 25, .156 25, .312 5, .312 5, .156 25, .031 25

**12.** $P(3 \text{ are good}) = \dfrac{{}_4C_3}{{}_6C_3} = \dfrac{1}{5}$
This is not a binomial probability question because the trials are not independent. The probability changes as soon as one battery is selected.

**13.** $P(\text{temperature above } 20° \text{ at least 40 times})$
$= {}_{50}C_{40}(0.8)^{40}(0.2)^{10} + {}_{50}C_{41}(0.8)^{41}(0.2)^9$
$+{}_{50}C_{42}(0.8)^{42}(0.2)^8+,,,+{}_{50}C_{50}(0.8)^{50}(0.2)^0$

OR

binompdf (50, 0.8, {40, 41, 42, 43, 44, 45, 46, 47, 48, 49, 50})

LIST $\rightarrow$ MATH $\rightarrow$ SUM $\rightarrow$ (ANS) $\rightarrow$ ENTER
$\doteq 0.584$

## *Practice Test*

### ANSWERS AND SOLUTIONS

**1.** Use a calculator as described in the notes:

**a)** Mean $\doteq 76.8$

**b)** Standard deviation $\doteq 3.2$

**c)** One standard deviation below the mean is approximately $76.8 - 3.2 = 73.6$. One standard deviation above the mean is approximately $76.8 + 3.2 = 80$.

There are 50 scores within one standard deviation of the mean.

**2.** The new mean will be 55% + 10% = 65%
The standard deviation will remain at 12% because the dispersion of marks has not changed.

**3.**

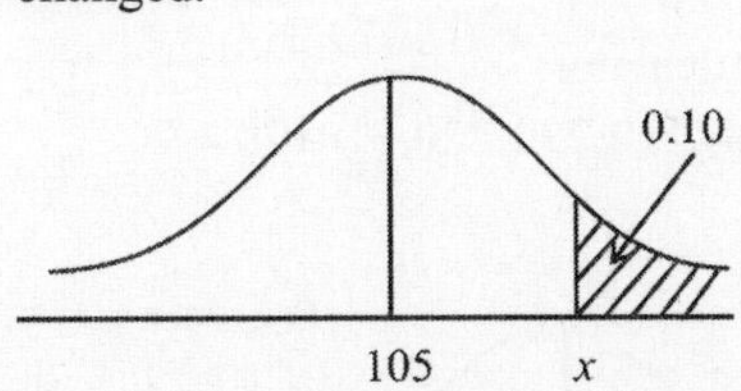


Using the table, the area under the curve is $(1 - 0.10) = 0.90$ for $z = 1.29$.

$1.29 = \dfrac{x-105}{15}$
$19.35 = x - 105$
$124.35 = x$

Thus, to be in the top 10% of all scores, Robert's score must be at least 125.

OR

Using ca alculator:
$\text{invNorm}(0.9, 105, 15) \doteq 124$

**4.**

$$z=\frac{x-\mu}{\sigma}$$

$$2=\frac{(12k-24)-6k}{(k+6)}$$

$$2k+12=6k-24$$

$$-4k=-36$$

$$k=9$$

Fred's actual score is $12(9)-24=84$.

**5.**

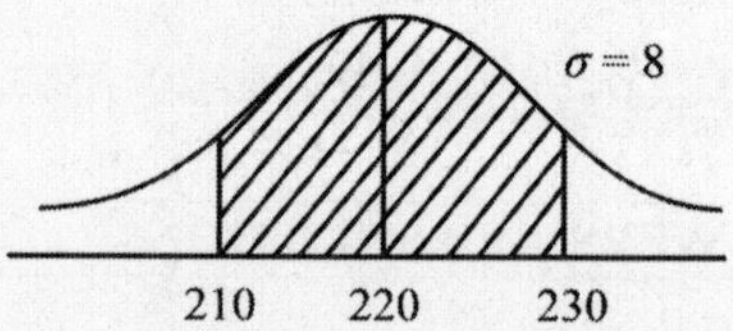


$$z=\frac{210-220}{8} \qquad z=\frac{230-220}{8}$$

$$z=-1.25 \qquad z=1.25$$

Using the table, the area between $z=-1.25$ and $z=1.25$ is $(0.894\ 4-0.105\ 6)=0.788\ 8$.

Thus, in a box 120 oranges, the number that can be expected to have a mass between 210 g and 230 g is $0.788\ 8\times 120 \doteq 94.656$, or approximately 94 oranges.

OR

Using a calculator:
$\text{normalcdf}(210,230,220,8)\times 120 \doteq 94$

**6.**

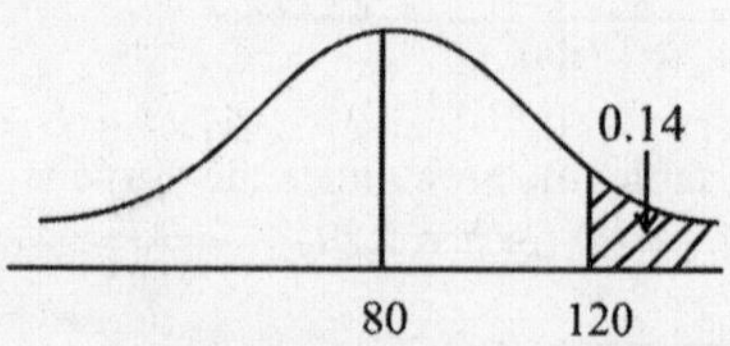


Using the table, the, area under the curve is $(1-0.14)=0.86$ for $z=1.09$.

$$1.09=\frac{120-80}{\sigma}$$

$$1.09\sigma=40$$

$$\sigma\doteq 36.7$$

The standard deviation is approximately 37 minutes.

OR

Using a calculator:
$z=\text{invNorm}(0.86,0,1)\doteq 1.08$

$$1.08=\frac{120-80}{\sigma}$$

$$\sigma\doteq 37$$

**7.**

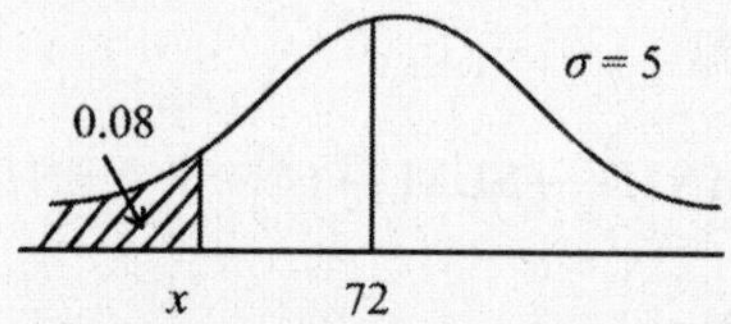


Using the table, the area under the curve is 0.08, when $z=-1.41$.

$$-1.41=\frac{x-72}{5}$$

$$-7.05=x-72$$

$$x=64.95$$

OR

Using a calculator:
$\text{invNorm}(0.08,72,5)\doteq 64.97$

The length of the warranty should be 64 months (because it is so close to 65, the manufacturer will probably use 65 months).

**8.**

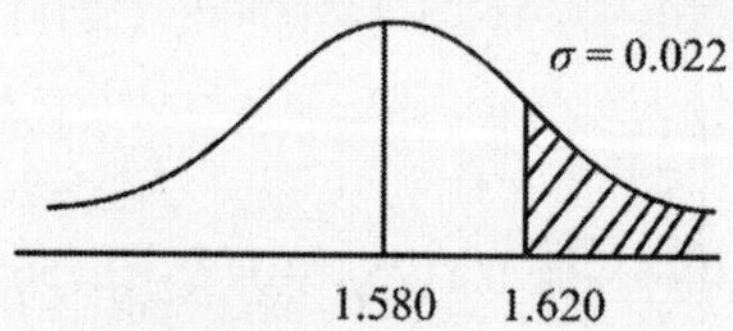


$$z=\frac{1.62-1.58}{0.022}$$

$$z\doteq 1.82$$

Using the table, the area to the right $z=1.82$ is $(1-0.965\ 6)\doteq 0.034\ 4$.

Thus, out of 1 000 golf balls, the expected number of balls that will be too heavy is $0.034\ 4\times 1\ 000=34.4$, or approximately 34.

OR

Using a calculator:
$\text{normal}(1.62,1\times 10^{99},1.58,0.022)\times 1000\doteq 34$

 

**9.** $P(3\ \ 6s) = {}_6C_3 \left(\frac{1}{6}\right)^3 \left(\frac{5}{6}\right)^3 \doteq 0.05$

OR

$P(3\ \ 6s) = \text{binompdf}\left(6, \frac{1}{6}, 3\right) \doteq 0.05$

**10.** $P(\text{at least 5 bull's-eyes})$
$= {}_{10}C_5 (0.5)^5 (0.4)^5 + {}_{10}C_6 (0.6)^6 (0.4)^4$

$+ {}_{10}C_7 (0.6)^7 (0.4)^3$
$+ {}_{10}C_8 (0.6)^8 (0.4)^2$
$+ {}_{10}C_9 (0.6)^6 (0.4)^1$
$+ {}_{10}C_{10} (0.6)^{10} (0.4)^0$
$\doteq 0.834$

OR

$P(\text{at least 5 bull's-eyes}) =$
$\text{binompdf}(10, 0.6, \{5, 6, 7, 8, 9, 10\})$

LIST → MATH → SUM → (ANS) →
ENTER
$\doteq 0.834$

**11.** $P(\text{effective at least once}) = 1 - {}_6C_0\ (0.4)^0 (0.6)^6$
$\doteq 0.953$

OR

$P(\text{effective at least once}) = 1 - \text{binompdf}(6, 0.4, 0)$
$\doteq 0.953$

**12.** $P(\text{at least one cracked}) = 1 - P(\text{no cracked})$
$= 1 - \frac{{}_{10}C_3}{{}_{12}C_3}$
$\doteq 1 - 0.545$
$\doteq 0.455$

This is not a binomial probability question because the trials are not independent. The probability changes as soon as one egg is selected.

**13.** $P(\text{at least 40 take accelerated math})$
$= {}_{50}C_{40} (0.7)^{40} (0.3)^{10}$
$+ {}_{50}C_{41} (0.7)^{41} (0.3)^9$
$+ {}_{50}C_{42} (0.7)^{42} (0.3)^8$
$+,,, + {}_{50}C_{50} (0.7)^{50} (0.3)^0 \doteq 0.079$

OR

$P(\text{at least 40 take accelerated math}) =$
binompdf (50, 0.7{40, 41, 42, 43, 44,
45, 46, 47, 48, 49, 50})
LIST → MATH → SUM → (ANS) →
ENTER $\doteq 0.079$

**14.**

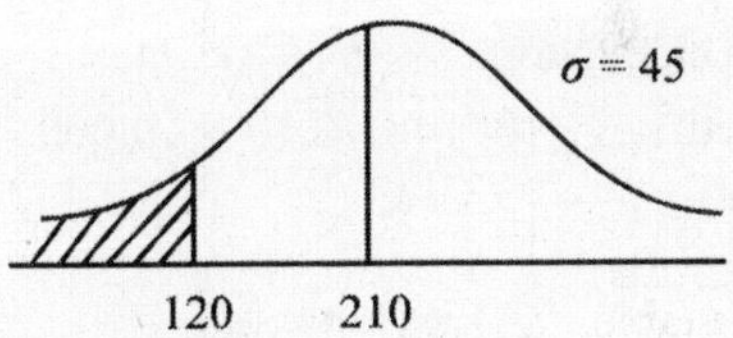


$z = \frac{120 - 210}{45}$

$z = -2$

Using the table, the area to the left of $z = -2$ is 0.022 8.

Thus, out of 1 400 students, the number expected to spend at most 2 hours per week using a computer is
$0.022\,8 \times 1\,400 = 31.92$, or 31 students.

OR

Using a calculator:
$\text{normalcdf}(-1 \times 10^{99}, 120, 210, 45) \times 1\,400 \doteq 31$

**15.**

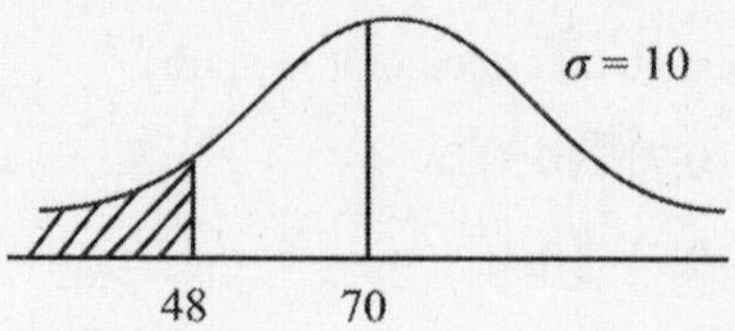


$$z = \frac{48-70}{10}$$
$$z = -2.2$$

Using the table, the area to the left $z = -2.2$ is 0.013 9.

Thus, out of 10 000 batteries, the number they would expect to replace is 0.013 9 × 10 000 = 139.

OR

Using a calculator:
$\text{normalcdf}\left(-1\times10^{99}, 48, 70, 10\right)\times 10\,000 \doteq 139$

| | | | |
|---|---|---|---|
| 1 | 3 | 9 | |

# CONICS

## *Lesson 1 Sections of Cones*

### CLASS EXERCISES ANSWERS AND SOLUTIONS

1. Half the vertex angle is 36°. If $\theta$ is the angle between the cutting plane and the axis then, to create a hyperbola, $0 \le \theta < 36°$.

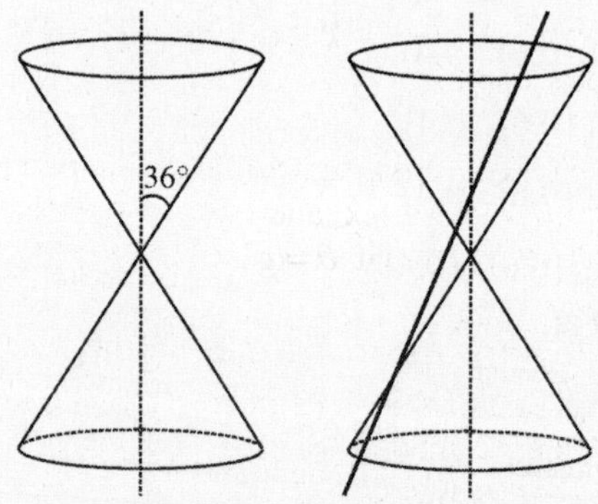


### PRACTICE EXERCISES ANSWERS AND SOLUTIONS

1. Point (degenerate circle)
2. Hyperbola
3. Parabola
4. One line (degenerate parabola)
5. Ellipse
6. Circle
7. One line (degenerate parabola)
8. Ellipse
9. Two parallel lines (degenerate parabola)
10. See lesson notes, p. 524
11. The plane passes through the vertex of a double-napped cone and intersects the axis at an angle $\theta$, such that $0 \le \theta < g$, where $g$ is the generator angle.
12. The plane passes through the vertex of a double-napped cone and intersects the axis at an angle $\theta$, such that $g < \theta \le 90°$, where $g$ is the generator angle.

**13.** The plane passes through a double-napped cone, not through the vertex, and intersects the axis at an angle $\theta$, such that $g < \theta < 90°$, where $g$ is the generator angle. The plane passes through a cylinder at an angle $\theta$ to the axis of the cylinder, such that $0° < \theta < 90°$.

**14.** The plane is parallel to the axis of the cylinder and intersects the cylinder.

**15.** The plane passes through a double-napped cone, not through the vertex, and intersects the axis at an angle $\theta$, such that $0 \le \theta < g$, where $g$ is the generator angle.

**16.** The plane passes through a double-napped cone, not through the vertex, and intersects the axis at an angle $\theta$, such that $\theta = g$, where $g$ is the generator angle.

**17.** If the angle that the plane makes with the axis of the cone is $\theta$, then to form
**a)** a circle: $\theta = 90°$
**b)** an ellipse: $40° < \theta < 90°$
**c)** a hyperbola: $0° \le \theta < 40°$
**d)** a parabola: $\theta = 40°$

**18.** If the angle that the plane makes with the axis of the cone is $\theta$, then to form
**a)** a circle: $\theta = 90°$
**b)** an ellipse: $53.5° < \theta < 90°$
**c)** a hyperbola: $0° \le \theta < 53.5°$
**d)** a parabola: $\theta = 53.5°$

**19.** If the angle that the plane makes with the axis of the cone is $\theta$, then to form
**a)** a circle: $\theta = 90°$
**b)** an ellipse: $14° < \theta < 90°$
**c)** a hyperbola: $0° \le \theta < 14°$
**d)** a parabola: $\theta = 14°$

**20.** The vertex angle is 46°.

**21.** Since an angle of 50° forms an ellipse, then the possible values of the vertex angle, $\theta$, are $0° < \theta < 100°$. A diagram might help you determine this.

**22.** Since an angle of 20° forms a hyperbola, then the possible values of the vertex angle, $\theta$, are $40° < \theta < 180°$. A diagram might help you determine this.

**23.** Ellipse

**24.** The light from the flashlight will produce a circle, ellipse, parabola, or hyperbola (one section). The shape produced is determined by the angle the flashlight is held in relation to the wall.

**25.** Notice that the angle at the centre of the cone of light is 60°.

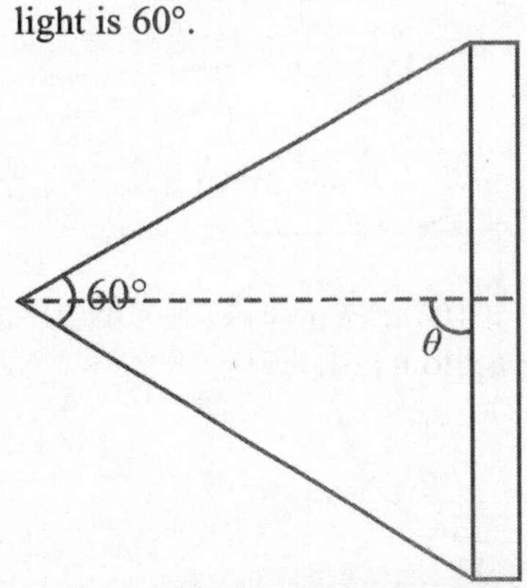


If $\theta$ is the angle that the flashlight is held in relation to the wall, then, to form
**a)** an ellipse: $30° < \theta < 90°$
**b)** a parabola: $\theta = 30°$
**c)** a hyperbola: $-30° < \theta < 30°$ (Notice that the flashlight can rotate up to 30° beyond parallel to the wall and still create a hyperbolic shape of light on the wall.)

**26.** C

**27.** B
Two parallel lines are formed when a plane intersects a cylinder with the plane parallel to the axis of the cylinder.

**28.** D

**29.** 74

| 7 | 4 | | |
|---|---|---|---|

## *Lesson 2 Equations of Conics—The Circle*

### CLASS EXERCISES ANSWERS AND SOLUTIONS

**1.** The graph of $(x+5)^2 + (y-2)^2 = 36$ is the graph of $x^2 + y^2 = 36$ translated 5 units to the left (since $x$ was replaced by $(x + 5)$) and 2 units up (since $y$ was replaced by $(y - 2)$). The radius is unchanged at 6.

**2.** The standard form of the equation is
$(x-3)^2+(y+5)^2=8^2$ or
$(x-3)^2+(y+5)^2=64$.

**3.** The centre is at (1, –6) and the radius, $r$, is
$r=\sqrt{49}=7$.

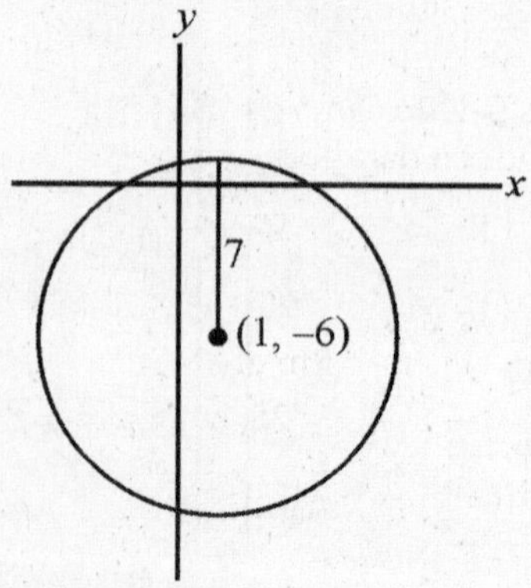


From the equation and graph, you can now determine the
**a)** domain
**b)** range
**c)** $x$-intercepts
**d)** $y$-intercepts

**a)** and **b)** For the domain and range, use the centre (1, –6) and the fact that the radius is 7 to find the endpoints of a horizontal diameter and a vertical diameter. The endpoints of the horizontal diameter are (1 –7, 6) = (–6, –6) and (1 +7, –6) = (8, –6). Thus, the domain is $-6\le x\le 8$.

Similarly, the endpoints of the vertical diameter are (1, –6 –7) = (1, –13) and (1, –6 +7) = (1, 1). Thus, the range is $-13\le y\le 1$.

**c)** To determine the $x$-intercepts, let $y = 0$.

$$(x-1)^2+(0+6)^2=49$$
$$(x-1)^2+36=49$$
$$(x-1)^2=13$$
- Subtract 36 from both sides

$$x-1=\pm\sqrt{13}$$
- Take the square root of both sides

$$x=1\pm\sqrt{13}$$
- Add 1 to both sides to solve for $x$

Thus, the $x$-intercepts are $1+\sqrt{13}$ and $1-\sqrt{13}$ (approximately 4.61 and –2.61).

**d)** To determine the $y$-intercepts, let $x = 0$.

$$(0-1)^2+(y+6)^2=49$$
$$1+(y+6)^2=49$$
$$(y+6)^2=48$$
- Subtract 1 from both sides

$$y+6=\pm\sqrt{48}$$
- Take the square root of both sides

$$y+6=\pm 4\sqrt{3}$$
- Simplify $\sqrt{48}$

$$y=-6\pm 4\sqrt{3}$$
- Subtract 6 from both sides to solve for $y$

Thus, the $y$-intercepts are $-6-4\sqrt{3}$ and $-6+4\sqrt{3}$ (approximately –12.93 and 0.93).

**4.** $$(x-3)^2+(y+1)^2=25$$
$$(x^2-6x+9)+(y^2+2y+1)=25$$
- Expand the binomials
Note: The trinomials in the brackets are perfect square trinomials.

$$x^2+y^2-6x+2y-15=0$$
- Collect like terms and make RHS = 0

**5.** Again, since $A = C$, this should be a circle.
$$3x^2+3y^2+12x-24y+60=0$$

Divide both sides of the equation by 3
$$x^2+y^2+4x-8y+20=0$$

Add –20 to both sides of the equation
$$x^2+y^2+4x-8y=-20$$

Rearrange and leave spaces to complete the squares
$$(x^2+4x+\_)+(y^2-8y+\_)=-20$$

Complete the squares by adding 4 and 16 to both sides
$$(x^2+4x+\underline{4})+(y^2-8y+\underline{16})=-20+4+16$$

Factor the perfect square trinomials produced and add the numbers on the RHS
$$(x+2)^2+(y-4)^2=0$$

 

This appears to be a circle with centre (–2, 4), but you can see that the radius is 0. This is a degenerate form of a circle since it represents a point. The only point that satisfies the equation is (–2, 4). The physical representation is the plane intersecting the vertex of the cone.

**6.** Again, since $A = C$, this should be a circle.

Following the steps described before:

$$x^2 + y^2 - 2x + 10y = -30$$

$$\left(x^2 - 2x + \_\_\right) + \left(y^2 + 10y + \_\_\right) = -30$$

$$\left(x^2 - 2x + \underline{1}\right) + \left(y^2 + 10y + \underline{25}\right) = -30 + 1 + 25$$

$$(x-1)^2 + (y+5)^2 = -4$$

This appears to be a circle with centre (1, –5), but the –4 on the RHS makes the equation impossible to solve. Substituting any numbers for $x$ and $y$ makes the LHS positive; it cannot equal –4. It is also clear that the square of the radius cannot be a negative number. Thus, there is no graph and it is a degenerate form of a circle.

## PRACTICE EXERCISES ANSWERS AND SOLUTIONS

**1.** The centre is (0, 0) and the radius is 4. Therefore, the standard form is $x^2 + y^2 = 4^2$. In general form, this equation is $x^2 + y^2 - 16 = 0$.

**2.** The centre is (0, 0) and the radius is 5. Therefore, the standard form is $x^2 + y^2 = 5^2$. In general form, this equation is $x^2 + y^2 - 25 = 0$.

**3.** The centre is (0, 0) and the radius is 1. Therefore, the standard form is $x^2 + y^2 = 1^2$. In general form, this equation is $x^2 + y^2 - 1 = 0$.

**4.** The centre is (2, –3) and the radius is 4. Therefore, the standard form is $(x-2)^2 + (y-(-3))^2 = 4^2$ or $(x-2)^2 + (y+3)^2 = 4^2$. Converting to general form gives $x^2 - 4x + 4 + y^2 + 6y + 9 = 16$, which is $x^2 + y^2 - 4x + 6y - 3 = 0$.

**5.** The centre is (–4, 2) and the radius is 3. Therefore, the standard form is $(x-(-4))^2 + (y-2)^2 = 3^2$ or $(x+4)^2 + (y-2)^2 = 3^2$. Converting to general form gives $x^2 + 8x + 16 + y^2 - 4y + 4 = 9$, which is $x^2 + y^2 + 8x - 4y + 11 = 0$.

**6.** The centre is (–1, 1) and the radius is 5. Therefore, the standard form is $(x-(-1))^2 + (y-1)^2 = 5^2$ or $(x+1)^2 + (y-1)^2 = 5^2$. Converting to general form gives $x^2 + 2x + 1 + y^2 - 2y + 1 = 25$, which is $x^2 + y^2 - 2y + 2x - 23 = 0$.

**7.** The graph is translated 2 units right and 8 units down.

**8.** The graph of $(x-4)^2 + (y+5)^2 = 16$ is the graph of $x^2 + y^2 = 16$ translated 4 units right and 5 units down.

**9.** The centre is (–2, 4) and the radius is 5.

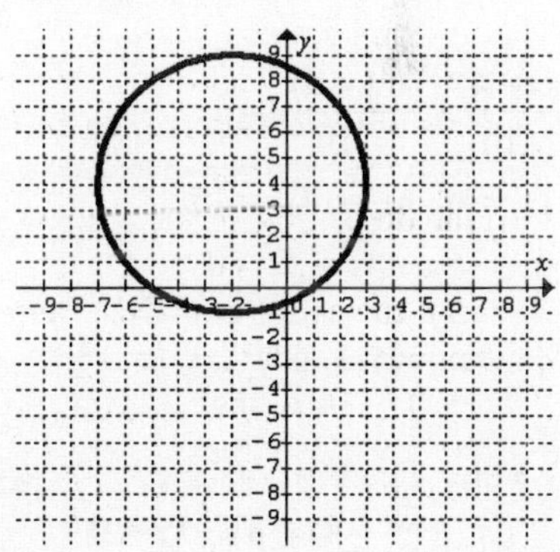


**a)** Domain: $-7 \le x \le 3$

**b)** Range: $-1 \le y \le 9$

**c)** $x$-intercepts:

$$(x+2)^2 + (0-4)^2 = 25$$

$$(x+2)^2 + 16 = 25$$

$$(x+2)^2 = 9$$

$$x + 2 = \pm 3$$

$$x = -5 \text{ or } 1$$

**d)** $y$-intercepts:

$$(0+2)^2+(y-4)^2=25$$
$$4+(y-4)^2=25$$
$$(y-4)^2=21$$
$$y-4=\pm\sqrt{21}$$
$$y=4+\sqrt{21} \text{ or } 4-\sqrt{21}$$

Note: For questions 10–12 the signs of the values of $A$, $C$, $D$, $E$, and $F$ may be reversed for any particular question.

**10.**

$$(x-2)^2+(y+5)^2=3^2$$
$$x^2-4x+4+y^2+10y+25=9$$
$$x^2+y^2-4x+10y+20=0$$

$A=1, C=1, D=-4, E=10, F=20$

**11.**

$$(x+3)^2+(y-1)^2=49$$
$$x^2+6x+9+y^2-2y+1=49$$
$$x^2+y^2+6x-2y-39=0$$

$A=1, C=1, D=6, E=-2, F=-39$

**12.**

$$\frac{(x-4)^2}{10}+\frac{(y+4)^2}{10}=1$$
$$(x-4)^2+(y+4)^2=10$$
$$x^2+y^2-8x+8y+22=0$$

$A=1, C=1, D=-8, E=8, F=22$

**13.**

$$x^2+y^2-4x+8y+19=0$$
$$\left(x^2-4x+\quad\right)+\left(y^2+8y+\quad\right)=-19$$
$$\left(x^2-4x+4\right)+\left(y^2+8y+16\right)=-19+20$$
$$(x-2)^2+(y+4)^2=1^2$$

The centre is (2, –4) and the radius is 1.

**14.**

$$x^2+y^2-6x+14y+54=0$$
$$\left(x^2-6x+\quad\right)+\left(y^2+14y+\quad\right)=-54$$
$$\left(x^2-6x+9\right)+\left(y^2+14y+49\right)=-54+58$$
$$(x-3)^2+(y+7)^2=2^2$$

The centre is (3, –7) and the radius is 2.

**15.**

$$2x^2+2y^2-8x+12y+14=0$$
$$x^2+y^2-4x+6y+7=0$$
$$\left(x^2-4x+\quad\right)+\left(y^2+6y+\quad\right)=-7$$
$$\left(x^2-4x+4\right)+\left(y^2+6y+9\right)=-7+13$$
$$(x-2)^2+(y+3)^2=\left(\sqrt{6}\right)^2$$

The centre is (2, –3) and the radius is $\sqrt{6}$.

**16.**

$$30x+3x^2+3y^2+51=0$$
$$10x+x^2+y^2+17=0$$
$$\left(x^2+10x+\quad\right)+y^2=-17$$
$$\left(x^2+10x+25\right)+y^2=-17+25$$
$$(x+5)^2+(y-0)^2=\left(2\sqrt{2}\right)^2$$

The centre is (–5, 0) and the radius is $2\sqrt{2}$.

**17.**

$$x^2+y^2-2x-6y-6=0$$
$$\left(x^2-2x+\quad\right)+\left(y^2-6y+\quad\right)=6$$
$$\left(x^2-2x+1\right)+\left(y^2-6y+9\right)=6+10$$
$$(x-1)^2+(y-3)^2=4^2$$

The centre is (1, 3) and the radius is 4.

**a)** Domain: $-3 \le x \le 5$

**b)** Range: $-1 \le y \le 7$

**c)** To find the $x$-intercepts, let $y=0$.

$$(x-1)^2+(0-3)^2=4^2$$
$$(x-1)^2+9=16$$
$$(x-1)^2=7$$
$$x-1=\pm\sqrt{7}$$
$$x=1\pm\sqrt{7}$$

**d)** To find the $y$-intercepts, let $x = 0$.

$(0-1)^2 + (y-3)^2 = 4^2$

$1 + (y-3)^2 = 16$

$(y-3)^2 = 15$

$y - 3 = \pm\sqrt{15}$

$y = 3 \pm \sqrt{15}$

**e)**

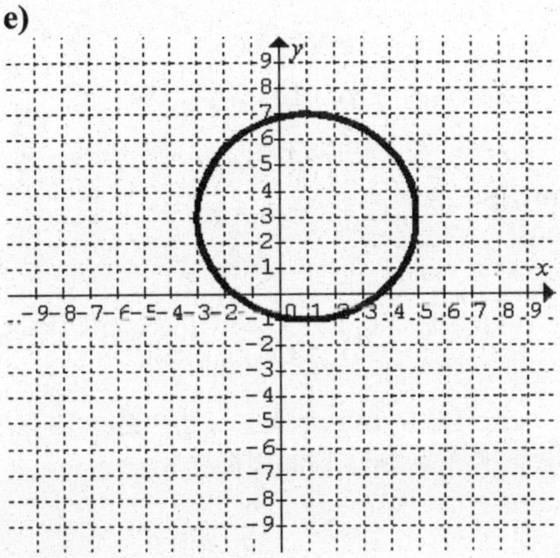

**18.** The new equation is

$\left(\frac{1}{2}x\right)^2 + \left(\frac{1}{2}y\right)^2 = 20$

$\frac{x^2}{4} + \frac{y^2}{4} = 20$

$x^2 + y^2 = 80$

The centre is unaffected. The radius changes from $\sqrt{20} = 2\sqrt{5}$ to $\sqrt{80} = 4\sqrt{5}$. The radius is doubled.

**19.** The new equation is

$(2x)^2 + (2y-2)^2 = 4$

$4x^2 + 4(y-1)^2 = 4$

$x^2 + (y-1)^2 = 1$

The centre moves from (0, 2) to (0, 1). The radius changes from 2 to 1. The radius is halved.

**20.** $x^2 + y^2 - 6x + 8y + F = 0$

$x^2 - 6x + y^2 + 8y = -F$

$\left(x^2 - 6x + \quad\right) + \left(y^2 + 8y + \quad\right) = -F$

$\left(x^2 - 6x + 9\right) + \left(y^2 + 8y + 16\right) = -F + 25$

For the equation to represent a non-degenerate conic, the value of $F$ must be less than 25 ($F < 25$).

**21.** $x^2 + y^2 - 10x - 12y + F = 0$

$x^2 - 10x + y^2 - 12y = -F$

$\left(x^2 - 10x + 25\right) + \left(y^2 - 12y + 36\right) = -F + 61$

For the equation to represent a degenerate conic the value of $F$ must be greater than or equal to 61 ($F \geq 61$).

**22.** To find the $x$-intercepts, let $y = 0$.

$x^2 + (0)^2 + 6x - 4(0) - 16 = 0$

$x^2 + 6x - 16 = 0$

$(x+8)(x-2) = 0$

$x = -8$ or $x = 2$

The value of the largest $x$-intercept is 2.

| 2 | | | |
|---|---|---|---|

**23.** For $x^2 + y^2 = 100$, $r = \sqrt{100} = 10$.

For $x^2 + y^2 = 25$, $r = \sqrt{25} = 5$.

Since the new radius is half the original radius, the stretch factor is $\frac{1}{2}$ or 0.5.

| 0 | . | 5 | |
|---|---|---|---|

## *Lesson 3 Equations of Conics—The Ellipse*

### CLASS EXERCISES ANSWERS AND SOLUTIONS

**1.** **a)** Here, $a^2 = 16$, thus, $a = 4$, since $a > 0$. The $x$-intercepts are +4 and –4.

Here, $b^2 = 64$, thus, $b = 8$, since $b > 0$. The $y$-intercepts are +8 and –8.

Using the intercepts to graph gives:

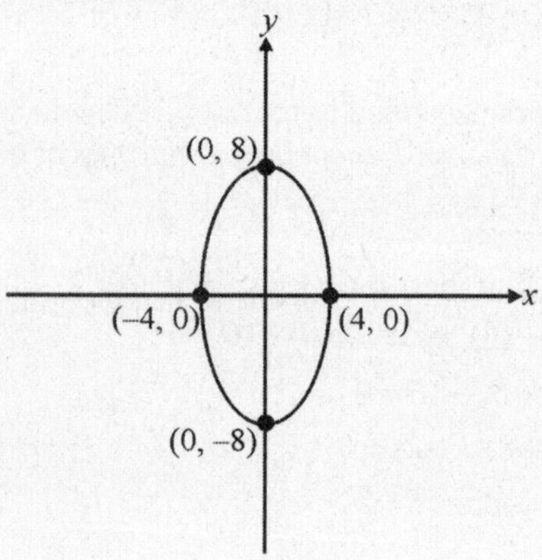


From the graph:
Domain: $-4 \leq x \leq 4$
Range: $-8 \leq y \leq 8$

**2.** $\frac{x^2}{16} + \frac{y^2}{64} = 1$

64 is the least common multiple of 16 and 64.

$4x^2 + y^2 = 64$ • Multiply both sides of the equation by 64

$4x^2 + y^2 - 64 = 0$ • Subtract 64 from both sides of the equation

Here, $A = 4$, $C = 1$, $D = 0$, $E = 0$, and $F = -64$.

**3.** You should recognize that this is the equation of an ellipse with centre at (–1, 2).

$a^2 = 49$ $\quad$ $b^2 = 16$
$a = 7$ $\quad$ $b = 4$

The endpoints of the major axis are 7 units from the centre and the endpoints of the minor axis are 4 units from the centre.

The graph is:

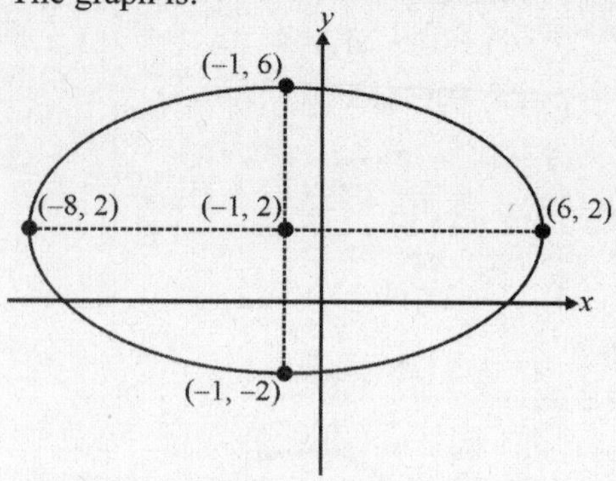


Domain: $-8 \leq x \leq 6$
Range: $-2 \leq y \leq 6$

**4.** **a)** Since $AC > 0$ and $A \neq C$, the conic is an ellipse or a degenerate ellipse.

**b)** Add –39 to both sides of the equation, rearrange the LHS

$3x^2 - 6x + 4y^2 + 24y = -39$

Factor out the coefficient of $x^2$ from the terms involving $x$ and the coefficient of $y^2$ from the terms involving $y$, and leave spaces to complete the squares

$3(x^2 - 2x + \_) + 4(y^2 + 6y + \_) = -39$

Since the squares were completed by adding 1 and 9 to the spaces and these numbers are in brackets, 3 and 36 were actually added to the LHS and thus 3 and 36 are added to the RHS

$3(x^2 - 2x + \underline{1}) + 4(y^2 + 6y + \underline{9}) = -39 + 3 + 36$

Factor the perfect square trinomials that were created and collect like terms on the RHS

$3(x-1)^2 + 4(y+3)^2 = 0$

Since the right-hand side of the equation is 0, the equation will not fit the form $\frac{(x-h)^2}{a^2} + \frac{(y-k)^2}{b^2} = 1$. Notice that the only ordered pair that satisfies the equation is (1, –3). Thus, this is a degenerate ellipse consisting only of one point. The physical representation is the plane intersecting the vertex of the cone.

## PRACTICE EXERCISES ANSWERS AND SOLUTIONS

**1.** $\frac{x^2}{4}+\frac{(y-2)^2}{16}=1$

$4x^2+(y-2)^2=16$

$4x^2+y^2-4y+4=16$

$4x^2+y^2-4y-12=0$

$A=4,\ C=1,\ D=0,\ E=-4,\ F=-12$

**2.** $\frac{(x-3)^2}{8}+\frac{(y+1)^2}{12}=1$

$3(x-3)^2+2(y+1)^2=24$

$3(x^2-6x+9)+2(y^2+2y+1)=24$

$3x^2-18x+27+2y^2+4y+2=24$

$3x^2+2y^2-18x+4y+5=0$

$A=3,\ C=2,\ D=-18,\ E=4,\ F=5$

**3.** **a)** Centre: (0, 0)

**b)** Vertices: (0, –4), (0, 4)

**c)** Domain: $-3\le x\le 3$, Range: $-4\le y\le 4$

**d)**

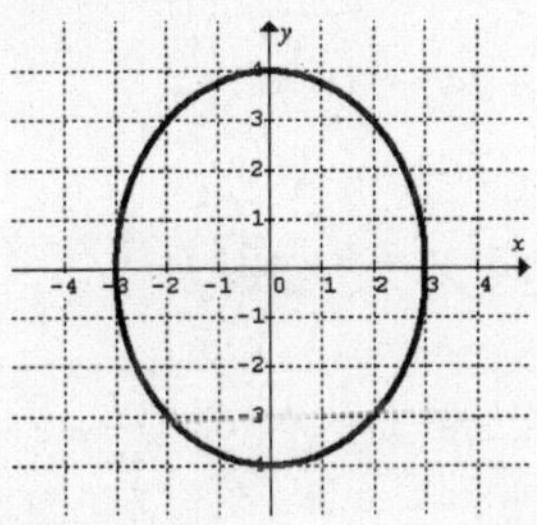

**4.** **a)** Centre: (0, 0)

**b)** Vertices: (0, –2), (0, 2)

**c)** Domain: $-1\le x\le 1$, Range: $-2\le y\le 2$

**d)**

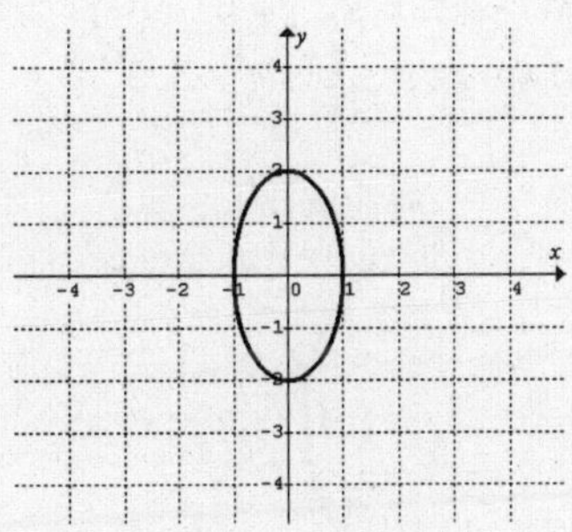

**5.** Convert to standard form:

$9x^2+4y^2=36$

$\frac{x^2}{4}+\frac{y^2}{9}=1$

**a)** Centre: (0, 0)

**b)** Vertices: (0, –3), (0, 3)

**c)** Domain: $-2\le x\le 2$, Range: $-3\le y\le 3$

**d)**

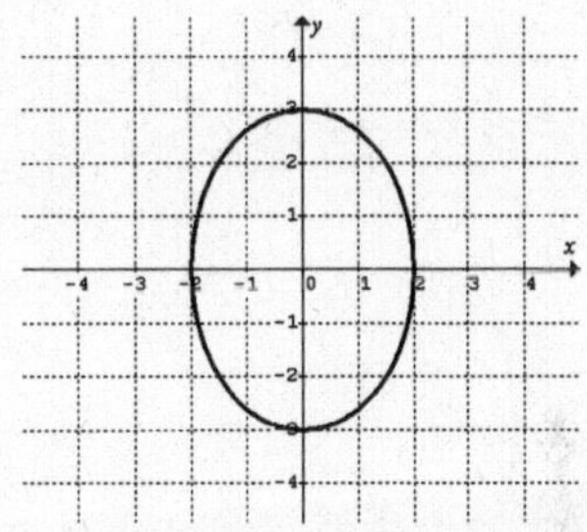

**6.** **a)** Centre: (2, –1)

**b)** Vertices: (–2, –1), (6, –1)

**c)** Domain: $-2\le x\le 6$, Range: $-3\le y\le 1$

**d)**

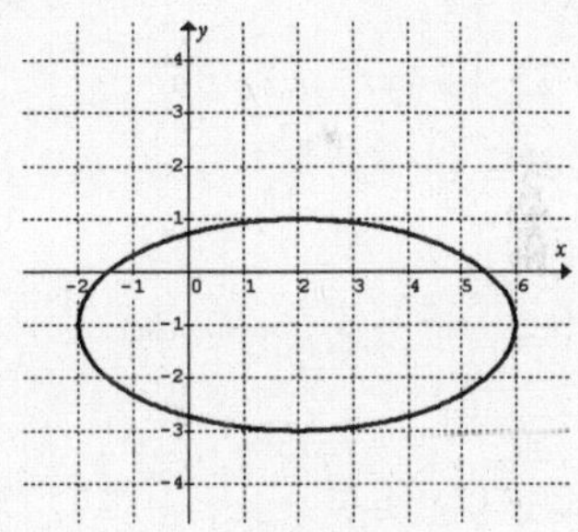

**7.** Convert to standard form:

$6(x+4)^2+24(y+3)^2=24$

$\frac{(x+4)^2}{4}+\frac{(y+3)^2}{1}=1$

**a)** Centre: (–4, –3)

**b)** Vertices: (–6, –3), (–2, –3)

**c)** Domain: $-6\le x\le -2$, Range: $-4\le y\le -2$

**d)**

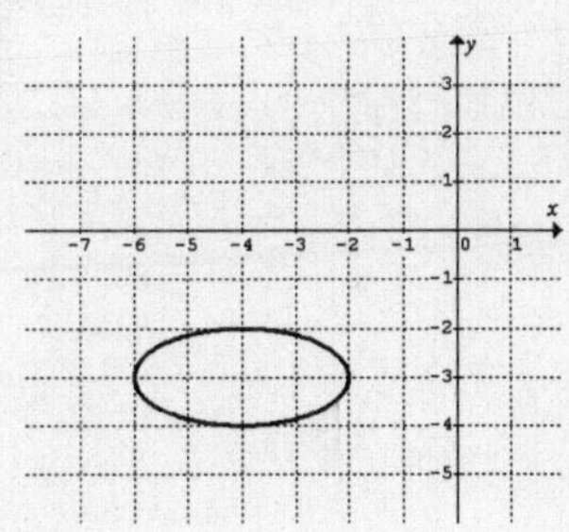

**8.** Convert to standard form:

$9(x-4)^2+2(y+2)^2=18$

$\frac{(x-4)^2}{2}+\frac{(y+2)^2}{9}=1$

**a)** Centre: (4, –2)

**b)** Vertices: (4, –5), (4, 1)

**c)** Domain: $4-\sqrt{2}\le x\le 4+\sqrt{2}$
Range: $-5\le y\le 1$

**d)**

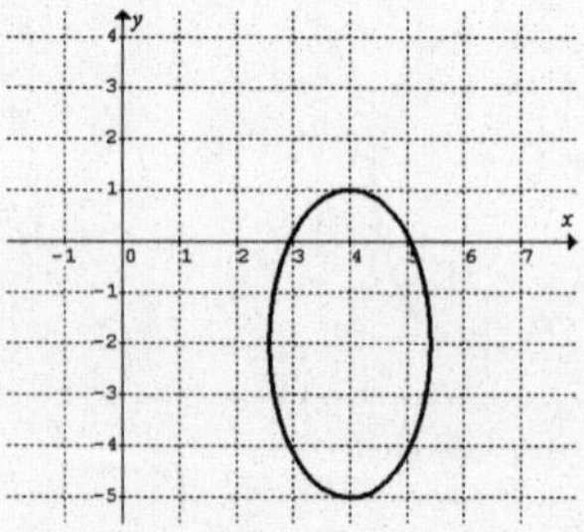

**9.** Convert to standard form by completing the square:

$x^2+4y^2+10x-11=0$

$x^2+10x+4y^2=11$

$\left(x^2+10x+25\right)+4y^2=11+25$

$(x+5)^2+4y^2=36$

$\frac{(x+5)^2}{36}+\frac{y^2}{9}=1$

**a)** Centre: (–5, 0)

**b)** Vertices: (–11, 0), (1, 0)

**c)** Domain: $-11\le x\le 1$, Range: $-3\le y\le 3$

**d)**

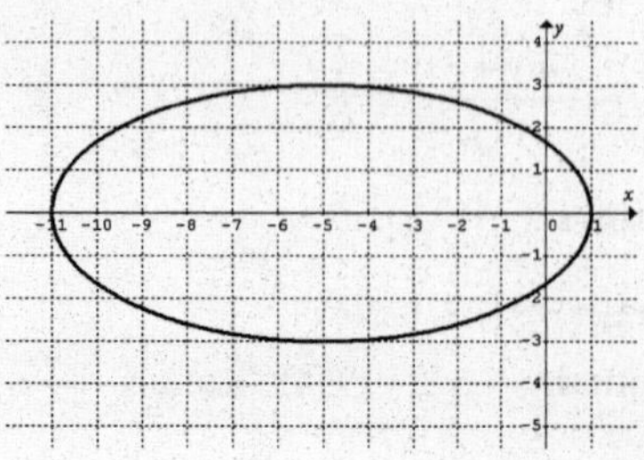

**10.** Convert to standard form by completing the square:

$25x^2+4y^2-150x+125=0$

$25x^2-150x+4y^2=-125$

$25\left(x^2-6x+9\right)+4y^2=-125+225$

$25(x-3)^2+4y^2=100$

$\frac{(x-3)^2}{4}+\frac{y^2}{25}=1$

**a)** Centre: (3, 0)

**b)** Vertices: (0, –5), (0, 5)

**c)** Domain: $1\le x\le 5$, Range: $-5\le y\le 5$

**d)**

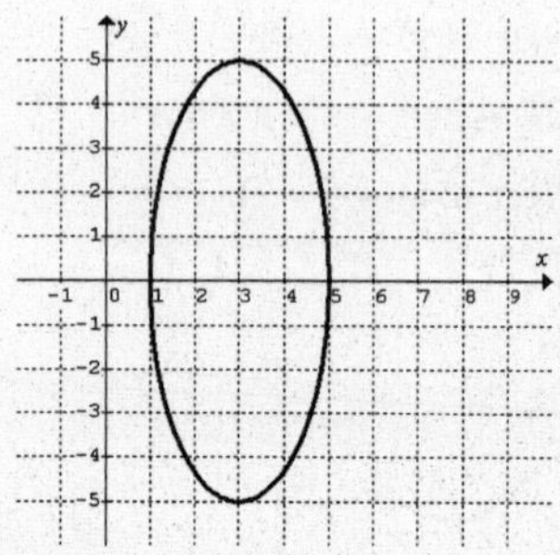

**11.** Convert to standard form by completing the squares:

$4x^2+3y^2+8x-24y+16=0$

$4x^2+8x+3y^2-24y=-16$

$4\left(x^2+2x+1\right)+3\left(y^2-8y+16\right)$

$=-16+4+48$

$4(x+1)^2+3(y-4)^2=36$

$\frac{(x+1)^2}{9}+\frac{(y-4)^2}{12}=1$

**a)** Centre: (–1, 4)

**b)** Vertices: $(-1,4-2\sqrt{3})$, $(-1,4+2\sqrt{3})$

**c)** Domain: $-4\le x\le 2$
Range: $4-2\sqrt{3}\le y\le 4+2\sqrt{3}$

**d)**

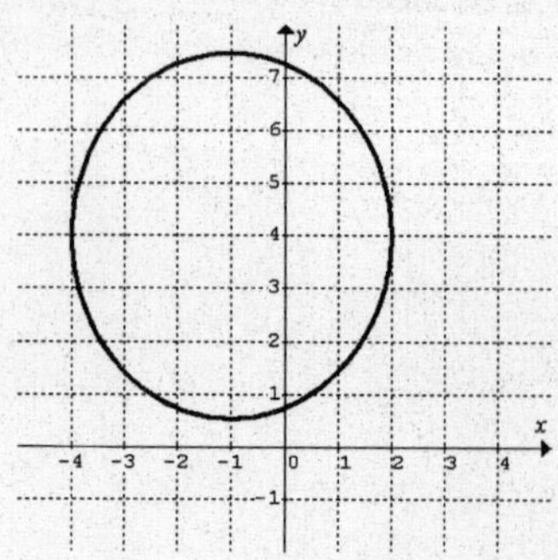

**12.** Convert to standard form by completing the squares:

$$2x^2+y^2-12x-6y-5=0$$
$$2x^2-12x+y^2-6y=5$$
$$2\left(x^2-6x+9\right)+\left(y^2-6y+9\right)=5+18+9$$
$$2(x-3)^2+(y-3)^2=32$$
$$\frac{(x-3)^2}{16}+\frac{(y-3)^2}{32}=1$$

**a)** Centre: (3, 3)

**b)** Vertices: $(3,3-4\sqrt{2}),\ (3,3+4\sqrt{2})$

**c)** Domain: $-1\le x\le 7$,
Range: $3-4\sqrt{2}\le y\le 3+4\sqrt{2}$

**d)**

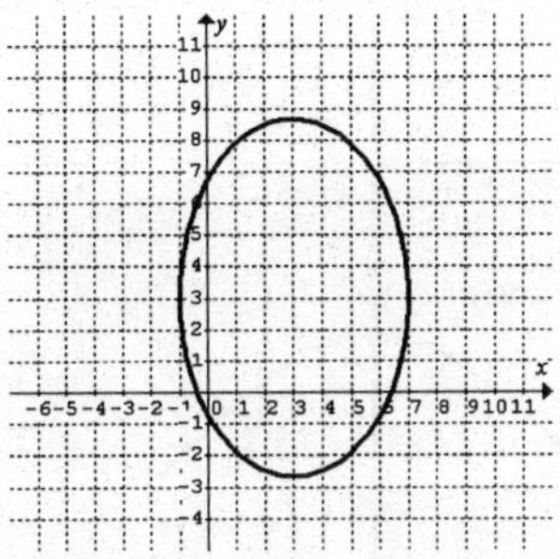

**13.** $\dfrac{x^2}{36}+\dfrac{y^2}{4}=1$

**14.** Centre: (0, 0)
$2a=2$, thus, $a=1$
$2b=10$, thus, $b=5$
Therefore, the equation is
$\dfrac{x^2}{1}+\dfrac{y^2}{25}=1.$

**15.** Centre: (2, –1)
$2a=8$, thus, $a=4$
$2b=6$, thus, $b=3$
Therefore, the equation is
$\dfrac{(x-2)^2}{16}+\dfrac{(y+1)^2}{9}=1.$

**16.** The equation of the transformed graph is

$$3(x-3)^2+2(y+2)^2-8=0$$
$$3(x^2-6x+9)+2(y^2+4y+4)-8=0$$
$$3x^2-18x+27+2y^2+8y+8-8=0$$
$$3x^2+2y^2-18x+8y+27=0$$

**17. a)** Changing the 9 to 81 would change the length of the horizontal axis from 6 to 18. The graph would be stretched horizontally about the $y$-axis by a factor of 3.

**b)** Changing the 36 to 4 would change the length of the vertical axis from 12 to 4. The graph would be stretched vertically about the $x$-axis by a factor of $\dfrac{1}{3}$.

**18. a)** $x^2+16y^2-16=0$

$$\frac{x^2}{16}+\frac{y^2}{1}=1$$

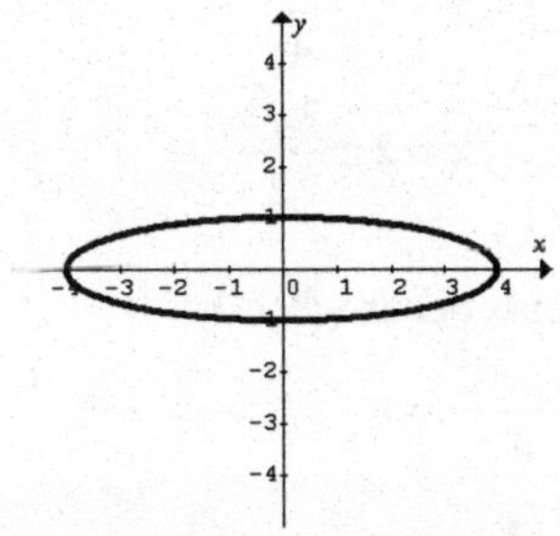

**b)** The result of the horizontal and vertical stretches would be the equation

$$\frac{\left(\frac{1}{2}x\right)^2}{16}+\frac{(3y)^2}{1}=1.$$

The result of the translations would be

$$\frac{\left(\frac{1}{2}(x+2)\right)^2}{16}+\frac{(3(y-5))^2}{1}=1.$$

Simplifying gives:

$$\frac{\frac{1}{4}(x+2)^2}{16}+\frac{9(y-5)^2}{1}=1$$
$$\frac{(x+2)^2}{64}+\frac{(y-5)^2}{\frac{1}{9}}=1$$

**c)** Domain: $-10\le x\le 6$,
Range: $4\frac{2}{3}\le y\le 5\frac{1}{3}$

**19. a)** Centre: (2, 1)
$2a=5-(-1)=6$ thus, $a=3$
$2b=3-(-1)=4$ thus, $b=2$

The equation is $\dfrac{(x-2)^2}{9}+\dfrac{(y-1)^2}{4}=1$

**b)** The $x$-intercepts occur when $y = 0$.

$$\frac{(x-2)^2}{9}+\frac{(0-1)^2}{4}=1$$
$$\frac{(x-2)^2}{9}+\frac{1}{4}=1$$
$$\frac{(x-2)^2}{9}=\frac{3}{4}$$
$$(x-2)^2=\frac{27}{4}$$
$$x-2=\pm\sqrt{\frac{27}{4}}$$
$$x=2\pm\frac{3\sqrt{3}}{2}$$

The $y$-intercepts occur when $x = 0$.

$$\frac{(0-2)^2}{9}+\frac{(y-1)^2}{4}=1$$
$$\frac{4}{9}+\frac{(y-1)^2}{4}=1$$
$$\frac{(y-1)^2}{4}=\frac{5}{9}$$
$$(y-1)^2=\frac{20}{9}$$
$$y-1=\pm\sqrt{\frac{20}{9}}$$
$$y=1\pm\frac{2\sqrt{5}}{3}$$

**c)**

$$4(x-2)^2+9(y-1)^2=36$$
$$4(x^2-4x+4)+9(y^2-2y+1)=36$$
$$4x^2-16x+16+9y^2-18y+9=0$$
$$4x^2+9y^2-16x-18y+25=0$$

**20. A**
The centre of the ellipse is (0, 3), which is not on the $x$-axis.

**21. C**
The largest value of the domain will be 3 + 4 = 7.

## *Lesson 4 Equations of Conics—The Hyperbola*

### CLASS EXERCISES ANSWERS AND SOLUTIONS

**1.** You should recognize this as a hyperbola with a vertical orientation with centre at (0, 0). Since $b^2=9$, $b=3$, since $b>0$, and thus, the vertices are at (0, 3) and (0, –3). A sketch of the graph is

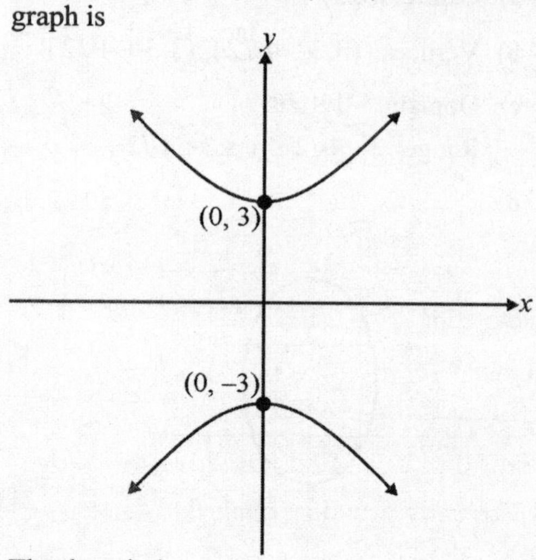


The domain is $x\in R$.
The range is $y\le -3$ or $y\ge 3$.

**2. a)** $36x^2-y^2=9$

$4x^2-\frac{y^2}{9}=1$ • Divide both sides of the equation by 9

$(2x)^2-\left(\frac{1}{3}y\right)^2=1$ • Writing with the coefficients of $x$ and $y$

Since $x$ has been replaced by $2x$, the graph has been stretched horizontally about the $y$-axis by a factor of $\frac{1}{2}$.

Since $y$ has been replaced by $\frac{1}{3}y$, the graph has been stretched vertically about the $x$-axis by a factor of 3.

**b)** From part (a) the equation can be written as

$$4x^2-\frac{y^2}{9}=1.$$

This is $\frac{4x^2}{1}-\frac{y^2}{9}=1.$

 

$$\frac{x^2}{\frac{1}{4}} - \frac{y^2}{9} = 1$$ • Because $4 = \frac{1}{\frac{1}{4}}$

$$\frac{x^2}{\left(\frac{1}{2}\right)^2} - \frac{y^2}{3^2} = 1$$ • Write denominators as squares

Thus, $a = \frac{1}{2}$ and $b = 3$.

**3.** **a)** You should recognize this as a hyperbola with centre (–1, 3) and having a vertical orientation, and thus, a transformation of $x^2 - y^2 = -1$.

$$\frac{(x+1)^2}{25} - \frac{(y-3)^2}{9} = -1$$

can be written as

$$\frac{1}{5^2}(x+1)^2 - \frac{1}{3^2}(y-3)^2 = -1$$

and then

$$\left(\frac{1}{5}(x+1)\right)^2 - \left(\frac{1}{3}(y-3)\right)^2 = -1$$

The transformations applied to the graph of $x^2 - y^2 = -1$ are:

1. Stretch horizontally about the $y$-axis by a factor of 5 and vertically about the $x$-axis by a factor of 3
2. Translate 1 unit left and 3 units up

**b)** This graph will be the same as the graph for $\frac{x^2}{25} - \frac{y^2}{9} = -1$, except it will be translated 1 unit left and 3 units up.

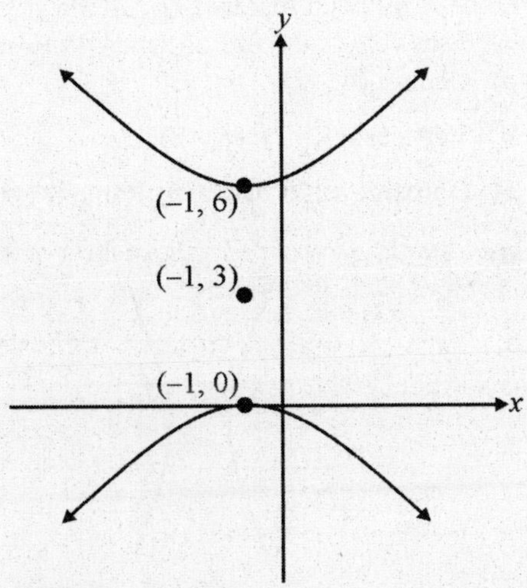


Domain: $x \in R$
Range: $y \geq 6$ or $y \leq 0$

The vertices will be 3 units from the centre, since $b^2 = 9$

$b = 3$, and because $b > 0$.

**4.**

$$16(x-2)^2 - 9(y+1)^2 = 144$$ • Multiply both sides of the equation by 144

Complete the algebraic processes:

$16(x^2 - 4x + 4) - 9(y^2 + 2y + 1)$ $= 144$ • Expand the binominals

$16x^2 - 64x + 64 - 9y^2 - 18y - 9$ $= 144$ • Multiply to remove the brackets

You should get
$16x^2 - 9y^2 - 64x - 18y - 89 = 0$, where
$A = 16$, $C = -9$, $D = -64$, $E = -18$, and $F = -89$.

**5.** It can be seen on the graph that the centre is (4, –1) and the vertices are (4, 3) and (4, –5). The distance from the centre to a vertex is 4 units. This is a vertical hyperbola and thus, $b = 4$. It is given that $\frac{4}{a} = 2$ thus, $a = 2$.

Therefore, the equation in standard form is $\frac{(x-4)^2}{4} - \frac{(y+1)^2}{16} = -1$.

## PRACTICE EXERCISES ANSWERS AND SOLUTIONS

**1.**

$$\frac{x^2}{9} - \frac{(y-1)^2}{4} = 1$$
$$4x^2 - 9(y-1)^2 = 36$$
$$4x^2 - 9(y^2 - 2y + 1) = 36$$
$$4x^2 - 9y^2 + 18y - 9 = 36$$
$$4x^2 - 9y^2 + 18y - 45 = 0$$

$A = 4$, $C = -9$, $D = 0$, $E = 18$, $F = -45$

**2.**
$$\frac{(x-2)^2}{8}-\frac{(y+2)^2}{12}=1$$
$$3(x-2)^2-2(y+2)^2=24$$
$$3(x^2-4x+4)-2(y^2+4y+4)=24$$
$$3x^2-12x+12-2y^2-8y-8=24$$
$$3x^2-2y^2-12x-8y-20=0$$

$A=3,\ C=-2,\ D=-12,\ E=-8,\ F=-20$

**3.** **a)** Centre: (0, 0)

**b)** Vertices: (–4, 0), (4, 0)

**c)** Domain: $x \le -4$ or $x \ge 4$, Range: $y \in R$

**d)**

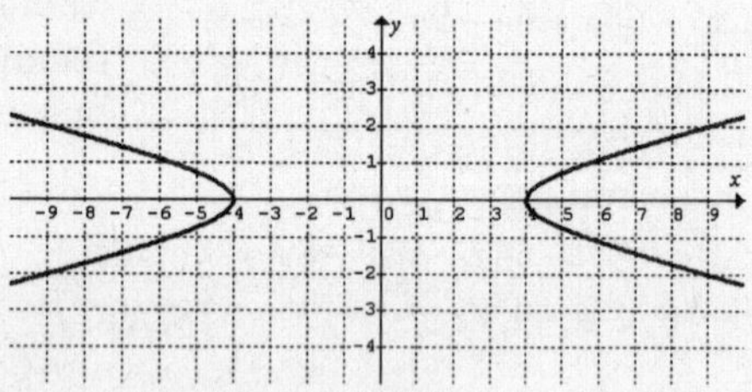

**4.** **a)** Centre: (0, 0)

**b)** Vertices: (0, –2), (0, 2)

**c)** Domain: $x \in R$, Range: $y \le -2$ or $y \ge 2$

**d)**

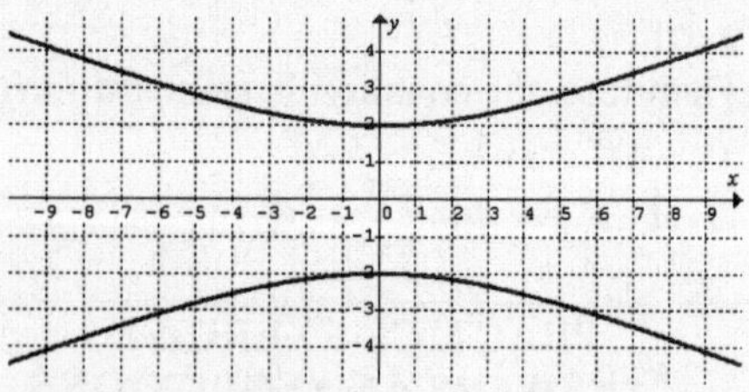

**5.** Convert to standard form:

$9x^2-4y^2=36$

$\frac{x^2}{4}-\frac{y^2}{9}=1$

**a)** Centre: (0, 0)

**b)** Vertices: (–2, 0), (2, 0)

**c)** Domain: $x \le -2$ or $x \ge 2$, Range: $y \in R$

**d)**

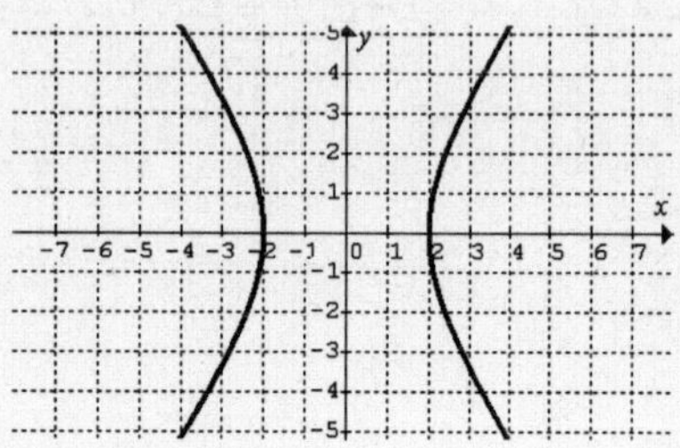

**6.** **a)** Centre: (2, –1)

**b)** Vertices: (–3, –1), (7, –1)

**c)** Domain: $x \le -3$ or $x \ge 7$, Range: $y \in R$

**d)**

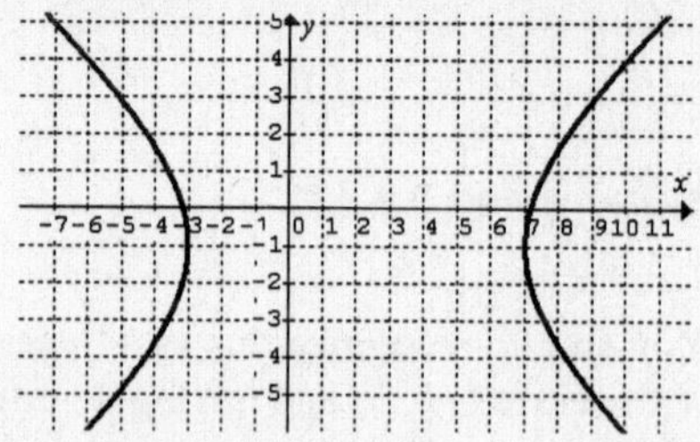

**7.** Convert to standard form:

$3(x-4)^2-12(y+3)^2=-12$

$\frac{(x-4)^2}{4}-\frac{(y+3)^2}{1}=-1$

**a)** Centre: (4, –3)

**b)** Vertices: (4, –4), (4, –2)

**c)** Domain: $x \in R$, Range: $y \le -4$ or $y \ge -2$

**d)**

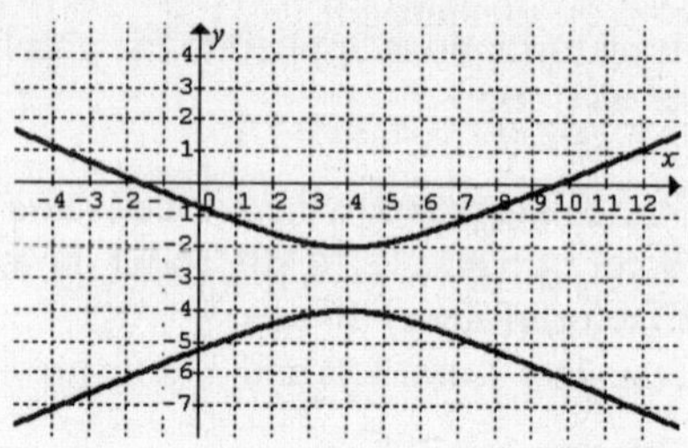

**8.** Convert to standard form:

$18(x-4)^2-2(y+2)^2=18$

$\frac{(x-4)^2}{1}-\frac{(y+2)^2}{9}=1$

**a)** Centre: (4, –2)

**b)** Vertices: (3, –2), (5, –2)

**c)** Domain: $x \le 3$ or $x \ge 5$, Range: $y \in R$

**d)**

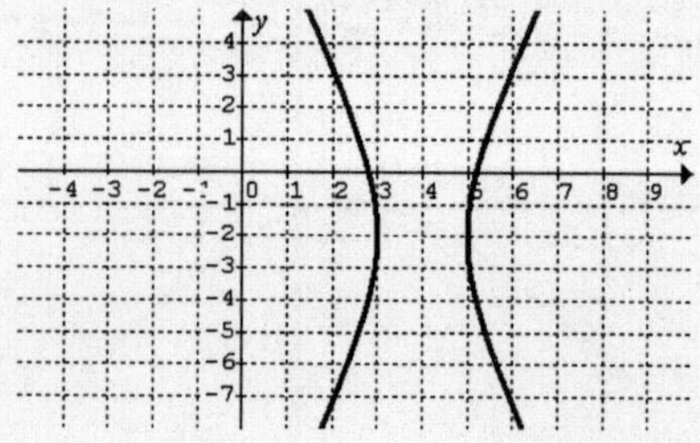

**9.** Convert to standard form by completing the square:

$4x^2 - y^2 - 4y - 8 = 0$

$4x^2 - y^2 - 4y = 8$

$4x^2 - \left(y^2 + 4y + 4\right) = 8 - 4$

$4x^2 - (y+2)^2 = 4$

$\dfrac{x^2}{1} - \dfrac{(y+2)^2}{4} = 1$

**a)** Centre: (0, –2)

**b)** Vertices: (–1, –2), (1, –2)

**c)** Domain: $x \le -1$ or $x \ge 1$, Range: $y \in R$

**d)**

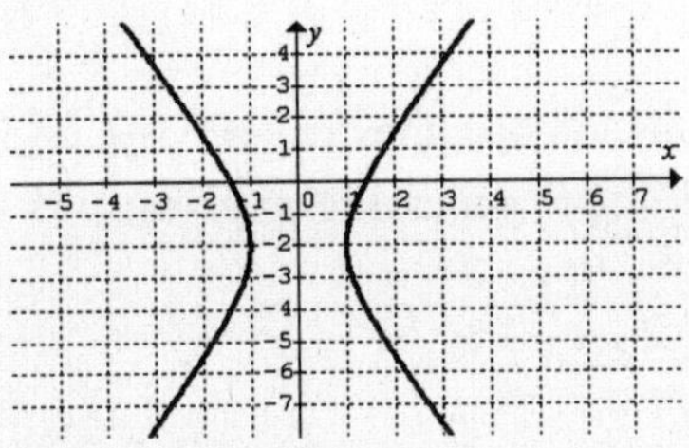

**10.** Convert to standard form by completing the square:

$x^2 - 9y^2 + 6x = 0$

$x^2 + 6x - 9y^2 = 0$

$\left(x^2 + 6x + 9\right) - 9y^2 = 0 + 9$

$(x+3)^2 - 9y^2 = 9$

$\dfrac{(x+3)^2}{9} - \dfrac{y^2}{1} = 1$

**a)** Centre: (–3, 0)

**b)** Vertices: (–6, 0), (0, 0)

**c)** Domain: $x \le -6$ or $x \ge 0$, Range: $y \in R$

**d)**

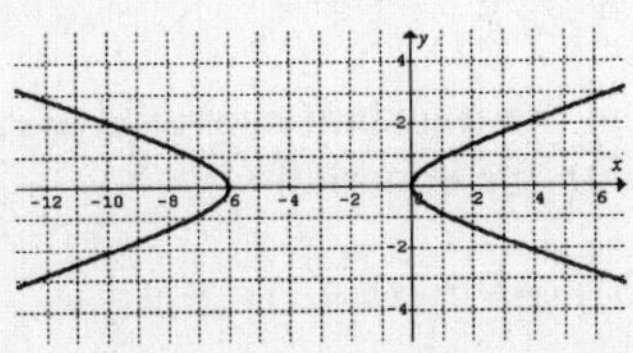

**11.** Convert to standard form by completing the squares:

$4x^2 - 9y^2 + 8x - 36y - 41 = 0$

$4x^2 + 8x - 9y^2 - 36y = 41$

$4\left(x^2 + 2x + 1\right) - 9\left(y^2 + 4y + 4\right) = 41 + 4 - 36$

$4(x+1)^2 - 9(y+2)^2 = 9$

$\dfrac{4(x+1)^2}{9} - \dfrac{(y+2)^2}{1} = 1$

$\dfrac{(x+1)^2}{\frac{9}{4}} - \dfrac{(y+2)^2}{1} = 1$

**a)** Centre: (–1, –2)

**b)** Vertices: (–2.5, –2), (0.5, –2)

**c)** Domain: $x \le -2.5$ or $x \ge 2.5$, Range: $y \in R$

**d)**

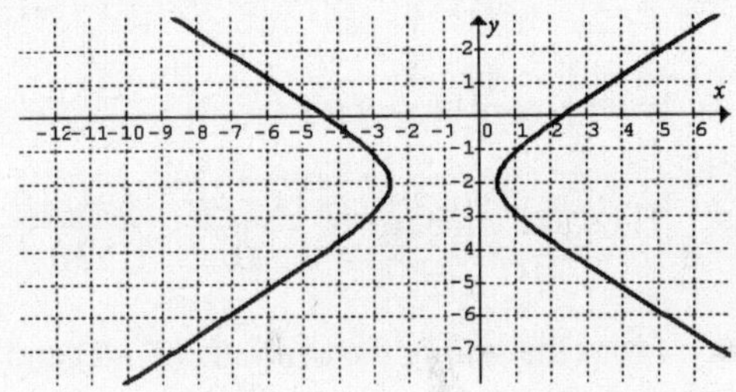

**12.** Convert to standard form by completing the squares:

$4x^2 - y^2 + 40x + 4y + 80 = 0$

$4x^2 + 40x - y^2 + 4y = -80$

$4\left(x^2 + 10x + 25\right) - \left(y^2 - 4y + 4\right)$

$= -80 + 100 - 4$

$4(x+5)^2 - (y-2)^2 = 16$

$\dfrac{(x+5)^2}{4} - \dfrac{(y-2)^2}{16} = 1$

**a)** Centre: (–5, 2)

**b)** Vertices: (–7, 2), (–3, 2)

**c)** Domain: $x \le -7$ or $x \ge -3$, Range: $y \in R$

**d)**

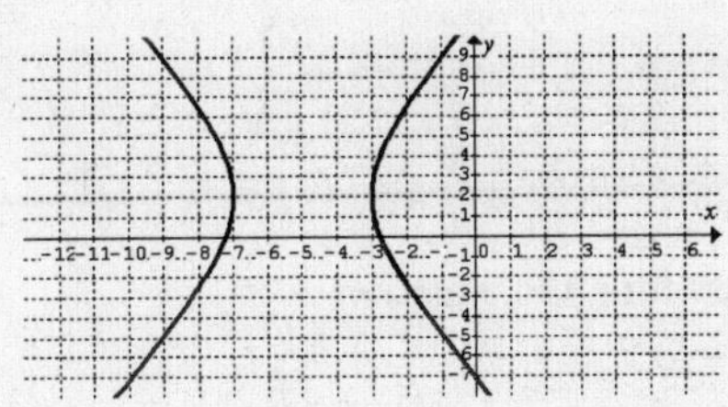

**13.** Centre: (0, 0)
The hyperbola has a horizontal orientation.
$a=6$

$\frac{b}{a}=\frac{1}{6} \to \frac{b}{6}=\frac{1}{6} \to b=1$

Thus, the equation is $\frac{x^2}{36}-\frac{y^2}{1}=1$.

**14.** Centre: (0, 0)
The hyperbola has a vertical orientation.
$b=4$

$\frac{b}{a}=\frac{2}{3} \to \frac{4}{a}=\frac{2}{3} \to a=6$

Thus, the equation is $\frac{x^2}{36}-\frac{y^2}{16}=-1$.

**15.** Centre: (3, –2)
Since one vertex is at (–1, –2), the hyperbola has a horizontal orientation and
$a=3-(-1)=4$.

$\frac{b}{a}=\frac{3}{2} \to \frac{b}{4}=\frac{3}{2} \to b=6$

Thus, the equation is $\frac{(x-3)^2}{16}-\frac{(y+2)^2}{36}=1$.

**16.** From the graph, the centre is (3, –2) and the vertices are (3, –6) and (3, 2).
The graph has a vertical orientation with
$b=2-(-2)=4$.
From the graph it can be seen that the positive slope of the asymptote is $\frac{2}{1}$.

Thus, $\frac{b}{a}=\frac{2}{1} \to \frac{4}{a}=\frac{2}{1} \to a=2$.
Therefore, the equation is
$\frac{(x-3)^2}{4}-\frac{(y+2)^2}{16}=-1$.

**17.** From the graph, the centre is (5, 1) and the vertices are (2, 1) and (9, 1).
The graph has a horizontal orientation with
$a=9-5=4$.
From the graph it can be seen that the positive slope of the asymptote is $\frac{5}{3}$.

Thus, $\frac{b}{a}=\frac{5}{3} \to \frac{b}{3}=\frac{5}{3} \to b=5$.
Therefore, the equation is
$\frac{(x-5)^2}{9}-\frac{(y-1)^2}{25}=1$.

**18.** Since the centre is at (2, –1), $h=2$ and $k=-1$.
Thus, the equation is $\frac{(x-2)^2}{a^2}-(y+1)^2=-1$.
Substitute the coordinates (–2, 2) and solve for $a$:

$\frac{(-2-2)^2}{a^2}-(2+1)^2=-1$

$\frac{16}{a^2}-9=-1$

$\frac{16}{a^2}=8$

$a^2=2$

Thus, in standard form, the equation is
$\frac{(x-2)^2}{2}-\frac{(y+1)^2}{1}=-1$.

**19.** The equation becomes $(x-4)^2-9(y+2)^2=9$
Converting to general form gives:
$(x-4)^2-9(y+2)^2=9$
$x^2-9y^2-8x-36y-29=0$

**20.** **a)** Convert to standard form by completing the square:
$x^2-9y^2+9=0$
$x^2-9y^2=-9$
$\frac{x^2}{9}-\frac{y^2}{1}=-1$

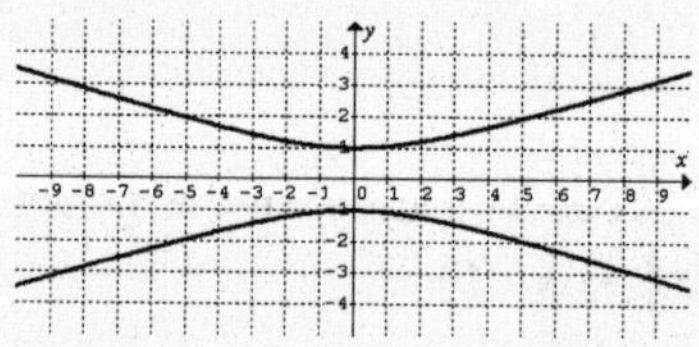

**b)** The horizontal and vertical stretches would create a graph with the equation

$(4x)^2-9\left(\frac{1}{2}y\right)^2+9=0$

Applying the translations would result in a graph with the equation

$\left(4(x-3)\right)^2-9\left(\frac{1}{2}(y+1)\right)^2+9=0$.

Simplifying gives:

$16(x-3)^2-\frac{9}{4}(y+1)^2=-9$

$\frac{16(x-3)^2}{9}-\frac{1}{4}(y+1)^2=-1$

$\frac{(x-3)^2}{\frac{9}{16}}-\frac{(y+1)^2}{4}=-1$

**c)** Domain: $x \in R$,
Range: $y \leq -3$ or $y \geq 1$

**21. a)** If the 4 was changed to 25, the vertices would change from (–2, 0) and (2, 0) to (–5, 0) and (5, 0). The graph would be stretched horizontally about the $y$-axis by a factor of $\frac{5}{2}$. Notice that $\frac{x^2}{25} - \frac{y^2}{9} = 1$ is equivalent to $\left(\frac{4}{25}\right)\frac{x^2}{4} - \frac{y^2}{9} = 1$, which is

$$\frac{\left(\frac{2}{5}x\right)^2}{4} - \frac{y^2}{9} = 1.$$

**b)** If the 9 was changed to 1, the graph would be stretched vertically about the $x$-axis by a factor of $\frac{1}{3}$. Notice that $\frac{x^2}{4} - \frac{y^2}{1} = 1$ is equivalent to $\frac{x^2}{4} - \frac{y^2}{9}\left(\frac{9}{1}\right) = 1$, which is

$$\frac{x^2}{4} - \frac{(3y)^2}{9} = 1.$$

**22. B**
The equation represents a hyperbola with a vertical orientation. The centre is (–2, 3) and $b = 3$. Thus, the vertices are (–2, 0) and (–2, 6). The range is $y \geq 6$ or $y \leq 0$.

**23. D**
The $x$-intercepts occur when $y = 0$. This gives:
$x^2 - 4(0)^2 - 6x + 5 = 0$
$x^2 - 6x + 5 = 0$
$(x-5)(x-1) = 0$
$x = 5$, or $x = 1$
The $y$-intercepts occur when $x = 0$. This gives:
$0^2 - 4y^2 - 6(0) + 5 = 0$
$-4y^2 = -5$
$y = \pm\sqrt{\frac{5}{4}}$
Thus, the total number of intercepts is 4.

## *Practice Quiz*

### ANSWERS AND SOLUTIONS

**1.** If the angle that the plane makes with the axis of the cone is $\theta$, then to form
**a)** a circle: $\theta = 90°$
**b)** an ellipse: $58° < \theta < 90°$
**c)** a parabola:, $\theta = 58°$
**d)** a hyperbola: $0° \leq \theta < 58°$

**2.** Since an angle of 48° forms an ellipse, then the possible values of the vertex angle, $\theta$, are $0° < \theta < 96°$. A diagram might help you determine this.

**3. a)** $x^2 + y^2 - 6x + 8y - 11 = 0$
$(x^2 - 6x + \quad) + (y^2 + 8y + \quad) = 11$
$(x^2 - 6x + 9) + (y^2 + 8y + 16) = 11 + 9 + 16$
$(x-3)^2 + (y+4)^2 = 6^2$
The original equation represents a circle with centre at the origin with radius = 1. The new equation represents a circle with centre at (3, –4) and with radius = 6.
The transformations, in order, are
1. Stretch the graph horizontally about the $y$-axis and vertically about the $x$-axis by a factor of 6.
2. Translate the graph 3 units right and 4 units down.

**b)** Domain: $-3 \leq x \leq 9$,
Range: $-10 \leq y \leq 2$

**c)** To determine the $x$-intercepts, let $y = 0$.
$(x-3)^2 + (0+4)^2 = 6^2$
$(x-3)^2 = 36 - 16$
$(x-3) = \pm\sqrt{20}$
$x = 3 \pm 2\sqrt{5}$
To determine the $y$-intercepts, let $x = 0$.
$(0-3)^2 + (y+4)^2 = 6^2$
$(y+4)^2 = 36 - 9$
$(y+4) = \pm\sqrt{27}$
$y = -4 \pm 3\sqrt{3}$

**4. a)** $\frac{(x-1)^2}{4} + (y+2)^2 = 9$
$\frac{(x-1)^2}{36} + \frac{(y+2)^2}{9} = 1$
The conic is an ellipse.

**b)**

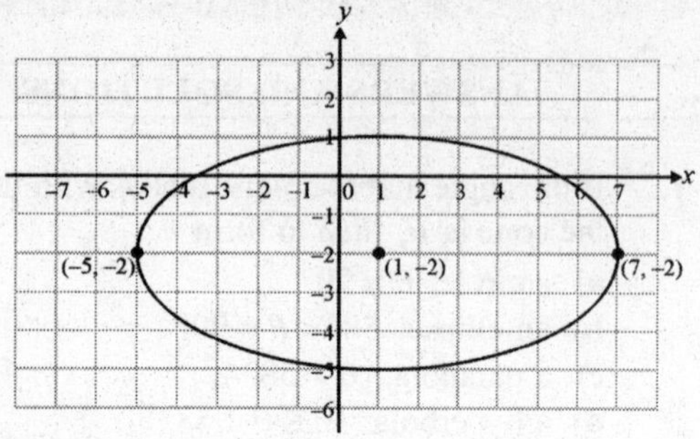


**c)** Domain: $-5 \le x \le 7$,
Range: $-5 \le y \le 1$

**d)** $\frac{(x-1)^2}{4} + (y+2)^2 = 9$

$(x-1)^2 + 4(y+2)^2 = 36$

$x^2 - 2x + 1 + 4y^2 + 16y + 16 = 36$

$x^2 + 4y^2 - 2x + 16y - 19 = 0$

$A = 1, C = 4, D = -2, E = 16, F = -19$

**5.** **a)** $4x^2 + y^2 - 36 = 0$

$\frac{x^2}{9} + \frac{y^2}{36} = 1$

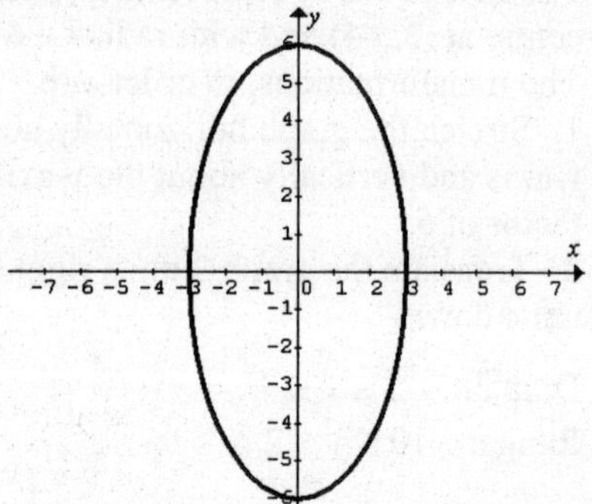


**b)** The transformations produce the following equations.

1. Stretches: $\frac{(2x)^2}{9} + \frac{\left(\frac{1}{3}y\right)^2}{36} = 1$

2. Translations:

$\frac{(2(x-1))^2}{9} + \frac{\left(\frac{1}{3}(y+4)\right)^2}{36} = 1$

Simplifying and writing in standard form gives:

$\frac{4(x-1)^2}{9} + \frac{\frac{1}{9}(y+4)^2}{36} = 1$

$\frac{(x-1)^2}{\frac{9}{4}} + \frac{(y+4)^2}{4} = 1$

**c)** Domain: $1 - \frac{3}{2} \le x \le 1 + \frac{3}{2}$, which is

$-\frac{1}{2} \le x \le \frac{5}{2}$

Range: $-4 - 2 \le y \le -4 + 2$, which is
$-6 \le y \le -2$

**6.** **a)** $4x^2 - y^2 = -100$

$\frac{x^2}{25} - \frac{y^2}{100} = -1$

$a = 5,\ b = 10$

**b)** Since $\frac{x^2}{25} - \frac{y^2}{100} = -1$ is $\left(\frac{1}{5}x\right)^2 - \left(\frac{1}{10}y\right)^2 = -1$, the graph of $x^2 - y^2 = -1$ has been stretched horizontally about the $y$-axis by a factor of 5 and vertically about the $x$-axis by a factor of 10.

**7.** **a)** $4x^2 - 25y^2 - 24x - 50y + 111 = 0$

$4x^2 - 24x - 25y^2 - 50y = -111$

$4\left(x^2 - 6x + 9\right) - 25\left(y^2 + 2y + 1\right) = -111 + 36 - 25$

$4(x-3)^2 - 25(y+1)^2 = -100$

$\frac{(x-3)^2}{25} - \frac{(y+1)^2}{4} = -1$

**b)**

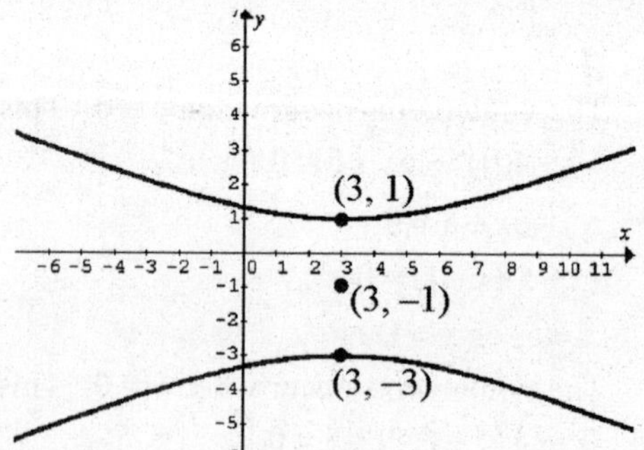


**c)** Domain: $x \in R$, Range: $y \le -3$ or $y \ge 1$

**8.** **a)** The graph is an ellipse with the centre at $(2, -1)$, $a = 2$, and $b = 6$. Therefore, the equation is $\frac{(x-2)^2}{4} + \frac{(y+1)^2}{36} = 1$.

**b)** The graph is a hyperbola with a horizontal orientation and with the centre at (2, 1) and $a = 3$. Since $\frac{b}{a} = 1$, then $b = 3$. Therefore, the equation is $\frac{(x-2)^2}{9} - \frac{(y-1)^2}{9} = 1$.

## *Lesson 5 Equations of Conics—The Parabola*

### CLASS EXERCISES ANSWERS AND SOLUTIONS

**1. a)** First, $y$ would change to $\frac{1}{3}y$, to give

$\frac{1}{3}y = x^2$ or $y = 3x^2$.

For the translations, $x$ is replaced with $(x-2)$ and y with $(y+4)$.

This gives $y+4 = 3(x-2)^2$. In Mathematics 20P, this was written in the equivalent form of $y = 3(x-2)^2 - 4$.

**b)**

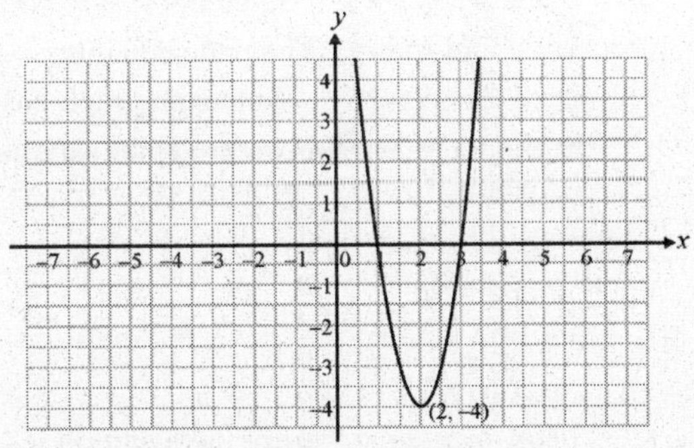


The vertex is (2, –4)
The domain is $x \in R$, the range is $y \geq -4$.

**2.** First, $x$ is replaced with $\frac{1}{4}x$, to give $\frac{1}{4}x = y^2$ or $x = 4y^2$ (do the stretches first).

Now, replace $x$ with $x - 3$ and $y$ with $y + 2$.

This gives $x - 3 = 4(y+2)^2$.

**3.**

$$y - 2 = 3(x^2 - 2x + 1)$$
$$y - 2 = 3x^2 - 6x + 3$$
$$0 = 3x^2 - 6x - y + 5$$

Thus, the general form is $3x^2 - 6x - y + 5 = 0$.

Here, $A = 3$, $C = 0$, $D = -6$, $E = -1$, and $F = 5$.

**4.**

$$4x^2 + 8x = -y + 7$$

• Add $-y + 7$ to both sides of the equation

$$4(x^2 + 2x + \_\_) = -y + 7$$
$$4(x^2 + 2x + \underline{1}) = -y + 7 + 4$$

• Add 1 in the brackets on the LHS to complete the square, and add 4 on the RHS to balance the equation

$$4(x+1)^2 = -y + 11$$

• Factor the perfect square trimonial and collect like terms

$$-4(x+1)^2 = y - 11$$

• Multiply both sides of the equation by –1. In standard form the coefficient of the first degree term must be +1.

Now you can see that the vertex is (–1, 11) and since $a = -4$, the parabola opens down. Since $|a| = |-4| = 4$, compared to the graph of $y = x^2$, the graph has been stretched vertically by a factor of 4. Since $a < 0$, the graph is reflected in the $x$-axis. The domain is $x \in R$ and the range is $y \leq 11$.

**5.**

| | |
|---|---|
| $2y^2-12y=-3x+6$ | • Add $-3x+6$ to both sides of the equation |
| $2(y^2-6y+\_\_)=-3x+6$ | • Factor out a common factor of 2 on the LHS and leave a space to complete the square |
| $2(y^2-6y+\underline{9})=-3x+6+18$ | • Add 9 in the space to complete the square. Since 9 is in brackets, you actually added 18 to the LHS and thus must add 18 to the RHS. |
| $2(y-3)^2=-3x+24$ | • Factor the perfect square trinomial that was created and collect like terms on the RHS |
| $2(y-3)^2=-3(x-8)$ | • Factor out a common factor of –3 on the RHS |
| $-\frac{2}{3}(y-3)^2=x-8$ | • Divide both sides of the equation by –3. In Standard form, the coefficient of the first-degree term must be +1. |

Since $a=-\frac{2}{3}$, the parabola opens left (because $a<0$).

The vertex is at (8, 3).
The domain is $x\le 8$ and the range is $y\in R$.
The $x$-intercept occurs when $y=0$.

$$-\frac{2}{3}(0-3)^2=x-8$$
$$-\frac{2}{3}(9)=x-8$$
$$-6=x-8$$
$$2=x$$

Thus, the $x$-intercept is 2.

The $y$-intercepts occur when $x=0$.

$$-\frac{2}{3}(y-3)^2=0-8$$
$$(y-3)^2=12$$
$$y-3=\pm\sqrt{12}$$
$$y=3\pm2\sqrt{3}$$

Thus, the $y$-intercepts are $3\pm2\sqrt{3}$.

Sketch of the graph:

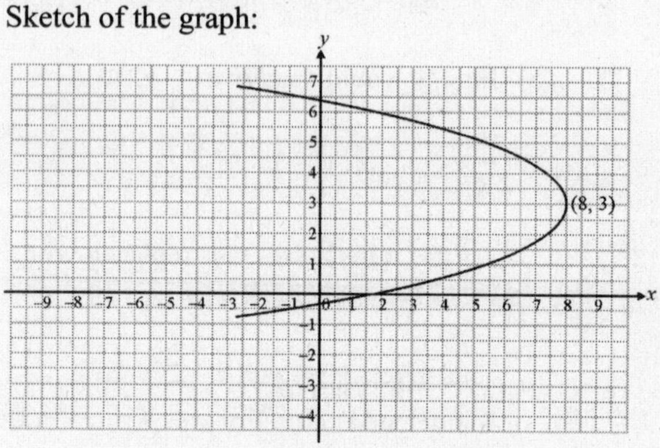


**6.** In standard form, the equation fits the form, $x-h=a(y-k)^2$. Since the vertex is (2, –1), $h=2$ and $k=-1$. Thus, the equation is $x-2=a(y+1)^2$. Since the graph passes through (–4, 2), this ordered pair satisfies the equation.

$$-4-2=a(2+1)^2$$
$$-6=9a$$
$$-\frac{2}{3}=a$$

Thus, in standard form, the equation is

$$x-2=-\frac{2}{3}(y+1)^2$$

Converting to general form:

| | |
|---|---|
| $3x-6=-2(y+1)^2$ | • Multiply both sides of the equation by 3 |
| $3x-6=-2(y^2+2y+1)$ | • Expand $(y+1)^2$ |
| $3x-6=-2y^2-4y-2$ | • Multiply to remove brackets |
| $2y^2+3x+4y-4=0$ | • Add and subtract terms to make RHS = 0 |

 

## PRACTICE EXERCISES ANSWERS AND SOLUTIONS

**1.** The vertical stretch gives $y = 5x^2$.
The translations give $y - 2 = 5(x + 4)^2$.

**2.** The horizontal stretch gives $y = (3x)^2$.
The translations give $y + 1 = \left(3(x - 6)\right)^2$.
Writing in standard form gives
$y + 1 = 9(x - 6)^2$.

**3.** The horizontal stretch gives $y = (2x)^2$.
The vertical stretch gives $\frac{1}{4}y = (2x)^2$ or
$y = 4(2x)^2$.
The translations give $y + 5 = 4\left(2(x - 2)\right)^2$.
Writing in standard form gives
$y + 5 = 16(x - 2)^2$.

**4.** The horizontal stretch gives $y = \left(\frac{1}{3}x\right)^2$.
The vertical stretch gives $\frac{2}{3}y = \left(\frac{1}{3}x\right)^2$ or
$y = \frac{3}{2}\left(\frac{1}{3}x\right)^2$.
The translations give $y + 5 = \frac{3}{2}\left(\frac{1}{3}(x - 2)\right)^2$.
Writing in standard form gives
$y + 5 = \frac{1}{6}(x - 2)^2$.

**5.** Vertical stretch about the *x*-axis by a factor of 6 (or horizontally about the *y*-axis by a factor of $\frac{1}{\sqrt{6}}$), translations of 7 units left and 2 units up.

**6.** Vertical stretch about the *x*-axis by a factor of $\frac{1}{4}$ (or horizontally about the *y*-axis by a factor of 2), translations of 1 unit right and 2 units down.

**7.** First, factor to convert to standard form:
$y + 6 = (9x - 18)^2$
$y + 6 = \left(9(x - 2)\right)^2$
$y + 6 = 81(x - 2)^2$
Transformations are: vertical stretch about the *x*-axis by a factor of 81 (or horizontally about the *y*-axis by a factor of $\frac{1}{9}$), translations of 2 units right and 6 units down.

**8.** First, convert to standard form:
$y = (4x + 12)^2 - 5$
$y + 5 = \left(4(x + 3)\right)^2$
$y + 5 = 16(x + 3)^2$
Transformations are: vertical stretch about the *x*-axis by a factor of 16 (or horizontally about the *y*-axis by a factor of $\frac{1}{4}$), translations of 3 units left and 5 units down.

**9.** Since the equation represents a parabola that is opening to the right, the first transformation is a reflection across the line $y = x$. Next, the graph is translated 2 units left and 4 units up.

**10.** Since the equation represents a parabola that is opening to the right, the first transformation is a reflection across the line $y = x$. Next, the graph is stretched horizontally about the *y*-axis by a factor of $\frac{1}{4}$ (or vertically about the *x*-axis by a factor of 2). Finally, the graph is translated 5 units left and 3 units up.

**11.** **a)** Vertex: (2, 3)

**b)** Domain: $x \in R$,
Range: $y \le 3$

**c)** To determine the *x*-intercepts, let $y = 0$.
$0 - 3 = -\frac{1}{4}(x - 2)^2$
$12 = (x - 2)^2$
$x - 2 = \pm\sqrt{12}$
$x = 2 \pm 2\sqrt{3}$
To determine the *y*-intercept, let $x = 0$.
$y - 3 = -\frac{1}{4}(0 - 3)^2$
$y - 3 = -\frac{9}{4}$
$y = -\frac{9}{4} + \frac{12}{4} = \frac{3}{4}$

**12.** **a)** Vertex: (3, –2)

**b)** Domain: $x \geq 3$,
Range: $y \in R$

**c)** To determine the $x$-intercept, let $y = 0$.

$x-3=2(0+2)^2$

$x-3=8$

$x=11$

To determine the $y$-intercepts, let $x = 0$.

$0-3=2(y+2)^2$

$\frac{-3}{2}=(y+2)^2$

There is no solution, thus no $y$-intercepts.

**13.** $3y^2-x-12y+11=0$

$3(y^2-4y+\ \ )=x-11$

$3(y^2-4y+4)=x-11+12$

$3(y-2)^2=x+1$

The vertex is at (–1, 2).

**14.** Complete the square:

$x^2-6x+3y=0$

$(x^2-6x+\ \ )=-3y$

$(x^2-6x+9)=-3y+9$

$(x-3)^2=-3(y-3)$

$-\frac{1}{3}(x-3)^2=y-3$

**a)** Vertex: (3, 3)

**b)** Domain: $x \in R$,
Range: $y \leq 3$

**c)** To determine the $x$-intercepts, let $y = 0$.

$-\frac{1}{3}(x-3)^2=0-3$

$(x-3)^2=9$

$x-3=\pm 3$

$x=0, 6$

To determine the $y$-intercept, let $x = 0$.

$-\frac{1}{3}(0-3)^2=y-3$

$-3=y-3$

$y=0$

**d)**

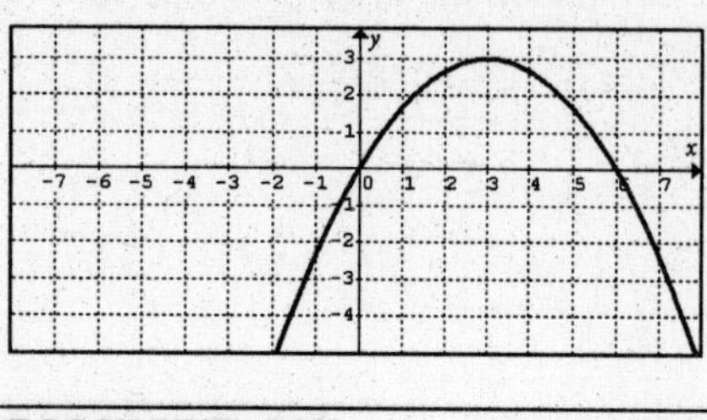

**15.** Complete the square and change to standard form:

$2y^2+16x=16y-64$

$2y^2-16y=-16x-64$

$2(y^2-8y+16)=-16x-64+32$

$2(y-4)^2=-16x-32$

$2(y-4)^2=-16(x+2)$

$-\frac{1}{8}(y-4)^2=x+2$

**a)** Vertex: (–2, 4)

**b)** Domain: $x \leq -2$,
Range: $y \in R$

**c)** To determine the $x$-intercept, let $y = 0$.

$-\frac{1}{8}(0-4)^2=x+2$

$-2=x+2$

$x=-4$

To determine the $y$-intercepts, let $x = 0$.

$-\frac{1}{8}(y-4)^2=0+2$

$-\frac{1}{8}(y-4)^2=2$

$(y-4)^2=-16$

There is no solution, thus no $y$-intercepts.

**d)**

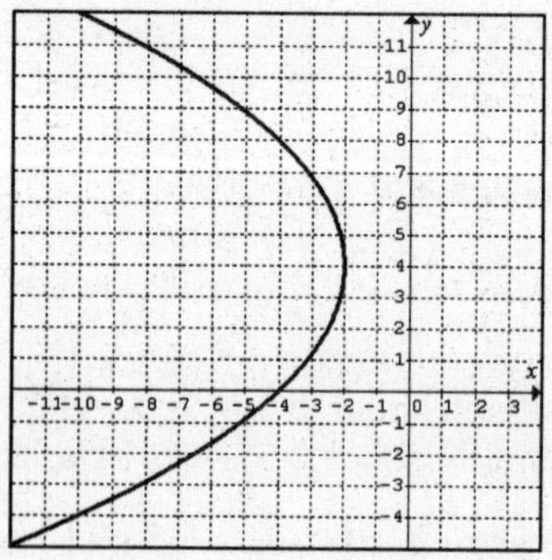

**16.** The equation in standard form has the form $y-1=a(x+2)^2$.

Substitute (2, –4) into the equation.

$-4-1=a(2+2)^2$

$-5=16a$

$a=-\frac{5}{16}$

Thus, the equation is $y-1=-\frac{5}{16}(x+2)^2$.

 

Convert to general form.

$y-1=-\frac{5}{16}(x+2)^2$

$16y-16=-5(x^2+4x+4)$

$16y-16=-5x^2-20x-20$

$5x^2+20x+16y+4=0$

$A=5,\ C=0,\ D=20,\ E=16,\ F=4$

**17.** The equation in standard form has the form $x-3=a(y+2)^2$.
Substitute (–3, 5) into the equation.

$-3-3=a(5+2)^2$

$-6=49a$

$a=-\frac{6}{49}$

Thus, the equation is $x-3=-\frac{6}{49}(y+2)^2$.

Convert to general form.

$x-3=-\frac{6}{49}(y+2)^2$

$49x-147=-6(y^2+4y+4)$

$49x-147=-6y^2-24y-24$

$6y^2+49x+24y-123=0$

$A=0,\ C=6,\ D=49,\ E=24,\ F=-123$

**18.** Since the graph is congruent to $x=\frac{3}{2}y^2$. the vertical stretch factor is $\frac{3}{2}$. Since the graph opens left and has its vertex at (4, –3), the equation in standard form is $x+2=-\frac{3}{2}(y-6)^2$.
Convert to general form.

$x+2=-\frac{3}{2}(y-6)^2$

$2x+4=-3(y^2-12y+36)$

$2x+4=-3y^2+36y-108$

$3y^2+2x-36y+112=0$

$C=3,\ D=2,\ E=-36,\ F=112$

**19.** First, complete the square to convert the equation to standard form.

$2x^2-4x-4y-6=0$

$2(x^2-2x+1)=4y+6+2$

$2(x-1)^2=4y+8$

$2(x-1)^2=4(y+2)$

$\frac{1}{2}(x-1)^2=y+2$

To be congruent to the graph of the given equation, the vertical stretch factor must be $\frac{1}{2}$.
With a vertex at (4, –3), the equation in standard form is $y+3=\frac{1}{2}(x-4)^2$.
Convert to general form.

$y+3=\frac{1}{2}(x-4)^2$

$2y+6=x^2-8x+16$

$x^2-8x-2y+10=0$

$A=1,\ D=-8,\ E=-2,\ F=10$

**20.** **B**
With the vertex at (–4, 6), the equation is $y-6=a(x+4)^2$. Since the parabola has an $x$-intercept of –2, then (–2, 0) satisfies the equation.

$0-6=a(-2+4)^2$

$-6=4a$

$-\frac{3}{2}=a$

**21.** **B**
Since $y$-intercepts occur when $x=0$, $0-h=-2(y-k)^2$ must have two solutions.
Thus, $\frac{h}{2}=(y-k)^2$ must have two solutions.
This will be true only if $h>0$.

**22.** The new equation is $y-5=3\left((x+2)^2-4\right)$.
The $y$-intercept occurs when $x=0$.

$y-5=3\left((0+2)^2-4\right)$

$y-5=3(0)$

$y=5$

| 5 | | | |
|---|---|---|---|

## *Lesson 6 Conics Summary, and Further Questions*

### CLASS EXERCISES ANSWERS AND SOLUTIONS

1. The general form is $3x^2+2y^2+12x-20y-10=0$. Because $AC>0$ and $A\neq C$, you should recognize the equation as that of an ellipse.

   Converting to standard form by completing the squares gives:

$$3x^2+2y^2+12x-20y-10=0$$
$$3(x^2+4x+\_)+2(y^2-10y+\_)=10$$
$$3(x^2+4x+4)+2(y^2-10y+25)=10+12+50$$
$$\frac{(x+2)^2}{24}+\frac{(y-5)^2}{36}=1$$

   The centre is at (–2, 5). Since $a^2<b^2$, the ellipse has a vertical orientation and thus, the vertices are at (–2, 5 + 6) and (–2, 5 – 6), and (–2, 11) or (–2, –1).

### PRACTICE EXERCISES ANSWERS AND SOLUTIONS

1. The graph is a circle with the centre at (3, –2) and radius = 5. Therefore, the equation is $(x-3)^2+(y+2)^2=5^2$.

2. The graph is an ellipse with centre the at (–1, –2), $a=2$, and $b=1$. Therefore, the equation is $\frac{(x+1)^2}{4}+\frac{(y+2)^2}{1}=1$.

3. The graph is a parabola opening up with its vertex at (3, –2). Thus, the equation is of the form $y+2=a(x-3)^2$. Since the $y$-intercept is 1, (0, 1) satisfies the equation.

$$1+2=a(0-3)^2$$
$$3=9a$$
$$a=\frac{1}{3}$$

   Thus, the equation is $y+2=\frac{1}{3}(x-3)^2$.

4. The graph is a parabola opening left with its vertex at (1, 3). Thus, the equation is of the form $x-1=a(y-3)^2$. Since the graph passes through the point (–1, 1), (–1, 1) satisfies the equation.

$$-1-1=a(1-3)^2$$
$$-2=4a$$
$$a=-\frac{1}{2}$$

   Thus, the equation is $x-1=-\frac{1}{2}(y-3)^2$.

5. The graph is a hyperbola with a horizontal orientation and its centre at (1, 0) and vertices at (–1, 0) and (3, 0). Since the distance from the centre to a vertex is 2, $a=2$. It is given that $b=1$. Thus, the equation is $\frac{(x-1)^2}{4}-\frac{y^2}{1}=1$.

6. The graph is a hyperbola with a vertical orientation and centre at (3, –1) and vertices at (3, –4) and (3, 2). Since the distance from the centre to a vertex is 3, $b=3$. It is given that $a=4$. Thus, the equation is $\frac{(x-3)^2}{16}-\frac{(y+1)^2}{9}=-1$.

7. Complete the square to convert to standard form.

$$2y^2-8y=x-7$$
$$2(y^2-4y+4)=x-7+8$$
$$2(y-2)^2=x+1$$
$$x+1=2(y-2)^2$$

   The graph is a parabola opening to the right with its vertex at (–1, 2). Notice that the $x$-intercept can easily be determined by letting $y=0$.

$$2(0)^2-x-8(0)+7=0$$
$$x=7$$

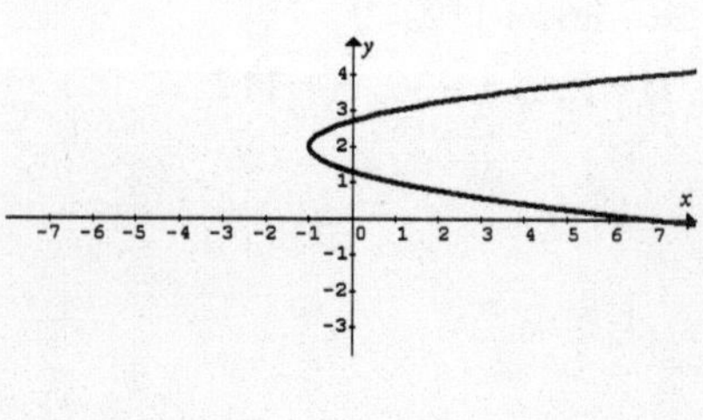

**8.** Complete the squares to convert to standard form.

$4x^2-8x-9y^2-36y=68$

$4(x^2-2x+1)-9(y^2+4y+4)=68+4-36$

$4(x-1)^2-9(y+2)^2=36$

$\frac{(x-1)^2}{9}-\frac{(y+2)^2}{4}=1$

The graph is a hyperbola with a horizontal orientation, its centre at (1, –2), and vertices at (–2, –2) and (3, –2).

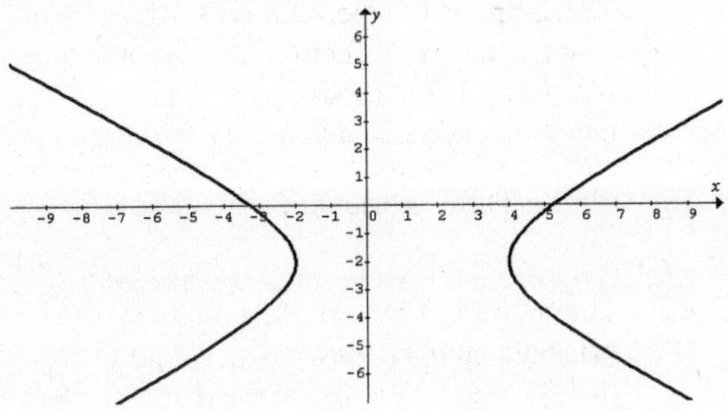

**9.** Complete the squares to convert to standard form.

$16x^2+96x+y^2+4y=-132$

$16(x^2+6x+9)+(y^2+4y+4)$

$=-132+144+4$

$16(x+3)^2+(y+2)^2=16$

$\frac{(x+3)^2}{1}+\frac{(y+2)^2}{16}=1$

The graph is an ellipse with a vertical orientation, its centre at (–3, –2), and vertices at (–3, –6) and (–3, 2).

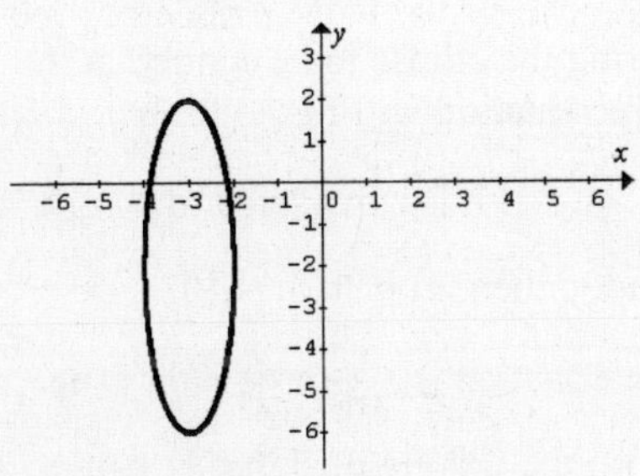

**10.** The results of the transformations are:

$(x-2)^2-\frac{\left(\frac{1}{2}y\right)^2}{9}=-1$

(vertical stretch by a factor of 2)

$((x+5)-2)^2-\frac{\left(\frac{1}{2}y\right)^2}{9}=-1$

(translation of 5 units left)

Simplify and write in standard form.

$(x+3)^2-\frac{\frac{1}{4}y^2}{9}=-1$

$\frac{(x+3)^2}{1}-\frac{y^2}{36}=-1$

**11.** Since the centre is at (–3, 2), the equation has the form $\frac{(x+3)^2}{a^2}+\frac{(y-2)^2}{b^2}=1$. The coordinates (2, 2) and (–3, 5) satisfy the equation. Thus, $\frac{(2+3)^2}{a^2}+\frac{(2-2)^2}{b^2}=1$ and $\frac{(-3+3)^2}{a^2}+\frac{(5-2)^2}{b^2}=1$. Solving the first equation for $a^2$ gives $\frac{25}{a^2}=1\rightarrow a^2=25$. Solving the second equation for $b^2$ gives $\frac{9}{b^2}=1\rightarrow b^2=9$. Thus, the equation is $\frac{(x+3)^2}{25}+\frac{(y-2)^2}{9}=1$.

**12.** **a)** Because of the two squared terms separated by the subtraction sign, the conic is a hyperbola.

**b)** Notice that $9(x+4)^2-4(y+3)^2=-1$ can be written as $(3(x+4))^2-(2(y+3))^2=-1$. From this form, it can be seen that the transformations applied to the graph of $x^2-y^2=-1$, in order, are:

**1.** A horizontal stretch about the $y$-axis by a factor of $\frac{1}{3}$ and a vertical stretch about the $x$-axis by a factor of $\frac{1}{2}$.

**2.** Translations of 4 units left and 3 units down

**c)** In standard form, the equation is $\frac{(x+4)^2}{\frac{1}{9}} - \frac{(y+3)^2}{\frac{1}{4}} = -1$. Notice that the equation is of a hyperbola with a vertical orientation and with $a = \frac{1}{3}$ and $b = \frac{1}{2}$.

**i)** Centre: (–4, –3)

**ii)** Vertices: $(-4, -3-0.5) \to (-4, -3.5)$ and $(-4, -3+0.5) \to (-4, -2.5)$

**iii)** Domain: $x \in R$,
Range: $y \leq -3.5$ or $y \geq -2.5$

**iv)** Slopes of the asymptotes are

$$\pm\frac{b}{a} = \pm\frac{\frac{1}{2}}{\frac{1}{3}} = \pm\frac{3}{2}$$

**13.** The equation $y^2 - x + 4y + F = 0$ represents a parabola opening to the right. Completing the square gives

$(y^2 + 4y + 4) = x - F + 4$

$(y+2)^2 = x - (F-4)$

The vertex of the parabola is $(F-4, -2)$.

If the 4 is changed to a 2, the equation is $y^2 - x + 2y + F = 0$.

$(y^2 + 2y + 1) = x - F + 1$

$(y+1)^2 = x - (F-1)$

The vertex of the parabola is $(F-1, -1)$.

Thus, the transformation effect is that the parabola has been translated 3 units right and 1 unit up.

**14.** The equation $x^2 + y^2 + 4x = 0$ is the equation of a circle. Completing the square gives:

$(x^2 + 4x + 4) + y^2 = 0 + 4$

$(x+2)^2 + y^2 = 4$

The centre is at (–2, 0) and the radius is 2.

If the 4 is changed to a 6, the equation is

$(x^2 + 6x + 9) + y^2 = 0 + 9$

$(x+3)^2 + y^2 = 9$

The centre is at (–3, 0) and the radius is 3. Thus the transformation effect, in order is

1. The graph has been stretched both horizontally about the *y*-axis and vertically about the *x*-axis by a factor of $\frac{3}{2}$ (the radius has changed from 2 to 3).
2. The circle has been translated 1 unit left.

**15.** The original circle has its centre at (0, 0) and a radius of 2. Thus, the equation is $x^2 + y^2 = 4$. First, the circle must be stretched horizontally about the *y*-axis by a factor of 2. The equation becomes $\left(\frac{1}{2}x\right)^2 + y^2 = 4$. Second, the circle must be translated 6 units left and 2 units up. The equation becomes

$$\left(\frac{1}{2}(x+6)\right)^2 + (y-2)^2 = 4.$$

Writing the equation in standard form gives:

$$\frac{1}{4}(x+6)^2 + (y-2)^2 = 4$$

$$\frac{(x+6)^2}{16} + \frac{(y-2)^2}{4} = 1$$

**16.** Ellipse with a horizontal orientation

**17.** Hyperbola with a vertical orientation

**18.** Parabola opening down

**19.** Circle

**20.** Parabola opening to the left

**21.** Ellipse with a vertical orientation

**22.** Parabola opening to the right

**23.** Parabola opening up

**24.** Hyperbola with a vertical orientation

## *Lesson 7 Applications of Conics Questions*

### CLASS EXERCISES
### ANSWERS AND SOLUTIONS

**1.** Considering the ellipse to be centred at the origin, the equation will fit the form $\frac{x^2}{a^2} + \frac{y^2}{b^2} = 1$. If it is considered to be oriented horizontally, then $2a = 80$, $a = 40$.

Thus, the equation is $\frac{x^2}{40^2} + \frac{y^2}{b^2} = 1$. The position on the skating oval that is shown can be represented by the point (30, 8). These coordinates must satisfy the equation.

$$\frac{30^2}{40^2} + \frac{8^2}{b^2} = 1.$$

 

Solving for $b^2$ gives

$$\frac{30^2}{40^2}+\frac{8^2}{b^2}=1$$

$$\frac{64}{b^2}=1-\frac{900}{1600}$$
• Subtract $\frac{30^2}{40^2}$ from both sides of the equation

$$\frac{64}{b^2}=\frac{7}{16}$$
• Simplify

$$b^2=\frac{64\times 16}{7}=\frac{32^2}{7}$$
• Apply the fact that cross-products are equal

Thus, the equation is $\frac{x^2}{40^2}+\frac{y^2}{\frac{32^2}{7}}=1$, which is

$$\frac{x^2}{40^2}+\frac{7y^2}{32^2}=1.$$

## PRACTICE EXERCISES ANSWERS AND SOLUTIONS

**1.** **a)** Since the vertex is at the origin, the equation has the form $x=ay^2$. A point that is on the graph is (4.6, 9.2). Thus, these coordinates satisfy the equation.

$$4.6=a(9.2)^2$$

$$\frac{4.6}{(9.2)^2}=a$$

$$a=\frac{1}{18.4}$$

Thus, the equation is $x=\frac{1}{18.4}y^2$.

**b)** The diameter at the front is twice the radius at the front. The radius at the front is the positive $y$-value when $x=11$.

$$11=\frac{1}{18.4}y^2$$

$$y^2=202.4$$

$$y\doteq\pm 14.23$$

Thus, the diameter at the front is approximately $14.23\times 2\doteq 28.5$ cm.

**2.** **a)** Since the centre is at the origin, the equation of the ellipse will have the form $\frac{x^2}{a^2}+\frac{y^2}{b^2}=1$. Since the tunnel is 12.0 m wide and 4.0 m high, the corresponding values for $a$ and $b$ are $a=6$ and $b=4$. Therefore, the equation is

$$\frac{x^2}{36}+\frac{y^2}{16}=1,\ y\geq 0.$$

**b)** The height of the tallest vehicle corresponds to the positive $y$-value when $x=4$.

$$\frac{4^2}{36}+\frac{y^2}{16}=1,\ y\geq 0$$

$$\frac{4}{9}+\frac{y^2}{16}=1$$

$$y^2=16\left(\frac{5}{9}\right)$$

$$y=\frac{4\sqrt{5}}{3}\doteq 2.981$$

The maximum height of the vehicle is approximately 2.9 m. Notice that 3.0 m is a little too much.

**3.** **a)**

**i)** If the parabola is considered to have its vertex at (0, 10) and opens downward, the equation has the form $y-10=ax^2$. Since the width is 80 m, the coordinates (40, 0) satisfy the equation.

$$0-10=a(40)^2$$

$$-\frac{10}{40^2}=a$$

$$a=-\frac{1}{160}$$

Thus, the equation is $y-10=-\frac{1}{160}x^2$, $-40\leq x\leq 40$.

**ii)** Considering that the centre of the ellipse is at the origin, the equation has the form $\frac{x^2}{a^2}+\frac{y^2}{b^2}=1$. Since the width is 80 m and the height is 10 m, $a=40$ and $b=10$. Therefore, the equation is

$$\frac{x^2}{1600}+\frac{y^2}{100}=1,\ y\geq 0.$$

**iii)** Consider the lower branch of the hyperbola to have its vertex at (0, 10). Since the value of $b$ is 5, the centre would be at (0, 15) and the equation has the form

$$\frac{x^2}{a^2}-\frac{(y-10)^2}{25}=-1,\ 0\leq y\leq 10.$$

Since the width is 80 m, the coordinates (40, 0) must satisfy the equation.

$$\frac{40^2}{a^2}-\frac{(0-10)^2}{25}=-1,\ 0\le y\le 10$$

$$\frac{1\,600}{a^2}-\frac{100}{25}=-1$$

$$\frac{1\,600}{a^2}=3$$

$$a^2=\frac{1\,600}{3}$$

Therefore, the equation is

$$\frac{x^2}{\frac{1600}{3}}-\frac{(y-10)^2}{25}=-1,\ 0\le y\le 10.$$

**b)** Since the barge reaches a height of 15 m, the height above the base of the support would be 5 m. If the barge is in the centre, the top right-hand corner would correspond to the coordinates (15, 5) on the graph. Determine the $y$-coordinate for an $x$-value of 15 for each of the conics from part (a).

**i)** For the parabola:

$$y-10=-\frac{1}{160}(15)^2$$

$$y=-\frac{225}{160}+10$$

$$y=-\frac{225}{160}+\frac{1600}{160}=\frac{1\,375}{160}$$

$$y\doteq 8.59$$

Thus, the clearance would be approximately $8.6-5=3.6$ m.

**ii)** For the semi-ellipse:

$$\frac{(15)^2}{1\,600}+\frac{y^2}{100}=1,\ y\ge 0$$

$$\frac{y^2}{100}=\frac{1\,600}{1\,600}-\frac{225}{1\,600}$$

$$y^2=\frac{137\,500}{1\,600}$$

$$y=\frac{50\sqrt{55}}{40}\doteq 9.27$$

Thus, the clearance would be approximately $9.3-5=4.3$ m.

**iii)** For the lower branch of the hyperbola:

$$\frac{(15)^2}{\frac{1\,600}{3}}-\frac{(y-10)^2}{25}=-1,\ 0\le y\le 10$$

$$-\frac{(y-10)^2}{25}=-1-\frac{(3)(225)}{1\,600}$$

$$(y-10)^2=25+\frac{(3)(225)(25)}{1\,600}$$

$$y-10\doteq\pm\sqrt{35.55}$$

$$y\doteq 10\pm 5.96$$

$y\doteq 10-5.96\doteq 4.04$ (for the lower branch)

Thus, there would not be enough clearance; the barge would be approximately $5-4.0=1.0$ m too tall.

**c)** Each of the conics would be translated 40 units to the right. The equations would be:

**i)** Parabola:

$$y-10=-\frac{1}{160}(x-40)^2, 0\le x\le 80$$

**ii)** Semi-ellipse: $\frac{(x-40)^2}{1\,600}+\frac{y^2}{100}=1,\ y\ge 0$

**iii)** Hyperbola:

$$\frac{(x-40)^2}{\frac{1\,600}{3}}-\frac{(y-10)^2}{25}=-1,\ 0\le y\le 10$$

**4.** The ellipse has its centre at (0, 0). At the widest part, the width is 100 cm. If the ellipse is considered to have a horizontal orientation, $2a=100$ and thus, $a=50$. Therefore, the equation has the form $\frac{x^2}{50^2}+\frac{y^2}{b^2}=1$. Since the box is 80 cm wide and 20 cm tall, the coordinates (40, 10) must satisfy the equation. You can now solve for $b^2$.

$$\frac{40^2}{50^2}+\frac{10^2}{b^2}=1$$

$$\frac{10^2}{b^2}=1-\frac{40^2}{50^2}$$

$$\frac{10^2}{b^2}=\frac{900}{2\,500}$$

$$b^2=\frac{2\,500\times 10^2}{900}=\frac{2\,500}{9}$$

Therefore, the equation is $\frac{x^2}{50^2}+\frac{y^2}{\left(\frac{50}{3}\right)^2}=1.$

**5.** **a)** Since the arch is 30 m wide and 20 m tall the coordinates (15, 20) can be considered to be the coordinates of the vertex of a parabola that opens down. Thus, the equation has the form $y-20=a(x-15)^2$. The coordinates (0, 0) satisfy the equation.

$0-20=a(0-15)^2$

$-\frac{20}{225}=a$

$a=-\frac{4}{45}$

Thus, the equation is

$y-20=-\frac{4}{45}(x-15)^2, \quad 0\le x\le 30$

Domain: $0\le x\le 30$

Range: $0\le y\le 20$

**b)** For the semi-ellipse, the centre would be at (15, 0) and the vertices would be at (0, 0) and (30, 0). Since the value of $a$ is the distance from the centre to a vertex, $a=15$. Thus, the equation has the form $\frac{(x-15)^2}{15^2}+\frac{y^2}{b^2}=1$. The coordinates (15, 20) represent the top of the arch and satisfy the equation.

$\frac{(15-15)^2}{15^2}+\frac{20^2}{b^2}=1$

$b^2=20^2$

Thus, the equation is

$\frac{(x-15)^2}{15^2}+\frac{y^2}{20^2}=1,\ 0\le y\le 20.$

Domain: $0\le x\le 30$

Range: $0\le y\le 20$

**c)** If the truck is located in the right lane so that the left side of its load is at the centre, then the right side would be 15 + 8 = 23 m from the left "foot" of the arch. This corresponds to $x=23$ in the equation. The corresponding $y$-value is the maximum height for the load.

For the parabolic arch:

$y-20=-\frac{4}{45}(23-15)^2$

$y=20-\frac{4}{45}(64)$

$y\doteq 14.31$

For the semi-elliptical arch:

$\frac{(23-15)^2}{15^2}+\frac{y^2}{20^2}=1,\ 0\le y\le 20$

$\frac{y^2}{400}=1-\frac{64}{225}$

$\frac{y^2}{400}=\frac{161}{225}$

$y=\sqrt{\frac{400\times 161}{225}}$

$y\doteq 16.92$

The maximum load heights are approximately 14.3 m for the parabolic arch and 16.9 m for the semi-elliptical arch.

Thus, the semi-elliptical arch would accommodate a taller load.

## *Practice Test*

### ANSWERS AND SOLUTIONS

**1.** If the angle that the plane makes with the axis of the cone is $\theta$, then to form a hyperbola, $0\le\theta<26°$.

**2.** Circle, ellipse, two parallel lines, one line, no graph

**3.** Since $AC=0$, with $A=0$, and $CD<0$, the equation represents a parabola opening to the right.

Convert the equation to standard form:

$2y^2-x-16y+35=0$

$2y^2-16y=x-35$

$2(y^2-8y+16)=x-35+32$

$2(y-4)^2=x-3$

Domain: $x\ge 3$,

Range: $y\in R$

**4.** **a)** $x^2-4y^2+6x-8y-11=0$

$x^2+6x-4y^2-8y=11$

$(x^2+6x+9)-4(y^2+2y+1)=11+9-4$

$(x+3)^2-4(y+1)^2=16$

$\frac{(x+3)^2}{16}-\frac{(y+1)^2}{4}=1$

**b)** The graph is a hyperbola oriented horizontally with a centre at (–3, –1) and vertices at (–7, –1) and (1, –1).

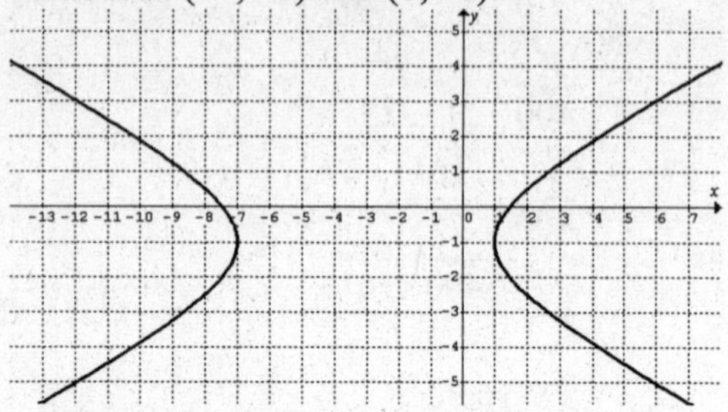

**c)** Domain: $x \le -7$ or $x \ge 1$,

Range: $y \in R$

**5.** **a)** Since $AC = 0$, with $C = 0$, and $AE < 0$, the equation represents a parabola opening up.

**b)** Change the equations to standard form to determine the effect of changing the 6 to a 10.

$x^2 - 6x - y + F = 0$

$\left(x^2 - 6x + 9\right) = y - F + 9$

$(x-3)^2 = y - (F - 9)$

$x^2 - 10x - y + F = 0$

$\left(x^2 - 10x + 25\right) = y - F + 25$

$(x-5)^2 = y - (F - 25)$

The parabola remains congruent to the original parabola and is translated 2 units right and 16 units down.

**6.** **a)** $x^2 + y^2 = 13 \rightarrow \left(\frac{1}{3}x\right)^2 + (2y)^2 = 13$

Write in general form:

$\frac{1}{9}x^2 + 4y^2 = 13$

$x^2 + 36y^2 = 117$

$x^2 + 36y^2 - 117 = 0$

**b)** $P(2, 3)$ would be transformed to

$P'(2 \times 3, 3 \times \frac{1}{2}) = P'(6, \frac{3}{2})$

**7.** The hyperbola has a horizontal orientation with its centre at $(\frac{-2+6}{2}, 4) = (2, 4)$. Since the distance from the centre to a vertex is 4 units, $a = 4$. Since the slope of an asymptote is $-\frac{4}{3}$,

$\frac{b}{4} = \frac{4}{3}$, and thus, $b = \frac{16}{3}$.

The equation of the hyperbola in standard form is $\frac{(x-2)^2}{4^2} - \frac{(y-4)^2}{\left(\frac{16}{3}\right)^2} = 1$.

**8.** Since the distance from the centre to a vertex is $8 - 1 = 7$ units, the value of $a$ is 7.

**9.** The equation has the form

$\frac{(x+1)^2}{a^2} + \frac{(y-4)^2}{b^2} = 1.$

The points (2, 4) and (–1, –1) must satisfy the equation and thus,

$\frac{(2+1)^2}{a^2} + \frac{(4-4)^2}{b^2} = 1$, and

$\frac{(-1+1)^2}{a^2} + \frac{(-1-4)^2}{b^2} = 1.$

From the first equation, $\frac{9}{a^2} = 1$, and thus,

$a^2 = 9$.

From the second equation, $\frac{25}{b^2} = 1$ and thus,

$b^2 = 25$.

Therefore, the equation is

$\frac{(x+1)^2}{9} + \frac{(y-4)^2}{25} = 1.$

**10.** $16(x+1)^2 + 25y^2 - 400 = 0$

Convert to standard form:

$\frac{(x+1)^2}{25} + \frac{y^2}{16} = 1$

Since the equation represents an ellipse with a horizontal orientation with its centre at (–1, 0) and $a = 5$, the vertices are (–6, 0) and (4, 0).

**11.** Since the graph is a hyperbola with a vertical orientation and a centre at (2, 1), the equation has the form $\frac{(x-2)^2}{a^2} - \frac{(y-1)^2}{b^2} = -1$.

 

From the graph, it can be seen that the distance from the centre to a vertex is 2 units, thus, $b = 2$. It is given that $a = 3$. Therefore, the equation is $\frac{(x-2)^2}{9} - \frac{(y-1)^2}{4} = -1$.

**12. a) i)** The graph will be a parabola opening down with its vertex at (8, 6). Thus, the equation has the form $y - 6 = a(x-8)^2$. Since (0, 0) is a point on the graph, (0, 0) satisfies the equation.

$$0 - 6 = a(0-8)^2$$
$$-\frac{6}{64} = a$$
$$a = -\frac{3}{32}$$

Therefore, the equation is

$$y - 6 = -\frac{3}{32}(x-8)^2,\ 0 \le x \le 16.$$

**ii)** The graph will be a semi-ellipse with a vertex at (8, 0), $a = 8$, and $b = 6$. Thus, the equation is $\frac{(x-8)^2}{64} + \frac{y^2}{36} = 1,\ 0 \le y \le 6.$

**b)** If the vehicle is in the right hand lane, then the maximum height of the vehicle is the $y$-value when $x = 8 + 4 = 12$.

For the parabolic tunnel opening:

$$y - 6 = -\frac{3}{32}(12-8)^2$$
$$y = 6 - \frac{3}{32}(16)$$
$$y = 6 - \frac{3}{2} = 4.5$$

For the semi-elliptical tunnel opening:

$$\frac{(12-8)^2}{64} + \frac{y^2}{36} = 1$$
$$\frac{16}{64} + \frac{y^2}{36} = 1$$
$$\frac{y^2}{36} = \frac{3}{4}$$
$$y = \sqrt{27} \doteq 5.196$$

Therefore, for the top of the vehicle to enter the tunnel, the maximum heights, to the nearest tenth, are 4.4 m for the parabolic opening and 5.1 m for the semi-elliptical opening.

# SCHOOL ORDER FORM

| THE KEY | QUANTITY |
|---|---|
| Biology 30 | |
| Chemistry 30 | |
| English 30-1 | |
| English 30-2 | |
| Math30 (Pure) | |
| Math 30 (Applied) | |
| Physics 30 | |
| Social Studies 30 | |
| Social Studies 33 | |
| Biology 20 | |
| Chemistry 20 | |
| English 20-1 | |
| Math 20 (Pure) | |
| Physics 20 | |
| Social Studies 20 | |
| English 10-1 | |
| Math 10 (Pure) | |
| Science 10 | |
| Social Studies 10 | |
| Language Arts 9 | |
| Math 9 | |
| Science 9 | |
| Social Studies 9 | |
| Math 8 | |
| Math 7 | |
| Language Arts 6 | |
| Math 6 | |
| Science 6 | |
| Social Studies 6 | |
| Math 4 | |
| Math 3 | |
| Language Arts 3 | |
| | |
| | |

| SNAP WORKBOOKS Notes and Problems/ Student Notes and Problems | QUANTITY | |
|---|---|---|
| | Workbooks | Solutions Manuals |
| Chemistry 30 | | |
| Chemistry 20 | | |
| Physics 30 | | |
| Physics 20 | | |
| Math 30 Pure | | |
| Math 30 Applied | | |
| Math 31 | | |
| Math 20 Pure | | |
| Math 20 Applied | | |
| Math 10 Pure | | |
| Math 10 Applied | | |
| Science 10 | | |
| Science 9 | | |
| Math 9 | | |
| Math 8 | | |
| Math 7 | | |
| Math 6 | | |
| Math 5 | | |
| Math 4 | | |
| Math 3 | | |

| TOTALS | |
|---|---|
| KEYS | |
| WORKBOOKS | |
| SOLUTION MANUALS | |

### *Learning Resources Centre*

**Castle Rock Research is pleased to announce an exclusive distribution arrangement with the Learning Resources Centre (LRC). Under this agreement, schools can now place all their orders with LRC for order fulfillment. As well, these resources are eligible for applying the Learning Resource Credit Allocation (LRCA), which gives schools a 25% discount off LRC's selling price. Call LRC for details.**

**Orders may be placed with LRC by**
**telephone: (780) 427-5775**
**fax: (780) 422-9750**
**internet: www.lrc.learning.gov.ab.ca**
**or mail: 12360 - 142 Street NW**
**Edmonton, AB T5L 4X9**

Learning Resources Centre

*PAYMENT AND SHIPPING INFORMATION*

Name: ______

School Telephone: ______

**SHIP TO**

School: ______

Address: ______

City: ______ Postal Code: ______

**PAYMENT**

☐ by credit card

VISA/MC Number: ______ Expiry Date: ______

Name on Card: ______

☐ enclosed cheque

☐ invoice school P.O. number: ______

Castle Rock Research Corp

#2340, 10180 – 101 Street, Edmonton, AB T5J 3S4 Tel: 780.448.9619 Fax: 780.426.3917
email: learn@castlerockresearch.com Toll-free: 1.800.840.6224

www.castlerockresearch.com

SNAP